RFID AND SENSOR NETWORKS

Architectures, Protocols, Security, and Integrations

WIRELESS NETWORKS AND MOBILE COMMUNICATIONS

Dr. Yan Zhang, Series Editor
Simula Research Laboratory, Norway
E-mail: yanzhang@ieee.org

Broadband Mobile Multimedia:
Techniques and Applications
Yan Zhang, Shiwen Mao, Laurence T. Yang,
and Thomas M. Chen
ISBN: 978-1-4200-5184-1

Cooperative Wireless Communications
Yan Zhang, Hsiao-Hwa Chen,
and Mohsen Guizani
ISBN: 978-1-4200-6469-8

Distributed Antenna Systems:
Open Architecture for Future Wireless
Communications
Honglin Hu, Yan Zhang, and Jijun Luo
ISBN: 978-1-4200-4288-7

The Internet of Things:
From RFID to the Next-Generation
Pervasive Networked Systems
Lu Yan, Yan Zhang, Laurence T. Yang,
and Huansheng Ning
ISBN: 978-1-4200-5281-7

Millimeter Wave Technology in Wireless
PAN, LAN and MAN
Shao-Qiu Xiao, Ming-Tuo Zhou
and Yan Zhang
ISBN: 978-0-8493-8227-7

Mobile WiMAX: Toward Broadband
Wireless Metropolitan Area Networks
Yan Zhang and Hsiao-Hwa Chen
ISBN: 978-0-8493-2624-0

Resource, Mobility, and Security
Management in Wireless Networks
and Mobile Communications
Yan Zhang, Honglin Hu, and Masayuki Fujise
ISBN: 978-0-8493-8036-5

RFID and Sensor Networks:
Architectures, Protocols, Security
and Integrations
Yan Zhang, Laurence T. Yang,
and JimIng Chen
ISBN: 978-1-4200-7777-3

Security in RFID and Sensor Networks
Yan Zhang and Paris Kitsos
ISBN: 978-1-4200-6839-9

Security in Wireless Mesh Networks
Yan Zhang, Jun Zheng and Honglin Hu
ISBN: 978-0-8493-8250-5

Unlicensed Mobile Access Technology:
Protocols, Architectures, Security,
Standards, and Applications
Yan Zhang, Laurence T. Yang, and Jianhua Ma
ISBN: 978-1-4200-5537-5

WiMAX Network Planning and
Optimization
Yan Zhang
ISBN: 978-1-4200-6662-3

Wireless Ad Hoc Networking: Personal-
Area, Local-Area, and the Sensory-Area
Networks
Shih-Lin Wu, Yu-Chee Tseng, and Hsin-Chu
ISBN: 978-0-8493-9254-2

Wireless Mesh Networking:
Architectures, Protocols, and Standards
Yan Zhang, Jijun Luo, and Honglin Hu
ISBN: 978-0-8493-7399-2

Wireless Quality-of-Service: Techniques,
Standards, and Applications
Maode Ma, Mieso K. Denko, and Yan Zhang
ISBN: 978-1-4200-5130-8

AUERBACH PUBLICATIONS

www.auerbach-publications.com
To Order Call: 1-800-272-7737 • Fax: 1-800-374-3401
E-mail: orders@crcpress.com

RFID AND SENSOR NETWORKS

Architectures, Protocols, Security, and Integrations

Edited by

Yan Zhang · Laurence T. Yang · Jiming Chen

CRC Press
Taylor & Francis Group
Boca Raton London New York

CRC Press is an imprint of the
Taylor & Francis Group, an **informa** business
AN AUERBACH BOOK

CRC Press
Taylor & Francis Group
6000 Broken Sound Parkway NW, Suite 300
Boca Raton, FL 33487-2742

© 2010 by Taylor and Francis Group, LLC
CRC Press is an imprint of Taylor & Francis Group, an Informa business

No claim to original U.S. Government works

Printed in the United States of America on acid-free paper
10 9 8 7 6 5 4 3 2 1

International Standard Book Number: 978-1-4200-7777-3 (Hardback)

Library of Congress Cataloging-in-Publication Data

RFID and sensor networks : architectures, protocols, security, and integrations / editors,
    Yan Zhang, Laurence T. Yang, and Jiming Chen.
        p. cm. -- (Wireless networks and mobile communications)
    "A CRC title."
    Includes bibliographical references and index.
    ISBN 978-1-4200-7777-3 (alk. paper)
        1. Radio frequency identification systems. 2. Sensor networks. I. Zhang, Yan, 1977-
    II. Yang, Laurence Tianruo. III. Chen, Jiming, 1978-

    TK6570.I34R47 2009
    681'.2--dc22
                                                                                2009018070

Visit the Taylor & Francis Web site at
http://www.taylorandfrancis.com

and the CRC Press Web site at
http://www.crcpress.com

Contents

PART I: RFID

v

PART II: WIRELESS SENSOR NETWORKS

PART III: INTEGRATED RFID AND SENSOR NETWORKS

Preface

Radio frequency identification (RFID) technology is witnessing a recent explosion of development in both industry and academia. A number of applications include supply chain management, electronic payments, RFID passports, environmental monitoring and control, office access control, intelligent labels, target detection and tracking, port management, food production control, animal identification, and so on. RFID is also an indispensable foundation to realize the pervasive computing paradigm—"Internet of things." It is strongly believed that many more scenarios will be identified when the principles of RFID are thoroughly understood, cheap components available, and when RFID security is guaranteed.

Wireless sensor networks (WSNs) are also attracting significant interest due to recent advances of the enabling technologies, including digital electronics, embedded systems, signal processing, and wireless communications. A WSN consists of a large number of small sensors with sensing, control, data processing, as well as communication and networking capabilities. WSNs are characterized by dense node deployments; unreliable sensors; frequent topology changes; and severe power, computation, and memory constraints. These unique characteristics pose considerable challenges for the design of WSNs. Because sensor networks usually transmit data and operate in hostile, unattended environments, the requirements and design of sensor networks are significantly different from other wireless networks like cellular networks, and ad hoc networks or mesh networks.

In practice, there is an increasing trend in integrating RFID and WSNs due to their complementary natures, and a flexible combination and demand for ubiquitous computing. A variety of applications are under development or in practical usage, e.g., smart homes, surveillance systems, and in personal healthcare. The integration of the two complementary technologies can exponentially enhance the visibility and monitoring capability. However, compared with either RFID or sensor networks alone, integrating RFID and WSNs has more technical, operational, business, and policy challenges.

RFID and Sensor Networks: Architectures, Protocols, Security, and Integrations provides a comprehensive technical guide covering introductory concepts; fundamental techniques; recent advances; and open issues in RFID, WSNs, and integrated RFID and

WSNs. This book contains illustrative figures and allows for complete cross-referencing. It also details information on the particular techniques for efficiently improving the performance of a RFID and a sensor network, and their integration.

This book is organized into three parts:

- Part I: RFID
- Part II: Wireless Sensor Networks
- Part III: Integrated RFID and Sensor Networks

Part I introduces the fundamentals and principles of RFID. This part provides readers with a knowledge of RFID, e.g., tags, readers, middleware, security, and services. Part II introduces the fundamentals and principles of WSNs. This part provides readers with a knowledge of WSNs, e.g., routing, medium access control, localization, clustering, mobility, security, and cross-layer optimization. Part III explores the principles and the applications of integrated RFID and WSNs.

This book has the following salient features:

- Serves as a comprehensive and essential reference on RFID, WSNs, and integrated RFID and WSNs
- Covers basics, a broad range of topics, and future development directions
- Introduces architectures, protocols, standards, security, and applications
- Assists professionals, engineers, students, and researchers to understand RFID and WSNs
- Provides a unique content on integrated RFID and WSNs

This book can serve as a useful reference for students, educators, research strategists, scientists, researchers, and engineers in the field of wireless communications and networking. In particular, this book has an instant appeal to students, researchers, developers, and consultants in developing RFID, WSNs, and integrated RFID and WSNs.

We would like to acknowledge the effort and time invested by all contributors for their excellent work. All of them are extremely professional and cooperative. Special thanks go to Richard O'Hanley, Stephanie Morkert, and Joette Lynch of the Taylor & Francis Group for their support, patience, and professionalism from the beginning until the final stage of the book. We are very grateful for Sridharan Sathyanarayanamoorthy for his painstaking efforts during typesetting. Last but not least, a special thank you to our families and friends for their constant encouragement, patience, and understanding throughout this project.

Yan Zhang
Simula Research Laboratory, Norway

Laurence T. Yang
St. Francis Xavier University, Canada

Jiming Chen
Zhejiang University, China

Editors

Yan Zhang received his BS in communication engineering from the Nanjing University of Post and Telecommunications, China; his MS in electrical engineering from the Beijing University of Aeronautics and Astronautics, China; and his PhD from the School of Electrical & Electronics Engineering, Nanyang Technological University, Singapore.

Dr. Zhang is an associate editor and also serves on the editorial boards of the *International Journal of Communication Systems* (*IJCS*, Wiley); the *International Journal of Communication Networks and Distributed Systems* (*IJCNDS*); the *Journal of Ambient Intelligence and Humanized Computing* (*JAIHC*, Springer); the *International Journal of Adaptive, Resilient and Autonomic Systems* (*IJARAS*); *Wireless Communications and Mobile Computing* (*WCMC*, Wiley); *Security and Communication Networks* (Wiley); the *International Journal of Network Security*; the *International Journal of Ubiquitous Computing*; *Transactions on Internet and Information Systems* (*TIIS*); the *International Journal of Autonomous and Adaptive Communications Systems* (*IJAACS*); the *International Journal of Ultra Wideband Communications and Systems* (*IJUWBCS*); and the *International Journal of Smart Home* (*IJSH*).

He is currently serving as the editor for the book series "Wireless Networks and Mobile Communications" (Auerbach Publications, CRC Press, Taylor & Francis Group). He also serves as a guest coeditor for the *WCMC* special issue for best papers in the IWCMC 2009; *Multimedia Systems Journal* (ACM/Springer) special issue on wireless multimedia transmission technology and application; *Journal of Wireless Personal Communications* (Springer) special issue on cognitive radio networks and communications; the *IJAACS* special issue on ubiquitous/pervasive services and applications; the *EURASIP Journal on Wireless Communications and Networking* (*JWCN*) special issue on broadband wireless access; the *IEEE Intelligent Systems* special issue on context-aware middleware and intelligent agents for smart environments; *Security and Communication Networks* (Wiley) special issue on secure multimedia communication; *Wireless Personal Communications* (Springer) special issue on selected papers from ISWCS 2007; *Computer Communications* (Elsevier) special issue on adaptive multicarrier communications and networks; *IJAACS* special issue on cognitive radio systems; the *Journal of Universal Computer Science* (*JUCS*) special issue on multimedia security in communication; the *Journal of Cluster Computing* (Springer) special issue on algorithm and distributed computing in wireless sensor networks; the *JWCN* special issue on OFDMA architectures, protocols, and applications; and the *Journal of Wireless Personal Communications* (Springer) special issue on security and multimodality in pervasive environments.

He is currently serving as a coeditor for the following books: *Resource, Mobility, and Security Management in Wireless Networks and Mobile Communications*; *Wireless Mesh Networking: Architectures, Protocols and Standards*; *Millimeter-Wave Technology in Wireless PAN, LAN and MAN*; *Distributed Antenna Systems: Open Architecture for Future Wireless Communications*; *Security in Wireless Mesh Networks*; *Mobile WiMAX: Toward Broadband Wireless Metropolitan Area Networks*; *Wireless Quality-of-Service: Techniques, Standards and Applications*; *Broadband Mobile Multimedia: Techniques and Applications*; *Internet of Things: From RFID to the Next-Generation Pervasive Networked Systems*; *Unlicensed Mobile Access Technology: Protocols, Architectures, Security, Standards and Applications*; *Cooperative Wireless Communications*; *WiMAX Network Planning and Optimization*; *RFID Security: Techniques, Protocols and System-On-Chip Design*; *Autonomic Computing and Networking*; *Security in RFID and Sensor Networks*; *Handbook of Research on Wireless Security*; *Handbook of Research on Secure Multimedia Distribution*; *RFID and Sensor Networks*; *Cognitive Radio Networks*; *Wireless Technologies for Intelligent Transportation Systems*; *Vehicular Networks: Techniques, Standards and Applications*; *Orthogonal Frequency Division Multiple Access (OFDMA)*; *Game Theory for Wireless Communications and Networking*; and *Delay Tolerant Networks: Protocols and Applications*.

Dr. Zhang serves as a program cochair for IWCMC 2010, as a program cochair for WICON 2010, as a program vice chair for CloudCom 2009, as a publicity cochair for IEEE MASS 2009, as a publicity cochair for IEEE NSS 2009, as a publication chair for PSATS 2009, as a symposium cochair for ChinaCom 2009, as a program cochair for BROADNETS 2009, as a program cochair for IWCMC 2009, as a workshop cochair for ADHOCNETS 2009, as a general cochair for COGCOM 2009, as a program cochair for UC-Sec 2009, as a journal liasion chair for IEEE BWA 2009, as a track cochair for ITNG 2009, as a publicity cochair for SMPE 2009, as a publicity cochair for COMSWARE 2009, as a publicity cochair for ISA 2009, as a general cochair for WAMSNet 2008, as a publicity cochair for TrustCom 2008, as a general cochair for COGCOM 2008, as a workshop cochair for IEEE APSCC 2008, as a general cochair for WITS-08, as a program cochair for PCAC 2008, as a general cochair for CONET 2008, as a workshop chair for SecTech 2008, as a workshop chair for SEA 2008, as a workshop co-organizer for MUSIC'08, as a workshop co-organizer for 4G-WiMAX 2008, as a publicity cochair for SMPE-08, as an international journals coordinating cochair for FGCN-08, as a publicity cochair for ICCCAS 2008, as a workshop chair for ISA 2008, as a symposium cochair for ChinaCom 2008, as an industrial cochair for MobiHoc 2008, as a program cochair for UIC-08, as a general cochair for CoNET 2007, as a general cochair for WAMSNet 2007, as a workshop cochair for FGCN 2007, as a program vice cochair for IEEE ISM 2007, as a publicity cochair for UIC-07, as a publication chair for IEEE ISWCS 2007, as a program cochair for IEEE PCAC 2007, as a special track cochair for "Mobility and Resource Management in Wireless/Mobile Networks" in ITNG 2007, as a special session co-organizer for "Wireless Mesh Networks" in PDCS

2006, and as a member of the Technical Program Committee for numerous international conferences, including ICC, GLOBECOM, WCNC, PIMRC, VTC, CCNC, AINA, ISWCS, etc. He received the best paper award in the IEEE 21st International Conference on Advanced Information Networking and Applications (AINA-07).

Since August 2006, he has been working with the Simula Research Laboratory, Lysaker, Norway (http://www.simula.no/). His research interests include resource, mobility, spectrum, and data, energy, and security management in wireless networks and mobile computing. He is a member of the IEEE and the IEEE ComSoc.

Laurence T. Yang is a professor at St. Francis Xavier University, Antigonish, Canada. His research interests include high performance, embedded and ubiquitous/ pervasive computing. He has published around 300 papers (including around 100 international journal papers such as in *IEEE* and *ACM Transactions*) in refereed journals, conference proceedings, and book chapters in these areas. He has been involved in more than 100 conferences and workshops as a program/general/steering conference chair and more than 300 conference and workshops as a program committee member. He served as the vice chair of the IEEE Technical Committee of Supercomputing Applications (TCSA) until 2004. Currently, he is the chair of the IEEE Technical Committee of Scalable Computing (TCSC) and the chair of the IEEE Task force on Ubiquitous Computing and Intelligence. He is also in the steering committee of the IEEE/ACM supercomputing conference series.

In addition, Dr. Yang is the editor in chief of several international journals and a few book series. He serves as an editor for around 20 international journals. He has been an author/coauthor or an editor/coeditor of 25 books from Kluwer, Springer, Nova Science, American Scientific Publishers, and John Wiley & Sons. He has won five best paper awards (including the IEEE 20th International Conference on Advanced Information Networking and Applications [AINA-06]); two IEEE best paper Awards in 2007 and 2008; two IEEE outstanding paper awards in 2007 and 2008; one best paper nomination in 2007; the Distinguished Achievement Award in 2005; the Canada Foundation for Innovation Award in 2003; and a University Research Award (1999–2002), a University Publication Award (2002–2005), and a University Teaching Award (2005–2008).

Jiming Chen received his PhD in control science and engineering from Zhejiang University, Hangzhou, China in 2005. He was a visiting scholar at the University of Waterloo, the French National Institute for Research in Computer Science and Control, and the National University of Singapore. He is currently an associate professor with the Institute of Industrial Process Control, State Key Lab of Industrial Control Technology, Zhejiang University. He leads the Networked Sensing and Control Group, Zhejiang University. Dr. Chen has published over 50 peer-reviewed papers. He currently serves as an associate editor for the *International Journal of Communication System* (Wiley); *Ad Hoc & Sensor Wireless Networks, an International Journal*; and the *Journal of Computers*, and was a guest editor of *Wireless Communication and Mobile Computing* (Wiley). He also serves as a general symposia cochair of IWCMC 2009, and track cochair of WiCON 2010 MAC.

Contributors

Nadjib Achir
Institut Galilée
University of Paris 13
Villetaneuse, France

Nadjib Aitsaadi
LiP6 Laboratory
University of Pierre and Marie Curie
Paris, France

Ali Hammad Akbar
Al-Khwarizmi Institute of
  Computer Science
University of Engineering and
  Technology
Lahore, Pakistan

Kashif Ali
School of Computing
Queen's University
Kingston, Ontario, Canada

Ebtisam Amar
LiP6 Laboratory
University of Pierre and Marie Curie
Paris, France

John Attia
Department of Electrical and
  Computer Engineering
Prairie View A&M University
Texas A&M University System
Prairie View, Texas

Abdalkarim Awad
Department of Computer Science
University of Erlangen
Erlangen, Germany

LeRoy A. Bailey
Department of Computer Science
  and Engineering
Michigan State University
East Lansing, Michigan

Jalel Benothman
PRiSM Laboratory
University of Versailles
Versailles, France

Andrea Boni
Department of Information Engineering
University of Parma
Parma, Italy

Khaled Boussetta
Institut Galilée
University of Paris 13
Villetaneuse, France

Shafique Ahmad Chaudhry
Department of Computer Science
National University of Ireland
Cork, Ireland

Jiming Chen
State Key Laboratory of Industrial
 Control Technology
Zhejiang University
Hangzhou, China

Paolo Ciampolini
Department of Information Engineering
University of Parma
Parma, Italy

Sebastian Dengler
Department of Computer Science
University of Erlangen
Erlangen, Germany

Christos Douligeris
Department of Informatics
University of Piraeus
Piraeus, Greece

Falko Dressler
Department of Computer Science
University of Erlangen
Erlangen, Germany

Alessio Facen
Department of Information Engineering
University of Parma
Parma, Italy

Damianos Gavalas
Department of Cultural Informatics
University of the Aegean
Mytilene, Greece

Reinhard German
Department of Computer Science
University of Erlangen
Erlangen, Germany

Matteo Grisanti
Department of Information Engineering
University of Parma
Parma, Italy

Hossam S. Hassanein
School of Computing
Queen's University
Kingston, Ontario, Canada

Peter J. Hawrylak
Radio Frequency Identification
 Center of Excellence
Swanson School of Engineering
Pittsburgh, Pennsylvania

Pekka Jäppinen
Department of Information Technology
Lappeenranta University of Technology
Lappeenranta, Finland

Jehn-Ruey Jiang
Department of Computer Science
 and Information Engineering
National Central University
Jhongli, Taiwan

Deepak Kataria
HCL America
Florham Park, New Jersey

Ki-Hyung Kim
Department of Information
 and Communication
Ajou University
Suwon, South Korea

Charalampos Konstantopoulos
Department of Informatics
University of Piraeus
Piraeus, Greece

Lin Li
Department of Computer Science and
 Engineering
University of Electronic Science
 and Technology of China
Chengdu, China

Xiangfang Li
Department of Electrical and
 Computer Engineering
Texas A&M University
College Station, Texas

Alex X. Liu
Department of Computer Science
 and Engineering
Michigan State University
East Lansing, Michigan

Eugene Lutton
Faculty of Science and Information
 Technology
The University of Newcastle
Newcastle, New South Wales, Australia

Basilis Mamalis
Department of Informatics
Technological Educational Institution
 of Athens
Athens, Greece

Di Miao
State Key Laboratory of Industrial
 Control Technology
Zhejiang University
Hangzhou, China

Marlin H. Mickle
Electrical and Computer Engineering
University of Pittsburgh
Pittsburgh, Pennsylvania

Aikaterini Mitrokotsa
Faculty of Electrical Engineering,
 Mathematics, and Computer Science
Delft University of Technology
Delft, the Netherlands

Melody Moh
Department of Computer Science
San Jose State University
San Jose, California

Teng-Sheng Moh
Department of Computer Science
San Jose State University
San Jose, California

Carlo Morandi
Department of Information Engineering
University of Parma
Parma, Italy

Hamid Mukhtar
Department of Information
 and Communication
Ajou University
Suwon, South Korea

Ilaria De Munari
Department of Information Engineering
University of Parma
Parma, Italy

Nidal Nasser
Department of Computing and
 Information Science
University of Guelph
Guelph, Ontario, Canada

Suat Ozdemir
Computer Engineering Department
Gazi University
Ankara, Turkey

Grammati Pantziou
Department of Informatics
Technological Educational Institution
 of Athens
Athens, Greece

Santiago Pujol
Department of Civil Engineering
Purdue University
West Lafayette, Indiana

Guy Pujolle
LiP6 Laboratory
University of Pierre and Marie Curie
Paris, France

Lijun Qian
Department of Electrical and
  Computer Engineering
Prairie View A&M University
Texas A&M University System
Prairie View, Texas

Sk. Md. Mizanur Rahman
Department of Computing and
  Information Science
University of Guelph
Guelph, Ontario, Canada

Brian Regan
Faculty of Science and Information
  Technology
The University of Newcastle
Newcastle, New South Wales, Australia

Andrea Ricci
Department of Information Engineering
University of Parma
Parma, Italy

Pedro M. Ruiz
Faculty of Informatica
University of Murcia
Murcia, Spain

Tarek El Salti
Department of Computing and
  Information Science
University of Guelph
Guelph, Ontario, Canada

Juan A. Sánchez
Faculty of Informatica
University of Murcia
Murcia, Spain

Geoff Skinner
Faculty of Science and Information
  Technology
The University of Newcastle
Newcastle, New South Wales, Australia

Abd-Elhamid M. Taha
School of Computing
Queen's University
Kingston, Ontario, Canada

Denis Trček
Laboratory of E-Media
Faculty of Computer and
  Information Science
University of Ljubljana
Ljubljana, Slovenia

Rolland Vida
Department of Telecommunications
  and Media Informatics
Budapest University of Technology
  and Economics
Budapest, Hungary

Attila Vidács
Department of Telecommunications
  and Media Informatics
Budapest University of Technology
  and Economics
Budapest, Hungary

Zachary Walker
Department of Computer Science
San Jose State University
San Jose, California

Yu Wang
Department of Computer Science
University of North Carolina
  at Charlotte
Charlotte, North Carolina

Bashir Yahya
PRiSM Laboratory
University of Versailles
Versailles, France

David K. Y. Yau
Department of Civil Engineering
Purdue University
West Lafayette, Indiana

Ming-Kuei Yeh
Department of Information Management
Nanya Institute of Technology
Chungli, Taiwan

Seung-Wha Yoo
Department of Information
  and Communication
Ajou University
Suwon, South Korea

RFID

I

Chapter 1

Medium Access Control in RFID

Kashif Ali, Abd-Elhamid M. Taha, and Hossam S. Hassanein

CONTENTS

A radio frequency identification (RFID) system overcomes challenges of other identification systems including barcode systems, optical character recognition systems, smart cards, and biometrics (voice, fingerprinting, retina scanning) because it does not require line-of-sight communication, sustains harsh physical environments, maintains a cost- and power-efficient operation, and allows for simultaneous tag identification. As with other radio systems, RFID system requires Medium Access Control (MAC) protocols to bypass different types of collisions as they waste network resources and slow down the reading procedure. Further efficiencies are required in the case of RFID due to the tags' limited capabilities in terms of processing and memory. The objective of this chapter is to provide an overview of state-of-the-art MAC procedures for an RFID system. We will highlight the requirements of the different procedures in terms of signaling, memory and processing, and differentiate between the different proposals based on nominal merits. We will also identify how MAC protocols behave and affect the overall RFID system performance under assorted practical scenarios: for example, mobility and high reader or tag densities. Finally, we will dispense some thoughts on the potential integration of RFID and wireless sensor networks (WSNs).

1.1 Introduction

Radio frequency identification (RFID) is a prominently emerging automated identification technology. It has an edge over other identification systems such as barcode systems, optical character recognition systems, smart cards, and biometrics (voice, fingerprinting, retina scanning) because it requires no line of sight for communication, sustains harsh physical environments, allows for simultaneous identification, and is cost and power efficient. An RFID effortlessly turns everyday objects into mobile network nodes which can be tracked, traced, monitored, trigger actions, or respond to action requests.

An RFID system is typically composed of an application host, a reader, and a set of tags. A tag is designed to store certain information, the size of which varies between 32 bits and 32,000 bytes. Tags can be either passive or active. A passive tag has no physical power source. It gathers energy from a reader's generated radio waves, sufficient to carry out processing and communication objectives. While most common in the RFID market, the passive tag has limited functionality for processing and communication. It processes simple state machines and has no medium sensing capabilities. However, the active tag has a power source for its processing needs and it may possess certain sensing capabilities for temperature or pressure. An RFID reader acts as a master for the tag and a slave for the application host. This master–slave concept is depicted in Figure 1.1. Tags targeted by a certain reader are said to be within that reader's interrogation zone. Specifically, an interrogation zone is the physical distance within which the strength of the electromagnetic waves, generated by the reader, is able to power the tags, receive the tags' signals, and successfully decode them—a process known as singulation. At any arbitrary instant, a reader can only read one tag within its interrogation zone and a tag can only be read by one reader. Tags within the interrogation zone of a reader, and

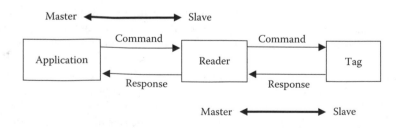

Figure 1.1 Master–slave architecture of the RFID system.

readers with overlapping interrogation zones, may simultaneously attempt to access the wireless medium for data communication.

However, this simultaneous wireless medium access results in collisions that undermine an RFID system's overall performance. To sustain system operability, efficient mechanisms for Medium Access Control (MAC) are required. As with other radio-based systems, the main objective of mechanisms aimed at regulating medium access, is to reduce collisions, be it in a proactive or a reactive manner. Proactively, collisions are avoided by distributing sufficient information about the access requirements of elements sharing the medium. Reactive mechanisms respond to collisions and attempt to speed the system's recovery from a collision stall. Conventional collision avoidance methods, such as Carrier Sense Multiple Access (CSMA) cannot be adopted for RFID systems, especially when passive tags are used due to power limitations and its basis of reflection-based communication, that is, use of backscattering modulation. Avoidance mechanisms also carry the risk of increasing the overall cost of tags and shrinking the potential interrogation zone for a reader. Proposals for RFID systems have therefore favored reactive approaches to alleviate collisions. Specific to RFID systems, collisions can be classified based on the type of entities involved:

Tags-to-reader collisions: Occur when more than one tag within a reader's interrogation zone attempts to reply to the reader's requests at the same time, as depicted in Figure 1.2. Tags-to-reader collisions are the most devastating, especially when passive tags are involved. They result in reduced reading rates, wasted resources, and increased delay.

Readers-to-tag collisions: Occur when one tag is interrogated by more than one reader, as shown in Figure 1.3. In such a scenario, multiple readers try to singulate a single tag which results in corruption of the tag's internal state. As a result, the tag may not be detectable.

Reader-to-reader collisions: Are result of the conventional frequency interferences, that is, multiple readers within each other's interference zones are locked on the same frequencies. Existing mechanisms such as frequency-hopping, dynamic frequency allocation, and dynamic power adjustment are utilized to hinder these collisions.

Our objective in this chapter is to survey mechanisms that have been proposed in the literature to alleviate the different types of collisions in RFID systems. We also provide a comprehensive and up-to-date classification of the MAC protocols, compared

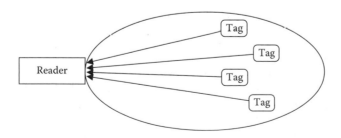

Figure 1.2   Multiple tags-to-reader collisions.

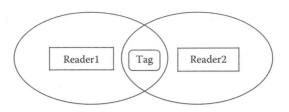

Figure 1.3 Multiple readers-to-tag collisions.

to other literature surveys [1,2]. Toward this end, we review preliminary aspects relative to MAC in RFID systems and identify the reasons for the popularity of TDMA-based solutions. This review is offered is Section 1.2. We then turn our focus to tag collisions in Section 1.3 and offer a taxonomy for the different solutions proposed based on the memory and processing required from a tag. We follow this by a survey of proposals for reader collisions in Section 1.4. In both sections, we also qualitatively compare different proposals based on general solution requirements. Finally, we conclude the chapter in Section 1.5 with some insights into the challenges and opportunities relative to medium access in current and future RFID systems.

1.2 Preliminaries on MAC in RFID Systems

In a radio-based system, MAC protocols provide mechanisms for channel access control, allowing multiple devices to share the same physical medium. The most prominent MAC techniques are based on the CSMA, Multiple Access with Collision Avoidance (MACA), or the Aloha protocol. RFID systems use half-duplex, point-to-point communication links in which communication between the reader and the tag is based on backscatter modulation. In backscattering, the tag sends its serial number by adjusting the antenna reflectivity to modulate the reflected signal. The nature of reflection-based communication results in failure of conventional MAC techniques as tags cannot sense the medium, detect collisions, or sense the presence of other channel traffic. This means that no collision avoidance or resolution mechanisms could be implemented at the tags. Hence, MAC in RFID systems is confined to resolving of collisions only at the reader, resulting in anticollision algorithms for both tag collisions and reader collisions, mainly employing Aloha-based contention avoidance schemes.

RFID systems exhibit very short durations of very high bursty activity followed by relatively long durations of inactivity. For a reader, the number of tags within its interrogation zone is usually unknown. This adds critical design requirements of scalability and protocols adaptability. Generally speaking, four different approaches exist in the literature in handling multi access issues in radio technologies. The approaches regulate access based on time, space (location), frequency, or code. Code Division Multiple Access (CDMA) supports high-rate data multiplexing using the spread-spectrum (SS) technique. In conventional SS, each user encodes data packets using an orthogonal spreading code, allowing successful transmission by multiple users. However, complex receiver design and higher computational and energy requirements restrained adaptation of CDMA technology in RFID systems. A combination of TDMA- and CDMA-based schemes has been investigated recently [3]. However, the CDMA scheme is a largely unexplored area in RFID systems.

Space Division Multiple Access (SDMA) [4,5] spatially reuses the channel. The philosophy behind the SDMA procedure is that at practically any given time there will be a limited number of RFID tags spatially colocated at the same position, independent of the total number of tags that may be present in the overall interrogation zone. Therefore, by spatially isolating the tags, the interference caused by other tags can be minimized. Numerous spatial isolation techniques exist with the most notable ones being adjusting the reader power levels (power control) [6–8], using adaptive arrays and multiple input multiple output (MIMO) antennae [9], and the use of electronically controlled directional antenna [5].

The power control cluster-based algorithm [6,8] divides the interrogation zone into smaller clusters based on the distance to the reader by adjusting the reader's power levels. Tags in each cluster are read separately. Since the number of tags in a cluster is less than that in the whole interrogation zone, collisions are minimized. An example of such a partitioning is shown in Figure 1.4, where the interrogation zone is divided into three clusters: d, d', and d''. When the reader sends a request, only those tags in the current cluster respond. For instance, assume the reader has read the tags from cluster d and has just sent a request to cluster d'. In that instance only tags from cluster d', that is, tags marked as T', will respond to that request. After all tags marked as T' have been read, they are put into a sleep mode. Then, the reader's zone is increased to d'' and a new request is sent, and so on. Similar transmission control schemes have been used

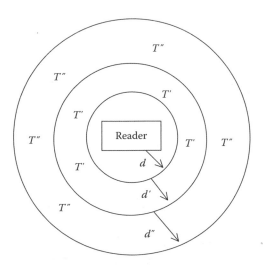

Figure 1.4 Clustering based on the distance between the reader and the tags, using reader's power levels to adjust its interrogation zone.

for reader collision avoidance [7], where reader's power levels are adjusted to reduce the interrogation overlaps.

Smart antenna techniques such as adaptive array antenna and MIMO antenna [9] are used to maximize system throughput by reducing the collisions. The antenna array adaptively nullifies interferences from other array elements, that is, one element receives the desired signal whereas other array elements are used to remove the interference to maximize the received signal strength. With MIMO, however, each antenna element receives a superposition of the multiple transmitted streams with different spatial signatures. These differences are used to separate multiple streams with signal processing at the receiver. These two techniques lower tag collisions, but with significant increase in the reader cost.

Electronically controlled directional antennas [5] are used to adaptively adjust the directional beam of the RFID reader at subset of tags, one at a time. Such techniques are commonly known as adaptive SDMA and are depicted in Figure 1.5. To read a specific tag or all the tags within the interrogation zone, the reader scans the area using the directional beam until the desired tag is detected or, in the latter case, when all the tags have been read by the reader. The SDMA technique is effective in reducing collisions but with a relatively high implementation cost because of sophisticated antenna systems, and is therefore restricted to a few specialized applications. One exceptional approach is the aforementioned power control scheme [6], which is low in cost and is effective in reducing both reader and tag collisions.

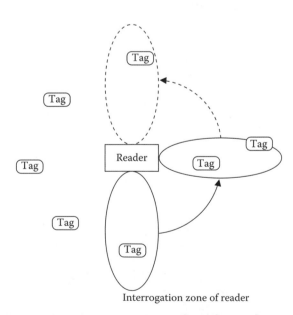

Figure 1.5   Adaptive SDMA with an electronically controlled directional antenna.

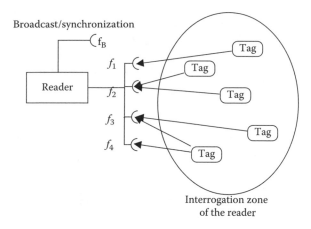

Figure 1.6 FDMA procedure, several frequency channels are available for communication between the tags and the reader.

Frequency Domain Multiple Access (FDMA) proposes the use of multiple channels, each with distinct carrier frequencies for communication. In the context of RFID systems, this implies the availability of broadcasting frequency for the reader and several frequencies for the tags that they can lock on to. The reader, using the broadcasting frequency, synchronizes and issues the interrogation commands. The tags reply to the reader using one of the many available frequencies, shown as f_1, f_2, f_3, and f_4, in Figure 1.6. The advantage of FDMA is the availability of noninterfering frequencies for concurrent communication by multiple tags and readers. The FDMA technique has not been significantly utilized in RFID systems. The reason is the impracticality for tags and relatively high cost of the readers as a dedicated receiver must be provided for every reception channel, restricting the use of FDMA for limited and specialized applications.

Time Domain Multiple Access (TDMA) relates to techniques where the available channel is divided, along the time dimension, between potential participants. TDMA, in a modified and mostly hybrid manner, is used in other prominent networks, for example, Global System Mobile (GSM), Bluetooth, and IEEE 802.16 (WiMax). The TDMA technique is by far the most dominant medium access protocols in RFID systems. This is because of TDMA simplicity, low processing overhead for passive tags, and low complexity (computational, processing, and monetary cost) compared with other available procedures such as FDMA, CDMA, OFDMA, etc. In the context of the RFID system, the TDMA procedure is further classified into tag-driven and reader-driven, as is shown in Figure 1.7. The tag-driven procedure operates in an asynchronous fashion as the reader does not control the data transfer. Tag-driven procedures are naturally very slow and inflexible with limited applicability. Most of the TDMA procedures are therefore based on the reader-driven approach. The reader-driven approach is a synchronous mechanism, with all the tags' data transfer handled by the reader, that is, control when the tags transfer data and select which ones, amongst the tag population, will transmit

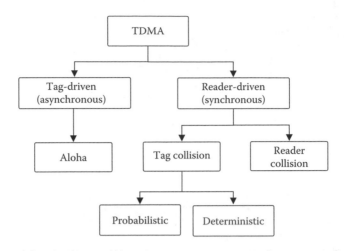

Figure 1.7   Classification of the time division multiple access procedures.

its data at any given time. This selection process, that is, isolating an individual tag from a large group of tags, is known as the singulation process. The reader-driven approach is used to tackle the collisions in RFID systems.

Tag collision schemes usually involve a logical partitioning of the tag population using tree-based algorithms or probabilistic-framed Aloha schemes, into a more manageable set of tags. Reader collisions employ conventional contention resolution approaches such as scheduling, interference learning, and coloring schemes to tackle reader collisions, that is, reader-to-reader and readers-to-tags collisions. The various collision resolution schemes for tags and readers' collisions are classified based on their underlying techniques, and are discussed next, in a sequential manner, in the following sections.

1.3   Tag Collisions

Tag collisions are the most common source of collisions in RFID systems. They take place when multiple tags are within the interrogation zone of a reader and simultaneously reply to the reader's commands. The passive RFID tag is by design reader-driven as it does not have carrier sensing or intertag communication capabilities. Therefore, tag collisions are resolved by the reader utilizing techniques collectively known as anticollisions schemes. These anticollisions schemes are classified in literature, as depicted in Figure 1.7, to be either deterministic or probabilistic [1,5].

In deterministic mechanisms the reader splits and identifies a set of tags to respond in a given time. Splitting is based on contention information obtained from the previous interrogation cycle and attempts to reduce contention for the next cycle. Deterministic anticollision mechanisms fall under the general algorithmic classification of tree-based algorithms because of their splitting approach. The deterministic mechanisms utilize either tags' serial numbers (identification codes) or randomly generated numbers to be used for splitting of the tree branches. Under certain circumstances, the deterministic approach may take a considerably longer time. However, it does not suffer from the tag starvation problem. In tag starvation, a tag may not be identifiable for a long time, and in the worst case, might not be able to be read at all.

In probabilistic mechanisms, the reader communicates the frame length, and the tag, randomly transmits a particular slot in the frame. The frame size may be adjusted, based on the information from the previous interrogation cycle, encouraging adaptability according to tag density and distribution. The frame process is repeated until all the tags have been identified. The probabilistic approach is fast, due to its low overhead, but suffers from tag starvation syndrome.

1.3.1 Deterministic Anticollision Mechanisms

Deterministic anticollision mechanisms are essentially tree-based anticollision algorithms. The tree-based algorithm is a two-way handshake algorithm involving sequences of interaction between the reader and the tags, known as the interrogation cycle. The objective of these interrogation cycles is to split the tags, using their serial numbers (IDs) or randomly generated numbers, into a more manageable set of tags. The splitting of the tree, mostly binary trees, into two branches (leaves) is based on the bit collisions and their respective positions; the location of which is obtained from the previous interrogation cycle. Obtaining collision information at the bit level requires that the precise bit position of collision is identifiable by the reader. For this particular reason, normally, either the Manchester coding or the NRZ (non-return-to-zero) bit coding is used by the RFID reader.

To understand the deterministic tree-based algorithm, we trace the execution of the conventional binary search tree algorithm [10]. The objective of using the binary search algorithm with multiple interrogation cycles is to singulate a tag from a larger set. The two-way handshakes during each interrogation cycle (iteration) are the commands that the reader broadcasts based on its previous iteration for the following iteration. The set of commands constitutes four main commands, which are as follows.

1. REQUEST: Carries a serial number with the tag as a parameter. If a tag's own serial number is less than or equal to the received serial number, the tag sends its own serial number back to the reader. Otherwise, the tag will not respond.
2. SELECT: Carries a serial number with the tag as a parameter. Only a tag with an identical serial number is selected for the processing of other commands, for example, reading and writing data. Only the selected tag will continue to respond to the reader commands.
3. READ_DATA: The selected tag sends stored data to the reader.
4. UNSELECT: Cancels the selection of an up-until-now active tag, that is, *mutes* the tag. Beyond this, the unselected tag becomes completely inactive and does not respond to further REQUEST commands until it is reset by the reader.

Let us assume that there are three RFID tags within the interrogation zone of the reader with 4-bit IDs of 1010_b, 1011_b, and 1110_b. The singulation process starts with the RFID reader sending the highest possible serial number. The purpose is to make all

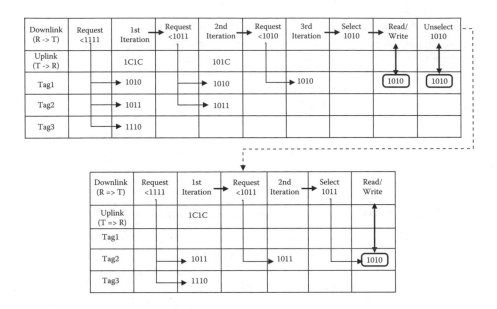

Figure 1.8 **Execution trace of the binary search tree anticollision algorithm using three tags and a single reader.**

tags respond so that the exact bit collisions among all the tags' ID's could be determined. In our example, shown in Figure 1.8, the first iteration of the algorithm begins with the reader transmitting the REQUEST command with the serial number 1111_b as an argument. The serial number 1111_b is the highest possible in this case. As the serial number of the tags in the interrogation zone is less than the requested serial number, all three tags reply back with their IDs. This first iteration results in collisions (C) at position 0 and 2, starting from the least significant bit (LSB), that is, 1C1C.

The third bit is the highest valued bit at which the collision has occurred during the first iteration. This implies that there is at least one tag between 1100_b and 1011_b. The binary search algorithm at this point splits the search into two subsets in an attempt to limit the search zone for subsequent interaction. The algorithm sets the third bit of the request to 0, that is, 1011_b. The LSBs after the third bit are all set to 1 in an attempt to capture all the tags whose two most significant bits (MSBs) are 10. The reader now broadcasts the command REQUEST with argument 1011_b. Two tags fulfill the command's criteria, that is, tag1 and tag2. The response back from these two tags causes collision at the bit position 0, that is, 101C. The reader, repeating the same splitting procedure, now chooses 1010_b as the requesting string for the subsequent iteration, that is, the second iteration. This request is fulfilled by only one tag tag1 which is now singulated. For now, no further iterations are required, as the reader has successfully detected a single tag without collision. Using subsequent SELECT commands, tag1 is selected using the detected tag ID address and can now be read or written by the reader without interference from other tags. At this point, other tags are silent as the READ_DATA command is selective. The tree, from the request point of view, is shown

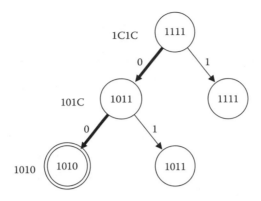

Figure 1.9 Singulation tree for the example RFID system.

in Figure 1.9. The string shown in the tree nodes are requested and sent by the reader, whereas the string next to each circle is the received response. The node splits into two children nodes, appending 0 and 1 at the MSB collision, to the left and right child node, respectively. This process continues until tag 1010_b is singulated, which is distinguished in the figure by a double concentric circle.

After the completion of the required read/write operation, the selected tag (tag1) is muted by the use of the UNSELECT command. The muted tag is completely inactive and does not reply to any further REQUEST commands. Muting the tags reduces the number of responding tags in the subsequent iteration of the singulation process, therefore resulting in lower collisions and less iterations. Referring back to Figure 1.8, the singulation of the tag (tag2) requires one less iteration, compared with tag1's singulation. The muting of tag1 has created a positive impact by yielding low collision as its serial number; otherwise, it would have collided with the tag2 serial number.

We further classify the deterministic anticollision mechanisms, as shown in Figure 1.10, based on whether they use a collision tracking or a collision detection approach.

Collision tracking: In the collision tracking method, both the reader and the tag maintains a certain amount of collision information from the previous interrogation cycle. The information, mostly as pointer to the most recent query, collision bit, or node in the tree, is used to send a subsequent query for the next interrogation cycle. The binary search tree algorithm (explained above using exemplary RFID system) and its variants [11–22], fall under the collision tracking class of the deterministic anticollision algorithms. The variants of conventional binary tree algorithms differ in their enhancement objectives, that is, shortening execution time, reducing memory usage, and reducing processing overhead or eliminating the unnecessary iterations.

In the conventional binary tree algorithm, when a tag is successfully singulated, read, and muted, the singulation process initiates from the root for the subsequent tag. This was seen, for instance, in our example above where the reader sends out the initial query of 1111_b, that is, the highest possible serial number, after the first tag, that is, tag1, is singulated and unselected. This results in unnecessary overhead, both

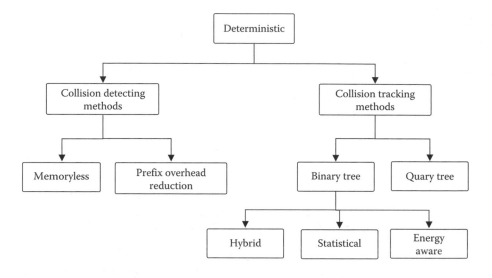

Figure 1.10 Classification of the deterministic TDMA schemes for tag anticollision.

from communication and processing perspectives. The overhead can be eliminated, after successful singulation, by setting the starting query point to the queued query from the last interrogation cycle. This translates to selecting either the parent node of the current query node or its sibling node [20,21]. This simple modification speeds up the interrogation process and reduces overhead by utilizing the saved (queued) information. The conventional binary tree-based algorithm recognizes only a single tag at a time through setting the most significant collision bit to either 1 or 0 during the interrogation process. The work in Refs. [14,17] notes that this results in low performance and claims that without any fundamental change in the tree navigation process two tags are identifiable at one time. This takes place during the last bit of the collision sequence, that is, when the LSB collision has been resolved. This is evident, as the LSB collision indicates the presence of two tags, that is, with collision bits of 0 and 1, therefore, speeding the interrogation process. The use of other mechanisms such as interference cancellations [18], utilizing the similarities among the tags' IDs [23], using height-oriented, depth-first search algorithms with tag state support [24], and secure variation of the tree algorithms [12,22] have also been investigated. All these approaches enhance the binary tree algorithm without deterring significantly from the conventional algorithm.

We further classify the collision tracking approach to be either statistical, hybrid, or energy aware, as is shown in Figure 1.10. The statistical approach exploits information stored in previous interrogation cycles (iterations) to assist in the forthcoming cycle. Adaptive binary splitting algorithms [25–28] initiate the conventional tree splitting procedure by using the information obtained from the last reading cycle. The fixed tags, that is, static tags, which have been identified in the previous cycles will be reidentified again by the reader in the next reading cycle. As the reader knows about the fixed tags it can avoid collisions, whereas conventional methods are used to identify the mobile tags. The mobile tag is defined as the tag just coming in, and therefore was not present during the last reading cycle, or has moved out of the interrogation zone of the reader. These adaptive splitting algorithms significantly reduce the number of collisions while supporting mobility but at the cost of substantial storage requirements. In such scenarios, where a significant portion of tags are mobile, the adaptive splitting algorithms will perform no better than the conventional binary algorithm.

The hybrid tree approach [29] is a combination of the deterministic and probabilistic approach. It combines the tree-based protocol with a slotted back-off mechanism. The algorithm uses a 4-ary tree instead of the conventional binary tree, with slotted back-off aimed at reducing the number of unwanted and unnecessary query commands. Upon responding to the reader, a tag sets its back-off timer using a part of its ID. If there is a collision the reader can partially deduce how the IDs of the tags are distributed and potentially reduce the unwanted idle cycles. Like the statistical approach of Ref. [26], this approach utilizes the information from previous reading cycles to avoid collision for the static tags and starts the querying process from the root to accommodate the tag mobility. In addition to inheriting the drawbacks of the statistical approaches, the hybrid approach increases the tags' cost as it requires additional functionality, for example, back-off timer.

The MAC scheme's objective of reducing collisions usually comes at the expense of increasing energy consumption. This energy consumption occurs because of additional processing and communication overhead imposed by the MAC protocols. The energy-aware approaches [15,30–32] address the energy issues, mainly the tag energy consumption, while maintaining low collisions. The approach in Ref. [15], while utilizing the conventional binary tree algorithm, involves the reader taking action if bit collision is detected. This translates to the reader sending a special symbol to the tags to stop sending data back. Fewer bits are thus sent by the tags, indirectly reducing energy consumption, by reducing communication overhead. More efficient schemes [32] use multiple time slots per tree node and three different anticollision protocols. The motive is that to detect the tags, the binary tree algorithm relies on the collision as it transmits redundant queries to the tags. This is important in the conventional binary tree algorithm as it needs to determine which subtree to query in the subsequent interrogation cycle. The tags are allowed to transmit responses within a slotted time frame, thereby avoiding collisions. The queries are used to find colliding subtrees as well as to read the tags. These two mechanisms help to read more tags with fewer queries, thereby reducing the energy consumption at the tags.

Collision detection: Collision detection methods are stateless algorithms in that they do not exercise tracking. Relative to collision tracking, detection algorithms may take longer time to read all tags within an interrogation zone. Their main advantage is in their nonstorage requirements, especially for the tags and their comparatively low communication overhead. We classify the collision detection methods into memoryless and prefix reduction, as is shown in Figure 1.10.

Query tree protocol (QT) [33] and variants [15,28,34] are prominent examples of the memoryless approach in which each tag does not need additional memory beyond storing its serial number. This means that no storage is required for random serial numbers, pointers or states, as it was required by the binary tree of the collision tracking algorithms. The QT uses the tag's serial number for the tree splitting mechanism. The reader broadcasts the query, using a string of bits as an argument. The tag receiving the query matches the MSB of its own serial number with the broadcasted serial number. If there is a match the tag transmits the remaining least significant portion of its serial number, otherwise, it remains silent. The reader maintains the queue for the transmitted strings of bits. At the beginning of the frame, the queue is initialized with two 1-bit strings, 0 and 1. The reader pops a bit string from the queue and broadcasts to all tags in its interrogation zone. If the tag responses collide, the reader pushes two 1-bit longer bit strings, compared to the last bit string transmitted, into the queue. By expanding the query until there is either a response or no response—all the tags will respond eventually—the query bit string traverses from the MSB to the LSB for the possible serial numbers. Contrary to binary trees, a QT imposes simple functions and requires no state or pointer maintenance by the tag. However, the tag reading delay for the QT protocol is significantly affected by the numerical distribution of the tag serial number, in that the reading delay is increased when tags have similar serial numbers.

The QT and its variants has noticeable overhead in terms of the number of bits that are sent as query argument, that is, the string bits. The reason is the string bits have to traverse, from most significant to least significant, equal to the length of the serial number. This results in communication overhead as redundant bits are transmitted, both by tags and readers, during the interrogation cycles.

Prefix overhead reduction algorithms [16,18,19,35] maintain prefix and iteration overhead reduction approaches which enhance the performance of the memoryless mechanisms. Prefix Randomized Query Tree (PRQT) [18], falls under the prefix reduction mechanism and overcomes the limitation of the QT-based algorithm, that is, create longer reading cycles due to tag length and its distributions. PRQT similar to QT is a stateless approach, however, and it differs from QT in that it uses prefixes chosen randomly by tags (rather than using their ID-based prefixes), thereby providing additional memory space. Each tag generates a random length prefix which is then used during the interrogation process. The optimal length prefix depends on an estimate of the number of tags within the interrogation zone. Tag estimation is proposed in the adaptive optimal incremental prefix length algorithm [35]. The algorithm begins by setting a small initial prefix length followed by the polling of all possible prefixes. The initial prefix length is then increased repeatedly until the collision ratio satisfies a prescribed condition. By reducing the prefix overhead, the collision detection mechanisms enjoy a faster reading rate and lower overheads with marginal need of additional memory at the tags.

1.3.2 Probabilistic Anticollision Mechanisms

In the probabilistic mechanisms the reader communicates the frame length and the tag picks a particular slot in the frame for transmission. The reader repeats this process until all tags have been successfully transmitted at least once. Reader-controlled synchronization is necessary as the tag has to transmit within its slotted time frame. The frame size may be adjusted based on the collision, idle, and occupied frame information from a previous interrogation cycle for the subsequent cycle. This encourages frame adaptability according to tag density and distribution, thereby reducing idle and collision frames.

The probabilistic approach is faster in comparison to the deterministic approach because of its low overhead. However, it suffers from tag starvation syndrome.

We will now illustrate the details of the conventional slotted Aloha anticollision procedure Section 1.3.1 using the example used from [5]. Consider again three tags with IDs of 1010_b, 1011_b, and 1110_b. Similar to the deterministic approach, the probabilistic approach is a chit-chat protocol that uses a sequence of interactive commands. The set of commands are as follows:

1. REQUEST: Synchronizes and prompts all tags to transmit their serial number to the reader using one of the time slots that follow. For our sampler RFID system, there are three available time slots.
2. SELECT: Sends a specific serial number to the tag as a parameter. Only the matching tag, that is, the tag whose serial number matches with the passed parameter, is flagged for further operation, allowing it to read and write. Tags with conflicting serial numbers, however, are still responsive to the REQUEST command.
3. READ_DATA: The selected tag sends stored data to the reader. In real RFID systems, other commands will also be available but they are omitted here for the purpose of simplicity.

The execution trace of the slotted Aloha protocol is shown in Figure 1.11. After a certain time interval, the reader periodically transmits the REQUEST command. After receiving the REQUEST command, the tag randomly selects one of the three available slots. The tag uses the slot to transmit its serial number back to the reader. Due to a random slot selection method, there is a collision between `tag1` and `tag2` in slot1. Only tag3 is successful in transmitting its serial number back to the reader. The successful tag is selected for subsequent reading and writing after it is been selected using the SELECT command. If no transmission was successful, the REQUEST command is iteratively repeated until the serial number can be received successfully and no collision is observed in the frames. As we have witnessed, in the probabilistic anticollision algorithms, the tag randomly selects a slot number in the frame and responds to the reader using the slot it

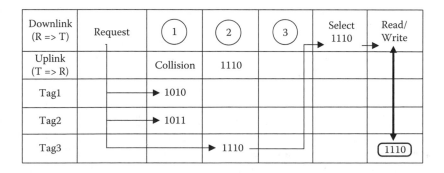

Figure 1.11 Execution trace of the conventional slotted Aloha protocol, using example RFID system of three tags and a single reader.

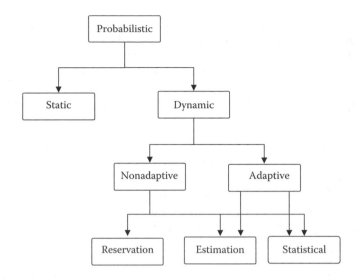

Figure 1.12 Taxonomy of the probabilistic TDMA-based schemes for tag anti-collision.

selected. When the number of tags is small, the probability of tag collision is low, and the time used to identify all tags is relatively short. As the number of tags increases, however, the probability of collision becomes higher and the time used to identify the tags increases rapidly. Therefore, the probabilistic algorithm performance depends significantly on the number of tags within the interrogation zone of the reader.

Generally speaking, and as shown in Figure 1.12, we classify the probabilistic mechanisms into two main groups: static and dynamic. The number of slots in the static approach is fixed, making it mostly used in instances of low tag density. An example of static algorithms includes those based on Aloha and framed slotted Aloha. While a fixed frame size leads to a simple implementation, static algorithms risk certain inefficiencies. For example, long frames used with a small number of tags, or vice versa, results in delays and resource underutilization. Such issues are addressed in dynamic approaches in which the frame size is adjusted based on the collision rate from tags within an interrogation zone. For example, the Dynamic Framed Slotted Aloha (DFSA) algorithm [5] determines the frame size based on information such as the number of slots used to identify the tag and the number of the slots with collisions.

Variations of the DFSA algorithm differ in frame size by changing approach. In one such variation, the frame size is adjusted based on some predefined threshold. For instance, the frame size is increased when the number of collision slots exceeds a preset threshold. As the collisions decrease, the frame size is reduced to its previous value. This dynamic adjustment of frame size allows readers to change its frame size depending on the tag density. That is to say, the frame size is increased as with an increase in number of tags increases the collisions, and vice versa.

In another variation, the interrogation cycle begins by setting the initial frame size to two or four. The frame size is incrementally increased with unsuccessful transmissions until at least one tag transmits successfully. If at least one tag is successfully identified, the current reading process is aborted and reinitiated from the beginning at its initial frame size. Despite the general performance of the DFSA and its variants, changing the frame size only may not be sufficient given the number of tags, as the frame size cannot be infinitely increased. For example, in the second variation above, when the number of tags is small the reader can identify all tags without much collision. If the number of tags is large, however, the number of slots needs to exponentially increase because the reader always starts with the initial minimum frame size.

We further classify the dynamic algorithms as adaptive or nonadaptive (see Figure 1.12). Adaptive algorithms use statistical information for frame size adjustment. Adaptive algorithms accommodate tag mobility more than nonadaptive algorithms because they reduce the probability of tag collisions while simultaneously expediting the identification of the RFID tags. Dynamic algorithms (both adaptive and nonadaptive) can be further classified based on their use of reservation, estimation, or statistical techniques. Estimation-based approaches are the most important Aloha-based schemes [10,37–43] and utilize tag count estimation techniques to adjust the frame accordingly. The Advanced Framed Slotted Aloha (AFSA) algorithm [10] estimates the number of tags, prior to initiating the reading process, and adjusts the frame size based on the estimation. The Chebyshev's inequality estimation function is utilized for tag count estimation. However, AFSA has to increase the frame size indefinitely, which is impractical with high tag density.

The Enhanced Dynamic Framed Slotted Aloha (EDFSA) [42] overcomes the limitation of AFSA by bounding the estimation based on maximum frame size, that is, setting an upper bound on the frame size. EDFSA initially estimates the number of unread tags using the AFSA estimation function and then partitions the unread tags into a number of groups. Only one group of tags is allowed to respond at one time. The number of groups that give output to the maximum throughput for the subsequent reading cycle is calculated and adjusted after every interrogation cycle. The reader transmits the number of tag groups and a random number to the tags when it broadcasts a request. A tag receiving the request generates a new number from the received random number and the tag's own serial number and divides the new number by the number of tag groups. Only the tags having the remainder of zero respond to the request. The estimation and grouping of the unread tags are performed after each read cycle until all the tags have been read. Although EDFSA overcomes the limitation of AFSA protocol, it has longer reading cycles and additional functionality at the tag.

Statistical algorithms [16,41,44,45] exploit statistical information to improve the read time of the RFID systems. Adaptive Slotted Aloha Protocol (ASAP) [41] utilizes information relative to the tag population, from previous interrogation cycles and reading processes, to estimate the number of tags presently within the interrogation zone of the reader. The Maximum Likelihood (ML) based estimation algorithm is used for this purpose. The frame size is adjusted optimally to reflect the tag estimation. The mobility is supported by accounting for tag arrival and departure rate, while initiating the estimation and frame adjustment at the beginning of every interrogation cycle. The statistical algorithm resembles the deterministic anticollision category and shares the same pros and cons.

1.3.3 Discussion

Most of the anticollision protocols discussed so far in this chapter attempt to maximize specific gains at the expense of other performance merits. In minimizing collision, for example, an algorithm increases the reading rate. This increase, however, comes at the expense of additional memory, overhead communication and processing requirements that may increase the cost of a tag. In what follows, we discuss some of the performance merits traded off in the design of anticollision algorithms.

1. Speed: The rate at which tags can be read. This is a general objective that is sought by most proposals.
2. Overhead: The communication and processing overhead for the tags and the readers. The communication overhead includes sending additional bits which otherwise would not be sent. The processing overhead includes additional interrogation cycles or data crunching beside the nominal.
3. State: The amount of state that can be reliably stored on the tag. Additional storage means additional memory. A reader may maintain statistical information but the storage cost at the reader is significantly less than at the tag.
4. Mobility: The ability to accommodate tags which enter and leave an interrogation zone during the interrogation process.
5. Scalability: The ability to accommodate high tag deployment densities.
6. Cost: In terms of additional functionality and memory requirements at both tags and reader. Increase in the cost also reflects an increase in the device monetary value.

Tables 1.1 and 1.2 show the scheme improvements, depicted by a checkmark ($\checkmark$), and the trade-offs, depicted by a down arrow ($\downarrow$), for the deterministic and probabilistic anticollision algorithms, respectively. A scheme may not affect certain metric but may do so indirectly. This shown by –. For instance, the prefix reduction scheme's main objective is to reduce the communication overhead which indirectly reduces the time it requires to interrogate the tags, thereby accelerating the tag reading rate. It is evident from the tables

Table 1.1 Comparison of Deterministic Anticollision Algorithms

Approach	Speed	Overhead	Statefull	Mobility	Scalability	Cost
Energy-aware [15,31,32]	$\checkmark$	$\checkmark$	–	$\downarrow$	$\downarrow$	$\checkmark$
Hybrid [29]	$\checkmark$	$\checkmark$	$\downarrow$	$\checkmark$	–	$\downarrow$
Memoryless [15,28,33,34]	$\checkmark$	$\downarrow$	$\checkmark$	–	–	$\checkmark$
Prefix [18,19,35, 36,46]	–	$\checkmark$	$\downarrow$	–	–	–
Statistical [25–28]	$\checkmark$	–	$\downarrow$	$\checkmark$	$\checkmark$	$\downarrow$

Table 1.2   Comparison of Probabilistic Anticollision Algorithms

Approach	Speed	Overhead	Statefull	Scalability	Cost
Static [8]	↓	✓	–	↓	–
Reservation [47]	✓	↓	↓	–	↓
Estimation [10,37–42]	✓	↓	↓	–	–
Statistical [16,41,44,45]	✓	–	↓	✓	↓

that each of the schemes, either deterministic or probabilistic, come with a trade-off and the suitability of each varies under different circumstances. Therefore, with diverse sets of RIFD applications, no single existing scheme fulfills all performance metrics. For example, in a typical warehouse scenario, the reader deployed at the docking doors or the conveyor belts is to interrogate hundreds of mobile tags, or possibly thousands with item-level tagging, as they move from one section (for example, docking doors) to another (for example, sorting) section of the warehouse. Therefore, it is of utmost importance to be able to read all the tags encompassing a certain performance metric, as speed, mobility and scalability matter more than overhead reduction. For such scenarios, the statistical approach is the most promising, compared to other high-speed reading schemes such as hybrid, with its limited scalability or memoryless, with its low support for mobility. An RFID-based access control system does not require scalability and high reading rates, therefore any generic tree-based algorithm is sufficient.

1.4   Reader Collisions

Readers with intersecting interrogation zones may interfere with each other's operation to the extent of barring a reader from identifying tags within its reach. As with other radio systems, a reader's operation may also be affected by other readers even if their interrogation zone does not overlap. In either case, a reader collision results [48,49]. While the use of multiple frequencies may effectively nix reader collisions, RFID tags are limited functionality devices that are incapable of differentiating between multiple readers and they cannot be mass produced to communicate using multiple frequencies.

We classify the reader collision schemes, shown in Figure 1.13. Multiple mechanisms based on scheduling, coverage, and learning are adopted to resolve collisions.

In the scheduling-based approach, frequency and an associated time slot is scheduled for the reader. Scheduling can be centralized or distributed, and can support both static and dynamic frequency assignments. Occasionally, the scheduling is enhanced with tag support, that is, additional data is stored within tags at the expense of additional memory requirement. However, scheduling without tag support is the most common approach. Algorithms such as colorwave [50,51] fall under this category. Learning-based approaches, on the other hand, are based on hierarchical online learning, genetic algorithms, and neural network methods. The learning approach attempts to minimize

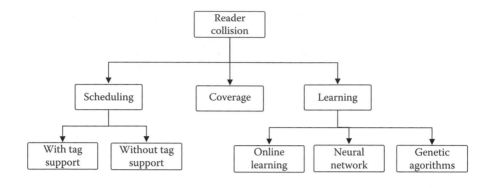

Figure 1.13   Taxonomy of reader collisions schemes.

collision by learning the collision patterns and assigning frequencies based on the learned pattern. Algorithm such as HiQ [52] falls under this category. Alternate approaches such as the use of beacon channels [53], central cooperators [54], and handling reader collisions as coverage problems [55] also exist in literature.

To achieve optimal frequency channel assignment, a pair of distributed algorithms called Distributed Color Selection (DCS) and Variable-Maximum Distributed Color Selection (VDCS) are introduced [50,51]. Both DCS and VDCS fall under the scheduling category. An objective of both algorithms, simply known as colorwave, is to color each node of the multiple reader networks using multiple colors in such a manner that the possibility of having two adjacent nodes with the same color is minimized. With color representing different frequencies, the difference between an adjacent node's color could indicate the availability of noninterfering frequencies. Colorwave optimizes the minimum number of colors and therefore frequencies required, up to a predefined maximum value while maintaining a configured successful transmission rate by the readers. Colorwave is a TDMA-based approach, wherein each reader randomly selects a time slot to transmit using its assigned frequency. Upon collision, the reader selects a new time slot and informs its adjacent nodes of the selection. If any neighbor is scheduled to use the same frequency (color), it reinitiates itself, selects a new frequency and informs its neighbor of its actions. The process of selecting frequency and scheduling time slot is repeated, as required. The maximum allowable color is increased if the successful transmission rate drops below the configured value. The colorwave reduces the reader collision up to certain network scales, as in practice only a limited number of frequencies are available that can be utilized.

The Hierarchical Q-Learning (HiQ) algorithm [52] is a hierarchical, online learning algorithm that finds the dynamic solution to reader collision problems. HiQ, similar to Colorwave, tries to minimize the reader-to-reader collisions by assigning frequencies to each reader over time. The frequency assignment, however, is based on learning the collision patterns among neighbors and then assigning the optimal frequency. The optimal or near-optimal frequency assignment is achieved by repetitive environment interaction, that is, collision patterns and estimation. HiQ utilizes three basic hierarchical tiers, in some form already present in the existing RFID systems, namely readers, R-servers, and Q-servers. At the lowest tier are the RFID readers. The RFID reader communicates using prescheduled frequency in the reserved time slots. The reader detects the collisions amongst the neighboring readers, that is, the readers with overlapping interrogation zones. This statistical information, as explained earlier, is utilized for optimal frequency assignments; the decision being made by upper tiers, that is, the reader-level server (R-server). A one-on-one relation exists between the low-level reader and the R-server. The Q-learning server (Q-server) is the highest tier and is responsible for resource allocation, by finding an optimum scheduling policy based on the underlying readers' learning experience or collision information. The HiQ learning algorithm is distributed at lower level (readers) with centralized scheduling at the top-most tier. HiQ provides optimal scheduling, resulting in higher throughput, but at the cost of a centralized solution.

Other learning-based approaches include neural networks and genetic algorithms. A neural network approach is a fixed channel solution where each neuron represents an RFID reader. The constraint and optimization criteria of the system are translated into an energy function, which the neural network tries to minimize. The outcome of these neurons determines if the channel may be usable for the represented RFID reader. Genetic algorithms are a form of blind local search. In the genetic algorithm approach to the channel assignment, a single solution is an assignment scheme for all RFID readers. The genetic algorithm takes a set of solutions as the population and begins the evolutionary process, which is repeated until a solution is found with no interference.

The learning and scheduling approaches are mainly targeted toward reader-to-reader collision. Alternate procedures such as beacon channels [53] and central cooperators [54] are used to tackle multiple readers-to-tag collisions. The distributed protocol, Pulse [53], is based on the beaconing mechanism. The reader, while reading the tags within its interrogation zone, periodically broadcasts a beacon using the control channel. Any other reader that wants to read the tags, that is, tags within the overlapping interrogation zone, has to ensure that no beacon has been transmitted on the control channel before it can initiate its interrogation process. The latter reader has to wait until the first reader has executed its interrogation process because the control channel is idle for a specific time period before triggering its own beacon. The pulse protocol, however, is inclined toward eliminating the collision caused by mobile RFID readers. In the central cooperative approach [54], a centralized device is responsible for the multiplexing of tags' data to multiple readers. Using a control cooperator, the present "multiple points to multiple points" collision problem (caused when multiple readers are trying to access multiple tags) translates into the classical problem of "multiple point to one point" collision. The central cooperator multiplexes multiple reader requests into a single request for the tags. The responses from the tags are demultiplexed to the individual readers separately. The central cooperator provides an affective approach in handling multiple readers-to-tag collisions. However, it requires a device with all the functionality of a conventional RFID reader.

The reader collision problem avoidance schemes include frequency assignment scheduling and online learning approaches to avoid reader collisions. Table 1.3 shows a contrast between various algorithms for resolving the collision problem. The algorithms vary and have fundamental differences. For instance, some are centrally controlled, that is, using central authority for channel assignment and multiplexing the reader queries.

Table 1.3   Comparison of Reader Collision Algorithm

Method	Approach	Centralized Control	Distributed Control	Fixed Channel	Dynamic Channel
DCS [50]	Scheduling	–	✓	✓	–
VDCS [51]	Scheduling	–	✓	–	✓
HiQ [52]	Learning	✓	✓	–	✓
Neural	Learning	–	✓	✓	–
Annealing [56]	Learning	✓	–	✓	–
Genetic	Learning	✓	–	✓	–
Redundancy elimination [55]	Coverage	–	✓	–	–
CC [54]	Coverage	✓	–	✓	–

Reader collision, especially reader-to-reader interference, is similar to a frequency assignment problem in conventional wireless communication. However, the fundamental difference is that RFID tags, especially passive tags, are unable to differentiate between various readers. Therefore, two readers communicating with the tag must communicate at different time slots or make use of sessions. According to the EPC Gen2 standard, the tag shall provide support for up to four sessions, allowing two or more readers to independently interrogate the tag. The readers, step-by-step, selectively singulate the tags into their respective sessions and move them into other respective sessions upon completion. This allows multiple readers to interrogate common populations of tags. Although effective, this requires extra memory at the tag, increasing the cost.

1.5   Future Outlook

Our objective in this chapter was to offer an up-to-date review for MAC for RFID systems. In doing so, we highlighted the different approaches that have been exercised to resolve collisions for both tags and readers. The general motivation for the relative research area has been the unique characteristics of RFID systems as a radio-based technology. This includes special traffic patterns, and the energy and functionality constraints at the tags.

In the evolution of RFID systems, it is expected that Wireless Sensor Networks (WSNs) will play a large role. WSNs differ from traditional wireless voice or data networks in several ways but have much in common with the RFID systems. Most sensor nodes in WSNs, for example, are likely to be battery powered, bearing similarity to the active RFID tags. Occasionally, sensor nodes utilize energy harvesting, akin to passive RFID tags. In addition, many applications employ large numbers of nodes, and WSN density vary in place and time. This characteristic resembles that of RFID systems employed in pallet-level and item-level tagging. More importantly, WSNs are trigger-driven networks, and may exhibit extremely high level of activity, as is the case with RFID systems.

There are, however, differences between WSNs and RFID systems. In WSNs, sensor nodes are often deployed in an ad hoc fashion rather than with careful preplanning, and it is generally left to the nodes to self-organize to oversee and maintain the different network functionalities. This is not the case in RFID systems. RFID tags, especially passive tags, cannot communicate in an ad-hoc manner. It is for this reason that general basis for MAC in WSNs is CSMA, the application of which is not feasible in RFID systems.

A growing interest has been shown in the literature in integration WSNs and RFID systems [57–59]. Possibilities include integrating tags with sensors, integrating tags with wireless sensor nodes, integrating readers with wireless sensor nodes and wireless devices, and a mix of RFID and sensors [59]. However, such integrations raise new challenges from the interference standpoint. Much effort has been made to reduce the interference in large RFID networks and WSNs in their own domains. However, with the inevitable integration of the WSNs and RFIDs, the situation becomes worse as the number of devices increases (wireless sensor nodes, RFID tags, and RFID readers) demanding collaborative schemes to tackle the interference amongst these devices. Possible directions are distributed frequency scheduling between hetrogenous device types, RFID tags aiding in MAC wakeup procedures, power-saving mode operation, and context-aware MAC scheduling for WSNs. Item-level tagging, mobile tags and mobile sensors, and exciting new areas in RFIDs and WSNs, all further demand sophisticated MAC protocols

References

1. D. Shih, P.L. Sun, D.C. Yen, and S.M. Huang, Taxonomy and survey of RFID anti-collision protocols, *Computer and Communications*, 29, 2150–2166, 2006.
2. Z. Tang and Y. He, Research of multi-access and anti-collision protocols in RFID systems, *2007 IEEE International Workshop on Anti-Counterfeiting, Security, Identification*, pp. 377–380, April 16–18, Xiamen, Fujian, China, 2007.
3. C. Mutti and C. Floerkemeier, CDMA-based RFID systems in dense scenarios: Concepts and challenges, *2008 IEEE International Conference on RFID*, pp. 215–222, April 16–17, Los Vegas, NV, 2008.
4. P. Vandenameele, *Space Division Multiple Access for Wireless Local Area Networks*, Kluwer Academic Publishers, Norwell, MA, 2001.
5. K. Finkenzeller, *RFID Handbook: Fundamentals and Applications in Contactless Smart Cards and Identification*, John Wiley & Sons, Inc., England, U.K., 2003.
6. K. Ali, H. Hassanein, and A.M. Taha, RFID anti-collision protocol for dense passive tag environments, *LCN '07: Proceedings of the 32nd IEEE Conference on Local Computer Networks*, pp. 819–824, Dublin, Ireland, 2007.
7. J. Kim, W. Lee, E. Kim, D. Kim, and K. Suh, Optimized transmission power control of interrogators for collision arbitration in UHF RFID systems, *IEEE Communications Letters*, 11, 22–24, 2007.

8. W. Alsalih, K. Ali, and H. Hassanein, Optimal distance-based clustering for tag anti-collision in RFID systems, *33rd IEEE Conference on Local Computer Networks*, pp. 266–273, Montreal, QC, Canada, 2008.

9. J. Lee, T. Kwon, Y. Choi, S.K. Das, and K. Kim, Analysis of RFID anti-collision algorithms using smart antennas, *SenSys '04: Proceedings of the 2nd International Conference on Embedded Networked Sensor Systems*, pp. 265–266, Baltimore, MD, 2004.

10. H. Vogt, Efficient object identification with passive RFID tags, *Pervasive '02: Proceedings of the First International Conference on Pervasive Computing*, pp. 98–113, Zurich, Switzerland, 2002.

11. B. Feng, J. Li, J. Guo, and Z. Ding, ID-binary tree stack anticollision algorithm for RFID, *ISCC '06: Proceedings. 11th IEEE Symposium on Computers and Communications*, pp. 207–212, Pula-Cagliari, Sardinia, Italy, 2006.

12. L. Bolotnyy and G, Robins, Randomized pseudo-random function tree walking algorithm for secure radio-frequency identification, *Fourth IEEE Workshop on Automatic Identification Advanced Technologies*, pp. 43–48, Buffalo, NY, 2005.

13. J. Capetanakis, Tree algorithms for packet broadcast channels, *IEEE Transactions on Information Theory*, 25, 505–515, 1979.

14. J.H. Choi, D. Lee, H. Jeon, J. Cha, and H. Lee, Enhanced binary search with time-divided responses for efficient RFID tag anti-collision, *ICC '07: IEEE International Conference on Communications*, pp. 3853–3858, Glasgow, Scotland, 2007.

15. F. Zhou, D. Jin, C. Huang, and M. Hao, Optimize the power consumption of passive electronic tags for anti-collision schemes, *Proceedings of 5th International Conference on ASIC*, vol. 2, pp. 1213–1217, Beijing, China, 2003.

16. C. Floerkemeier, Transmission control scheme for fast RFID object identification, *PerCom Workshops 2006: Fourth Annual IEEE International Conference on Pervasive Computing and Communications Workshops*, pp. 457–462, Pisa, Italy, 2006.

17. L. Liu, Z. Xie, J. Xi, and S. Lai, An improved anti-collision algorithm in RFID system, *2nd International Conference on Mobile Technology, Applications and Systems*, pp. 137–142, Guangzhou, Guangdong, China, 2005.

18. N. Zhang and B. Vojcic, Binary search algorithms with interference cancellation RFID systems, *MILCOM 2005: IEEE Military Communications Conference*, vol. 2, pp. 950–955, Atlantic City, NJ, 2005.

19. M. Nanjundaiah and V. Chaudhary, Improvement to the anticollision protocol specification for 900 MHz class 0 radio frequency identification tag, *AINA '05: Proceedings of the 19th International Conference on Advanced Information Networking and Applications*, pp. 616–620, Taipei, Taiwan, 2005.

20. T. Hwang, B. Lee, Y.S. Kim, D.Y. Suh, and J.S. Kim, Improved anti-collision scheme for high speed identification in RFID system, *ICICIC '06: First International Conference on Innovative Computing, Information and Control*, vol. 2, pp. 449–452, Beijing, China, 2006.

21. T. Wang, Enhanced binary search with cut-through operation for anti-collision in RFID systems, *IEEE Communications Letters*, 10, 236–238, 2006.

22. S. Weis, S.E. Sarma, R.L. Rivest, and D.W. Engels, Security and privacy aspects of low-cost radio frequency identification systems, *Security in Pervasive Computing*, 2802, 201–212, 2004.

23. G. Khandelwal, A. Yener, and M. Chen, OPT: Optimal protocol tree for efficient tag identification in dense RFID systems, *ICC'06: IEEE International Conference on Communications*, vol. 1, pp. 128–133, Istanbul, Turkey, 2006.

24. S.H. Kim and P. Park, An efficient tree-based tag anti-collision protocol for RFID systems, *IEEE Communications Letters*, 11, 449–451, 2007.

25. W. Chen, S. Horng, and P. Fan, An enhanced anti-collision algorithm in RFID based on counter and stack, *ICSNC '07: Proceedings of the Second International Conference on Systems and Networks Communications*, pp. 21–24, Cap Esteral, French Riviera, France, 2007.

26. J. Myung, W. Lee, and J. Srivastava, Adaptive binary splitting for efficient RFID tag anti-collision, *IEEE Communications Letters*, 10, 144–146, 2006.

27. J. Myung and W. Lee, Adaptive binary splitting: A RFID tag collision arbitration protocol for tag identification, *Mobile Network Applications*, 11, 711–722, 2006.

28. J. Myung, W. Lee, J. Srivastava, and T.K. Shih, Tag-splitting: Adaptive collision arbitration protocols for RFID tag identification, *IEEE Transaction on Parallel and Distributed System*, 18, 763–775, 2007.

29. J. Ryu, H. Lee, Y. Seok, T. Kwon, and Y. Choi, A hybrid query tree protocol for tag collision arbitration in RFID systems, *ICC '07: IEEE International Conference on Communications*, pp. 5981–5986, Glasgow, Scotland, 2007.

30. F. Zhou, C. Chen, D. Jin, C. Huang, and H. Min, Evaluating and optimizing power consumption of anti-collision protocols for applications in RFID systems, *ISLPED'04: Proceedings of the 2004 International Symposium on Low Power Electronics and Design*, pp. 357–362, Newport, CA, 2004.

31. V. Namboodiri and L. Gao, Energy-aware tag anti-collision protocols for RFID systems, *PerCom '07: Fifth Annual IEEE International Conference on Pervasive Computing and Communications*, pp. 23–36, White Plains, NY, 2007.

32. N. Pastos and R. Viswanathan, A modified grouped-tag TDMA access protocol for radio frequency identification networks, *WCNC 2000: IEEE Wireless Communications and Networking Conference*, vol. 2, pp. 512–516, Chicago, IL, 2000.

33. C. Law, K. Lee, and K. Siu, Efficient memoryless protocol for tag identification, *DIALM '00: Proceedings of the 4th International Workshop on Discrete Algorithms and Methods for Mobile Computing and Communications*, pp. 75–84, Boston, MA, 2000.

34. J. Choi, D. Lee, and H. Lee, Query tree-based reservation for efficient RFID tag anti-collision, *IEEE Communications Letters*, 11, 85–87, 2007.

35. K.W. Chiang, C. Hua, and P. Yum, Prefix-length adaptation for PRQT protocol in RFID systems, *GLOBECOM'06: IEEE Global Telecommunications Conference*, pp. 1–5, San Francisco, CA, 2006.

36. K.W. Chiang, C. Hua, and T.P. Yum, Prefix-randomized query-tree protocol for RFID systems, *ICC '06: IEEE International Conference on Communications*, vol. 4, pp. 1653–1657, Istanbul, Turkey, 2006.

37. J. Cha and J. Kim, Dynamic framed slotted ALOHA algorithms using fast tag estimation method for RFID system, *CCNC 2006: 3rd IEEE Consumer Communications and Networking Conference*, vol. 2, pp. 768–772, Los Vegas, NV, 2006.

38. J. Cha and J. Kim, Novel anti-collision algorithms for fast object identification in RFID system, *Proceedings 11th International Conference on Parallel and Distributed Systems*, vol. 2, pp. 63–67, Cambridge, MA, 2005.

39. J. Park, M.Y. Chung, and T. Lee, Identification of RFID tags in framed-slotted ALOHA with robust estimation and binary selection, *IEEE Communications Letters*, 11, 452–454, 2007.

40. J. Park, M.Y. Chung, and T. Lee, Identification of RFID tags in framed-slotted ALOHA with tag estimation and binary splitting, *ICCE '06: First International Conference on Communications and Electronics*, pp. 368–372, San Diego, CA, 2006.

41. G. Khandelwal, A. Yener, K. Lee, and S. Serbetli, ASAP: A MAC protocol for dense and time constrained RFID systems, *ICC '06: IEEE International Conference on Communications*, vol. 9, pp. 4028–4033, Istanbul, Turkey, 2006.

42. S. Lee, S. Joo, and C. Lee, An enhanced dynamic framed slotted ALOHA algorithm for RFID tag identification, *MobiQuitous 2005: The Second Annual International Conference on Mobile and Ubiquitous Systems: Networking and Services*, pp. 166–172, San Diego, CA, 2005.

43. W.J. Shin and J.G. Kim, Partitioning of tags for near-optimum RFID anti-collision performance, *WCNC 2007: IEEE Wireless Communications and Networking Conference*, pp. 1673–1678, Hong Kong, China, 2007.

44. C. Floerkemeier, Bayesian transmission strategy for framed ALOHA based RFID protocols, *IEEE International Conference on RFID 2007*, pp. 228–235, Grapevine, TX, 2007.

45. J. Choi, D. Lee, and H. Lee, Bi-slotted tree based anti-collision protocols for fast tag identification in RFID systems, *IEEE Communications Letters*, 10, 861–863, 2006.

46. J. Kim, J. Yu, J. Myung, and E. Kim, Effect of localized optimal clustering for reader anti-collision in RFID networks: Fairness aspects to the readers, *ICCCN 2005: 14th International Conference on Computer Communications and Networks*, pp. 497–502, San Diego, CA, 2005.

47. C.P. Wong, Grouping based bit-slot ALOHA protocol for tag anti-collision in RFID systems, *IEEE Communications Letters*, 11, 946–948, 2007.

48. D.W. Engels and S.E. Sarma, The reader collision problem, *2002 IEEE International Conference on Systems, Man and Cybernetics*, vol. 3, pp. 641–646, Hammamet, Tunisia, 2002.

49. K.S. Leong, M.L. Ng, and P.H. Cole, The reader collision problem in RFID systems, *MAPE 2005: IEEE International Symposium on Microwave, Antenna, Propagation and EMC Technologies for Wireless Communications*, vol. 1, pp. 658–661, Beijing, China, 2005.

50. J. Waldrop, D.W. Engles, and S.E. Sarma, Colorwave: An anticollision algorithm for the reader collision problem, *ICC '03: IEEE International Conference on Communications*, vol. 2, pp. 1206–1210, Anchorage, AK, 2003.

51. J. Waldrop, D.W. Engles, and S.E. Sarma, Colorwave: A MAC for RFID reader networks, *WCNC 2003: 2003 IEEE Wireless Communications and Networking*, vol. 3, pp. 1701–1704, New Orleans, LA, 2003.

52. J. Ho, D.W. Engels, and S.E. Sarma, HiQ: A hierarchical Q-learning algorithm to solve the reader collision problem, *SAINT Workshops 2006: International Symposium on Applications and the Internet Workshops*, pp. 88–91, Phoenix, AZ, 2006.

53. S.M. Birari and S. Iyer, Mitigating the reader collision problem in RFID networks with mobile readers, *13th IEEE International Conference on Networks*, pp. 463–468, Kuala Lumpur, Malaysia, 2005.

54. D. Wang, J. Wang, and Y. Zhao, A novel solution to the reader collision problem in RFID system, *WiCOM 2006: International Conference on Wireless Communications, Networking and Mobile Computing*, pp. 1–4, Wuhan, Hubai, China, 2006.

55. B. Carbunar, M.K. Ramanathan, M. Koyuturk, C. Hoffmann, and A. Grama, Redundant reader elimination in RFID systems, *IEEE SECON 2005: 2005 Second Annual IEEE Communications Society Conference on Sensor and Ad Hoc Communications and Networks*, pp. 176–184, Santa Clara, CA, 2005.

56. C. Lin and F. Lin, A simulated annealing algorithm for RFID reader networks, *WCNC'2007: IEEE Wireless Communications and Networking Conference*, pp. 1669–1672, Hong Kong, China, 2007.

57. J. Cho, Y. Shim, T. Kwon, Y. Choi, and S. Pack, SARIF: A novel framework for integrating wireless sensor and RFID networks, *IEEE Wireless Communications*, 14, 50–56, 2007.

58. L. Zhang and Z. Wang, Integration of RFID into wireless sensor networks: Architectures, opportunities and challenging problems, *Fifth International Conference on Grid and Cooperative Computing Workshops*, pp. 463–469, Changsha, Huan, China, 2006.

59. H. Liu, M. Bolic, A. Nayak, and I. Stojmenovic, Integration of RFID and wireless sensor networks, *Proceedings of The First ACM Workshop on Convergence of RFID and Wireless Sensor Networks and Their Applications*, Sydney, Australia, 2007.

Chapter 2

Anti-Collision Algorithm in RFID

Jehn-Ruey Jiang and Ming-Kuei Yeh

CONTENTS

In the radio-frequency identification (RFID) system, tags store unique identifications and are attached to objects; a reader performs the tag interrogation procedure to recognize an object by issuing wireless RF signals to interrogate the identification of the attached tag. Like other wireless communication systems, the RFID system also suffers from the signal interference problem. There are two types of signal interference. One is called the reader collision, which occurs when multiple readers issue signals to same tags simultaneously. The other is called the tag collision, which occurs when multiple tags respond to a reader simultaneously. Collisions hinder and slow down the tag interrogation procedure. Therefore, reader anti-collision and tag anti-collision protocols are required to respectively reduce reader collisions and tag collisions to improve interrogation procedure performance. In this chapter, we introduce existing reader anti-collision and tag anti-collision protocols. We intend to provide not only an extensive survey of the protocols, but also new research directions of them.

2.1 Introduction

The front end of an radio-frequency identification (RFID) system is composed of two components: readers and tags [1]. Tags store unique identifications and are attached to objects; a reader performs the tag interrogation procedure to recognize an object by issuing wireless RF signals to interrogate the identification (ID) of the attached tag. Since tags are designed for an attempt of worldwide deployment in commercial or alike applications, they are supposed to be tiny, low cost, and equipped with a simple circuit of limited computation and communication capabilities [2]. Most RFID tags are passive; they do not have on-tag power source and derive energy from the RF field generated by the reader to drive the circuit. When a tag and a reader are close enough, they can communicate with each other. For such a situation, we say that the tag is in the interrogation zone of the reader. Like other wireless communication systems, the RFID system also suffers from the signal interference problem [3]. There are two types of signal interference. One is called the reader collision, which occurs when multiple readers issue signals to same tags simultaneously. The other is called the tag collision, which occurs when multiple tags respond to a reader simultaneously. Collisions hinder and slow down the tag interrogation procedure. Therefore, reader anti-collision and tag anti-collision protocols are thus required to respectively reduce reader collisions and tag collisions to improve interrogation procedure performance. In this chapter, we introduce existing reader anti-collision and tag anti-collision protocols. We intend to provide not only a comprehensive survey of the protocols, but also new research directions of them.

Because the tag is energized by the reader, the tag's response range (also called the interrogation range) is much less than the reader's RF transmission range (also called the interference range) [4]. Furthermore, tags and readers have very different computation and communication capabilities. Due to all the asymmetries, we cannot rely on common collision avoidance mechanisms, such as the RTS/CTS mechanism used in wireless local area network [3], to solve the collision problem.

Some reader anti-collision protocols are proposed to reduce reader collisions based on the concepts of Time Division Multiple Access (TDMA), Frequency Division Multiple Access (FDMA) or Carrier Sense Multiple Access (CSMA) [5]. TDMA-based reader anti-collision protocol divides the transmission time into intervals and a reader can only transmit messages in its assigned intervals. The assignment of intervals can be done in a distributed or a centralized way [6]. Waldrop et al. [7] propose two distributed TDMA-based reader anti-collision protocols, called Distributed Color Selection (DCS) and Colorwave. A reader graph is first derived, where any two readers are defined to be adjacent and have an edge between them if they may interfere with each other. Each reader is assigned a color which stands for a reservation of a specific time slot for transmitting signals. If all the adjacent readers are with different colors, the reader collision is avoided. In DCS protocol, the maximum number of colors (max_colors) is fixed, and a reader transmits only in its assigned color (time slot). On the contrary, Colorwave protocol has dynamic values of max_colors; it is a dynamic color assignment mechanism to minimize the required number of colors in the reader graph. With the reduction in the number of used colors, the efficiency of message transmission is increased.

FDMA-based protocols divide all available frequency bands into several noninterfering frequency channels. If a frequency channel is only assigned to a transmitter at a time, transmitters can transmit messages simultaneously without causing any interference. Ho et al. propose HiQ [8], which is a both TDMA-based and FDMA-based protocol. It attempts to minimize reader collisions by learning the collision patterns of the readers and by effectively assigning frequencies over time. HiQ depends on a distributed, hierarchical, and online learning scheme called Q-learning for determining frequency and time assignments. By interacting repeatedly with the system, Q-learning attempts to discover an optimum frequency assignment over time. EPCGlobal Gen 2 [9] is a famous protocol that adopts FDMA technology to solve the reader collision problem. Readers can choose separate transmission channels to avoid interference by the frequency hopping spread spectrum technique.

CSMA is another mechanism used to solve the reader collision problem. In CSMA mechanism, each reader needs to check before transmitting messages whether the carrier (the shared communication channel) is free or not. If the carrier is sensed to be idle, the reader sends out messages at once. Otherwise, the reader delays a random period of time and then starts sensing carrier again. The European Telecommunications Standards Institute (ETSI) EN 302 208 Standard [10] utilizes "Listen Before Talk (LBT)" mechanism that is based on the concept of CSMA to solve the reader collision problem.

Several tag anti-collision protocols are proposed to reduce tag collisions. They can be categorized mainly into three classes: ALOHA-based, tree-based, and counter-based protocols [11]. The ALOHA [12], slotted ALOHA [13], and frame slotted ALOHA [14] protocols are ALOHA-based protocols. In ALOHA protocol, a reader first sends a command to make tags transmit their IDs. On receiving the reader's interrogation signal, each tag in the interrogation zone independently waits for a random back-off time and then responds with its tag ID to the reader. If no collision occurs during a tag's ID response, its ID can be identified properly. In slotted ALOHA protocol, the random back-off time must be a multiple of a prespecified slot time. Frame slotted ALOHA protocol is similar to slotted ALOHA protocol except that the whole interrogation procedure is divided into a set of frames with each having a fixed number of time slots, and a tag can send its ID to the reader only in one randomly chosen slot during a frame period. ALOHA-based protocol is simple but has the tag starvation problem that a tag may never be identified properly for the reason that its responses always collide with others'.

The basic idea of tree-based protocols [15–19] is to repeatedly split the tags encountering collisions into subgroups according to tag IDs until there is only one tag in a subgroup to be identified successfully. The protocols can be applied to tags with or without writable memory. Tags with memory have higher cost. However, protocols for such a kind of tags have better performance. The Query Tree (QT) protocol [16] is applicable to tags without on-tag writable memory. In the protocol, the reader broadcasts a request bit string S with a variable length to tags. A tag with an ID prefix matching S will respond its ID to the reader. When collisions occur, the reader broadcasts again with a longer bit string S0 or S1 to split colliding tags into two subgroups. The bit-by-bit binary tree [15] is applicable for tags with writable memory. In the protocol, a reader broadcasts a request command first and each tag will respond with the first bit of its tag ID. If collisions occur, the reader will acknowledge the tags with 0 (or 1). Only the tag with the first bit being 0 (or 1) will respond with the next bit to the reader. In this way, the tags are continuously split into two groups. The other tree-based protocols, such as EPCglobal Class 0 [19], the tree-slotted ALOHA (TSA) [17], Bi Slotted Query Tree Algorithm (BSQTA) [18] and Bi Slotted Collision Tracking Tree Algorithm (BSCTTA) [18] protocol, also utilize similar concept to split tags to solve the tag collision problem. The main drawback of tree-based protocols is that their performance is affected by the length or the distribution of tag IDs. In general, the tree-based protocol has longer identification time latency than that of the ALOHA-based [20], but it does not have the tag starvation problem.

The concept of counter-based protocols [11,20–23] is similar to that of tree-based protocols. The major difference between these two kinds of protocols is that the former rely on static tag IDs for the splitting, and the latter rely on dynamically changing counters for the splitting. ISO/IEC 18000-6B [22] is a standard adopting the counter-based tag anti-collision protocol. In ISO/IEC 18000-6B, each tag has a counter initially set to 0. When a reader sends request to tags, every tag with counter value 0 can transmit its tag ID to the reader. When a collision occurs, the tags with counters of values greater than 0 then increase their counters by 1, while the tags with counter value 0 randomly generate a random bit, 0 or 1, and add it to their counters. In this way, the tags with counter value 0 are split into two subgroups. Other counter-based protocols, such as Adaptive Binary Splitting (ABS) protocol [24], utilize similar concept to split tags encountering collisions.

The counter-based protocols do not have the starvation problem. Furthermore, they have the stable property that their performance is not affected by the length of tag IDs or the distribution of tag IDs.

Table 2.1 shows the anti-collision protocols that will be discussed in the chapter and their classification. In summary, the reader anti-collision protocols are classified as TDMA, FDMA, and CSMA protocols, while the tag anti-collision protocols are classified as ALOHA-, tree-, and counter-based protocols. All protocols in Table 2.1 will be described in detail in the following context.

The rest of this chapter is organized as follows. In Section 2.2, collision problems are defined first. And in Section 2.3, reader anti-collision protocols, like TDMA, FDMA, and CSMA protocols, are described in detail. And tag anti-collision protocols, including ALOHA-, tree-, and counter-based protocols, are elaborated in Section 2.4. In Sections 2.3 and 2.4, examples are further given for some protocols to make them easy to understand. At last, we give a summary and suggestions of new research directions in Section 2.5.

Table 2.1   The Protocols That Will Be Discussed in This Chapter

Collision Type	Category	Protocol
Reader collision	TDMA	DCS [7]
		Colorwave [7]
	FDMA	HiQ [8]
		EPCglobal Gen 2 [9]
	CSMA	ETSI 302 208 Standard [10]
Tag collision	ALOHA-based	ALOHA [12]
		Slotted ALOHA [13]
		Frame slotted ALOHA [14]
		ISO/IEC 18000-6A [22]
	Tree-based	QT [16]
		Bit-by-bit binary tree [15]
		EPCglobal Class 0 [19]
		TSA [17]
		BSQTA [18]
		BSCTTA [18]
		AQS (Adaptive Query Splitting) [30]
	Counter-based	ISO/IEC 18000-6B [22]
		ABS [30]

2.2 Collision Problems in the RFID System

When a reader (or called interrogator) transmits a request to a tag, it also provides energy to power up a passive tag. If the reader and the passive tag are close enough, the reader can receive the signal reflected from the tag. For such a situation, we say that the tag is in the interrogation zone of the reader. When two or more readers are too close or many tags appear in one reader's interrogation zone, there arise interference problems, which are mainly classified as the reader collision problem and the tag collision problem. Below, we describe the two types of problems.

- The reader collision (or reader interference) problem:
  Because the tag is energized by the reader, the tag's response zone (i.e., the interrogation zone) is much less than the reader's transmission zone (also called interference zone). When a tag is within the interrogation zone of a reader A and within the interference zone of another reader B, due to the interference of readers, either the tag cannot receive the request command from reader A correctly or reader A cannot interpret the response from the tag properly. This is called the reader collision problem. For example, in Figure 2.1, tag T is within the interrogation zone of reader A and within the interference zone of reader B. The reader collision problem occurs for such a situation.

- Tag collision problem:
  To identify tags within the interrogation zone, a reader sends a request to ask tags to send back their IDs. When multiple tags within the reader's interrogation zone respond to the request simultaneously, collision occurs and the reader cannot identify any tag properly. This is called the tag collision problem. For example, in Figure 2.1, tags S and T are within the interrogation zone of reader A. If tags S and T send their IDs for responding to reader A's request simultaneously, the tag collision problem occurs and neither tag can be recognized by reader A.

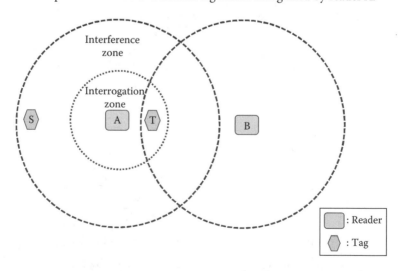

Figure 2.1 The relation between the interrogation zone and the interference zone.

2.3   Reader Anti-Collision Protocols

Several reader anti-collision protocols are proposed to solve the reader collision problem. They are classified into three classes: TDMA, FDMA, and CSMA protocols [5]. Below, we describe some reader anti-collision protocols class by class.

2.3.1   TDMA Protocols

The basic idea of TDMA-based reader anti-collision protocols is to divide the whole time period into intervals and to allow a reader to transmit message only within its allocated intervals. In this way, the reader collision can be avoided. Below, we introduce two TDMA-based reader anti-collision protocols: DCS and Colorwave algorithms [7].

2.3.1.1   DCS Algorithm

DCS is a reader anti-collision protocol proposed by Waldrop et al. in Ref. [7]. Time slots are assumed to be colored by colors 0, 1, maxColors cyclically. DCS solves the reader collision problem by first deriving a reader graph, where readers are represented as nodes and two nodes (readers) are defined to be adjacent and have an edge between them if they may interfere with each other. It then assigns each reader a color which stands for a reservation of a specific time slot for transmitting signals. If all the adjacent readers are of different colors, the reader collision is avoided.

DCS is a distributed algorithm that allows each reader to randomly and locally choose a color (time slot) from color set {0, 1, maxColors}, where maxColors is an input parameter whose value will never change. When a reader wants to send a message to the tags, it will queue the message until the time slot of the chosen color arrives. If a reader transmits a message in the time slot of its chosen color but finds that collisions occur, it will rechoose a new color and notify all its neighbors to change their chosen colors accordingly. Note that, DCS algorithm needs to synchronize the timing of time slots but does not need to synchronize the value of colors among all readers in the system.

2.3.1.2   Colorwave Algorithm

Colorwave algorithm, or Variable-Maximum Distributed Color Selection (VDCS) algorithm, is an extension of the DCS algorithm. In Colorwave, a mechanism is proposed to optimize the number of colors (i.e., maxColors) required to color the reader graph. If the used colors are reduced, the efficiency of signal transmission can be improved.

When a reader observes by itself or is notified by neighboring readers that the successful transmission rate is below an addition_maxColors threshold, it will increase its local maxColors value and broadcasts the new maxColors to its neighboring readers to make them reselect colors to reduce the transmission collisions. On the contrary, a reader will decrease its local maxColors value to decrease the transmission waiting time when the successful transmission rate is above a subtraction_maxColors threshold.

2.3.2 FDMA Protocols

FDMA protocols divide all available frequency bands into several noninterfering channels. Readers can use different channels to communicate with tags simultaneously. Below, we introduce two protocols, HiQ [8] and the EPCglobal Gen 2 [9], which adopt the FDMA mechanism to solve the reader collision problem.

2.3.2.1 HiQ Protocol

HiQ [8] is a hierarchical, distributed, and online learning algorithm based on TDMA and FDMA to solve the reader collision problem. The designed goal is to maximize the number of concurrent communication channels between readers and tags while minimizing the number of reader collisions by learning the collision patterns of readers to assign frequencies to each time slot to the readers effectively.

The hierarchical control structure of HiQ consists of readers, R-servers, and Q-servers, as shown in Figure 2.2. RFID readers are at the lowest tier and each server in R-server tier manages several readers. When a reader needs to send messages to the tags in its interrogation zone, it must request resources, namely the frequency channel and the time slot, from its master R-server. The reader can send messages at a specific frequency channel in a time slot only after the channel and time slot are granted by its master R-server.

With the distributed architecture, the neighboring readers can send messages in the same time slot or in the same frequency channel to cause collisions. It is the responsibility

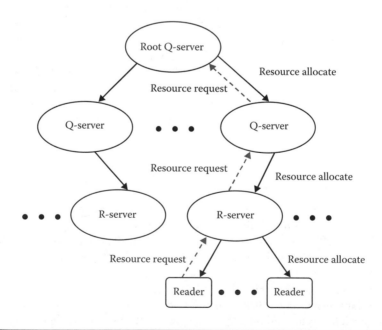

Figure 2.2   The hierarchical control structure of HiQ protocol.

of readers to detect collisions with neighboring readers. Each reader should report the number of collisions, type of collisions, and the number of successful reads to its master R-server. The R-server can then determine which slave readers are interfering mutually by the feedback reports and reallocates the resources dynamically to avoid the collisions.

The resources that the R-server can allocate are from its master Q-server (Q-learning server) in the hierarchical structure. For greater flexibility and scalability, Q-Servers may themselves work in a hierarchical architecture. But there is always only one root Q-Server in the whole system that has the power of full control over the allocation of all frequency channels and time slots.

2.3.2.2   EPCglobal Gen 2 Protocol

The Class 1 Generation 2 UHF standard [9] proposed by EPCglobal uses FDMA technology to reduce reader interference. The entire allocated frequency band is divided into channels. A reader will only use a certain channel for communication. The carrier frequency used by readers and tags are separate. That is, readers (resp., tags) will collide with readers (resp., tags) only. Readers use frequency hopping spread spectrum technique to avoid interference. In Europe, a bandwidth of 200 kHz is regulated for frequency allocation [25]. It is suggested that readers use even-numbered channels while tags backscatter signals in odd-numbered channels. In the United States, a wider bandwidth of 500 kHz is regulated for frequency allocation. All channels are available for reader interrogation but the tag can backscatter signals at the boundaries of these channels. EPCglobal Gen 2 protocol can solve the reader collision problem. Because most low-cost tags do not have frequency selection capability, the tag collision problem still exists [3].

2.3.3   CSMA Protocols

CSMA is a common mechanism used in wired or wireless systems to avoid collisions. In this mechanism, each device needs to check whether the media channel is free before transmitting messages. If the media is occupied, the device will wait until it is released.

ETSI 302 208 is a European regulation that adopts a CSMA mechanism called "LBT" to solve the reader collision problem. It allocates the frequency band of 865 to 868 MHz for RFID applications [10,25] and divides the band into 15 channels, each of 200 kHz bandwidth. With the maximum effective radiation power (ERP) of 2 W, only 10 channels are available for communication and 5 channels are defined as guard bands or reserved for lower power readers. The receiver module of a reader is first activated to monitor selected channel for a specified time period (5 ms) before transmission. If it senses that the channel is idle over the specified time period, the reader can send the message directly for up to 4 s and then the reader activates the receiver module to detect signal interference. If the channel is occupied by other readers, the reader will search for another free channel for transmitting messages.

2.4 Tag Anti-Collision Protocols

Several tag anti-collision protocols are proposed for reducing tag collisions. They can be categorized into three classes: ALOHA-based, tree-based, and counter-based protocols [11]. Below, we introduce some of the protocols class by class.

2.4.1 ALOHA-Based Protocols

ALOHA-based tag anti-collision protocols [21,26–28] are based on a backoff mechanism that operates in a probabilistic manner. They try to stagger the response times of tags in the interrogation zone. Below, we introduce several ALOHA-based protocols: ALOHA [12], slotted ALOHA [13], and frame slotted ALOHA [14]. In general, ALOHA-based protocols are simple and have fair performance. However, they have the tag starvation problem that a tag may never be identified because its responses always collide with others'.

2.4.1.1 ALOHA Protocol

ALOHA protocol [12] is the simplest ALOHA-based tag anti-collision protocol. When a reader requests tags to respond to their IDs, each tag in the interrogation zone chooses a random back-off time individually and responds with its tag ID to the reader after the back-off time. If no collision occurs during the transmission of a tag ID, this ID is identified successfully and acknowledged by the reader. A tag with acknowledged ID will stop responding to the reader. And a tag will repeatedly select a random back-off time and send its ID until the ID is identified and acknowledged by the reader.

2.4.1.2 Slotted ALOHA Protocol

In slotted ALOHA protocol [13], the random back-off time must be a multiple of a prespecified slot time. Note that a slot time is usually set to be a time period that is long enough for a tag to send out its ID and for the reader to recognize the ID and acknowledge the ID. The reader needs to synchronize the slot times for all the tags in the interrogation zone. If only one tag transmits its ID in a period of a slot time, it can be identified and acknowledged by the reader properly. Tags not identified by the reader will repeatedly select a time slot randomly for transmitting their IDs. It is shown in Ref. [29] that the performance of slotted ALOHA protocol is twice that of the ALOHA protocol because there is no partial collision of tag ID responses in slotted ALOHA protocol.

2.4.1.3 Frame Slotted ALOHA Protocol

In frame slotted ALOHA protocol [14], the whole interrogation procedure is divided into a set of frames, each having several time slots. On receiving the reader's REQUEST command, each tag can respond just in one randomly chosen slot during a frame period. If there is only one tag response in a slot, the reader can identify the tag successfully. Tags not identified successfully will reselect a time slot in the next frame for retransmitting their IDs. At the time when no tag responds, all tags are identified successfully. The frame rounds continue until that time.

In Figure 2.3, we show an example of frame slotted ALOHA protocol in which each frame has four time slots. Suppose that there are six tags with unique 5-bit IDs in the interrogation zone of a reader. The execution procedure of the protocol is described as follows:

1. The reader sends REQUEST command first to synchronize the beginning of a frame.
2. Each tag randomly chooses one of the four available time slots in frame 0 to respond to its tag ID after receiving REQUEST command. In our example, in frame 0, only tag ID (01110) in time slot 1 can be identified successfully. Collisions occur in time slots 2 and 4, and no tag responds in time slot 3.
3. The identified tag can be selected by SELECT command for reading or writing data. It will stop responding to REQUEST commands in later frames.
4. The reader sends REQUEST commands repeatedly until all tags are identified successfully, as shown in frames 1 and 2 of Figure 2.3.

One drawback of frame slotted ALOHA protocol is that its performance will degrade when the number of slots in the frame does not match properly the number of tags in the interrogation zone. Dynamic frame slotted ALOHA protocols [26–28,30] try to eliminate the drawback by dynamically adjusting the frame size according to the estimated number of tags. Their performance is better than that of frame slotted ALOHA protocol.

2.4.1.4   ISO/IEC 18000-6A Protocol

ISO/IEC 18000-6 [22] is a standard that defines the air-interface communication at 860–960 MHz for the RFID system. There are three different types (A, B, and C) of communication protocols defined in this standard. Among them, types A and C are ALOHA-based protocols. Because the type C protocol is a derivation of the type A protocol, we only introduce the type A protocol.

In ISO/IEC 18000-6A protocol, a reader initiates a round of the identification procedure by sending out Init_round command. In this command, the number of slots in a round, namely the round size, is given. It is noted that the reader can dynamically determine a proper round size for the next round according to the number of collisions in the current round. After receiving the command, a tag randomly selects a time slot to respond its ID to the reader. The tag keeps a slot counter to track the current time slot. When the selected time slot arrives, the tag waits a random delay time in the range of 0 to 7 periods and responds with a randomly chosen four-bit tag signature. If there is only one responding tag whose signature is received by the reader properly, the reader will send Next_slot command containing the received signature to the tag as an acknowledgment; otherwise, Close_slot command is sent. The tag has the following behaviors:

■ The tag increases the slot counter by one if it does not respond in the current slot and the received command is Close_slot or Next_slot.

		Frame 0					Frame 1					Frame 2		
		Time slot 1	Times lot 2	Time slot 3	Time slot 4		Time slot 1	Time slot 2	Time slot 3	Time slot 4		Time slot 1	Time slot 2	Time slot 3	Time slot 4
Reader	Request					Request					Request				
Tag1			10010						10010						
Tag2		01110													
Tag3					00101		00101							00101	
Tag4			11011				11011						11011		
Tag5			10110					10110							
Tag6					01001					01001					
State		Success	Collision	Idle	Collision		Collision	Success	Success	Success		Idle	Success	Success	Idle

Figure 2.3 An example of frame slotted ALOHA protocol.

- The tag increases the slot counter by one if it responds in the current slot and the received command is Close_slot.
- The tag changes to Quiet state if it responds in the current slot and the received command is Next_slot with the same tag signature as its.

During a round, the reader can suspend the round by sending Standby_round command to tags. The suspension of the round allows the reader to conduct a dialogue with a selected tag for data reading/writing. When the slot count equals the round size specified in Init_round command, the round is finished and all tags not in Quiet state (i.e., tags not yet identified) will randomly select a new slot and a new random signature to enter a new round.

2.4.2 Tree-Based Protocols

The basic idea of the tree-based tag anti-collision protocol is to repeatedly split the tags encountering collisions into subgroups according to tag IDs until there is only one tag in a subgroup to be identified successfully. The protocols can be applied to tags with or without writable memory. Tags with memory have higher cost. However, protocols for such kinds of tags have better performance. In general, the tree-based protocol has longer identification time latency than that of the ALOHA-based protocol, but it does not have the tag starvation problem. A further drawback of the tree-based protocol is that its performance is affected by the length or the distribution of tag IDs. Below, we introduce some tree-based protocols: query tree [16], bit-by-bit binary tree [15], EPCglobal Class 0 [19], TSA [17], BSQTA [18], and BSCTTA [18] protocols.

2.4.2.1 Query Tree Protocol

In QT protocol [16], a reader first broadcasts a request bit string S to tags. A tag with an ID prefix matching S will respond its whole ID to the reader. If only one tag responds at an instance, the tag is identified successfully. But if multiple tags respond simultaneously, the responses collide. In such a case, the reader broadcasts again with a longer bit string that has one more bit, 0 or 1, appended to S, that is, S0 or S1. Obviously, the tags with prefix S are split into two subgroups S0 and S1. The splitting procedure will be performed repeatedly until every tag in the interrogation zone is identified successfully. The QT protocol is a memory-less protocol because it does not require tags to be equipped with additional writable on-chip memory. We can observe that QT protocol's identification delay is affected by the distribution and the length of tag IDs. Specifically, if the tags have continuous tag IDs, the request bit string will grow longer and longer for identifying them. The delay time of the identification procedure will then increase significantly.

Below, we show an example of QT protocol. We assume that there are six tags with unique IDs 0010, 0011, 1001, 1100, 1101, and 1110. The tag interrogation process of QT protocol is described step by step as follows:

1. The reader sends out a request bit string S = "0" first and pushes another request bit string "1" into the stack. The tags with IDs 0010 and 0011 have the first bit of tag ID matching the request bit string S. They respond their tag IDs to the reader simultaneously and collision occurs.

2. The reader then sends out a longer request bit string S = "00" and pushes "01" into the stack. The tags with IDs 0010 and 0011 respond to the request simultaneously and collision again occurs.
3. The reader sends out a still longer request bit string S = "000" and pushes "001" into the stack. None of the tags has an ID prefix matching S, so there is no response.
4. For the case of no response, the reader pops "001" from the stack and sends it out as a request bit string. The tags with IDs 0010 and 0011 respond to the request simultaneously and collision again occurs.
5. The reader sends out a request bit string S = "0010" and pushes "0011" into the stack. Only the tag with ID 0010 responds to the request and is identified successfully.
6. For the case of successful identification, the reader pops "0011" from the stack and sends it out as a request bit string. Only the tag with ID 0011 responds to the request and is identified successfully.

The identification procedure is executed repeatedly until the stack is empty. And then all tags can be identified successfully. The steps of the whole procedure and the associated tree diagram are shown in Table 2.2.

2.4.2.2   Bit-by-Bit Binary Tree Protocol

With the assistance of writable on-tag memory, bit-by-bit binary tree protocol [15] can reduce the tag collision efficiently. In this protocol, a reader broadcasts a request command first and each tag will respond to the request with the first bit of its tag ID. If collisions occur, the reader will acknowledge the tags with 0 (or 1). Only the tag with the first bit being 0 (or 1) will respond with the next bit to the reader. The above procedure repeats bit by bit until there is only one responding tag. The reader can then ask the tag to send out the remaining bits of its ID for the purpose of identification. With the on-tag memory, tags can keep track of the on-going status of the identification procedure and response. Unlike QT protocol, bit-by-bit binary tree protocol does not require a reader to send long ID prefixes; the reader only sends out one bit at a time. Consequently, the delay time of the identification procedure is reduced.

2.4.2.3   EPCglobal Class 0

In the EPCglobal Class 0 protocol [19], the tag will respond to the reader's request with its first bit of the tag ID. Each tag responds with a single bit through one of two subcarrier frequencies, one for binary 0 and the other for binary 1, so that the reader can recognize 0 and 1 at the same time. If the reader receives 0 and 1 simultaneously, it will acknowledge 0 to the tags; otherwise, the reader will instead acknowledge the receiving bit value. Only tags with the first bit matching the acknowledgment bit can respond with the next bit to the reader, while the other tags will enter a mute state and keep silent temporarily until the reader requests the tags to start over to respond in a new round of tag interrogation. The above procedure repeats bit by bit until one tag can respond with full bits of its ID to be identified successfully. The tag can then enter a dormant state

Table 2.2 The Steps of the Identification Procedure of QT Protocol

Step	Request Bit String S	Response
1	0	Collision
2	00	Collision
3	000	Null
4	001	Collision
5	0010	0010
6	0011	0011
7	01	Null
8	1	Collision
9	10	1001
10	11	Collision
11	110	Collision
12	1100	1100
13	1101	1101
14	111	1110

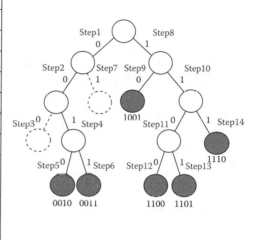

to sleep until the reader requests all tags to start over to launch the next interrogation procedure.

Below, we give an example to explain the details of EPCglobal Class 0 protocol. We assume there are three tags with unique IDs 001, 011, and 110, respectively. Some steps of the tag interrogation procedure are described as follows.

1. At the beginning, the reader sends a request command to ask tags to start a round of tag interrogation. On receiving the request, the tags respond with the first bit of their tag IDs. Specifically, tag1 (with ID 001) responds with "0," tag2 (with ID 011) responds with "0," and tag3 (with ID 110) responds with "1."
2. The reader receives both bits "0" and "1" from two separate subcarrier channels and acknowledges bit "0" to the tags. Tag1 and tag2 will respond with the second bit of its tag ID (i.e., tag1 responds with "0" and tag2 responds with "1"). Tag3 enters the mute state and will keep silent temporarily until a next request command is received.
3. The reader still receives both "0" and "1" at the same time. It acknowledges with "0" to the tags. Tag1 responds with the third bit "1" of its ID, while tag2 enters the mute state.
4. Since there is only tag1 responding "1" and the number of responding bits is equal to the ID length, the reader acknowledges with "1" and set tag1's ID (001) as

identified. On receiving the reader's acknowledgment to the sending of its last ID bit, tag1 keeps in the dormant state until the next interrogation procedure starts.

5. The reader requests tags to start a round of tag interrogation. All tags in the mute state start responding to the reader.

The steps of the interrogation procedure continue until all tags in the interrogation zone are identified successfully. The complete steps and the associated spitting tree diagram are shown in Table 2.3.

2.4.2.4 TSA Protocol

Tree slotted ALOHA (TSA) protocol [17] is a hybrid protocol which integrates the concepts of tree splitting and dynamic frame slotted ALOHA protocol. In TSA, the reader first decides an initial size S of the frame (the number of slots in the frame) and sends it to all tags to request their IDs. All tags randomly select a time slot numbered between 1 and S to transmit their tag IDs on receiving the request. If there is only one responding tag in the time slot, the tag is identified properly. However, if there are multiple responding tags in the time slot, the reader remembers the slot number and demands only those tags to respond in the next frame. It is noted that the frame size is calculated by using a particular estimation function defined in Ref. [14]. If there are still tags encountering collisions in the frame, the same action is performed for splitting the colliding tags level by level recursively. The action is similar to splitting colliding tags into subgroups according to a tree structure from the top level to the bottom level. This is why the protocol is called tree slotted ALOHA.

In TSA protocol, the reader includes in every request the frame size, the slot number for splitting colliding tags, and the level of the tag splitting tree. By memorizing the slot number selected and keeping a level variable of the tag splitting tree, tags can keep track of the status of the identification procedure. Therefore, the identification procedure can be performed properly and all tags can then be identified successfully.

2.4.2.5 BSQTA and BSCTTA Protocols

BSQTA and BSCTTA protocols are proposed by Choi et al. in Ref. [18] to improve the QT protocol. In the identification procedure of the QT protocol, when the reader sends the request bit string S of length k to the tags, the tag that has ID prefix matching S will respond with its partial tag ID of bits $k + 1, \ldots, n$ to the reader, where n is the length of the ID. If collision happens, the reader needs to send the request bit string S0 and S1 to tags latter. Choi et al. [20] observes that the request bit string S0 and S1 are the same in the first k bits and are different only in the last bit. On the basis of the observation, two methods, BSQTA and BSCTTA, are proposed to reduce the identification time with the help of two response time slots. Below we introduce the procedure of the two methods step by step.

1. A reader sends the request bit string S of length $h - 1$ to tags.
2. The tag in the interrogation zone of the reader will respond with its tag ID to the reader in one of two time slots if S matches with the first $h - 1$ bits of the tag ID. If the hth bit of the ID is "0," the tag responds in the first response time slot; otherwise, it responds in the second time slot.

Table 2.3 The Identification Procedure of EPCglobal Class 0 Protocol

Step	Command/ Ack. Bit	Response Bit	Status
1	NewRound	Tag 001:0 Tag 011:0 Tag110:1	Collision
2	0	Tag 001:0 Tag 011:1 Tag110:Mute	Collision
3	0	Tag 001:1 Tag 011:Mute Tag110:Mute	Success
4	1	Tag 001:Dormant Tag 011:Mute Tag110:Mute	Identified (Tag001)
5	NewRound	Tag 001:Dormant Tag 011:0 Tag110.1	Collision
6	0	Tag 001:Dormant Tag 011:1 Tag110:Mute	Success
7	1	Tag 001:Dormant Tag 011·1 Tag110:Mute	Success
8	1	Tag 001:Dormant Tag 011:Dormant Tag110:Mute	Identified (Tag011)
9	NewRound	Tag 001:Dormant Tag 011:Dormant Tag110:1	Success
10	1	Tag 001:Dormant Tag 011:Dormant Tag110:1	Success
11	1	Tag 001:Dormant Tag 011:Dormant Tag110:0	Success
12	0	Tag 001:Dormant Tag 011:Dormant Tag110:Dormant	Identified (Tag110)

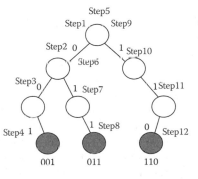

- For BSQTA, the tag responds with its ID from the $(h + 1)$th bit to the last bit.
- For BSCTTA, the tag responds with its ID from the $(h + 1)$th bit to the last bit until it receives an ACK command, which is sent by the reader to indicate the collision occurrence.

3. If there is no collision in a time slot, the tag can then be identified successfully.
4. If collisions occur in a response time slot (numbered with 0 or 1), then the reader should send a new request bit string to the tags.

   - For BSQTA, the new request bit string will be S appended by the time slot number (0 or 1).
   - For BSCTTA, the new request bit string will be S appended by the bits received before collisions occur.

The above procedure is repeated until all tags are identified successfully. As shown in Ref. [25], the performance of QT protocol can be improved significantly by BSQTA and BSCTTA.

2.4.2.6 AQS Protocol

AQS protocol is an adaptive tag anti-collision protocol proposed by Myung et al. [24] to improve QT protocol. The basic concept of this protocol is to reduce the collisions by referring the tag ID information obtained from the last identification round under the assumption that the tag population does not change greatly in consecutive rounds. The identification procedure of AQS protocol is the same as that of QT protocol except that the request bit strings in the ready-to-send string queue is copied from the last identification round. The queue includes not only the request bit strings of steps of successful tag identification but also those of steps without any tag response. If the population of tags in the interrogation zone remains the same, all tags can be identified successfully without modifying any request bit string in the queue. But if there are tags joining or leaving after the last identification round, the following actions must be done.

- Tags joining:
  If tag collisions occur for the request bit string S provided by the last identification round, there must be new tags moving into the interrogation zone of the reader after the last identification round. For such a case, the tree splitting procedure is performed and longer request bit strings are added to the queue.
- Tags leaving:
  If some tag leaves, there will be no response for some request bit string S provided by the last identification round. To improve the identification performance, the reader should merge the request bit string S with the one in the queue that has the same bit string as S except for the last bit.

2.4.3 Counter-Based Protocols

Counter-based protocols [11,22–24], like tree-based protocols, do not have the tag starvation problem. The basic idea of the two classes of protocols is to repeatedly split the tags encountering collisions into subgroups until there is only one tag in a subgroup to be identified successfully. The major difference between these two classes of protocols is that the tree-based protocol relies on static tag IDs for a deterministic splitting, but the counter-based protocol relies on dynamically changing counters for a probabilistic splitting. Since the counter-based protocol does not rely on tag IDs for the splitting, it has the stable property that its performance is not affected by the ID distribution or the ID length. In this section, we introduce two counter-based tag anti-collision protocols: ISO/IEC 18000-6B and ABS protocols.

2.4.3.1 ISO/IEC 18000-6B Protocol

ISO/IEC 18000-6B [22] is a standard adopting the counter-based tag anti-collision protocol. In ISO/IEC 18000-6B, each tag uses a dynamically changing counter and a random bit generator for tag identification. All tags' counters are initially set to 0 and every tag with counter value 0 can transmit its tag ID to respond to the request of a reader. When a collision occurs, the reader notifies all tags of the collision. The tags with counter values greater than 0 then increase their counters by 1, while the tags with counter value 0 randomly generate a random bit, 0 or 1, and add it to their counters. In this way, the tags with counter value 0 are split into two subgroups, one for tags with counter value 0 and the other for tags with counter value 1. The splitting procedure will be repeated until only one or no tag is of counter value 0. In the case of only one tag having counter value 0, this tag can be identified successfully and should keep silent until the end of the tag interrogation procedure. Either in the case of only one tag or in the case of no tag having counter value 0, the reader sends a command to inform all tags to decrease their counters by 1. The procedure will continue until all tags are identified successfully.

Below, we show an example to illustrate the procedure of ISO/IEC 18000-6B protocol. We assume there are four tags with unique IDs 0010, 0110, 1001, and 1110. The steps of the tag interrogation procedure are as follows:

1. At the beginning, the reader requests the tags to start a round of tag interrogation. On receiving the request, the tags reset their counters to 0. Tag1 (with ID 0010), tag2 (with ID 0110), tag3 (with ID 1001), and tag4 (with ID 1110) respond with their IDs to the reader simultaneously and collisions happen.
2. The reader sends a collision-notification command to make all tags randomly add 0 or 1 to their counters. Tags 1, 2, and 4 are with counter value 0. They respond with their IDs simultaneously and collisions occur again.
3. The reader sends a collision-notification command to make tags 1, 2, and 4 randomly add 0 or 1 to their counters, while tag3 increases 1 to its counter. Tag2 with counter value 0 responds with its ID to the reader and is identified successfully.
4. The reader acknowledges the identified ID with a success-notification command. The identified tag2 enters the silent state, and all unidentified tags 1, 3, and 4 decrease their counters by 1.

The identification procedure is repeated until all tags are identified successfully. The whole steps of the interrogation procedure and the associated tree diagram are described in Table 2.4.

Table 2.4 The Identification Procedure of ISO/IEC 18000-6B Protocol

Steps	Reader Command	Tag ID	Counter Value	Random Bit	New Counter Value	Response	Tree Diagram
1	REQUEST	1	–		0	0010	0 (0010, 0110, 1001, 1110)
		2	–		0	0110	
		3	–		0	1001	
		4	–		0	1110	
2	Collision	1	0	0	0	0010	0 (0010, 0110, 1110) 1 (1001)
		2	0	0	0	0110	
		3	0	1	1		
		4	0	0	0	1110	
3	Collision	1	0	1	1		2 (1001); (0110) 0 1 (0010, 1110)
		2	0	0	0	0110	
		3	1		2		
		4	0	1	1		
4	Success	1	1		0	0010	1 (1001); 0 (0010, 1110)
		2	0		–		
		3	2		1		
		4	1		0	1110	
5	Collision	1	0	0	0	0010	2 (1001); 0 (0010) 1 (1110)
		2	–		–		
		3	1		2		
		4	0	1	1		
6	Success	1	0		–		1 (1001); 0 (1110)
		2	–		–		
		3	2		1		
		4	1		0	1110	

Table 2.4 (continued)   The Identification Procedure of ISO/IEC 18000-6B Protocol

Steps	Reader Command	Tag ID	Counter Value	Random Bit	New Counter Value	Response	Tree Diagram
7	Success	1	–		–		
		2	–		–		
		3	1		0	1001	
		4	0		–		
8	Success	1	–				
		2	–				
		3	–				
		4	–				

2.4.3.2   ABS Protocol

ABS protocol [24] is proposed to improve ISO/IEC 18000 6B tag anti-collision protocol. A tag in ABS protocol keeps two counters, progressed slot counter (PSC) and allocated slot counter (ASC). PSC represents the number of tags identified successfully. It is initialized to 0 and is increased by 1 when a tag is successfully identified. With PSC and ASC, a tag can decide if it can transmit its ID to respond to a reader request. All tags with ASC equal to PSC can transmit their tag IDs. When there is no response, all tags with ASC larger than PSC decrease ASC by one. When collisions occur, the reader notifies all tags of the collisions. For such a case, the tags with ASC larger than PSC then increase ASC by 1, while the tags with ASC equal to PSC randomly add 0 or 1 to ASC. Note that tags with ASC less than PSC do not increase ASC; they do not even attempt to transmit their IDs until the tag interrogation procedure completes because they have already been identified.

After all tags are identified, tags have unique and successive ASC values. These values can be reserved for use in the next tag interrogation round to speed up the interrogation procedure. If there are tags joining or leaving after the last interrogation round, the following actions must be taken.

- Tags joining:
  When a new tag receives the reader's command to start a new interrogation round, it sets its PSC to 0 and sets its ASC to a random value R within a proper range passed by the reader. The new tag's response will collide with that of the old tag with ASC value R. The processes of ABS protocol mentioned above can deal with the collision properly by adjusting all tags' counters.

■ Tags leaving:
   If the reader detects that no tag responds to a request, it knows that there must be a leaving tag. All tags with ASC larger than PSC will decrease ASC by 1 to deal with the case.

As shown in Ref. [24], the performance of ISO/IEC 18000-6B tag anti-collision protocol is improved significantly by the ABS protocol. This justifies that the counter information obtained from the last interrogation round is very useful when the tag population does not change greatly in consecutive interrogation rounds.

2.5 Conclusion

How to identify the tags in the interrogation zone quickly and accurately is the fundamental design requirement for the reader/tag anti-collision protocols. There are many anti-collision protocols proposed to meet the requirement. We have introduced several reader collision protocols that are based on the concepts of TDMA, FDMA or CSMA in this chapter. We have also introduced some tag anti-collision protocols that are classified as ALOHA-, tree-, and counter-based protocols. Table 2.5 lists the properties of the protocols introduced in this chapter. Below, we further give summaries and new research directions for the protocols.

2.5.1 The Summary and New Directions for Reader Anti-Collision Protocols

Because a passive tag is tiny and is energized by the reader, it only has limited computation power and communication capability. The common mechanisms, such as RTS/CTS, used in wireless communications field to avoid collisions are not suitable for the RFID system. New mechanisms are thus needed to reduce collisions. In Section 2.3, we survey several reader anti-collision protocols for reducing reader collisions. They can be classified as TDMA, FDMA, and CSMA protocols. With the popularity of the RFID system, there exist the following new research directions for reader anti-collision protocols.

1. Reader anti-collision protocols for mobile reader environments:
   In a static reader environment, we can allocate resources like frequency channels and time slots to reduce as many collisions as possible. But if the reader can move around, the signal interference will be dynamic and unpredictable. No prior fixed plan is suitable for the dynamically changing environment of mobile readers. New reader anti-collision protocols are thus needed for such environments.
2. Reader anti-collision protocols for dense reader environments:
   How to fairly allocate the resources (for example, frequency channels or time slots) among the readers is more complex for environments of dense readers. More efficient anti-collision protocols are thus needed for such environments. In some cases, it is required for a reader to cooperate with others to track tags. Reader cooperation can extend the area where tags can be tracked.

Table 2.5   Properties of the Protocols Introduced in This Chapter

Protocol Type	Protocol	Properties
Reader anti-collision	DCS [7] Colorwave [7]	• In Colorwave (resp., DCS), the number of colors can (resp., cannot) be adjusted dynamically • The protocols assume time synchronization among readers • It is hard for a reader to detect collisions alone
	HiQ [8]	• It is inefficient to be applied to mobile environments • Due to the hierarchical architecture, more extra management overheads are required
	EPCglobal Gen 2 [9]	• It is not suitable for the low-cost tags having no frequency selection capability
	ETSI 302 208 Standard [10]	• It is hard for a reader to detect the collisions just by sensing the carrier alone
Tag anti-collision	ALOHA [12] Slotted ALOHA [13] Frame Slotted ALOHA [14] ISO/IEC 18000-6A [22]	• The algorithms are simple • They have the tag starvation problem • The performance of slotted ALOHA is twice better than ALOHA • Frame slotted ALOHA needs to estimate the frame size properly for better performance
	QT [16]	• The algorithm and the tag circuit are simple • No nonvolatile writable tag memory is needed • The identification delay is affected by the distribution and the length of tag IDs

Table 2.5 (continued)   Properties of the Protocols Introduced in This Chapter

Protocol Type	Protocol	Properties
	Bit-by-Bit Binary Tree [15] EPCglobal Class 0 [19]	• The traffic of message transmission is lower • The complexity depends on the length but not the distribution of tag IDs
	TSA [17]	• The tag needs a complex circuit
	BSQTA BSCTTA [18]	• The tag needs a complex circuit • Time synchronization is needed • In BSCTTA, the tag is hard to transmit and listen the channel on the same time
	AQS ABS [30]	• If the tag population does not change greatly, the protocols have excellent performance
	ISO/IEC 18000-6B [22]	• The performance is not affected by the length or the distribution of tag IDs • It has no tag starvation problem

3. Reader anti-collision protocols for environments with handheld devices and active tags:

   One of the fatal weak spots for the handheld device is its limited power capacity. Energy-efficient or energy-aware reader anti-collision protocols are thus needed for environments with handheld devices for the purpose of saving energy of handheld devices. The energy-saving and energy-aware considerations can also be extended to the operation of active tags to prolong their lifetimes.

2.5.2   The Summary and New Directions for Tag Anti-Collision Protocols

In Section 2.4, we have categorized tag anti-collision protocols into three classes, namely, ALOHA-, tree-, and counter-based. We have then introduced some of them class by class. ALOHA-based protocols are simple and have fair performance. However, they have the tag starvation problem that a tag may never be identified because its responses always collide with others'. Tree-based protocols usually have longer identification latency than ALOHA-based protocols, but they do not have the tag starvation problem. Tree-based protocols also have the drawback that their performance is affected by the length or the distribution of tag IDs. Like tree-based protocols, counter-based protocols do not have the starvation problem. And they have the stable property that their performance is not affected by the tag ID distribution or ID length. Table 2.6 gives a comparison of some representative tag anti-collision protocols in terms of time complexity, messages

Table 2.6 The Comparison of Representative Tag Anti-Collision Protocols in Terms of Time Complexity, Message Complexity and the Need of Nonvolatile Writable Memory in Tags

Protocol	Time Complexity	Message Complexity	The Need of Nonvolatile Writable Memory in Tags
QT [16]	Average case: $O(n)$ Worst case: $n \times (k + 2 - \log n)$ [31,32]	Worst case: $k \times (2.21 \log n + 4.19)$ [31,32]	No [33]
Bit-by-Bit Binary Tree [15]	Worst Case: $\theta(2^k)$ [32,34]	Average case: $O(n(k + 1))$ [32,34]	Yes (8 bit memory as a pointer to maintain the last bit sent [18])
ISO/IEC 18000-6B [22]	$O(n)$ [20,32]	Worst case: [32,35] • $\theta(n \log n)$ for the case that n is unknown • $\theta(n)$ for the case that n is known	Yes (8 or 16 bit memory as a counter to keep the tag response sequence and timing [30])
Framed Slotted ALOHA [14]	Upper bound: $t \times s +$ time required to estimate N [14,32]	Upper bound: $n \times s$ [32]	Yes (8 or 16 bit memory as a counter to keep which slot the tag can respond [14])

Note: n is the number of tags in the interrogation zone, k is tag ID length, t is the time duration of a frame, s is the number of frames needed to identify all tags, and N is the estimation value of n.

complexity, and the need of nonvolatile writable memory on tags. The information is derived from papers [14,18,20,31–35].

A good tag anti-collision protocol should have some characteristics. We list some characteristics that should be kept in mind when we develop new tag anti-collision protocols.

1. A reader need to recognize all the tags in its interrogation zone. If some tags cannot be identified properly for some reason (e.g., due to the tag starvation problem), this may cause problems in some applications. Therefore, a good tag anti-collision protocol should try to not miss any tag.

2. For many applications, tags are usually attached to mobile objects. Because a reader can only successfully identify the tags when the tags are within the interrogation zone, the reader needs to identify the tags as soon as possible so that mobile tags can be identified before they leave the interrogation zone.

3. Due to the limitation of communication and computation capabilities of tags, anti-collision mechanism should not be too complex. That is, we should keep the tag anti-collision protocol as simple as possible.

4. When a tag attached to an object is identified by readers of malicious people, the privacy of the person owning the object may be harmed. Therefore, there is a need to integrate anti-collision protocols with privacy-protection mechanisms so that tags can be identified efficiently without leaking privacy.

5. If the tag population does not change greatly in consecutive tag interrogation rounds, information obtained from the last interrogation round can be very useful. One can thus exploit the information to develop new protocols with faster interrogation procedures. However, the tag joining and leaving cases should be kept in mind when one develops such protocols.

References

1. P. Schaar. Working document on data protection issues related to RFID technology. In *Working Document Article 29—10107/05/EN*. European Union Data Protection Working Party, January 2005.
2. H. Chae, D. Yeager, J. Smith, and K. Fu. Maximalist cryptography and computation on the WISP UHF RFID tag. In *Proc. of the Conference on RFID Security*, Málaga, Spain, 2007.
3. S. M. Birari and S. Iyer. PULSE: A MAC protocol for RFID networks. In *Proc. of the EUC Workshops*, pp. 1036–1046, Nagasaki, Japan, 2005.
4. D. Y. Kim, B. J. Jang, H. G. Yoon, J. S. Park, and J. G. Yook. Effects of reader interference on the RFID interrogation range. In *Proc. the 37th European Microwave Conference (EuMC'07)*, pp. 728–731, Munich, Germany, 2007.
5. D. Y. Kim, H. G. Yoon, B. J. Jang, and J. G. Yook. Interference analysis of UHF RFID systems. *Progress in Electromagnetics Research*, 4:115–126, 2008.
6. Y. Tanaka and I. Sasase. Interference avoidance algorithms for passive RFID systems using contention-based transmit abortion. *IEICE Transactions*, 90-B(11):3170–3180, 2007.

7. J. Waldrop, D. W. Engels, and S. E. Sarma. Colorwave: An anti-collision algorithm for the reader collision problem. In *Proc. of IEEE International Conference on Communications*, Vol. 2, pp. 1206–1210, Anchorage, Alaska, 2003.

8. S. E. Sarma, J. Ho, and D. W. Engels. HiQ: A hierarchical Q-learning algorithm to solve the reader collision problem. In *Proc. of SAINT Workshops*, pp. 88–91, Phoenix, AZ, 2006.

9. EPCglobal. *EPCglobal Class-1 Generation-2 UHF RFID Protocol*, April 2004. Version 1.0.9.

10. ETSI. *EN 302 208-2 Protocol*, September 2004. Version 1.1.1.

11. M.-K. Yeh and J.-R. Jiang. Adaptive *k*-way splitting and pre-signaling for RFID tag anti-collision. In *Proc. of the 33rd Annual Conference of the IEEE Industrial Electronics Society (IECON'07)*, Taipei, Taiwan, 2007.

12. N. Abramson. The ALOHA system-another alternative for computer communications. In *Proc. of Fall Joint Computer Conference of AFIPS*, Vol. 37, pp. 281–285, Houston, TX, 1970.

13. L. Liu and S. Lai. ALOHA-based anti-collision algorithms used in RFID system. In *Proc. of Int'l Conf. on Wireless Communications, Networking and Mobile Computing 2006 (WiCOM 2006)*, pp. 1–4, Wuhan, China, 2006.

14. H. Vogt. Efficient object identification with passive RFID tags. In *Proc. of Pervasive Computing*, pp. 98–113, Berlin, 2002.

15. H. Choi, J. R. Cha, and J. H. Kim. Fast wireless anti-collision algorithm in ubiquitous ID system. In *Proc. of IEEE VTC*, Los Angeles, CA, 2004.

16. F. Zhou et al. Evaluating and optimizing power consumption of anti-collision protocols for applications in RFID systems. In *Proc. of the 2004 International Symposium on Low Power Electronics and Design*, New York, 2004.

17. M. A. Bonuccelli, F. Lonetti, and F. Martelli. Tree slotted Aloha: A new protocol for tag identification in RFID networks. In *Proc. of the 4th IEEE International Workshop on Mobile Distributed Computing (MDC'06)*, New York, 2006.

18. J. H. Choi, D. Lee, and H. Lee. Bi-slotted tree based anti-collision protocols for fast tag identification in RFID systems. *IEEE Communications Letters*, 10(12):861–863, 2006.

19. Draft protocol specification for a 900 MHz class 0 radio frequency identification tag, Auto-ID Center, Cambridge, MA, Technical report, 2003.

20. D. H. Shih, P. L. Sun, and D. C. Yen. Taxonomy and survey of RFID anti-collision protocols. *Computer Communications*, 29(11):2150–2166, 2006.

21. D. Krebs and M. J. Liard. *White Paper: Global Markets and Applications for Radio Frequency Identification*. Venture Development Corporation, 2001.

22. ISO/IEC. Information technology automatic identification and data capture techniques—radio frequency identification for item management air interface—part 6: parameters for air interface communications at 860–960 MHz. Final Draft International Standard ISO 18000-6, November 2006.

23. *Philips Semiconductors, UCODE,* http://www.semiconductors.philips.com, 2005.

24. J. Myung and W. Lee. Adaptive splitting protocols for RFID tag collision arbitration. In *Proc. of MobiHoc 2006*, pp. 202–213, Florence, Italy, 2006.

25. *Dense RFID Reader Deployment in Europe using Synchronization*, January 2008. Final draft ETSI EN 302 208-1 V1.2.1.

26. J. R. Cha and J. H. Kim. Novel anti-collision algorithms for fast object identification in RFID system. In *Proc. of the 11th International Conference on Parallel and Distributed Systems—Workshops (ICPADS'05)*, pp. 63–67, Fuduoka, Japan, 2005.

27. G. Khandelwal et al. ASAP: A MAC protocol for dense and time constrained RFID systems. In *Proc. of IEEE International Conference on Communications (ICC'06)*, Istanbul, Turkey, 2006.

28. S. Lee, S. D. Joo, and C. W. Lee. An enhanced dynamic framed slotted aloha algorithm for RFID tag identification. In *Proc. of Mobiquitous*, pp. 166–172, 2005.

29. L. G. Roberts. Extensions of packet communication technology to a hand held personal terminal. In *Proc. of AFIPS Spring Joint Computer Conf.*, Vol. 40, pp. 295–298, Montvale, NJ, 1972.

30. M. Kodialam and T. Nandagopal. Fast and reliable estimation schemes in RFID systems. In *Proc. of ACM Mobicom*, Los Angeles, CA, 2006.

31. C. Law, K. Lee, and K. Y. Siu. Efficient memoryless protocol for tag identification. In *Proc. of the 4th International Workshop on Discrete Algorithms and Methods for Mobile Computing and Communication*, pp. 75–84, Boston, MA, August 2000.

32. C. Abraham, V. Ahuja, A. K. Ghosh, and P. Pakanati. Inventory management using passive RFID tags: A survey. Technical report, Technical Report of Department of Computer Science, The University of Texas at Dallas, Dallas, TX, 2003.

33. K. Ali, H. Hassanein, and A. Taha. RFID anti-collision protocol for dense passive tag environments. In *Proc. of 32nd IEEE Conference on Local Computer Networks*, Dublin, Ireland, 2007.

34. M. Jacomet, A. Ehrsam, and U. Gehrig. Contact-less identification device with anti-collision algorithm. In *Proc. of IEEE International Conference on Circuits, Systems, Computers and Communications*, Sado Island, Niigata, July 1999.

35. D. Hush and C. Wood. Analysis of tree algorithms for rfid arbitration. In *Proc. of IEEE International Symposium on Information Theory*, Cambridge, MA, August 1998.

Chapter 3

Low-Power Transponders for RFID

Andrea Ricci, Alessio Facen, Matteo Grisanti, Andrea Boni,
Ilaria De Munari, Paolo Ciampolini, and Carlo Morandi

CONTENTS

Recently radio-frequency identification (RFID) systems have gained popularity in manufacturing units, inventory and logistics, as they represent an inexpensive and reliable solution for automatic identification. Moreover, RFID transponders are expected to become a key element in the ubiquitous computing scenario. Tags will likely be used to collect sensors data, enabling noninvasive environment monitoring. Low-cost passive ultra-high frequency (UHF) transponders are expected to play a major role in this context, due to extended read range capabilities. Within a passive tag, power harvested from the field irradiated by the reader during the communication should operate both digital control circuitry and potential sensing devices. Effective low-power operating strategies are to be implemented, to preserve large communication range and circuit functionalities. In this chapter, the state of the art in transponder design is examined. Insights for the design of a low-power tag, operating in the UHF band, are presented, in compliance with standard protocols (ISO 18000-6B and -6C), to support device interoperability. This chapter is concluded by identifying some open research issues in the realization of high-performance ubiquitous sensing transponders.

3.1   Introduction

Radio-frequency identification (RFID) systems are gaining increasing popularity in several fields (manufacturing, logistics, transportation, etc.) where unique identification and tracking of items is a major concern [1]. RFID tags represent a cost-effective solution for many problems in the full automation of supply chains and are candidates for the replacement of widespread optical barcode technologies, overcoming some of their inherent limitations (such as the reader–tag maximum distance) [2]. However, RFID promises to go well beyond the mere replacement of barcode functionalities. Research on ubiquitous and pervasive computing aims at the creation of "active spaces," integrating distributed computational infrastructure into the physical surroundings. RFID transponders may play a major role in noninvasive environment monitoring: the tag chip may embed local computing power, data storage, sensors, and communication devices, thus acting as "intelligent," yet cheap, network nodes. In particular, transponders operating in the UHF band, featuring large read ranges (up to several meters), can be profitably exploited. Aiming at mass-market penetration, cost figures in the order of few cents per tag should be achieved; to meet such a stringent constraint, a cheap fabrication process is needed. For most applications, this rules out active transponders (which include an expensive battery) and makes the choice of passive tag almost mandatory. A high-performance passive RFID tag just consists of a tiny integrated circuit chip, a flexible printed antenna, and an adhesive label substrate for application to items [1]. Power needed to operate the tag circuitry is obtained by harvesting energy from the field irradiated by the reader during the communication. Hence, effective low-power operating strategies are to be implemented, to preserve large communication range and circuit functionalities. Although a tag chip includes an analog RF front end and a memory as well, the largest power fraction is usually required by baseband processor [2]. Hence, exploiting an ultra-low-power baseband processor would provide sensing section with higher energy, thus improving measurement performance.

In this chapter, the state of the art in transponder design is presented. First a deep analysis of system requirements is introduced and then characteristics and design goals of transponder devices are reported. This is followed by presenting circuit-design solutions based on low-voltage operations and low-current design techniques, devised for both analog and digital tag building blocks. Open research issues, including sensor integration and security support, are discussed and some potential solutions are eventually discussed.

3.2 Survey on State-of-the-Art RFID Implementations

At present, several RFID systems operate in the industrial, scientific, and medical (ISM) 13.56 MHz band [1,3]. Such systems exploit near-field communication, relying on either capacitive or magnetic coupling between the tag and the reader. Although high frequency (HF) systems are reliable and suitable for many applications, the coupling methods substantially limit their achievable read range to a maximum of about 1 m. Alternative systems, which are currently under active development, exploit higher frequencies, such as the 868/915 MHz [4–13] and 2.4 GHz ISM bands [14,15]. UHF systems allow the reader-to-tag coupling by means of radiated energy in the far field, and hence may feature much larger read ranges. Moreover, a larger bandwidth is available, which, in turn, may be exploited to attain higher data rates.

Several works have been done on complete RFID chip, using proprietary communication protocols between a tag and a reader. Kocer et al. [4] reports a wireless telemetry device which just recovers power and a reference clock from a 450 MHz incident RF signal and returns an ID data by means of a 900 MHz binary-phase-shift keying (PSK) modulated carrier. Karthaus et al. [5] details a low-power RFID tag, operating in UHF band, which features EEPROM memory and a full communication protocol relying on pulsewidth modulation (PWM) on the forward link (reader-to-tag communication) and PSK on the return link (tag-to-reader communication). Curty et al. [15] details a fully integrated remotely powered RFID transponder, working at 2.45 GHz, featuring on–off keying (OOK) modulation on the forward link and amplitude-shift keying (ASK) backscatter modulation. Much work has been done on CMOS-based analog front end for RFID application [6–8].

The attention to established and widespread standards, to support devices interoperability, represents the founding element of further works. Gillen et al. [2] provides a high-level description of a passive RFID chip, implementing the EPC Class 0 protocol [16], whereas [10–12] focus on the optimization of a fully digital baseband processor. An ultra-low-power battery/passive RFID tag, ISO 18000-6B [17] compliant, was recently introduced in Ref. [9] which is able to operate in both high-frequency and microwave bands.

3.3 RFID System Requirements

Main RFID tag design targets can be summarized as follows:

- Wide read range, that is, effective exploitation of energy harvested from the incoming antenna field
- Low cost per die, that is, compatibility with standard VLSI low-cost fabrication processes and small device footprint
- Device interoperability, to allow for communication within etherogeneous readers networks

These requirements are addressed in the following.

3.3.1 EM Wave Propagation Basics and Tag Power Consumption

The propagation of an electromagnetic (EM) wave into open space from its source point is spherical. An antenna radiating a power P_T uniformly in all directions is called isotropic or spherical. The radiation density S measured at any point at a distance r from such an emitter is given by

$$S = \frac{P_T}{4\pi r^2} \tag{3.1}$$

This condition holds in the so-called far field, that is, r is higher than twice the wavelength of the EM field λ. On the contrary, in the near field it is very difficult to determine the relationships between the electrical and magnetic components of the EM wave, as well as the radiation density. We will assume for the purpose of analysis far-field conditions.

Actually, no physical isotropic antenna exists: every emitter has its radiation pattern, featuring a certain gain $G_T(\theta, \phi)$ for any direction described by the angles θ and ϕ. The directionality of an antenna is determined by its physical characteristics. Along the main radiation direction, therefore, the radiation density is G_T times the one of an isotropic emitter:

$$S = \frac{P_T G_T}{4\pi r^2} \tag{3.2}$$

The effective isotropic radiated power, or shortly EIRP, can be defined as the power that should be provided by a spherical antenna to experience the same radiation density at a distance r coming from the given emitter:

$$P_{T,EIRP} = P_T G_T \tag{3.3}$$

For example, a dipole antenna exhibits a maximum gain $G_T = 1.64$: hence, its $P_{T,EIRP}$ is equal to $1.64 P_T$. Actually, the power emitted by a dipole antenna is also known as effective radiated power (ERP), so $P_{T,EIRP} = 1.64 P_{T,ERP}$. On the other hand, a receiving antenna (in matching conditions) sensing a radiation density S delivers the following available power to its load [18]:

$$P_{AV} = S \frac{\lambda^2}{4\pi} G_{ANT} \tag{3.4}$$

where
  λ is the wavelength of the received EM wave
  G_{ANT} is the receiving antenna gain (assuming the antenna is oriented in the maximum
    gain direction)

From Equations 3.2 through 3.4 it follows that

$$P_{AV} = P_{T,EIRP} \left(\frac{\lambda}{4\pi r} \right)^2 G_{ANT} \tag{3.5}$$

The last equation, known as the Friis transmission equation [19], describes the amount of received power of an antenna with gain G_{ANT} at a distance r from a source emitting an EM wave with wavelength λ and power $P_{T,EIRP}$. The reader transmits an EM wave composed by a carrier, whose frequency is established by the regulations to be in the 860–960 MHz frequency range and the power is between 500 mW ERP and 4 W EIRP, depending on the local regulations [20]. Assuming the transponder has a unity-gain antenna (G_{ANT}), its available power (given Equation 3.5) versus the reading distance is reported in Figure 3.1: it is worth noticing that such a power rapidly drops into the microwatts range when the distance becomes higher than few meters. The power awareness and efficiency of a passive tag is, therefore, a must to meet the reading range requirements for the final product. In this framework, making the tag core circuitry operate slightly above the threshold voltage could represent a feasible solution to aggressively reduce power consumption while maintaining fair performance [21].

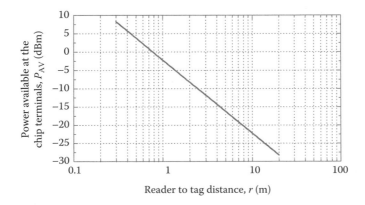

Figure 3.1 **Available power for a 0 dB matched antenna tag with a 500 mW ERP transmitted power versus reading distance.**

Figure 3.2 depicts a high-level block diagram of a passive transponder. As already stressed above, power is supplied to the digital core by rectifying and regulating voltage extracted at the antenna terminals (V_{ant}). An on-chip dumping capacitor (C_{dump}) is employed as an energy-storage element, needed to maintain an adequate power supply level during the interrogation and response phases. The dumping capacitor should be completely charged within the time interval that each standard reserves for the RFID powering up, just before the interrogation phase start ($t = t_0$). This initial condition can be expressed as follows:

$$V_{dd,t_0}^{unreg} = V_{dd,max}^{unreg} = \frac{Q_{max}}{C_{dump}} \tag{3.6}$$

Moreover, assuming a linear voltage regulator is used, a minimum input voltage,

$$V_{dd}^{unreg}(t) > V_{dd,min}^{unreg} \tag{3.7}$$

should be always maintained during the operating phase, to ensure the correct generation of regulated voltage (V_{dd}^{reg}). Theoretically, the tag's available instantaneous power ($dE_{out}(t)/dt$) is finite and limited by the incoming power ($dE_{in}(t)/dt$), while the total energy can be infinite as long as the tag harvests power from the field irradiated by the reader. The dumping capacitor just acts as a charge reservoir, increasing the available instantaneous power at tag circuitry with respect to the incoming one, thus facing a peak power request, which can be tolerated when the rectified voltage (V_{rect}) stays above the minimum value $V_{rect,min}$. This condition is fulfilled when the tag energy consumption, $E_{out}(t)$, is limited as follows:

$$E_{out}(t) < E_{in}(t) + \frac{C_{dump}}{2}\left(\left(V_{dd,t_0}^{unreg}\right)^2 - \left(V_{dd,min}^{unreg}\right)^2\right) \tag{3.8}$$

throughout the whole reading phase (i.e., for $t_0 < t < t_{end}$). In the equation above, the required energy $E_{out}(t)$ can be ascribed to different components of the tag:

$$E_{out}(t) = \int_{t_0}^{t} P_{out}(\tau)d\tau = \int_{t_0}^{t} \left[V_{drop}(\tau) + V_{dd}^{reg}(\tau)\right]I_{reg}(\tau)d\tau, \tag{3.9}$$

where
  V_{rect} is the rectified voltage
  V_{drop} is the voltage drop across the linear regulator
  regulator current I_{reg} comes from the sum of the baseband processor, memory, sensors, and oscillator currents

$$I_{reg} = I_{bp} + I_{mem} + I_{sens} + I_{osc}. \tag{3.10}$$

Incoming energy can be expressed as

$$E_{in}(t) = \int_{t_0}^{t} P_{in}(\tau)d\tau \cong \eta_{rect}G_{tag}\int_{t_0}^{t} P_{EIRP}(\tau)d\tau \tag{3.11}$$

where η_{rect} is the rectifier conversion efficiency.

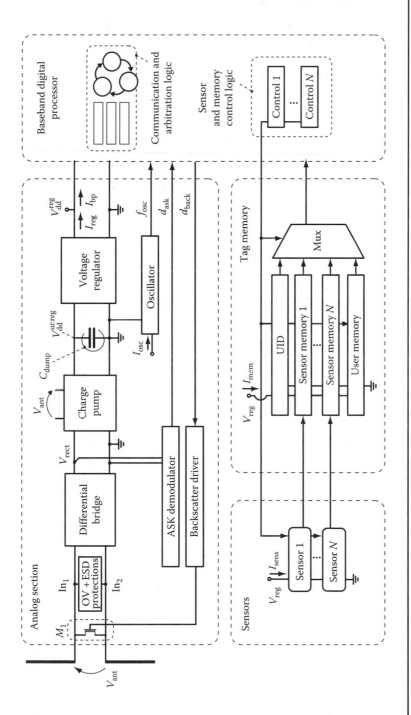

Figure 3.2 Block diagram of the UHF RFID transponder.

3.3.2 Fabrication Process

As already stressed in Section 3.1, extremely low costs are needed to foster RFID tag diffusion. At large production volumes, fabrication cost prevail over nonrecurring engineering costs, so that a careful evaluation of the actual fabrication process in terms of cost/performance figures is crucial. In terms of speed, RFID tags usually pose no critical constraint, whereas power is the main concern. A further issue is related to the technologies, maintenance, because long-running technologies are needed to preserve the investment costs. In addition, the price per tag can be highly reduced by adopting a low-cost, general purpose technology for the chip fabrication, although the design of the analog blocks may suffer from the devices' nonoptimal performances and the absence of specific process options: such limitations should be overcome with smart circuit-level solutions.

In the following, we refer to a 0.18 μm CMOS technology node when reporting details of prototypal tag implementations. Such a technology is currently mature enough to ensure proven reliability, while at the same time still ensuring long-term running perspectives and not yet presenting too severe leakage current problems. Although the selected technology provides designers with transistors featuring different threshold voltages (high-V_{th} = 0.5 V and low-V_{th} = 0.35 V), only high-V_{th} devices were exploited in the digital core design, to minimize the off-current leakage contribution to the power consumption. A digital core supply voltage was regulated close to the transistor threshold (i.e., V_{DD} = 0.6 V), to limit dynamic power consumption as well and take advantage of the relatively coarse speed requirement: simulations and measurements support the selected approach, even at the worst operating conditions. Nevertheless, a more scaled technology node (i.e., 0.13 nm) could be suitable for physical implementations as well [13].

3.3.3 Air Interface Standard

There are several standards regulating RFID transmissions for passive tags. Actually, the international standard that establishes the way tags and readers communicate in the UHF spectrum is the ISO 18000-6. There are currently three versions of the standard, the ISO 18000-6A, -6B, and -6C. Here, the -6B and -6C versions are analyzed, having a wider diffusion worldwide with respect to -6A standard. In addition, it is generally possible to reuse a considerable part of a chip designed for a certain standard while potentially migrating to another regulation.

Both the considered standards define the protocol for a passive backscatter RFID system, featuring the following capabilities: identification and communication with multiple tags in the field, selection of a subgroup of tags for identification or communication, reading from and writing to or rewriting data many times to individual tags, user-controlled permanently lockable memory, data integrity protection, interrogator-to-tag communication link with error detection, and tag-to-interrogator communication link with error detection.

According to the ISO 18000-6B standard [17], the reader modulates the amplitude of the RF carrier (i.e., ASK transmission) to convey data to the tag. Data are Manchester encoded to enable bit error detection by the tag receiver. The backward link air interface is based on classical passive backscatter modulation [20]: the tag transmits information

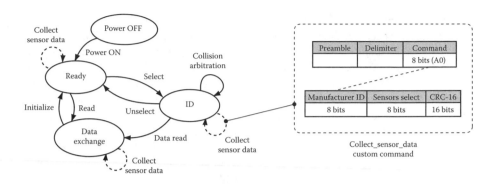

Figure 3.3   ISO 18000-6B tag state diagram and custom command format.

to the interrogator by modulating the incident energy and reflecting it back to the reader. The fastest transmission speed (40 kHz) allowed by the standard could be selected to maximize the tag read speed. The protocol deals with the collision arbitration by means of a probabilistic binary tree algorithm. In addition, command and data exchanged between the reader and the tag are contained in properly formatted frames, subdivided in several different fields. Each frame include a 16 bit cyclic redundancy check (CRC) field, to improve error detection capabilities. Figure 3.3 depicts the main states of an ISO 18000-6B compliant transponder. Further details are reported in Section 3.5.3, where sensor integration is discussed.

According to ISO 18000-6C standard, interrogator-to-tag communication is based on a packet-based scheme as well [23,24]. The 6C standard supports a higher transmission data rate, respect to 18000-6B regulation: transmission rate is defined by the interrogator, and could be changed at each communication round. To receive pulse-interval encoding (PIE)-encoded payload data (and to reply to the reader, if necessary) the tag measures reference time intervals (Tari, RTcal and, if needed, TRcal) included in the reader packet preamble. The measurement of Tari interval (i.e., the 0 symbol length, chosen in the {6.25, 12.5, 25 μs} set), and RTcal (i.e., data-0 plus data-1 length) can be reliably carried out with a reference clock frequency of at least 2 MHz. The interrogator specifies the tag's backscatter link frequency (LF, in the $40 \div 640$ kHz range) by means of different combinations of the TRcal interval and divides ratio parameters (DR, included in the payload). The actual LF and its tolerance FT, which definitely constrain the tag's minimum clock frequency, are deeply analyzed in Section 3.4.7. The encoding format, selected in response to interrogator commands, is either FM0 or Miller-modulated subcarrier. The tag inventory rounds are operated by means of a slotted random anticollision algorithm. Error detection capability is included in 6C protocol as well, based on either 16 or 5 bit CRC field, depending on interrogator command.

3.4   Analog Front End and Antenna Design Issues

As already stressed above, a passive UHF RFID tag harvests the necessary power for the analog and the digital core from the incident EM wave transmitted by the reader.

Therefore the sensitivity and the power efficiency of the RF front-end is a specification of primary importance, that directly impact on the reading range of the tag. The black-box schematic of the analog front end of the tag is shown in Figure 3.2. The front end is connected to the antenna terminals through pins in_1 and in_2, which are the only pins of the RFID chip. The differential bridge is a RF rectifier which is responsible for retrieving the maximum power from the antenna and provides a rectified voltage to the following blocks. Since the level of the rectified voltage is too low for powering the digital core of the chip (few hundreds of mV at the maximum distance from the reader) a voltage booster is necessary (referred in Figure 3.2 as charge-pump). This DC–DC up-converter generates a stable voltage (V_{dd}^{unreg}) across the on-chip filter capacitor (C_{dump}). Such a voltage level is strongly dependent on the available power at the antenna terminals, that is, on the reading distance. Therefore a series voltage regulator is introduced to provide a stable supply voltage, and almost independent on the reading conditions, to the digital core. Limiting the variation of the supply voltage allows to lower the tolerance of the reference frequency provided by the local oscillator, as discussed in Section 3.4.7.

An additional feature of the analog front end is the demodulation of the signal transmitted by the reader. As an example, the ISO 18000-6B/-6C standards dictate that the reader-to-tag communication is based on an ASK modulation of the UHF carrier with a modulation depth varying from 100 percent down to 18 percent [35]. Finally, the analog front end is responsible for the tag-to-reader communication, which is based on the backscatter technique [22]. This is a convenient method for transmitting back any information with a minimum power consumption: The tag sends either a 0 or a 1 to the reader by simply activating or deactivating a switch (M1) placed in parallel to the antenna. When the switch is closed the antenna is completely mismatched, thus leading to a strong reflection of power from the tag to the reader. Therefore, the logic value of the bit transmitted by the tag is identified by measuring the amount of reflected power.

3.4.1 Antenna Properties

To harvest the maximum available power from the RF field, the condition of impedance matching must be met between antenna and RF front end of the chip. Such a constraint means that the impedance of the antenna at the transmission frequency has to be equal to the complex conjugate of the front-end input impedance [26]. The latter is composed of a real (resistive) and a negative imaginary (capacitive) part, allowing hence to model the chip with a $R_{IN} - C_{IN}$ equivalent series circuit. Therefore, the power matching can be achieved by setting the antenna equivalent resistance R_{ANT} to be equal to R_{IN}, while C_{IN} has to be balanced with an equivalent inductor L, as shown in Figure 3.4. Since an accurate design allows to keep the inductance value as low as few dozens nH, there is no need to place an inductor as external component, that would basically add an undesired extra cost. As a matter of fact, an inductive antenna can be designed by setting its resonance frequency f_{ANT} slightly above the RF carrier used in the communication (i.e., $f = 869.5\,MHz$): a careful antenna design can thus grant the power-matching condition without adding extra components. If the matching condition is met ($R_{ANT} = R_{IN}$ and $\omega L = (\omega C_{IN})^{-1}$), the Friis equation in Equation 3.5 can be used without introducing any degrading factor due to the impedance mismatch, and the available

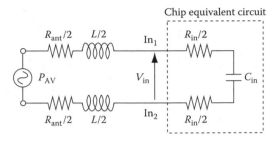

Figure 3.4   Passive tag equivalent schematic.

power at the chip inputs is the highest for the given reading distance. It is possible from here to calculate the input peak voltage: referring to Figure 3.4, it can be shown that

$$V_{IN} \approx \sqrt{2\frac{P_{AV}}{R_{IN}}\frac{1}{\omega C_{IN}}} \qquad (3.12)$$

From the above equation, it can be seen that the input peak voltage can be increased by designing a tag with low resistive and capacitive parts. Since, given the nonlinear behavior of the RF/DC power retriever, a higher rectification efficiency can be achieved with an elevate input peak level, the tag should be in some manner designed to minimize R_{IN} and C_{IN}. However, it should be noticed that the physical realization of a good antenna with a very low real impedance is a tough task, because the antenna gain would considerably decrease. Moreover, the parasitic effect of the ohmic losses in the antenna tracks would introduce a noticeable efficiency drop.

3.4.2   RF Rectifier

As already stressed above, passive RFID tags harvest power from the incoming EM field, by means of rectifying and regulating reader RF carrier.

The unmodulated carrier sent by the reader may be written as

$$v_{in}(t) = V_{IN}\sin(2\pi f t) = V_{IN}\sin\left(\frac{2\pi t}{T}\right) \qquad (3.13)$$

where
 f is the UHF carrier frequency (e.g., 869.5 MHz in Europe)
 T is the corresponding time period

Most RFID tags presented in the last years employ structures based on the well-known full-wave diode bridge, reported in Figure 3.5a. The voltage level at the *rect* terminal is one diode threshold (V_T) lower than the peak of the highest potential reached either by in_1 or in_2, assuming negligible, for the sake of simplicity, the discharge of the loading capacitor. Similarly, the *gnd* terminal sets V_T above the lowest peak hit by one of the two

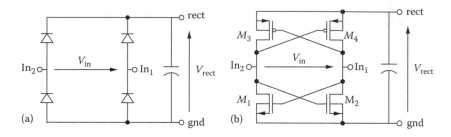

Figure 3.5 Schematic of (a) the full-wave bridge and (b) the low-voltage CMOS rectification stage.

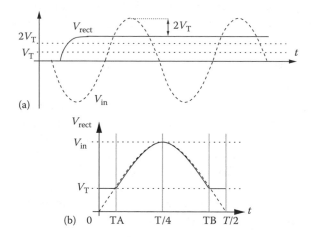

Figure 3.6 Full-wave bridge: (a) the output waveform and (b) rectifier output voltage during half input period.

input pins. Therefore, the resulting rectified voltage V_{rect}, as shown in Figure 3.6a, is

$$V_{rect} = V_{IN} - 2V_T \qquad (3.14)$$

It is immediate from Equation 3.14 that the minimum input peak value useful to achieve a nonnull rectified voltage is $V_{IN} = 2V_T$. The loss of two diode thresholds in the rectification process for a single stage calls for the use of low-V_T devices, especially in this kind of application where V_T is not negligible with respect to V_{IN}. As a matter of fact, most reported works resort to specific technology options. Most used are Schottky devices [5,27–29], which feature very low forward voltage, but also tags exploiting BiCMOS [30], Silicon-on-Insulator (SOI) [15] or ferroelectric [31] processes can be found in literature. Each one of these options brings considerable advantages in the design of power-efficient rectifiers; however they also feature production costs noticeably higher than the ones of a standard CMOS process.

In a CMOS technology, the limitation imposed by the $2V_T$ voltage drop by the circuit of Figure 3.5a heavily impacts on the maximum reading range. A RF rectifier fully compatible with a low-cost CMOS technology was proposed by [25,32] and, afterwards, in Ref. [33]. The schematic is shown in Figure 3.5b. To better understand the working principle of the circuit, initially the on-resistance of the four transistors is assumed to be negligible: hence, they behave like ideal switches. Moreover, the PMOS and NMOS thresholds are considered to be equal: $-V_{THp} = V_{THn} = V_T$. This approximation holds considering V_T as the maximum between $-V_{THp}$ and V_{THn}. During the time intervals where $|v_{in}(t)|$ is higher than V_T, either one of the NMOS transistors M_1 or M_2 experiences a V_{GS} high enough to be turned on, while the other, having an equivalent negative V_{GS} given the symmetries of the circuit and the input signal, is in off state. The closed switch shorts the *gnd* terminal to its source, that is, the pin with the lowest voltage: Therefore, *gnd* follows the minimum potential present at the inputs. Likewise, the *rect* terminal is connected to the input pin featuring the highest potential by either M_3 or M_4. During the positive phase of $v_{in}(t)$ the closed switches are M_1 and M_4, while during the negative phase it is the dual couple M_2–M_3 that conducts. It happens thus that $v_{rect}(t) = |v_{in}(t)|$. To reduce the ripple to an acceptable level for the application, $v_{rect}(t)$ has to be filtered: This operation can be done considering the MOS's on-resistance r_{sw}, that in the previous calculations had been neglected. In fact, the series of the r_{sw} of the two transistors and the loading capacitor forms a low-pass RC filter, that can be sized to suppress the UHF ripple. In this case, $v_{rect}(t) \simeq V_{RECT}$, the latter being the average level of the rectified waveform, which can be calculated as follows on the basis of Figure 3.6b. Analyzing half an input period (the one where $v_{in}(t) > 0$, for example), the rectifier output voltage can be written as

$$v_{rect}(t) = \begin{cases} V_{in} \sin\left(\frac{2\pi t}{T}\right) & \text{if } T_A < t < T_B \\ V_T & \text{if } 0 < t < T_A, T_B < t < \frac{T}{2} \end{cases} \tag{3.15}$$

where

$$T_A = \frac{T}{2\pi} \arcsin\left(\frac{V_T}{V_{IN}}\right) \tag{3.16}$$

and

$$T_B = \frac{T}{2} - \frac{T}{2\pi} \arcsin\left(\frac{V_T}{V_{IN}}\right) \tag{3.17}$$

The average value of the waveform can be computed by integrating Equation 3.15. However, given the symmetry of the function, it is simpler to calculate the integral only over half this interval (thus $T/4$) and then double the result:

$$V_{rect} = \frac{2}{\pi}\left[V_t + V_{in} \cos \arcsin\left(\frac{V_t}{V_{in}}\right)\right] \tag{3.18}$$

A plot of V_{rect} versus the V_{in}/V_t ratio is reported in Figure 3.7; in the same figure. A comparison is made between the DC output levels of the full-wave and the proposed

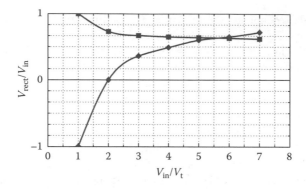

Figure 3.7 Comparison between full-wave (♦) and low-voltage CMOS (■) rectifier output.

rectifier, assuming the same V_T for both the diodes of Figure 3.5a and the MOS transistors of Figure 3.5b. It can be noticed that the adopted solution exhibits a lower activation threshold (V_t vs. $2V_t$) and a higher rectified output for low input voltages, that is, in the case where $V_{in} < 5.65V_t$: It is, therefore, preferable in the worst operating conditions, enhancing the reading range of the tag. In fact, the efficiency loss (with respect to the full-wave bridge) experienced when $V_{in} > 5.65V_t$ does not impact the correct behavior of the tag, because in those conditions there is plenty of power available to be transferred to the chip by the voltage booster driven by the rectifier.

Regarding the RC filter embedded in the RF rectifier, it must be sized to suppress the output ripple occurring at UHF frequencies, around 2×900 MHz, to preserve the amplitude modulation of the incoming RF carrier. To this aim, the filter pole frequency should thus be positioned somewhere between 50 kHz and 50 MHz to exhibit a fast tracking of the amplitude modulation and a good rejection of the UHF ripple. While this frequency span could seem wide at first sight, it proves crucial to carefully consider the choice of the components size, because of three main factors:

1. The on-resistance of the bridge switches is dependent on the input voltage. As a matter of fact, the higher the input voltage, the lower is the transistors's on-resistance: considering the low-threshold MOS devices used in the bridge (to further decrease the activation minimum level), for input peak voltages ranging from V_t (around 0.3 V) to the maximum $|V_{GS}|$ allowable on the devices (1.8 V), the value of a MOS r_{ds} may exhibit variations of more than two orders of magnitude.
2. The analysis of the small-signal equivalent of the MOS rectifier returns that the equivalent series impedance of the chip is partly influenced by the input resistors value. Since the input peak voltage V_{IN} is enhanced for high R_{in}, the higher r_{sw}, the higher R_{in}, thus potentially increasing the efficiency. Moreover, the capacitor should be big enough to exhibit a very low impedance in the UHF frequencies.

3. In the previous calculations, because the current flow is estimated to be a few µA, the ohmic losses on the MOS's r_{ds} have been utterly neglected. However, excessively increasing the transistors on-resistance might cause a considerable voltage drop across the switches, able to overwhelm the advantages of the previous point.

3.4.3 Voltage Booster

The output level of the bridge described in the previous section is usually too low to provide a supply voltage sufficient to bias the core circuitry, especially when the reading distance rises above some meters. Another rectification stage able to increase the DC supply voltage has to be introduced; among the different architectures available to perform such a task, a common Dickson charge pump [34] can be considered a good solution in terms of efficiency and area occupation. To keep the symmetry of the load on the two input pins, a pseudodifferential structure can be considered. A three-stage differential Dickson charge pump is shown in Figure 3.8. As mentioned before, Schottky diodes are not available in a low-cost technology. To reduce the voltage drop across the diodes, they can be implemented with diode-connected low-threshold MOS transistors; the efficiency of this solution is limited mainly by the body effect on the devices and by their parasitic capacitance toward the substrate.

Looking at the three-stages Dickson charge pump of Figure 3.8, in the half input period where in_2 is low the first diode is on and A_1 is set to $V_{RECT} - V_T$, while in the successive half period A_1 rises with in_2 and the diode is turned off. Therefore, the diode represents a device that stops the current flow when in_2 is high and conducts when in_2 is low; dually, it can be seen as a switch open if in_1 is low and closed (with a threshold loss) if in_1 is high. Given the fact that, according to Equation 3.18, V_{rect} is generally lower than the input peak voltage V_{IN}, it is also true that in_1 and in_2 exhibit a peak potential higher than *rect*. If the difference between these potentials was high enough to turn on a transistor, the first diode could be substituted with a device, that, acting as a switch driven by the input potential, performs the same task without losing the threshold voltage V_t. In a deep-submicron technology (0.25 or 0.18 µm)

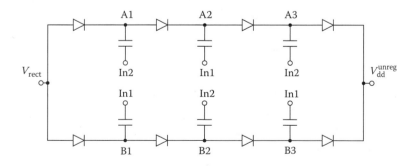

Figure 3.8 Schematic of a three-stage pseudodifferential Dickson voltage multiplier.

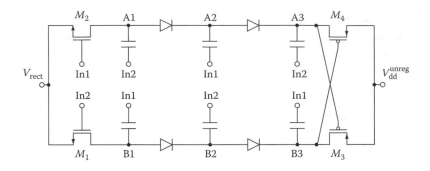

Figure 3.9   Schematic of the low-voltage multiplier.

native zero-threshold NMOS transistors (ZVT) are usually available. Such a device can successfully be employed for the task, allowing to gain a V_T in the rectification process: two ZVT NMOS's are put in place of the first couple of diodes, one driven by in_1 (between *rect* and A_1) and another driven by in_2 (between *rect* and B_1). The transistors' size is determined by balancing two constraints: They have to be large enough to neglect the voltage drop due to their on-resistance, without adding too much capacitance to the inputs. A second technique to improve the efficiency is the substitution of the last diodes with a cross-coupled PMOS peak detector similar to the one used in the bridge of Section 3.4.2. The ultra-low-voltage charge pump is reported in Figure 3.9.

During the reader-to-tag and tag-to-reader communication phases there are some time intervals when the input power is suddenly reduced or drops to almost zero. It occurs during the tag-to-reader communication phase when a logic 0 is detected and the amplitude modulation depth approaches 100 percent, or during the backscatter answering phase when a "high reflectivity" state may lead to the absence of any noticeable input power. As already stressed above, an energy storage element is mandatory to guarantee the required power supply to the digital core from the tag wakeup until the instant when the communication ends. The on-chip storage capacitor (C_{dump}) is sized on the basis of the maximum silicon area and considering the maximum time interval when the input power is absent. Furthermore the available standards requires a maximum wakeup time for the tag (from 400 μs to about 1 ms): This ultimately limits the maximum value of the on-chip capacitor.

3.4.4 Device Safety Protections

There are two kinds of electrical events that, despite having different origin, may lead to the physical damage of the chip: an electrostatic discharge (ESD) and an excessive potential difference between two transistor terminals due to a very short reading range. The destroying effects of ESD events are avoided if a low-resistive path is guaranteed between each couple of pads when an electrostatic discharge event occurs. Typical protection circuits are based on diodes or GC-MOS devices. However, such approach introduces a large capacitance at the RF input pads, leading to a strong reduction of

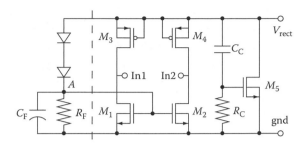

Figure 3.10 Schematic of the input protection circuit.

the peak value of the RF input signal. Furthermore, analytical analysis of the rectifier in Figure 3.5b shows that an added capacitance between the RF pads causes a reduction of the resistive part of the input impedance, raising issues in the design of the antenna. A second effect which may lead to a permanent damage of the RF front end is an overvoltage at the input pads, occurring when the tag is close to the reader. The maximum tolerated voltage is about 2 V if a 0.18 μm technology is used. This situation occurs at a tag-to-reader distance of about 1 m with 500 mW ERP at the reader. A suitable protection for the RFID analog front-end with minimum added capacitance is shown in Figure 3.10. The right part of the circuit is a modified GC-MOS protection. When an ESD event occurs, the input voltage $|in_1 - in_2|$ tends to rise abruptly. The diode-connected transistors $M_1 \ldots M_4$ allow *rect* and *gnd* to track promptly such a variation. The rising front is then reported by C_C to the gate of M_5, which is suddenly turned on. Under an ESD event a low-resistive path formed either by M_1–M_5–M_4 or M_2–M_5–M_3, depending on the polarity of the discharge, is provided. The left part of the circuit in Figure 3.10 provides input overvoltage protection. When the level of the input signal approaches the maximum tolerated value transistor M_1 and M_2 are progressively turned-on, thus limiting the input power by introducing a short circuit between the antenna terminal. A third effect leading to a permanent damage is the overvoltage at the output of the voltage booster. It occurs at a medium distance from the reader, when the input level is not enough to activate the input protection, but it causes the unregulated voltage to exceed the safety level. A suitable countermeasure is based on clamping diodes placed at the Dickson charge pump outputs. It is worth noticing that the added capacitance at those nodes does not affect the behavior of the circuit under normal operation.

3.4.5 Voltage Regulation

The unregulated supply voltage across the energy storage capacitor is strongly dependent on the available input power and thus on the reading distance. To provide a stable voltage to the digital core, almost independent of the reading conditions, a low-drop out regulator is required, Figure 3.11. The circuit should exhibit a very low-power consumption, in the hundreds of nW range, and a suitable regulation. A grounded gate ZVT NMOS is used as tail current generator, thus avoiding the need for a reference bias

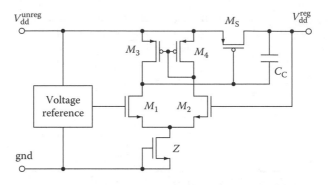

Figure 3.11 Schematic of the LDO.

current. The voltage reference block sets the value of the regulated voltage. The reference circuit should tolerate a large supply voltage variation (from 2 V down to 0.8 V) with an ultra-low-power consumption. An example of a suitable circuit is reported in Ref. [35].

3.4.6 ASK Demodulator

The RF rectifier described in Section 3.4.2, loaded with the small filtering capacitor behaves as an envelope detector. Therefore the low-frequency signal at the output of the rectifier is the modulating signal transmitted by the reader to the tag. Since the modulation depth can vary from 100 percent down to 18 percent depending on the standard of reference, a dedicated circuit translating the modulating signal into a CMOS digital signal is required. A suitable circuit is shown in Figure 3.12, where the common-mode level of the amplitude modulation is provided to the negative input of the comparator by means of a RC filter.

3.4.7 Clock Generation

An UHF RFID, compliant with both the ISO 18000-6B/C standards, needs a local oscillator providing a suitable reference clock to the logic core. Considering the requirements for a sub-microwatt power consumption, a small silicon area and a low output frequency, either a RC-oscillator or a ring oscillator are suitable architectures for such an application.

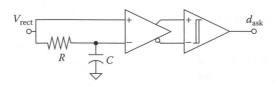

Figure 3.12 ASK demodulator to CMOS levels.

A relevant design issue is posed by the tolerance affecting the frequency of the backscatter signal (δ_f), which must be within ± 15 percent $\div \pm 22$ percent of nominal value, depending on selected standard and communication data rate. Such a relatively low value cannot be achieved with an untrimmed RC or relaxation oscillator. However, trimming at the wafer sort has a significant impact on the cost of the chip and, thus, should be avoided.

In a ring oscillator, the oscillation frequency is determined by the inverter delay which depends on MOS transistor parameters (threshold, oxide capacitance, and mobility) and on the supply voltage. If the oscillator is powered through a voltage regulator, Figure 3.11, the variation of the supply and, thus of the frequency, with the reading distance are sufficiently rejected. Furthermore, the issue of the shift of the frequency due to the process tolerance affecting the inverter delay can be addressed by introducing an autocalibration function in the RFID, which can be accomplished in the digital domain. It is worth noticing that such a solution avoids any trimming at the wafer sort. However, the application requires that the calibration technique has a negligible impact on silicon area and power consumption; moreover, calibration must not affect the normal operation of the tag. Finally a frequency reference is required to calibrate the oscillator.

Considering the ISO 18000-6B standards a time reference can be achieved from the interrogation phase (ASK modulation) starting with nine subsequent 0's, Manchester coded, at 40 kbps data rate (preamble). The local oscillator can be calibrated by measuring the time interval between two successive rising edges (T_{bit}) of the base-band signal (detected by the ASK demodulator) with the frequency reference provided by the ring oscillator itself. Therefore, we are measuring a known time interval, T_{bit}, with a frequency reference affected by a relevant tolerance. It is worth noticing that the frequency of the local oscillator should be higher than $1/T_{bit}$ to achieve a suitable resolution in the time measurement. The result of such a measurement (i.e., the number of periods of the local clock which fall within a T_{bit} time interval) is then used to program a divider placed between the local oscillator and the core logic.

It should be remarked that the proposed calibration strategy can be used for the latest standard, that is, ISO 18000-6C or EPC1-GEN2 also, with minimum modification. In particular the 0 (with PIE coding) at the beginning of the preamble can be used as time reference. Nevertheless, the higher precision dictated by the latest standard and the higher backscatter link frequency, ranging from 40 to 640 kHz, require a higher oscillator frequency (a few MHz) and a higher division ratio for the programmable divider (as reported in Section 3.3). Such modifications lead to a slightly larger power consumption due to the higher frequency and complexity, respect to ISO 18000-6B implementation.

Details of the digital implementation of a calibration circuitry (Digital Clock Manager, DCM) is reported in Section 3.5.

3.4.8 Backscatter Transmitter

The digital section, once elaborated, the instructions received in the interrogation phase provides a digital signal representing the answer to be sent to the reader. Such a data flow is transmitted through the backscatter modulation. With such an extremely low-power technique the RF front end of the tag is toggled between a "high reflectivity" to a "low reflectivity" state, according to the bit logical value to be transmitted to the reader [22].

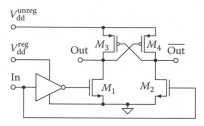

Figure 3.13   Level shifter from regulated to unregulated supply.

Standard sets the discrimination between high and low reflectivity state by referring to the variation of radar cross-sectional area. Such a parameter heavily depends on the antenna's physical features [22]; however a large variation of the radar cross section may be achieved by turning-on a MOS switch (referred as M_1 in Figure 3.2) placed across the antenna terminals, that is, the input pins of the RF front end of the chip. The aspect ratio of the switch should be minimized to provide a low on-resistance, leading to a large variation of the radar cross section. However the parasitic capacitance contributed by its drain-to-substrate and source-to-substrate junctions affects the level of the RF signal, as discussed in Section 3.4.1; therefore, the width of the device should be set according to the value of the maximum allowed parasitic capacitance. A relatively low on-resistance can be achieved with a low device width, if the switch is driven by a level shifter, powered by the unregulated supply. Therefore the maximum available voltage is used to drive the gate of the switch. The schematic of the shifter is shown in Figure 3.13.

3.5   Digital Baseband Processor

The schematic architecture of baseband processor is shown in Figure 3.2. The digital core accomplishes several functions, including:

1. Scaling-down of high-speed clock signal at slower sections and synchronization with incoming data
2. Demodulation of information coming from the reader (e.g., Manchester-coded data)
3. Encoding of reverse-link data according to either biphase space modulation or FM0/Miller for ISO 18000-6B and -6C, respectively
4. Execution of a frame-level, cyclic redundancy check on both direct and reverse link data
5. Collision arbitration, driven by reader commands
6. Management of sensor devices, if included in the tag

To implement all the above functions while shoving total power dissipation at a minimum, several solutions can be adopted, both at the architectural and circuitry level. First, an ad hoc architecture could be preferred with respect to a conventional general-purpose processor, to reduce switching activity. In addition, the baseband-processor design should be optimized to keep the driving clock frequency as low as possible, which allows to reduce dynamic power consumption. Extensive, fine-grained use of clock gating could be implemented to avoid unnecessary switching and thus to reduce energy waste. Finally, as already mentioned, making the tag core operating at a supply voltage slightly above the threshold voltage allows for aggressively reducing power consumption while maintaining fair performance.

The literature reports both full-custom [9] and semicustom [10–13] implementations of RFID baseband processors. The former methodology enables for maximum performance (i.e., minimum power consumption) by means of hand-crafted optimization of every processor building block, but requires a much higher design effort; the latter design style significantly reduces design costs, at the expense of less than optimal performance. As RFID technology evolves, more and more complex features are included into air interface standards, which makes full-custom digital core implementations increasingly difficult. Moreover, a standard cell-based design flow also fosters design reuse, making much simpler porting the processor design to future technology nodes. To the purpose of conjugating efficient design and extremely low-power performance, a compact library of standard cells has been designed and characterized, suitable for the design of power-limited systems operating at near-threshold supply voltage.

In the following, design of the standard-cell library is discussed. Then, partitioning and implementation of the processor circuit is described, taking into account major issues related to low-power constraints and openness to the integration of on-board sensors.

3.5.1   Low-Power Standard-Cell Design

The development of complex digital VLSI systems often relies on a standard-cell design style, which gives definite advantages for the physical implementation of the system. Large libraries of cells are commercially available, optimized to achieve high performance in terms of speed and area consumption. Recently, with integrated circuits power reduction becoming an increasingly critical issue, manufacturers started producing library devoted to low-power design.

Power-limited systems are often encountered in low-cost, low-performance applications. Power management and optimization technique significantly differ between energy- and power-limited systems. Passive RFID transponders represent a significant example of the latter family, for which low-power design techniques should be tuned to the peculiar application: In this case, long inactivity periods do not impact at all on the power budget, whereas, as stated above, performance (i.e., read range) closely depends on the peak available power.

Accounting for these issues, a standard-cell library has been developed, explicitly aimed at the design of power-limited systems, operating at near-threshold supply voltage.

To validate the proposed approach, a limited variety of cells could be actually implemented: It has been shown [36–38], that using a reduced set of properly selected cells does not critically affect performance (which is not the main concern, anyway) and may positively impact on the efficiency of the synthesis process. An example of suitable compact library is summarized in Table 3.1, which has been used for several tag implementations. The presence of both positive-edge and negative-edge triggered

Table 3.1 Contents of the Compact Standard-Cell Library

Category	Function	Cell Area [μm^2]
Flip-flops	D-FF positive edge	68.43
	D-FF negative edge	68.43
	D-FF asynchronous reset	131.10
	D-FF asynchronous set	131.10
Inverters and buffers	INVX1	17.10
	INVX2	22.81
	INVX4	22.81
	BUFX2	22.81
Two-input primitive gates	NAND2X1	22.81
	NOR2X1	22.81
	XOR2X1	42.80
Multiplexers	MUX2X2	45.86

flip-flops can be exploited to implement sections active at each semiperiod of the clock signal, thus locally doubling the operating speed without increasing the main clock frequency. Careful design and full characterization of each module over a full range of temperature and voltages should be carried out to properly integrate the library in a standard and reliable digital design flow.

As already stressed above, near-threshold operations could be selected as a way to attain both ultra-low-power consumption as well as fair performance. Referring to the 180 nm implementation, cells have been designed accounting for a supply voltage V_{dd} equal to 0.6 V, slightly larger than the threshold voltage of high-V_{th} devices, to tolerate some fluctuation of the voltage regulator output. For high-performance systems implemented in a 180 nm CMOS process, leakage currents play a relatively unimportant role; here, however, operating frequencies as low as 40 kHz can be purposely selected for most parts of the circuit, and thus, in a relative sense, the leakage power becomes relevant. This may impact on the optimal transistor sizing in a somehow unfamiliar fashion. To investigate such an aspect, simulations could help in finding out the optimal device sizing. Simulations of inverter cell total power consumption, as a function of channel lengths, are shown in Figure 3.14a. A 100 kHz switching frequency has been assumed and minimum width for n-channel and p-channel devices has been imposed to minimize input capacitance and hence dynamic power consumption. Nevertheless, for a 180 nm node moderate gradients of the power function are exhibited in the nearby

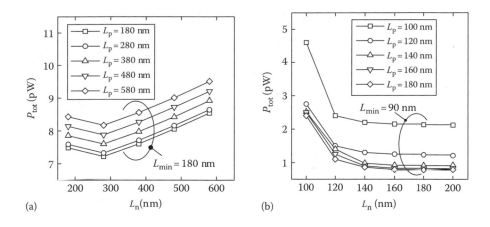

Figure 3.14 **Power consumption of an inverter cell: (a) 180 nm, V_{dd} = 0.6 V, V_{th} = 0.5 V, $W_n = W_p$ = 240 nm, f_{sw} = 100 kHz; (b) 90 nm, V_{dd} = 0.4 V, V_{th} = 0.35 V, $W_n = W_p$ = 300 nm, f_{sw} = 200 kHz.**

of the minimum point: This means that, with respect to the minimum-sized devices, selecting the "optimal" size would imply just a 1 percent power improvement, not sufficient to compensate an area penalty of 18 percent. Actual cell sizing is reported in Table 3.1 as well. In this case, minimum-size and minimum-power strategies appear to happily coexist. This is not the general case; however at lower frequencies or lower circuit activities, such a figure may significantly change and careful optimization should be made on these grounds. The problem may also become much more evident with future technology scaling: A similar optimization procedure has been carried out assuming a 90 nm technology. In this case, 0.4 V supply voltage and 0.35 V threshold voltage were assumed, and an effective switching frequency of 100 kHz was retained. The plot in Figure 3.14b illustrates the results: In this case, the minimum-power sizing lies far from the minimum-area sizing; in the analyzed range, a fair 70 percent saving in the overall power can be obtained (at the expense of area and speed, though). Since, as stressed above, reducing the overall power needed by the tag allows for increasing the read range, the importance of application-oriented power optimization becomes quite evident.

3.5.2 Baseband-Processor Building Blocks

The baseband logic circuitry should be designed aiming at keeping the operating frequency as low as possible, compatibly with functional constraints. Baseband-processor partitioning into several clock domains and even distribution of circuit activity in time represent the key elements for power reduction at the architectural design level.

A fundamental requirement consist of the need for real-time processing of incoming data; that is, the tag clock should be fast enough to allow for extracting data from incoming modulated RF carrier and performing all the related computations within the time slot allowed by the air-interface standard.

3.5.2.1 Solutions for ISO 18000-6B Implementation

With reference to the above mentioned ISO 18000-6B standard, thanks to clock calibration circuitry (DCM) described later on, the operating frequency can be lowered down to 40 kHz for the largest part of baseband circuitry. The digital core partitioning is reported in Figure 3.15.

A relatively "fast," 800 kHz nominal clock, ϕ_{osc}, is effectively generated by a small (three stages), power-efficient ring oscillator. A small digital block, referred to as the DCM, deals with clock prescaling and synchronization with incoming data ($d_{ask,smp}$). Large fluctuations of the primary clock frequency can thus be tolerated without compromising the proper functionality of the tag.

Clock scaling down DCM and synchronization are performed exploiting the frame preamble, typical of reader-to-tag commands, which consists of a field filled with nine consecutive Manchester-encoded zeroes (see Figure 3.16).

In principle, decoding Manchester-encoded data flowing in at a 40 kHz rate should call for a 80 kHz sampling frequency. By exploiting double-edge triggered sampling, however, operating frequency can be halved (thus saving energy). This, in turn, requires a close phase relationship between incoming data ($d_{ask,smp}$) and prescaled clock signal (ϕ_{dcm}) to be maintained throughout the whole packet analysis. This is accomplished by the DCM module, which includes two counters and a simple Finite State Machine (FSM).

Figure 3.16 illustrates the behavior of DCM module: ($d_{ask,smp}$) is the input signal, and the start section of a command coming from the reader is shown. Preamble and delimiter fields are evidenced. Once the start section of a new data packet is recognized, the actual ratio between the internal clock (nominally oscillating at 800 kHz) and the incoming data rate (40 kHz) is estimated by averaging the lengths of the zero symbols in the preamble. Actually, a weighted average is carried out, in which larger weights are associated to the last symbols: This allows for minimizing the influence of the demodulator turning-on transients (more sensible on the initial symbols). Just after seven symbols of the preamble have been recognized, a pulse is generated, which wakes up the following circuits and provides the arbitration logic with the low-speed clock signal (ϕ_{dcm}). From this moment on, the DCM continuously tracks the data input frequency analyzing Manchester transitions, ensuring clock synchronism, and maintaining a quarter-period phase shift between d_{in-smp} and ϕ_{dcm}, necessary to guarantee the correctness of the double-edge triggered sampling. Aligned markers M_1 and M_2 placed on the waveforms in Figure 3.16 illustrate such a behavior. An additional function carried out by the high-frequency section of DCM consists of a simple frame detection, capable of identifying the incoming command start and to enable subsequent DCM sections. In practice, this allows for an automatic clock-gating of all the section driven by the low-frequency clock. Simulations have been carried out to check for the reliability and robustness of this approach: Results have demonstrated that the DCM module correctly operates even for large, yet unrealistic, spreads of the local frequency. More precisely, synchronization was obtained for frequencies theoretical ranging from 480 kHz to 1.320 MHz. Figure 3.17 shows the dependance of DCM output frequency (ϕ_{dcm}) on local oscillator frequency (ϕ_{osc}). The solid line depicts the simulated behavior,

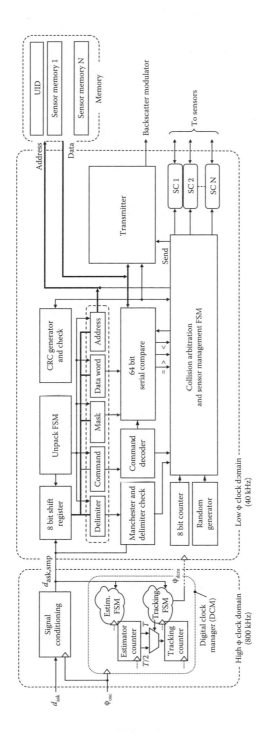

Figure 3.15 Architecture of the baseband processor (ISO 18000-6B protocol).

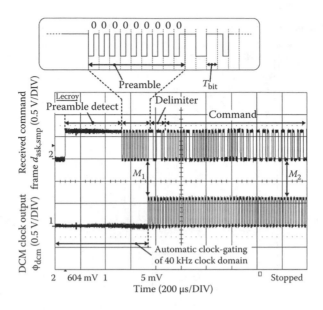

Figure 3.16   Architecture of the baseband processor (ISO 18000-6B protocol).

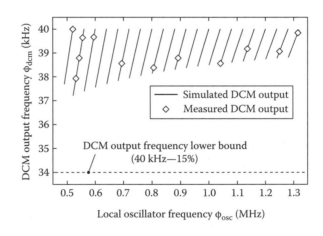

Figure 3.17   DCM output frequency: simulated and measured behavior.

according to

$$\phi_{dcm} = \phi_{osc}/n \tag{3.19}$$

and

$$n = \left[\frac{\phi_{osc}}{\phi_{in-smp}} \right] \tag{3.20}$$

where $\bar{\phi}_{in-smp}$ is the weighted average carried out on incoming data preamble. Circuitry architecture limits n range to $[12, \ldots, 32]$ interval. The lower bound is related to DCM data tracking FSM, which correctly operates with $\lfloor n/4 \rfloor > 3$, implying $n_{min} = 12$ (and hence $\phi_{osc} > 480\,\text{kHz}$). On the other hand, the n upper bound is related to average counter sizing. Six bits have been reserved for two preamble bit-period measure, to limit average counter power consumption, and leading to $n_{max} = 32$ (and hence $\phi_{osc} < 1.320\,\text{MHz}$). Measured performance exhibit a slightly narrow range of acceptable input frequency, $[520\,\text{kHz}, \ldots, 1.320\,\text{MHz}]$, due to gate delays and nonideality of incoming data preamble. Moreover, a full accordance of measured DCM output frequency (diamonds) with theoretical behavior is shown in Figure 3.17. The ISO 18000-6B standard defines the return link bit rate equal to $40\,\text{kHz} \pm 15$ percent. The DCM accuracy satisfies this further constraint, as highlighted in Figure 3.17 (dotted line). In the calibration technique discussed here, most of the digital core exploits the calibrated low-frequency clock; the oscillator and the divider operate at high frequency during normal operations, thus resulting in a lower power consumption, with respect to recent literature work [39].

Once the incoming data flow is tracked, validation of the data packet is carried out by the Manchester and Delimiter Check module, which rejects malformed packets. While waiting for the next valid datum, the clock still keeps triggering the collision arbiter, to preserve FSM state variables. The clock is briefly suspended only when a new command frame is identified, to allow for new clock-to-preamble locking. Valid data, instead, proceed to the shift register for serial-to-parallel conversion, which is performed in 8-bit segments. The input stream is continuously monitored by the Manchester tester; eventually, the frame-level CRC test is carried out to check for symbol errors.

The typical command frame encompasses several short fields, and a larger field, including a 64-bit word. Depending on the actual command to be executed, such a 64-bit word may have to be compared in different ways with the internal tag code.

From the power management point of view, processing of the 64-bit word is clearly the most critical issue, and some energy-saving features could be implemented in this section. First, bitwise shift of the incoming data would result in an intense switching activity at the 64-bit comparator inputs, inherently characterized by relatively large input capacitances; to reduce such effects, 8-bit buffers could be accounted for, and bytewise shift could be implemented instead (thus reducing related power dissipation by a factor 8). Second, we may take further advantage of the relatively weak timing constraints: 64-bit operations required by the standard, in fact, may result in a fairly demanding combinational logic network. The available time frame, however, allows for splitting the 64-bit comparison serially into eight 8-bit operations, thus permitting consistent saving in terms of area and switched capacitances (i.e., power). As a by-product, this also makes straightforward the implementation of the mask feature required from the standard for some comparison modes: Masked bytes just result in a no-op cycle in the serial comparison.

The collision arbitration FSM manages the overall collision arbitration process, according to the ISO standard specifications. It requires a status counter and a random number generator; subcircuits should be independently enabled (in a mutually exclusive

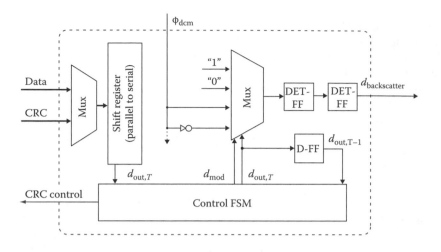

Figure 3.18   Details of transmitter module ISO 18000-6B compliant.

fashion, whenever possible) by the control unit, to keep the circuit activity as evenly distributed in time as possible.

The same FSM manages the transmission of tag response as well. As already introduced above, the standard requires backward-link data to be encoded according to biphase space modulation. According to such an encoding scheme, a transition occurs at the end of each bit period, and an additional transition at the middle of period is required to signal a logic 0. Thus, at 40 kHz data rate, biphase modulated symbol transitions may occur at 12.5 μs intervals, that is, at a frequency double than the available clock. To preserve low-frequency clock operation, a simple circuit solution could be devised, based on a four-input multiplexer. The encoder circuit is shown in Figure 3.18 and its operation can be straightforwardly interpreted. Parallel input data are serialized and fed to a simple finite state machine. Depending on the data bit sequence, transmitted signal is selected among constant high and low bit ("1" data bit) or clock and inverted clock signal ("0" data bit). Output double edge triggered (DET) flip-flops aim at removing potential glitches from the output signal.

Finally, the current status of the digital core could be communicated to the analog front end, to exploit energy-saving strategies at the whole tag level; for instance, the ASK demodulator could be safely turned off during the timeframe needed to work out the tag response.

As already mentioned above, test chips have been fabricated, onto which several measurements have been performed, to validate the proposed circuitry solutions. Figure 3.19a reports the baseband-processor power consumption, measured at room temperature (27 °C). The actual supply current has been measured at variable supply voltages, ranging in the {0.46, 0.5, 0.6, and 0.7 V} set. Adsorbed current is split into contributions required by the (relatively) high-frequency section (i.e., the local oscillator) and the lower frequency section, clocked by the DCM module. The high-frequency domain power consumption increases (as expected) almost linearly with the clock rate; as the clock

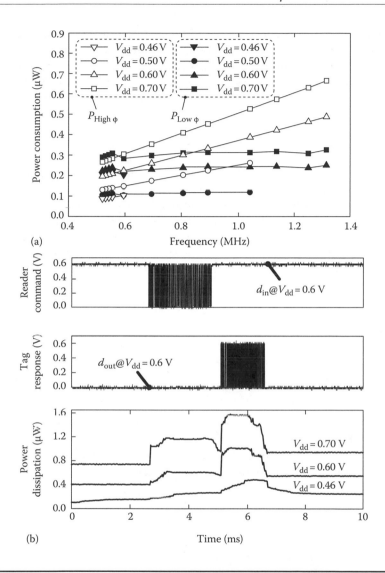

Figure 3.19 Examples of measurement results: (a) Power consumption of high φ and low φ clock domains and (b) Distribution of baseband-processor power consumption during data reception (i.e., GROUP_SELECT_EQ command) and tag ID transmission.

rate exceeds 600 kHz, it prevails on the low-frequency domain contribution, the power consumption of which is nearly constant over clock frequency, thanks to the DCM module action described above. At target operating conditions (i.e., 800 kHz clock frequency and 0.6 V supply voltage) the digital circuitry dissipates only 440 nW. The power distribution over the execution cycle is detailed in Figure 3.19b, which correlates the actual power with the input signal (i.e., the *GROUP_SELECT_NE* reader-packet shown in the upper plot) and the tag response (i.e., the ID transmission shown in the middle plot). The baseband-processor power consumption is reported in the lower plot, accounting for three different supply voltages (0.46, 0.6, and 0.7 V). The power dissipation reaches its maximum value during reply transmission (about 1 μW). When operated at the lower supply voltage ($V_{dd} = 0.46$ V), the peak power dissipation is as as low as 420 nW.

3.5.2.2 Solutions for ISO 18000-6C Implementation

As already stressed in Section 3.4.7, clock scaling for ISO 18000-6C compliant implementation is limited by the tag's reply data rate.

The interrogator specifies the tag's backscatter link frequency (LF, in the 40 kHz ÷ 640 kHz range) by means of different combinations of the TRcal interval and divide ratio parameters (DR, included in the payload). The actual LF and its tolerance FT constrain the tag minimum clock frequency. With reference to Equation 3.21, the estimation of such a frequency can be performed by means of simulations

$$|FT| = \frac{\left| T_{LF}^{nom} - \hat{n} T_{clk} \right|}{\hat{n} T_{clk}} \tag{3.21}$$

where

$$\hat{n} = \begin{cases} \text{round}\left(\left(3 \left\lfloor \frac{TR_{cal}}{T_{clk}/2} \right\rfloor \right) \Big/ (64.2) \right) & DR = 64/3 \\ \text{round}\left(\left\lfloor \frac{TR_{cal}}{T_{clk}/2} \right\rfloor \Big/ (8.2) \right) & DR = 8. \end{cases} \tag{3.22}$$

The plot in Figure 3.20 illustrates results for both DR = 64/3 and DR = 8: The percentage of backscatter frequencies (LF) that do not satisfy frequency tolerance (FT) constraints are depicted, as a function of local oscillator frequency. The clock frequency should be chosen in the 1.6–2.20 MHz safe range. A somehow larger figure (e.g., 2 MHz)

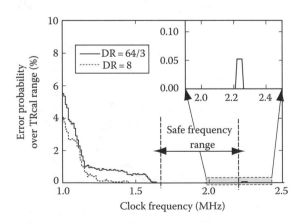

Figure 3.20   ISO 18000-6C baseband-processor clock frequency selection.

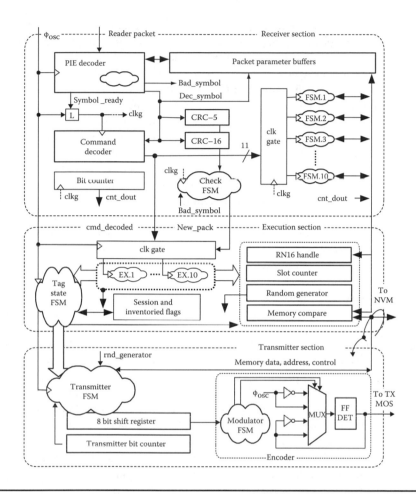

Figure 3.21   **Architecture of the baseband processor (ISO 18000-6C protocol).**

respect to minimum reliable frequency is to be adopted if a coarse-precision ring oscillator is used as low-power local clock generator.

Figure 3.21 reports a possible architecture of ISO 18000-6C baseband processor. ϕ_{osc} directly feeds a PIE decoder block, which samples incoming reader packets. Decoding operations are triggered by the packet delimiter identification. PIE Decoder embeds a binary counter and some buffers exploited for Tari, RTcal, and TRcal measurement. A simple FSM uses the same counter to interpret subsequent interrogator symbols, comparing counter output with RTcal/2. On receiving a new symbol, the PIE decoder issues a T_{osc} wide pulse which is used to gate clock signal feeded to remaining packet reception circuitry. Command Decoder block analyzes the first decoded symbols to identify mandatory commands, while a 6-bit counter (Bit Counter) computes total received symbols. Ten independent FSMs could be implemented to manage different mandatory standard commands: By keeping them separate, more effective fine-grained clock-gating strategies can be adopted, at the expense of a slight area increase. Received symbols flow through error-processing elements (CRC-16 or CRC-5, depending on actual received command) to report bad PIE-encoded data.

The tag state should be managed through an always active FSM, running at 2 MHz. Execution phase starts when the receiver Check FSM validates incoming packet. Again, different command-related tasks can be managed by independent small controllers. Execution data path includes four independent modules devoted to 16-bit random compare, slot counter operations, random generation, and memory comparison.

The available time between received packet and potential tag response could be exploited for splitting the memory comparison serially into 8-bit operations, thus permitting consistent saving in terms of area and switched capacitances (i.e., peak power).

Generation of 16-bit random numbers, required by the arbitration process, is accomplished by a linear feedback shift register (LFSR)-based random generator. Initialization is performed during power-up and while the tag remains in ready state. During arbitration and transmission the random generator is normally turned off to reduce power dissipation. Short reactivations (T_{osc} wide) are used for further random number extraction, during arbitration process. Memory comparison is performed in a 8-bit serial fashion, to save area and switched capacity.

Transmitter operations include FM0 and Miller encoding of different data (random numbers and memory content). According to such schemes, a signal switch occurs at the end of each bit period; symbols "0" (FM0) or "1" (Miller) are encoded by an additional signal edge at the middle of the period. Thus, actual symbol switching may occur at a doubled frequency with respect to the selected clock.

To avoid doubling the local clock frequency, a simple circuit solution could be devised, based on a four-input multiplexer (similar to ISO 18000-6B implementation). The encoder circuit is shown in Figure 3.21 and its operation can be straightforwardly interpreted: The symbol boundary transitions are accomplished inverting the value of the output DET flip-flop, whereas the mid-symbol inversions are obtained sampling a single pulse clock. A finite state machine properly drives the multiplexer, depending on the transmitted data and the symbol period.

3.5.3 Integration with Sensing Devices

Emerging RFID technology aims at on-chip sensor integration, to create wireless sensing devices. Sensor integration poses two major concerns, related to control/measurement communication between transponder and reader and sensing peripheral power dissipation. The ISO 18000-6B air interface standard supports the implementation of custom commands, to add manufacturer specific functions. The only fields that can be customized are the parameters and the data fields. Any custom command should contain as its first parameter the IC manufacturer code. This allows IC manufacturers to implement custom commands without risking duplication of command codes and thus misinterpretation. Such commands could be exploited to manage eventual sensor data [10]: A custom code can be reserved to store measured information into tag memory whereas a standard command could be applied to collect sensor data. Figure 3.3 depicts the main tag states, according to ISO 18000-6B standard: Dotted transition lines represent potential custom sensor-data collecting operations. According to 6B standard, an interrogator could issue a custom command both before and after arbitration session. In the former case ("Ready" state), all tags could be involved into potential monitoring task whereas, in the latter case ("ID" or "Data Exchange" states), only selected transponders participate in sensor-derived data collection. Custom command example, referred in Figure 3.3 as *COLLECT_SENSOR_DATA*, enables for either single or multiple sensor selection by means of an 8 bits wide committed field (Sensor Select).

Referring to ISO 18000-6C standard, an interrogator shall issue a custom command only after singulating a tag and reading (or having prior knowledge of) the tag manufacturer's identification (embedded into tag TID memory). This limits the flexibility of potential monitoring task, preventing from actual simultaneous collection of multiple-tags sensor-derived data.

From the power management point of view, power-supply gating could be used to place sensor devices in standby while the tag is performing communication tasks (i.e., during transponders arbitration or while retrieving memory content). By cutting off the sensor current, whenever it is possible, negative impact on the read range is minimized.

3.6 Open Issues

Today the problem of illicit RFID tag inventorying and tracking is still of limited concern. However, RFID transponders are very likely to replace the widespread optical barcode technologies in several fields within a few years, becoming a key element of supply chains and retail stores management. Hence, increasingly important data security issues are expected to raise. Moreover, as already discussed above, RFID transponders can host sensors aimed at monitoring environmental and personal parameters. This may involve transmission of private or sensitive data, and thus again implies privacy concerns. To guarantee data security and integrity in RFID systems, suitable features must be embedded into RFID devices, supporting data privacy and authentication [40]. The efficient design and implementation of cryptographic primitives, geared specifically at the very tight power constraints of RFID tags, could provide useful security tools.

Several works are focused on privacy-protecting schemes for basic tags, which do not include cryptography primitives, or explore implementations specifically devoted to RFID environment (see Ref. [40] for a survey). Recently, the European Network for Excellence in Cryptology (ECRYPT) [41] has identified a portfolio of promising new stream ciphers, geared toward resource-constrained hardware platforms and potentially suitable for RFID tags.

Implementation of on-tag security features, as well as performance improvement are still deserving additional research effort, to further expand potential application fields of RFID devices.

3.7 Conclusions

In this chapter, techniques adopted for the development of a passive RFID transponder were discussed. Special attention was devoted to power issues, which directly limits the tag performance. Several power optimization strategies have been discussed, including optimization of analog front-end circuitry, dedicated standard-cells design and digital system power-efficient implementation.

Design and development of test chips has been discussed, based on a 180 nm technology and fully compliant with ISO 18000-6B/-6C standards. By adopting a standard-cell-based design flow for the implementation of the digital core, design effort has been reduced, at the same time fostering an easier port toward future, more aggressively scaled technology nodes. Integration with sensor devices was also discussed, with respect to the air-interface standards.

The proposed approach has been validated by means of simulation and experimental tests, and remarkable performance figures have been obtained. By means of careful choices, the area-power balance is effectively exploited for the peculiar application at hand, tailoring the design to the target of read range maximization. Extremely low-power dissipation could actually be achieved, which holds promises for low-cost, high-performance passive devices, suitable for ubiquitous computing applications.

Acknowledgment

This work was supported by TECAL Lab., funded by Regione Emilia–Romagna (Italy), PRRIITT Misura 3.4 Azione A.

References

1. K. Finkenzeller, *RFID Handbook, Radio-Frequency Identifications Fundamentals and Applications*, 2nd ed. Wiley, New York, 2003.
2. R. Glidden et al., Design of ultra-low cost UHF RFID tags for supply chain applications, *IEEE Commun. Mag.*, 42(8): 140–151, August 2004.
3. S. Masui, E. Ishii, T. Iwawaki, Y. Sugawara, and K. Sawada, A 13.56-MHz CMOS RF identification transponder integrated circuit with a dedicated CPU, in *Dig. Tech. Papers Solid-State Circuits Conf. (ISSCC)*, pp. 162–163, San Francisco, CA, February 1999.
4. F. Kocer and M.P. Flynn, A long-range RFID IC with on-chip ADC in 0.25 μm CMOS, in *Dig. Papers IEEE Radio Freq. Integr. Circuits (RFID) Symp.*, pp. 361–364, Long Beach, CA, June 2005.
5. U. Karthaus and M. Fischer, Fully integrated passive UHF RFID transponder with 16.7 μW minimum RF input power, *IEEE J. Solid State Circ.*, 38(10): 1602–1608, October 2003.
6. B. Jamali, D.C. Ranasinghe, and P.H. Cole, Analysis of UHF RFID CMOS rectifier structures and input impedance characteristics, in *Proc. SPIE (Microelectronics: design, technology, and packaging II)*, vol. 6035, pp. 313–323, Brisbane, Australia, 2005.

7. T. Umeda et al., A 950-MHz rectifier circuit for sensor network tag with 10-m distance, *IEEE J. Solid State Circ.*, 40(1): 35–41, January 2006.
8. A. Facen and A. Boni, A CMOS analog frontend for a passive UHF RFID tag, in *Proc. Int. Symp. on Low Power Electronic and Design*, pp. 280–285, Tagernsee, Germany, October 2006.
9. V. Pillai et al., An ultra-low-power long range battery/passive RFID tag for UHF and microwave bands with a current consumption of 700 nA at 1.5 V, *IEEE Trans. Circuits Syst.*, 54(7): 1500–1512, July 2007.
10. A. Ricci and I. De Munari, Enabling pervasive sensing with RFID: An ultra low-power digital core for UHF transponders, in *Proc. IEEE Int. Symph. on Circuit and Systems (ISCAS)*, pp. 1589–1592, New Orleans, LA, May 2007.
11. A. Mann et al., Design and implementation of a low-power baseband-system for RFID tag, in *Proc. IEEE Int. Symp. on Circuit and Systems (ISCAS)*, pp. 1585–1588, New Orleans, LA, May 2006.
12. H. Yan, H. Jianyun, L. Qiang, and M. Hao Design of low-power baseband-processor for RFID tag, in *Proc. Int. Symp. on Applications and the Internet Workshops*, Phoenix, AZ, January 2006.
13. R. Barnett, G. Balachandran, S. Lazar, B. Kramer, G. Konnail, S. Rajasekhar, and V. Drobny, A passive UHF RFID transponder for EPC Gen 2 with −14 dBm sensitivity in 0.13 μm CMOS, in *2007 Solid-State Circuits Conference, Digest of Technical Papers*, pp. 582–583, San Francisco, CA, February 2007.
14. W.G. Yeoh, Y.B. Choi, K.Y. Tham, S.X. Diao, and Y.S. Li, A CMOS 2.45-GHz radio frequency identification tag IC with read/write memory, in *Dig. Papers IEEE Radio Freq. Integr. Circuits (RFIC) Symp.*, pp. 365–368, Long Beach, CA, June 2005.
15. J.-P. Curty, N. Joehl, C. Dehollain, and M.J. Declercq, Remotely powered address-able UHF RFID integrated system, *IEEE J. Solid State Circ.*, 40(11): 2193–2202, November 2005.
16. EPC Global, 860 MHz–930 MHz class 0 radio frequency identification tag protocol specification candidate recommendation, Version 1.0.0. MIT Auto-ID Center, June 2003.
17. International Standards Organization, Type B UHF RFID, ISO/IEC WD 18000 Part 6. August 2004.
18. R.J. Marhefka and J.D. Kraus, *Antennas*, McGraw-Hill, New York, 2002.
19. H.T. Friis, A note on simple transmission formula, *Proc. Inst. Radio Eng.*, 34: 254–256, May 1946.
20. ETSI, *TR 101 445*, v1.1.1 ed., April 2002. Electromagnetic compatibility and radio spectrum matters (ERM); short-range devices (SRD) intended for operation in the 862 MHz to 870 MHz band; system reference document for radio frequency identification (RFID) equipment.
21. J. Rabaey, Scaling the power wall, *Keynote presentation, 44th DAC*, June 2007.
22. K.V.S. Rao and P.V. Nikitin, Theory and measurement of backscattering from RFID tags, *IEEE Ant. Propag. Mag.*, 48(6): 212–218, December 2006.
23. EPC Global, EPC radio-frequency identity protocols class1 generation2 UHF, RFID protocol for communications at 860 MHz–960 MHz, Version 1.0.9, January 2005.

24. EPC Global, EPC radio-frequency identity protocols class1 generation2 UHF, RFID conformance requirements, Version 1.0.2, January 2005.
25. A. Facen and A. Boni, CMOS power retriever for UHF RFID tags, *IET Electron. Lett.*, 43(25): 1424–1425, December 2007.
26. K.V.S. Rao, P.V. Nikitin, and S.F. Lam, Impedance matching concepts in RFID transponder design, in *Fourth IEEE Workshop on Automatic Identification Advanced Technologies*, pp. 39–42, Buffalo, NY, 2005.
27. K. Seemann, F. Cilek, G. Hofer, and R. Weigel, Single-ended ultra-low-power multistage rectifiers for passive RFID tags at UHF and microwave frequencies, in *Radio and Wireless Symposium*, pp. 479–482, San Diego, CA, January 2006.
28. N. Tran, B. Lee, and J. Lee, Development of long-range UHF-band RFID tag chip using schottky diodes in standard CMOS technology, in *Radio Frequency Integrated Circuits (RFIC) Symposium*, pp. 281–284, Honolulu, HI, June 2007.
29. A. Navarro and J.L. Del Valle, Voltage generator for UHF RFID passive tags using Schottky diodes based on a 0.5 µm CMOS technology, in *3rd International Conference on Electrical and Electronics Engineering*, pp. 1–4, Veracruz, Mexico, November 2006.
30. G. De Vita and G. Iannaccone, Ultra low power RF section of a passive microwave RFID transponder in 0.35 µm BiCMOS, in *International Symposium on Circuit and Systems*, vol. 5, pp. 5075–5078, Kobe, Japan, May 2005.
31. H. Nakamoto, D. Yamazaki, T. Yamamoto, H. Kurata, S. Yamada, K. Mukaida, T. Ninomiya, T. Ohkawa, S. Masui, and K. Gotoh, A passive UHF RF identification CMOS tag IC using ferroelectric RAM in 0.35 µm technology, *IEEE J. Solid State Circ.*, 42(1): 101–110, January 2007.
32. A. Facen and A. Boni, Power supply generation in CMOS passive UHF RFID tags, in *PhD Research in Microelectronics and Electronics*, pp. 33–36, 2006.
33. S. Mandal and R. Sarpeshkar, Low-power CMOS rectifier design for RFID applications, *IEEE Trans. Circuits Syst. I: Fundamental Theory and Applications*, 54(6): 1177–1188, June 2007.
34. J.F. Dickson, On-chip high-voltage generation in NMOS integrated circuits using an improved voltage multiplier technique, *IEEE J. Solid State Circ.*, 11(3): 374–378, June 1976.
35. G. De Vita and G. Iannaccone, A sub-1V 10 ppm/C, nanopower voltage reference generator, *IEEE J. Solid State Circ.*, 42(7): 1536–1542, July 2007.
36. N.M. Duc and T. Sakurai, Compact yet high-performance (CyHP) library for short time-to-market with new technologies, in *Proc. 5th Asia and South Pacific Design Automation Conf. (ASP-DAC)*, pp. 475–480, Yokohama, Japan, January 2000.
37. J.M. Masgonty, S. Cserveny, C. Arm, P.D. Pfister, and C. Piguet, Low-power low-voltage standard cell libraries with a limited number of cells, in *Proc. Int. Workshop—Power And Timing Modeling, Optimization and Simulation (PATMOS)*, Yverdon-Les-Bains, Switzerland, September 2001.
38. A. Ricci, I. De Munari, and P. Ciampolini, An evolutionary approach for standard-cell library reduction, in *Proc. 17th ACM Great Lakes Symp. on VLSI*, pp. 305–310, Stresa (VB), Italy, March 2007.

39. F. Cilek, K. Seemann, G. Holweg, and R. Weigel, Impact of the local oscillator on baseband processing in RFID transponder, in *International Symposium on Signal, Systems and Electronics*, pp. 231–234, Montréal (Québec), Canada, 2007.
40. A. Juels, RFID security and privacy: A research survey, *IEEE J. Select. Areas Commun.*, 24(2): 381–394, February 2006.
41. Stream Cipher Project Web Page, ECRYPT (European network for excellence in cryptology), 2005, available: http://www.ecrypt.eu.org/stream/(online).

Chapter 4

EPC Gen-2 Standard for RFID

Peter J. Hawrylak and Marlin H. Mickle

CONTENTS

Numerous communication protocols exist for radio-frequency identification (RFID) systems. One of the most popular protocols is the EPC Gen-2 protocol. This protocol is widely used in RFID applications in the retail space. The Gen-2 protocol provides great flexibility in physical layer properties and link layer procedures to accommodate various environments. This flexibility is critical to maximizing throughput, or the number of tags read/accessed per second. This chapter presents an overview of the physical and link layers of Gen-2 and provides some insight into the different strategies to maximize throughput.

4.1 Introduction

Radio frequency identification (RFID) systems consist of two devices, a RFID reader (sometimes referred to as an interrogator), and a RFID tag. The RFID tag is attached to an item and contains information about that particular item. Typically there are some software applications between the reader and the central database or back-end system. This chapter will focus on one communication protocol used by the tags and readers to communicate. RFID systems may fall into one of two general categories; passive systems where tags do not have any onboard battery; and active systems where tags have an onboard battery. Within each of the previously mentioned two categories the frequency at which the readers and tags communicate with each other further subdivides the categories. There are many different standards governing the RFID reader-to-tag link and this chapter focuses on the EPC Gen-2 standard.

RFID offers significant benefits over a traditional bar code. Today bar codes are used to identify an object and require a line of sight between the bar code reader and the bar code to be read. On the other hand, a RFID tag can be read without a line of sight between the RFID reader and the RFID tag. Further, RFID tags can store more information than a traditional bar code, enabling identification to a unique item; for example, pair 1275 of blue jeans. Also, of importance, RFID tags support writing new or updating old data during their lifetime. Because of this ability, RFID tags offer many advantages for usage in ePedigree (ePedigree is an electronic log of the chain of custody for an asset, i.e., pharmaceutical product) applications.

4.1.1 EPC Gen-2 Background

The EPC Gen-2 (Gen-2) standard is one of many standards governing RFID systems in the RFID space. Recently the Gen-2 system was ratified as an International Standards Organization (ISO) standard, ISO 18000-Part 6C and is a leading standard for passive (battery-less tags) ultrahigh frequency (UHF) RFID system. Many of the major retailers that use RFID, that is, Wal*Mart and Metro, use Gen-2-based systems.

RFID readers and tags that comply with the Gen-2 standard communicate in the UHF range, in the frequency band of 860 to 960 MHz depending on the geographic location. The applicable regulatory board in a country is located, for example, the Federal Communications Commission (FCC) in the United States, governs the frequency band that can be used. For example in the European Union (EU) the allowed frequency range is between 868 and 870 MHz, while the allowable range in the United States is 902 to 928, and 960 MHz in Japan. The Gen-2 standard defines the communication between the reader and the tag regardless of the particular frequency range in which the system is operating.

4.1.1.1 Goals and the Need for the Gen-2 Standard

The primary goal of the Gen-2 standard was to provide a uniform method to read data from, write data to, and communicate with a RFID tag. The hope was to prevent some of the problems that have developed in the electronic article surveillance (EAS) market. In the EAS market there is no single standard for operation requiring manufacturers and retailers to purchase and maintain several different sets of EAS equipment and several different sets of EAS tags. This results in a very large cost overhead. Standards, such as Gen-2, overcome this issue by providing developers with a basic blueprint describing how the system (readers and tags) will interact. The developers are then able to produce products following the standard that are interoperable with products from other vendors that follow the same standard. Thus, users can purchase a single system that works with products from all vendors.

RFID standards face a number of challenges. First, the frequency range at which the devices operate must be determined. This issue is made more difficult because the use of radio frequencies (RFs) are governed by local regulatory bodies. Hence, an RFID standard must identify a frequency range that can be used free of charge throughout the world. Gen-2 uses a wide range (100 MHz) not a single frequency band due to different local regulations as described above. Second, the communication protocol must fit the general application while providing for specialized operation within that application space. The communication protocol must be simple enough to provide basic functionality, access, read, and write. The standard must also provide hooks that can be used by developers to put custom features on top of the basic level of functionality. Third, an RFID standard must define a unique identifier (ID) for each tag and provide a means to allow this unique ID to be scaled to very large numbers (one per tag). Fourth, the physics of the system must be taken into account when selecting the operating frequency range. Read range is one example of a physical parameter where a long read range is desirable in an inventory management system, while a short read range is preferred for transit fare collection systems.

Gen-2 is the leading standard used for RFID in the retail sector. The retail sector requires a license plate tag which simply provides a unique ID that is used as a key to access a centralized database containing information such as cost or expiration date. Gen-2 provides a robust mechanism, the EPC number, for the unique ID capable of handling the need for the trillions of unique IDs required by the retail sector. Further, the breakdown of EPC number into a set of fields allows users to quickly identify tags assigned to specific vendors and products. There are other RFID standards for other application spaces such as library check out and check in systems, public transit fee collection, and highway and bridge toll collection.

4.1.1.2 Goals and the Need for the EPC Numbering System

The primary goal of the EPC numbering system is to provide a universal system by which a particular item can be uniquely identified. The EPC number contains information identifying the manufacturer, type of item, and a serial number unique to that type of manufacturer and item. Thus, the EPC number is universal and enables tags to be addressed based on more than just a unique serial number. Using the EPC number all tags for a particular manufacturer can be identified, simplifying inventory management. The information encoded in the EPC number can be used to access information about that particular item in a larger database.

4.1.2 Overview of Commonly Used Features of Gen-2

The most commonly used features of the Gen-2 protocol are the inventory commands. The inventory commands consist of the following four commands: (1) Select; (2) Query;

(3) QueryRep; and (4) QueryAdjust. These commands are used to read the EPC numbers of all tags within range of the reader. The EPC number can be used to access a central database for more information about the asset to which the tag is attached. In the scenario where Gen-2 tags are affixed to merchandise the inventory commands enable the user to take an inventory of all items on hand at that particular time. In the retail environment the inventory is currently sufficient for the retailers needs. The second most commonly used feature of the Gen-2 protocol is the write and read commands. These two commands fall into the access command category. Write enables the user to write one word (16-bits) of data to a given location on the tag. Similarly Read provides the user the capability to read up to 256 words from the tag; however, the memory bank being read may not contain 256 words.

Not all retail facilities are RFID enabled and as a result RFID Gen-2 tags are often attached to a label that can be printed on. During the printing process the RFID tag is run through a printer which will print a bar code and other information on the label portion of the tag. This enables non-RFID enabled facilities to use the bar code to identify the item in question and provides a backup in the event that the RFID tag fails at some point. The *Write* command is used during the printing process to write the appropriate EPC number to the tag. During this printing process the tag is linked to a specific item, and the EPC number for that specific item must be written into the tag. The EPC number will also identify the manufacturer of the item. The *Read* command is used by the printer to read back the EPC number and verify that the correct data were written. Other data in addition to the EPC number, that is, food expiration data or lot number, can be written and verified by the printer.

4.2   Physical Layer Communication Features

The physical communication interface in Gen-2 is similar in concept to the physical (PHY) layer of the seven layer Open Systems Interconnection (OSI) stack. The reader controls all aspects of the physical layer in the Gen-2 protocol and encodes these aspects in the preamble portion of all commands it sends to the tags. There are two communication links in the Gen-2 protocol: (1) the reader-to-tag link and (2) the tag-to-reader link. Both links are independent and can have different data encodings, data rates, and data modulation schemes. The specifics of both communication links are controlled by the reader.

This enables the reader to adjust the communication links to account for changing environmental and situational circumstances. For example, the reader may use one of the Miller encodings in a very RF noisy environment to reduce the number of bit errors in the tag response. Another example is where the user wishes to take a quick inventory of a very large number of tags, and the reader can then use the fastest allowable data rates for the reader-to-tag and tag-to-reader links.

The data rate options are described in Section 4.2.1, the different modulation types are described in Section 4.2.2, and Section 4.2.3 describes the FM0 and Miller class of data encodings.

4.2.1 Data Rate

There are two communication links defined by the Gen-2 protocol. The first link is from the reader to the tag and is used to send commands (from reader) to the tag(s). The reader transmits the command and then maintains a carrier wave (CW). The CW is an unmodulated signal and is simply the reader transmitting energy to the tag. The second link is from the tag to the reader and is used to send tag replies (from a tag) to the reader. In Gen-2, tags communicate with the reader using backscattering.

When exposed to RF energy all antennas will absorb some of that energy and will reflect the remainder. Backscatter is defined as the reflected energy from any antenna. The tag has the ability to switch its antenna characteristics between two settings: (1) reflect very little energy and (2) reflect almost all the energy. When replying to a reader, Gen-2 tags use this antenna switching ability to alter (modulate) the CW transmitted by the reader and encode the data of the reply in the CW.

4.2.2 Modulation Type

Modulation defines how the data is physically encoded onto the CW signal. Gen-2 supports two types of modulation: amplitude shift keying (ASK) and phase shift keying (PSK).

As previously described, two independent communication links exist in Gen-2: the reader-to-tag link; and the tag-to-reader link. The modulation for the reader-to-tag link can be one of three types of ASK modulation: (1) single-sideband amplitude shift keying (SSB-ASK), (2) double-sideband amplitude shift keying (DSB-ASK), or (3) phase-reversal amplitude shift keying (PR-ASK) [1].

After each command the reader transmits the unmodulated CW. A tag then modulates the backscatter as described in Section 4.2.1 using either ASK or PSK modulation.

The Gen-2 protocol is a half-duplex protocol when a single reader is communicating with a single tag. Hence, the reader will never transmit a command while the tag is backscattering its reply. ASK modulation changes the amplitude of the sine wave to differentiate between a high and a low symbol. For example a sine wave may be defined as follows:

$$(m_{\mathrm{d}})(A)\sin(2\pi ft) \tag{4.1}$$

where
   m_{d} is the modulation depth
   A is the maximum amplitude of the sine wave
   f is the frequency of the unmodulated CW
   t is a discrete point in time

The modulation depth, m_{d} is the percentage of the maximum amplitude, A, of the sine wave that is generated and can range from 1.0 to 0.0 or maximum amplitude to minimum amplitude. When m_{d} is 1.0, the amplitude of the sine wave is at its maximum indicating a high symbol. When m_{d} is less than 1.0, the sine wave and transmitted signal is attenuated, amplitude is less than A, and this represents a low symbol, when m_{d} is below a set threshold, typically between 0.2 and 0.0. The maximum value of m_d such that the tag can differentiate between a symbol high and symbol low depends on the receiver sensitivity and receiver circuitry of the tag.

PSK changes the phase of the sine wave to encode a data 0 and a data 1.

4.2.3 Data Encoding

The data transmitted by the reader to the tag (reader to tag link) are encoded using pulse-interval encoding (PIE) [1]. PIE encoding employs two different length pulses to represent a logical or data 0 and a logical or data 1. In Gen-2 the data 0 pulse is shorter than the data 1 pulse [1]. The Tari is the basic reference time unit in the Gen-2 encoding and ranges from a minimum of 6.25 microseconds to a maximum of 25 microseconds [1]. Data 0 symbols are 1 Tari in length and consist of a transmitted CW followed by an attenuated CW as shown in Figure 4.1.

The length of the attenuated CW portion of the data 0 encoding is defined by the pulse-width (PW) parameter. The PW parameter is on the order of microseconds and is between [1]

$$\max(0.265 \times \text{Tari}, 2) \leq \text{PW} \leq 0.525 \times \text{Tari} \qquad (4.2)$$

Figure 4.1 PIE encoding of a data 0.

The data 1 symbol is longer in length than the data 0 symbol. Specifically the data 1 symbol can be between 1.5 and 2.0 times as long as the data 0 symbol [1]. The data 1 symbol is shown in Figure 4.2.

4.2.4 Message Preambles

In wireless communications a preamble is commonly used to signal the beginning of a wireless message. The preamble serves several purposes. First, it alerts the receiver that a message is being transmitted. Second, it enables the receiver circuitry to synchronize itself, or lock onto, the signal. Third, the preamble can contain information about the origin of the message (from a reader or from a tag) and information about the message encoding itself.

The Gen-2 protocol defines two sets of preambles. The first set of preambles precedes reader-to-tag commands. The type of preamble used for reader-to-tag commands depends on the command being sent. Both types of reader-to-tag preambles contain information about how the tag is to reply to the command. The second set of preambles precedes tag-to-reader replies. The type of preamble used for tag-to-reader replies depends on the data encoding used in the reply.

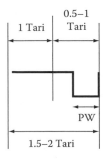

Figure 4.2 PIE encoding of a data 1.

4.2.4.1 Reader-to-Tag Preamble

The reader-to-tag preamble contains information on how the tag is to backscatter the reply to the reader. The two types of a reader to tag preamble are: (1) the preamble and (2) the frame sync. The command transmitted defines which type of preamble is used. The preamble is used when the reader sends a Query command [1]. The frame sync is used for any command other than the Query command [1].

The frame sync consists of three parts: (1) the delimiter; (2) a data 0 bit; and (3) the reader-to-tag calibration (RTcal) symbol [1]. Figure 4.3 shows the reader-to-tag frame sync.

The delimiter is the beginning of the frame sync and is of fixed length [1]. A single data 0 bit follows the delimiter [1]. The RTcal symbol follows the data 0 bit and is the final symbol in the frame sync [1]. The RTcal symbol is equal in length to the length of a data 0 bit plus the length of a data 1 bit [1]. The length of the RTcal symbol is measured in units of Tari symbols [1].

The RTcal symbol is used by the tag to differentiate between a data 0 bit and a data 1 bit. The tag uses the RTcal symbol to compute the pivot value. The pivot value is defined as [1]

$$\text{pivot} = \frac{\text{RTcal}}{2} \qquad (4.3)$$

Any data bit the tag receives that is shorter than the pivot is interpreted as data 0 while any received data bit that is longer than the pivot is interpreted as a data 1 [1]. The data 0 and data 1 bits have a distinct pattern as shown in Figures 4.1 (data 0) and 4.2 (data 1). The tag also uses the RTcal to distinguish between a valid data bit and an invalid, or bad, data bit. Any symbol (bit) received by the tag that is longer than four times the length of RTcal is interpreted as an invalid, or bad, data bit [1].

The frame sync precedes all commands except the Query command. When the reader transmits the Query command, the preamble, not the frame sync is used. The preamble contains four components: (1) the fixed length delimiter; (2) a single data 0 bit; (3) the RTcal symbol; and (4) the tag-to-reader calibration (TRcal) symbol [1]. Hence, the preamble consists of a frame sync followed by the TRcal symbol. The reader-to-tag preamble is shown in Figure 4.4.

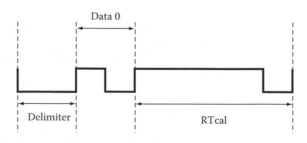

Figure 4.3   Reader-to-tag frame sync.

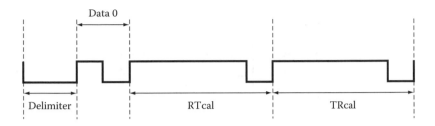

Figure 4.4 **Reader-to-tag preamble.**

The delimiter, data 0 bit, and the RTcal symbol are identical to those used in the frame sync. The pivot value is computed and used in the same fashion as in the frame sync. The additional component, TRcal, is used to instruct the tag which data encoding and data rate to use. Recall that the tag-to-reader communication link can use either FM0 or Miller encoding. Both encodings can operate at a variety of data rates.

The preamble is only used when the reader transmits a Query command to the tag. The Query command has a divide ratio (DR) field. The DR field in the query command and the TRcal symbol of the preamble are used to compute the data encoding rate of the tag-to-reader link. The tag measures the length of the TRcal and uses the value of the DR field in the query command to calculate the data encoding rate, termed the backscatter link frequency (BLF). The BLF is computed using the following equation [1],

$$BLF = \frac{DR}{TRcal} \tag{4.4}$$

With the BLF the tag knows the rate to encode the data of the response to the reader. The BLF is set once for each inventory round by the Query command interrogator-to-tag preamble [1]. To change the BLF a new Query command, and hence a new inventory round, must be started.

4.2.4.2 Tag-to-Reader Preamble

The tag precedes all replies with a preamble. The data encoding of the tag-to-reader link can be one of two options, FM0 or Miller encoding. The preamble used depends on the type of data encoding used in the reply. The reader controls which data encoding scheme to use by its choice of the *M* field of the Query command [1].

The tag preamble may or may not include a pilot tone. The pilot tone is always included in the tag preamble when a tag responds to a command writing to its memory, otherwise the inclusion of the pilot tone depends on the value of the TRext field of the Query command [1]. The pilot tone is included when the TRext field of the Query command is a 1 and is not included when the TRext field is a 0 [1].

The FM0 preamble consists of the optional FM0 pilot tone followed by the FM0 preamble. The FM0 pilot tone consists of 12 FM0 encoded zeros [1]. The FM0 preamble contains a set data sequence and 1 bit that is in violation of the FM0 encoding [1]. The FM0 preamble consists of a total of six symbols. The first four symbols of the preamble consist of an alternating sequence 1 and 0 bits, with the first bit being a 1, followed by

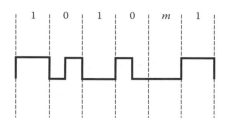

Figure 4.5   FM0 preamble without the pilot tone.

the single bit that violates the FM0 encoding, then a 1 bit terminates the preamble. The bit that violates the FM0 encoding is a data 0 bit without the required phase inversion in the middle of the bit time and is the second to last bit in the FM0 preamble [1].

The FM0 preamble, without the pilot tone is illustrated in Figure 4.5, with the bit violating the FM0 encoding labeled as "m." The FM0 preamble with the pilot tone is illustrated in Figure 4.6; again the bit that violates the FM0 encoding is labeled as "m."

The Miller encoding preamble may be preceded by a pilot tone based on the TRext field value and the command issued by the reader. The Miller pilot tone consists of 12 zeros (baseband Miller encoding) that have been mixed with the appropriate Miller subcarrier clock [1]. This is similar to the FM0 pilot tone with the exception that the 0 bits are encoded according to the Miller scheme rather than the FM0 scheme. The preamble is 10 bits in length. Unlike the FM0 preamble, none of the 10 bits in the Miller preamble violate the Miller encoding.

4.3   Tag State Machine

The behavior of the Gen-2 tag can be described using a finite state machine. This state machine has a large number of states when all flag bits and counters are taken into account. However, the tag finite state machine can be reduced to a finite state machine consisting of seven states. Each of these seven states encompasses many states when the flag bits and slot counters are taken into account. The reduced seven state finite state machine requires knowledge of these internal flag bits and slot counters. Thus, the reduced seven state finite state machine can be used to determine the behavior of the tag based on a given set of current settings and current input.

The next section describes each of the seven states of the reduced finite state machine. Then, the movement through the tag state machine during the inventory (query) procedure is explained. This section concludes with an explanation of movement through the tag state machine during an access command.

4.3.1   Overview of Different Tag States

4.3.1.1   Ready State

The ready state is one of two states in which a tag can enter upon power up. The other state a tag may enter upon power-up is the killed state. The tag will enter the killed state

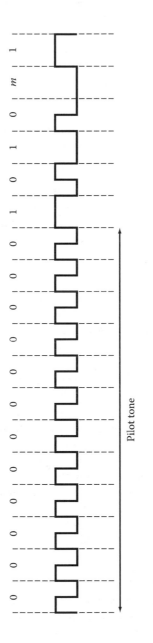

Figure 4.6   FM0 preamble with the pilot tone.

upon power up only if it has been killed prior to power-up. If the tag has not previously been killed, then it will enter the ready state on power-up.

A tag in the ready state is not participating in an inventory round. An inventory round begins when the reader issues a query command and is terminated with the next query command [1]. Thus, the query command signals the end of the current inventory round and the beginning of the next inventory round. The reader uses inventory rounds to obtain the EPC numbers of a population of tags.

The tag leaves the ready state when a query command is received. The query command contains parameters for the tag to select a random number to load into its slot counter. If the random number is zero, then the tag transits into the reply state. Otherwise, the tag transits into the arbitrate state (random number is nonzero).

4.3.1.2 Arbitrate State

Tags in the arbitrate state are participating in the current inventory round and their slot counter contains a nonzero value. Tags in this state are waiting for their slot counter to reach zero. When a tag's slot counter reaches zero the tag will transition into the reply state.

4.3.1.3 Reply State

The reply state is the first of two states the tag enters when it is time for that tag to transmit its EPC number (unique ID) to the reader. Tags in the reply state have a slot counter that is zero and backscatter their RN16 (RN16 is a 16-bit random number) number to the reader. Backscattering the RN16 is the first of two steps in the tag transmitting its EPC number to the reader.

If the reader receives the tag's RN16 number correctly, the reader will transmit an acknowledge (ACK) command with that RN16 back to the tag. The reader may not receive the RN16 due to: (1) a collision of two or more tags transmitting their RN16; (2) RF interference corrupting the RN16; or (3) the reader simply missed the RN16. The tag will then backscatter its EPC number, along with a PC (describes physical properties of the tag) and a CRC (for error detection). The receipt of the ACK command with the correct RN16 causes the tag to transition from the reply state into the acknowledged state.

The tag will only remain in the reply state for a limited time in the absence of receiving any command from the reader. After this time, the tag will automatically return to the arbitrate state.

4.3.1.4 Acknowledged State

The acknowledged state is entered after the tag sends its EPC number to the reader. The acknowledged state is the gateway state to the access commands (read and write commands). However, the tag cannot be killed in the acknowledged state.

The acknowledged state, like the reply state, has a timer that is reset every time a command is received from the reader. If no command from the reader is received within the specified time (timer expires) then the tag will automatically return to the arbitrate state.

4.3.1.5   Open State

The open state is the first of two states used exclusively by the access commands. There is no timer in the open state. Hence, if there is no reader command the tag will remain in the open state as long as the tag remains powered. The secured state can be entered from the open state.

4.3.1.6   Secured State

The tag may employ an optional password feature that requires the reader to supply the correct password to perform any of the access commands. The default value of the password is zero and indicates that password protection is either not implemented or activated. Setting the password to a nonzero value enables the password protection feature.

If the password feature of the tag is implemented and activated then all access commands must be performed from the secured state. The correct password must be provided to the tag before it will enter the secured state.

4.3.1.7   Killed State

RFID tags contain information identifying a particular object. The ID number is unique and corresponds to one particular object. Hence, a number of security and privacy concerns have been raised with the use of RFID. Some of these concerns include reading the RFID tags to track people; thieves reading RFID tags to determine which customers are carrying expensive merchandize; and eavesdroppers reading RFID tags on medicines to determine a person's medical condition.

One method to combat this problem is to provide the ability to turn off the tag. This is referred to as killing the tag in Gen-2. The kill operation cannot be undone and permanently destroys the tag. After being killed, the tag will not respond to any reader command. Hence, no malicious entity can read the tag's EPC number after the tag has been killed.

The killed state is entered after the tag has received the kill command along with the correct kill password. From this point on, the tag will always enter the killed state, instead of the ready state when it is powered up. A tag in the killed state will not respond to any reader command.

4.3.2   Overview of the Movement through the Tag State Machine during the Inventory Procedure

When the reader wishes to read the EPC numbers of a given subset of tags within range, it uses the commands described above starting with the Select command. Assuming that all tags have not been energized for at least 10 seconds and that no tags have been killed, all tags will enter the ready state when energized.

First, the Select command is used to condition the set of tags within the range of the reader to a particular state by altering the SL and inventoried flags. The tags will remain in the ready state.

Then, the Query command initiates the inventory round and specifies which subset of tags will participate in the inventory round. In response to the Query command each tag will select a Q-bit random slot number loading that value into its slot counter. A tag will backscatter its RN16 to the reader when its slot counter reaches zero; all tags that select the random slot number of zero will respond immediately [1]. Assuming that all tags match the query and that the tag's slot counter is nonzero, it will enter the Arbitrate state and will not backscatter reply [1]. However, if the tag's slot counter is zero it will backscatter its RN16 and transition into the Reply state [1].

A tag with a nonzero slot counter will remain in the Arbitrate state until its slot counter equals zero [1]. Tags in the Arbitrate state will respond to all matching Query, QueryRep, and QueryAdjust commands. When the slot counter reaches zero the tag will backscatter its RN16 and transition into the Reply state.

The reader may or may not receive the RN16 that the tag backscattered. A collision may have malformed the RN16 or the reader was simply not able to detect the tags response. If the reader successfully received the backscattered RN16, it will transmit an ACK (acknowledge) command using the received RN16 as the RN parameter of the ACK command. The tag will store the RN16 it backscattered when its slot counter reached zero and will compare its stored RN16 to the RN16 in the ACK command; if they match the tag will transmit its EPC number to the reader; otherwise it will ignore the command [1]. When the tag transmits its EPC number to the reader it will transition into the acknowledged state. The acknowledged state is the gateway state to the portion of the tag state machine covering the Access (read, write, and security) commands.

4.3.3 Overview of the Movement through the Tag State Machine during an Access Command

The access commands provide the reader the ability to perform advanced functions with the tag such as reading and writing data, changing passwords, locking memory on the tag, and completely disabling or killing the tag. Access commands can be used to read and write information about the item associated with the tag allowing that information to travel or reside with the item itself. One example of using this feature in the perishable goods area where writing temperature or other environmental conditions allows the receiver to determine if the item was kept under the proper environmental conditions during transit and storage. This can assist in preventing spoiled goods from reaching the market and causing health problems. Alternatively, the EPC number of the tag can be stored in a centralized database and pertinent information about the associated item could be stored in that database.

The tag must be in the acknowledged state before it will respond to any of the access commands. This means that the tag must be accessed almost immediately after its EPC number is read. Thus, reading a tag's EPC number is the first step (to put the tag in the acknowledged state) in issuing any of the access commands to the tag and this procedure is described in Section 4.3.2.

After reading the EPC number the reader must send the Req_RN command with the RN16 (16-bit random number that was used with the ACK command to read the tag's EPC number) to cause the tag to transition into either the open state or the secured state. Which state the tag transitions into from the Acknowledge state depends on the value of the access password. An access password of all zeros indicates that the access password feature is not implemented or enabled [1]. If the access password feature is not implemented or enabled (access password is zero) the tag will transition into the secured state in response to the Req_RN command. If the access password is nonzero then the access password feature is enabled and the tag will transition into the open state [1]. In response to a valid Req_RN command (correct RN16 with correct access password or access password is zero) the tag will backscatter a handle which is used by the other access commands [1]. When the tag is in the open state and has a nonzero access password the Access command must be used with the correct handle and access password for the tag to transition into the secured state.

The Lock command allows memory locations to be locked, or made read only, write only, or both read and write [1]. The Lock command can unlock memory locations making them readable, writable, or both if they were previously locked [1]. The Lock operation will only take effect when the tag is in the Secured state [1]. Other access commands, including the Kill command, can be performed while the tag is in the Open state.

In summary, there are three steps in performing an access command. First, the tag EPC number must be read (see Section 4.3.2). Second, the Req_RN command must be sent with the correct RN16 to the tag. Third, the tag must be put into the Secured state using the Access command if the access password is nonzero. At this point any access command, including the Lock command, can be performed.

4.4   Tag Inventorying Features

The Gen-2 protocol defines a set of commands. The lengths of the commands can vary. Thus, the Gen-2 protocol attempts to minimize the length of those commands which are most frequently used.

The most commonly used Gen-2 commands are used to collect an inventory of the tags in range of the reader at a given time. This process is referred to as inventorying the tags and the lengths of those commands used for inventorying the tags are minimized. Gen-2 employs a slotted Aloha collision avoidance algorithm to reduce the amount of data corruption resulting from a collision. In a slotted Aloha communication scheme time is divided into a number of discrete slots, called time slots. A device wishing to transmit will start to transmit at the beginning of a time slot. Further, the transmission must not overlap into the adjacent (next) time slots. Thus, when there is a collision, the data in only one time slot is corrupted.

Without the restrictions of the time slot it is possible for one device, Device A, to start transmitting in the middle of Device B's transmission and continue until the middle of Device C's transmission; where Device B's and Device C's transmissions do not overlap. In this case three messages are lost where in a slotted Aloha environment one message would have gotten through.

4.4.1 Overview of the Query Commands

Collecting the tag EPC numbers (unique IDs) is one of the most important operations in the Gen-2 protocol. There are four commands that are used to inventory the tags: (1) the Select command; (2) the Query command; (3) the QueryRep command; and (4) the QueryAdjust command. Each of these four commands performs a different function within the inventorying process.

4.4.1.1 Query

The Query command signals the beginning of the Gen-2 inventory round [1]. The Query command, along with the reader-to-tag preamble, defines all the tag-to-reader link characteristics. Three fields in the Query command are used by the tag in conjunction with the reader-to-tag preamble to determine the tag-to-reader communication link parameters. These three fields are the DR, M, and TRext fields. The DR field is used by the tag to compute the data rate of the tag-to-reader link as described in Section 4.2.4 and Equation 4.4. The M field is used to select the FM0 or one of the three Miller encoding schemes to encode the data backscattered by the tag. The TRext field of the Query command determines if the tag includes the tag preamble when replying. The TRext field is 1 bit and the tag preamble is included when TRext is a one, and the tag preamble is not included when TRext is a zero [1].

When a tag receives a Query command, the tag first checks to see if the flag values specified by the Query command match its stored flag values. The tag will participate in the inventory round if and only if all its stored flag values match the flag values specified by the Query command, otherwise the tag will ignore the Query command. The tag will then pick (if participating) a random number of length defined by Q field of the Query command. The value of the Q field can range from 0 to 15 and determines the total number of time slots available for the tags to choose from. The number of time slots, N_{Slots}, available for a given value of Q is defined by the following equation:

$$N_{Slots} = 2^Q - 1 \tag{4.5}$$

The tag will then wait for that slot and then respond by sending the RN16 to the reader. The RN16 is a 16-bit random number that is used to access a unique tag as the probability of two tags simultaneously selecting the same RN16 (16-bit random number) is extremely small. The parameter Q instructs the tag how many bits from its 16-bit random number generator to use for its time slot termed the slot counter [1]. The reader uses the RN16 value to read the tag's EPC number (unique ID) and as a handle for high-level functions (read, write, and security). The Query command is 22 bits in length.

4.4.1.2 QueryRep

The QueryRep command is used to advance to the next slot by causing all participating unread tags to decrement their slot counter by one. Any tag which selected that slot to reply in will send their RN16 value to the reader. The QueryRep command is probably the most frequently used Gen-2 command and is thus the shortest of the Gen-2 commands with a length of only 4 bits.

4.4.1.3   QueryAdjust

The Query command always starts a new inventory round, but sometimes the reader only needs to make a slight adjustment to the Q value (number of time slots) or to ask all noninventoried tags to select new slot numbers in which to respond. The QueryAdjust command provides a mechanism to increase the Q value by one, decrease the Q value by one, and to keep the Q value the same. In all three cases all noninventoried tags (tag that have not backscattered their EPC number) will select new time slots in which to respond. The QueryAdjust command is shorter than the Query command, being only 9 bits in length, and does not start a new inventory round. Starting a new inventory round has additional side effects on the state of the tags currently being inventoried that may adversely impact the time required to complete the inventorying process.

4.4.1.4   Select

The Select command defines the population of tags that will participate in the upcoming inventory round. Although all tags within the range of the reader may participate in the current inventory round, this is not necessary. Thus, only a subset of those tags within the range of the reader will participate in the current inventory round. The Select command provides a number of features to tell the tags whether or not they are to participate in the current inventory round. The usefulness of the Select command to define a subset of tags to participate in the current inventory round will be illustrated in the examples provided later in this chapter.

4.4.2   Use of Sessions

When two readers try to communicate with a single tag in the same time frame a number of consistency problems arise. This is true for both writing data to the tag and for reading data or the EPC number from the tag. RFID systems must allow multiple readers to communicate with a single tag. The Gen-2 standard provides some limited support for allowing up to four readers to communicate with a tag in the same time frame. This mechanism is termed sessions in Gen-2.

There are four sessions defined in Gen-2 and each session has its own inventoried flag [1]. The remaining tag state variables are shared among all four sessions [1]. The sessions are denoted as S0, S1, S2, and S3 [1]. These four sessions enable up to four readers to independently communicate with a tag for the purposes of readings its EPC number. The initial value of the inventoried flag depends on the session used by the reader and the previous value of the inventoried flag. When the tag powers up (energized by the reader) the inventoried flag of session S0 is always set to A, while the inventory flag of sessions S1, S2, and S3 are set to previous value (either A or B) if the tag is energized before the persistence time expires, otherwise the inventoried flag will default to A [1].

Because the tag has no onboard power supply each session has an associated persistence time which is the length of time that the inventoried flag for each session is maintained after the reader stops transmitting energy to the tag. The tag is capable of storing a small amount of energy using a capacitor and the amount of energy that can be stored on the capacitor determines the persistence time. Thus, each session has its own capacitor each having a different capacitance value (storing different amounts of energy) which determines the persistence time of the inventoried flag. The persistence time for the inventoried flag when the tag loses power in sessions S2 and S3 is a minimum of 2 seconds while the persistence time for session S1 varies between 500 milliseconds and 5 seconds, and there is no persistence time (persistence time is zero) for session S0 [1]. The S1 inventoried flag will revert to A after the persistence time expires, even if the tag is energized (powered) [1].

Sessions are useful because they can be used to set the inventoried flag of all tags to a known value. Then as the reader reads the EPC number of each tag that tag's inventoried flag will switch and that tag will no longer participate in the inventory round. Thus, the reader can verify that it has read all tags within range when there are no more tags left with a given inventoried flag. This requires the reader to select the proper session to use. Both sessions S2 and S3 have a persistence time and, as long as the reader can energize the tags before that persistence time expires, the above process will work. Session S0 has a persistence time of zero, so in this case the reader must continually energize the tags. This is achieved by the reader transmitting the CW between commands. When using session S0, the inventoried flag of any tag that loses power during this process will revert to A regardless of its previous value. Assuming that the reader is looking for only tags with an inventoried flag of A it may read the EPC number of the same tag multiple times due to this issue. When using session S1, care must be taken to ensure that the entire inventory round does not last longer than the persistence time. If the inventory round lasts longer than the persistence time, the inventoried flag will revert to A regardless of its previous value. When using session S1, it is possible for tags to continually reenter the process as the persistence time expires. Thus, session S1 should be used only for short inventory rounds. However, maintaining the CW between commands will prevent the resetting of the inventoried flag due to a persistence time timeout.

The sessions provide some limited support for up to four different readers to access the same tag to read its EPC number in the same period. To do this each reader must select a different session to avoid continually switching the inventoried flag between A and B every time one of the readers reads the EPC number. Because the higher level read, write, and access functions require more than just the inventoried flag, the use of sessions does not provide the ability for independent high-level commands to be executed by multiple readers. However, this is beneficial because issues such as data consistency and state consistency arise when multiple readers are performing high-level operations on the same tag.

4.4.3 Features of the Select Command

The Select command is used to select the subset of those tags within the range of the reader that will participate in the upcoming inventory round. This subset may include all, some, or none of the tags within the range of the reader. The Query command, which starts an inventory round, uses the selected flag (SL flag) and one of the four inventoried flags to select the desired subset of tags to participate in the inventory round. The Select command allows the reader to alter either the selected flag (SL flag) or one of the four inventoried flags stored in the tag.

The Select command contains six mandatory fields and one optional field that can be used to identify the subset of tags that will participate in the upcoming inventory round. The first mandatory field is the Target field which allows the reader to alter the SL flag, or the inventoried flag of one of the four sessions (S0, S1, S2, or S3), but not to alter the SL or inventoried flags [1]. The second mandatory field is the Action field which specifies the action that all tags matching the parameters of the Select command will take. The actions available are to set the value of the SL flag or to set the value of the inventoried flag to either A or B [1]. The action specified is taken only if a tag's parameters match the set of parameters specified in the MemBank, Pointer, Length, and Mask fields of the Select command [1]. The MemBank, Pointer, Length, and Mask fields are used to further specify the matching criteria for the tag beyond just the SL or inventoried flags. The Mask field is optional and has a variable length ranging from 0 bytes to 255 bytes [1]. The Mask field, when used, contains an N byte string that must match the contents of a specific location in the tag's memory (for instance the EPC number) for that tag to participate in the next inventory round. The Length field defines the length of the Mask field in bytes [1]. The Mask can be compared against one of four types of memory in the tag. Gen-2 divides the tag memory space into four sections: (1) the first section is the Reserved Memory bank and contains the passwords associated with that tag; (2) the second section is the EPC Memory bank containing the EPC number; (3) the third section is the TID Memory bank and contains tag and vendor specific information; and (4) the fourth section is the User Memory bank containing user-defined data [1]. The MemBank field specifies which of the four memory banks the Mask will be compared too [1]. The Pointer field defines the starting address in memory where the Mask is be to compared against [1]. The pointer provides a memory address to memory relative to the memory bank specified by the MemBank field [1]. Complex and multiple subsets of tags can be facilitated by issuing multiple Select commands to further define and refine the subsets of tags.

A seventh field, the Truncate field, is part of the Select command. The Truncate field is used in conjunction with the Mask field to reduce the number of bits of the EPC number (unique ID) that the tag backscatters to the reader in response to the ACK command [1]. Truncate is only used when the MemBank points to the EPC memory, when the Mask is specified, when the Target specifies those tags whose SL flag is asserted, and this particular Select command is the last Select command before the inventory round begins [1]. When truncating is enabled, the tag will send only the portion of the EPC number that follows the Mask [1]. Hence, the Truncate field can be used to shorten the length of the tag reply to the ACK command. Use of the Truncate field is explored with the case studies in Section 1.6.

4.4.4 Features of the Query Command

The Query command starts a new inventory round, identifies the subset of tags which will participate in the new inventory round, and selects the tag-to-reader data encoding and data rate [1]. The Select command is used to alter the SL flag and the four inventoried flags of the tags. The Query command specifies a value for the SL flag, a particular session, and a value for the inventoried flag for that particular session [1]. Those tags whose SL flag and inventoried flag for the particular session match the requested values in the Query command will participate in the new inventory round.

The Query command contains seven fields and a CRC for error detection. Three of the seven fields, the DR, M, and TRext fields are used to define the tag-to-reader data rate, data encoding respectively, and if the pilot tone is included in tag-to-reader preamble [1]. The DR and M fields are discussed in Section 4.4.1. The TRext field tells the tag whether or not to backscatter the pilot tone before backscattering the tag-to-reader preamble when replying to the reader [1].

Three of the remaining four fields, the Sel, Session, and Target fields define the SL flag and inventoried flag parameters needed by the tag to determine if it should participate in the inventory round [1]. The Sel field specifies the value of the SL flag that a tag must have to participate in the inventory round. The Sel field may specify that the SL flag be asserted, not asserted, or either one (all SL flag values match) [1]. The Session field defines one of the four sessions, S0, S1, S2, and S3 [1]. The Target field defines the value of the inventoried flag, either A or B, for the session specified by the Session field [1]. Using these three values each tag can determine if it should participate in the inventory round or not. All tags, whose SL flag and inventoried flag for the session specified by the Session field match the values specified in the Query command, will participate in the inventory round; those tags with different flag values will not participate in the inventory round.

4.4.5 Features of the QueryRep Command

The QueryRep command is the shortest command in the Gen-2 protocol. The QueryRep command instructs all tags to advance to the next slot. Each tag will decrement its slot counter in response to the QueryRep command. When a tags slot counter reaches zero (or selects zero as the random Q-bit number) it will respond by backscattering its RN16 to the reader.

The QueryRep command has one parameter, the Session field, which is used to indicate which Session the tag should decrement its slot counter for [1]. If a tag is participating in an inventory round for a given session, say S0, and receives a QueryRep command with another session (S1, S2, or S3) it will ignore that QueryRep command [1].

4.4.6 Features of the QueryAdjust Command

The QueryAdjust command is used to adjust the value of Q by one and then select a new slot counter, or to instruct the tags to select a new slot counter without changing Q. The QueryAdjust has two parameters, the Session parameter and the UpDn parameter [1]. The Session parameter indicates the session to which that QueryAdjust command is related [1]. As with the QueryRep command any tag participating in an inventory round in one session will ignore all QueryAdjust commands with a different Session parameter [1]. The UpDn parameter defines the action that the tag is to take in response to the QueryAdjust command.

The QueryAdjust command can cause the tag to take three actions. The first action is to keep the previous value of Q and then to select a new Q-bit random slot number [1]. The second action is to increase the previous value of Q by one and then to select a Q-bit random slot number [1]. The third action is to decrease the previous value of Q by one and then to select a Q-bit random slot number [1].

4.5 Tag Singulation

Tag singulation is the process by which a reader establishes communication for reading the EPC number or for advanced commands (read, write, or security commands) with a single tag among a population of tags. Singulating a tag depends on the current operating environment and on the makeup of the population of tags themselves. Although most methods will work in all cases, it is often beneficial to select the method that requires the least amount of time. This helps to remove RFID from being a bottleneck or at least mitigate any bottleneck effect in the larger system.

4.5.1 EPC Gen-2 Tag Data Encoding Classes

The EPCglobal organization, which developed and maintains the Gen-2 standard, also defines a set of EPC number encodings. These EPC encoding types are described by the Tag Data Standard maintained by EPCglobal [2]. The protocol-control (PC) bits that are backscattered by the tag when the tag is asked for its EPC number contain information defining the tag data encoding used and the length of the tag's EPC number in units of Gen-2 words. This standard defines 11 different data encodings EPC tags [2]. The maximum length of the EPC memory bank in a Gen-2 tag is 496 bits (31 Gen-2 words) [1]. The theoretic minimal length of the EPC number is one Gen-2 word, but the shortest EPC number defined by the Tag Data Standard is 6 Gen-2 words or 96 bits [2]. One Gen-2 word is equal to 16 bits [1]. Shorter encodings (64 bits) have reserved headers but are only included for backward compatibility with older tags, and eventually these headers will be recycled for new encodings.

Each of the 11 different tag data formats contain different information or the same information in different formats. One of the encoding formats is the General Identifier (GID-96) which is a generic 96-bit EPC number [2]. The GID-96 contains four fields: (1) the Header identifying that this EPC number is encoded according to the GID-96 specification; (2) the General Manager Number identifying the owner of the asset; (3) Object Class identifying the type of the asset; and (4) the Serial Number identifying a unique instance of an asset [2]. The Object Class and Serial Numbers are assigned by the owner and are unique within the owner's set of numbers [2].

The GID-96 data standard is very similar to an IP address assignment where large organizations are given a Class B address space containing a large number of individual IP addresses which they allocate to their internal organization. The Object Class field can be likened to a division within the organization that is allocated a Class C address space for use by that particular division. The Serial Number can be thought of as the individual IP address assigned to a particular device in an office.

4.5.2 Selecting a Single Tag

Selecting a single tag requires the most knowledge of the set of tags present in the reader field. A single tag can be selected by using that tag's EPC number as the mask field of the Select command. After the Select command the only tag selected (select flag is asserted) will be the tag having the EPC number specified as the mask field of the Select command just sent. Selecting a single tag is a shortcut to performing higher level commands (read and write commands) on that tag.

The Query command can be issued looking only for those tags with their select flag asserted and the Q value can be set to zero forcing the single selected tag to respond. Once that tag responds and is acknowledged the higher level commands can be performed.

4.5.3 Selecting a Group of Tags

Groups of tags can be selected on the basis of the value of their SL flag, their inventoried flag for a particular session, or based on the contents of their memory, that is, EPC number. The SL and inventoried flags do not provide a good assurance of selecting the correct group of tags. There is no guarantee that all tags desired to be in the subset will actually be in the subset and that no unwanted tags are included.

However, the ability of the Select command to put tags into the subset based on memory contents can be used to create the subset. This is possible through the use of the mask feature of the Select command that requires tags to match the Mask field with their specified memory data to take the action specified by the Select command [1]. The EPC number has a structure that standardizes this process. Thus, all tags that are wanted in the subset can be put into the subset, and tags that are not to participate in the inventory round will be in the subset.

The GID-96 encoding structure described in Section 4.5.1 will be used throughout this chapter for the EPC number encoding to illustrate the concepts presented in this chapter. The methods presented can be used with the other encodings. The GID-96 contains four fields where the Header field does not change for a GID-96 type encoding. The second and third fields, the General Manager Number and Object Class denote the owner and type of asset respectively. The fourth field is the Serial Number field and is unique within a given General Manager Number and Object Class combination. Hence, the General Manager Number can be used to select a group of tags based on their owner or manufacturer. This subset can be further refined by using the Object Class to select a particular type of asset within a given General Manager Number. Because multiple Select commands can be chained together it is possible to select a group of item types from a group of different owners.

4.5.4 Selecting All Tags

Sometimes it is beneficial to have all tags participate in the inventory round. This can be guaranteed by issuing a single Select command. To do this, session S0 should be used. The Select command will alter the session S0 inventoried flag, using "000" as the Target field, and will not use the mask or truncate features [1]. This will cause all tags that match the first Select command to set their session S0 inventoried flags to a value of *A* and all tags that do not match will set their session S0 inventoried flags to a value of *A* [1]. At this point all tags will have set their session S0 inventoried flag to *A* because there was no Mask (mask length was zero) specified in the Select command and this means that all tags match [1]. The inventory round is begun by issuing a Query command with the Sel field set to all (value of "00" or "01"), the Session field set to session S0 and the Target field set to *A* [1]. Now, all tags will participate in this inventory round.

4.6   Trade-Offs

Retail goods are the primary target of Gen-2 RFID tags. Pallets are often sent to customers containing a quantity of a single product. Frequently Gen-2 tags are attached to the items on the pallet. Although barcodes require line of sight to be read Gen-2 RFID tags can be read through other objects. Hence, when a pallet arrives it must be disassembled or broken down but if the items are tagged with RFID tags then the inventory of that pallet can be obtained using RFID without having to break down the pallet saving a significant amount of time.

There are a number of trade-offs in the Gen-2 protocol. One trade-off is in the construction of the subset of tags that will participate in a given inventory round. Creation of this subset results from the use of the Select and Query commands. This section explores this trade-off in more detail and gives example cases where different subsets of tags are useful. Selecting the correct set of values to optimize will decrease the amount of time required for the RFID system to read all the RFID tags on the pallet and increase throughput.

Another trade-off is in the selection of the reader to tag, tag to reader, reader data encoding, and tag data encoding. These values typically depend on the characteristics of the RF environment where the system operates. For example, in an environment with very low RF noise fast data rates and fast (simple) data encodings would be preferable to produce the faster reader–tag interactions. Conversely, in an environment with significant RF noise the slower data rates and more robust data encodings, such as Miller 4 or Miller 8 would be preferable to decrease bit errors (increase chance messages can be correctly decoded) but these data encodings result in a lower data rate.

In the example cases presented in this section the generic 96-bit EPC number (GID-96) format for the EPC number on the tag is assumed to be employed. Further, it is assumed that it has been at least 10 seconds because any tag on the pallet had been energized.

4.6.1   *Inventorying the Tags on a Pallet Containing One Type of Product*

This is the simplest case to visualize as the pallet contains only a single type of item from a single vendor. Recall that the Select command contains a Mask field that can be used to put only those tags whose memory matches the mask to participate in the current inventory round. Also recall that the Truncate field can be asserted to have the tag send back a shortened EPC number. In this case the Header, General Manager Number, and Object Class values will be the same for all tags and only the Serial Number will differ.

In this case, all tags must be included in the subset of tags participating in the inventory round. This can be ensured by issuing two Select commands. The first Select command will target all tags whose SL flags are asserted and use action code "110" which will cause no change for the matching tags and force those tags that do not match to assert their SL flags. The second Select command will target all tags with their SL flag asserted (this is the entire population of tags) again using action "110" (no change) in tags should result and with the Mask specified and the Truncate flag asserted. The Mask will be set to the Header, General Manager Number, and Object Class. With the Truncate flag asserted, each tag will return only its Serial Number. The Serial Number is 36 bits in length versus 96 bits for the entire EPC number [2]. This is a savings of 60 bytes or 62.5 percent reduction in bits backscattered for each tag read. From this point the reader simply starts the inventory round with the Query command and continues to read tag EPC numbers using the Query, QueryRep, and QueryAdjust commands until all tags have been read.

When the Header, General Manager Number, and Object Class values are not known ahead of time they can be obtained by reading an EPC number from a single tag. Once the Header, General Manager Number, and Object Class values are obtained from the single EPC number the process in the above paragraph can be used to read the remaining tags.

4.6.2 Accessing Tags on a Pallet Containing One Type of Product

Accessing a tag can be done when that tag's EPC number is read as described in Section 4.3.3 or after all tags have been read. To access a tag after all tags have been read requires using the Select command with the Mask set to the entire EPC number of that tag. Then a Query with Q set to zero (a single time slot) is issued. The tag will respond in the single time slot and the reader can acknowledge the tag and then access the tag. If the tag is not accessed immediately after being read it is important to wait until all tags have been read to access that tag using the above method. This is because the above method may alter the SL flags of the tags and cause problems in ensuring that all remaining unread tags have been read during the current inventory round.

4.6.3 Inventorying the Tags on a Pallet Containing Multiple Types of Products from a Single Vendor

In this case, the pallet contains a number of different types of items from the same vendor. Therefore, the Header and General Manager Number of all tags will be the same, but the Object Class and Serial Number values will be different. A similar procedure to that described in Section 4.6.1 can be used with the exception that only the Header and General Manager Number will be specified in the Mask field of the second Select command. If there are only a few different types of products on the pallet or a large percentage of the pallet is a single product then it may be beneficial to include the Object Class in the Mask to reduce the amount of the EPC that those tags backscatter to the reader. However, this trade-off must be carefully balanced because two Select commands and several inventory (Query, QueryRep, and QueryAdjust) commands must be issued for each type of product. As the number of items of each type of product decreases the benefit from having the tag reply with a shorter part of the EPC number is negated by increased costs in setting up and performing the inventory round.

4.6.4 Accessing Tags on a Pallet Containing Multiple Types of Products from a Single Vendor

In this case the tags can be grouped into smaller subgroups based on the type of product using the Object Class. Thus, only a particular subset of tags must then be read to find the desired tag. This reduces the time required to access a particular tag because the subgroup contains fewer tags than inventorying the entire pallet. The tag may be accessed immediately after being read or after all tags in the subgroup (tags with identical Object Class fields) have been read. The trade-off mentioned in Section 4.6.3 of the overhead of sending the sequence of Select commands to build the subgroup must be compared against the time required to simply read all tags on the pallet.

4.6.5 Inventorying the Tags on a Pallet Containing Multiple Types of Products from Multiple Vendors

In this case, the pallet contains multiple types of products from different vendors. Therefore, only the Header portion of the EPC number is the same for all tags. The Header field specifies the type of data encoding on the tag and is 8 bits in length [2]. The mask and truncate features of the Select command can used to reduce the amount of the EPC number that is sent back but that only saves 8 bits or 8.33 percent of the EPC information using the GID-96 encoding that the tag backscatters. The General Manager Number, Object Class, and Serial Number values will vary in this type of pallet.

Thus, using the mask and truncate features of the Select command must be balanced against the added expense of issuing the extra Select commands. If the quantities of similar items are small or the pallet contains items from a large number of different vendors then the cost of issuing the extra Select commands may increase the time required to take the inventory. In this case, it is preferable to use the mask and truncate feature of the Select command using just the Header and then to read the EPC numbers from all tags one at a time.

4.6.6 Accessing Tags on a Pallet Containing Multiple Types of Products from Multiple Vendors

In this case the tags can be subdivided based on manufacturer and product type using the General Manager Number and Object Class fields, respectively. If these values are known then the subgroup can be easily formed and potentially eliminate a large percentage of the tags on the pallet. The desired tag may be accessed immediately after being read or after all tags with matching General Manager Number and Object Class fields have been read. The trade-off of the increased overhead of the sequence of Select commands to build the subgroup of tags versus the time required to simply read all tags on the pallet must be evaluated. For pallets with only a few items of the same type and vendor, forming the subgroup is quicker.

4.7 Open Issues

Open issues with the efficiency of the Gen-2 protocol include determining the best physical layer settings (data encoding and data rate) to reduce bit errors and in optimizing the inventorying and access processes to minimize the time required to perform operations on the tags.

Determination of the best physical layer properties involves trading off more robust communication links reducing bit errors versus the decrease in data rate that this causes. Tractable and accurate models to predict the future RF environmental conditions are critical but are difficult to formulate. For wide commercial deployment, these models must address a wide range of different operating conditions, such as in a factory, in a store, or on a moving truck. Also, the ability of the RF signals to travel through the items is important because if the signal is attenuated (weakened) too much as it travels through the items the reader or tag will not receive the messages properly or at all.

Minimizing the time to read the EPC numbers of all tags requires making predictions based on the makeup of the set of tags being inventoried as well as the status of each tag. Knowing the type of pallet being inventoried a priori enables the reader to select which set of masks to use to reduce the size of the EPC number backscattered by each tag. However, this information may not be available and either a prediction or game theory type model must be developed to quickly determine the type of pallet or this information must be provided to the reader by the pallet.

4.8 Conclusions and Future Research Directions

The features of the Gen-2 protocol enables users to optimize the time required to interact with a group of tags. This is particularly useful in the retail and distribution sectors where minimizing the time to inventory a pallet of items will increase throughput. However, there are still a number of open research areas spanning from more error-resistant physical layers up to the models and methodologies used to determine how to communicate with the tags themselves.

Reducing the number of bit errors caused by the RF environment (interference or ambient noise) will decrease the time required to take an accurate inventory of the pallet. Signal processing techniques may provide some assistance in deciphering the correct data from a message garbled by noise. Also, of interest is identifying at least one of the RN16s backscattered by a tag during a collision. This would greatly improve the number of tags that is able to be read in a given time period.

The attenuation of the RF signals by the other items reduces the range at which readers and tags can communicate. This limits the size of the pallet to ensure that all tags can be read to take an accurate inventory. Changing the physical layer properties of the protocol is not wise as this could obsolete existing products based on Gen-2.

Using the right mix of commands to inventory a group of tags is important. Although simply trying to inventory all tags within a range at the same time will work, the benefits of the truncate and mask features of the Select command are not used. Properly using the Select command is important as time is added to define the subset of tags that will participate in the inventory round. Accurately predicting the type of pallet being inventoried is important to minimizing the time to take the inventory. Also, selecting the right mix of inventory commands is important to minimizing the time to read all the EPC numbers. Research into developing models or policies for both areas is needed.

References

1. EPC^{TM} radio-frequency identify protocols class-1 generation-2 UHF RFID protocol for communications at 860 MHz–960 MHz version 1.1.0, EPCglobal Inc., 2005.
2. EPCglobal tag data standards version 1.3.1, EPCglobal Inc., 2007.

Chapter 5

RFID Authentication and Privacy

Alex X. Liu and LeRoy A. Bailey

CONTENTS

Radio frequency identification (RFID) tags are cheap, simple devices that can store unique identification information and perform simple computation to keep better inventory of packages. This feature provides a significant advantage over barcodes, allowing them to be used in applications throughout various fields such as inventory tracking, supply chain management, theft-prevention, and the like. However, unlike barcodes, these tags have a longer range in which they are allowed to be scanned, subjecting them to unauthorized scanning by malicious readers and to various attacks, including cloning. Therefore, a security protocol for RFID tags is needed to ensure privacy and authentication between each tag and their reader. This chapter provides a general look over various security approaches created in recent years. These approaches include separate devices that were developed to protect an RFID tag and low-computation algorithmic protocols developed within the tag itself, two of which were developed by the authors of this chapter. The chapter is concluded by discussing the future direction of RFID security and some open research issues concerning its field of study.

5.1   Introduction

Radio frequency identification (RFID) tags are small electronic components that are used to identify and track objects. They have applications in various fields such as inventory tracking, supply chain management, theft-prevention, and the like. An RFID system consists of an RFID tag (i.e., transponder), an RFID reader (i.e., transceiver), and a back-end database. An RFID reader consists of an RF transmitter and receiver, a control unit, and a memory unit. These instruments work together to transfer and receive information stored on radio waves between the reader and an antenna attached to an RFID tag. This information interacts with stored items upon a back-end database that some readers are able to connect to. Depending on the type of the tag, they too have the capability to perform different functions with the information transferred from a reader.

There are three broad categories of RFID tags: passive, semipassive, and active. Passive tags are powered by the signal of an interrogating reader and can only work within short ranges (a few meters). Active tags maintain their internal state and power transmission using a battery. Semipassive tags are battery-assisted tags that use some battery power to maintain their internal volatile memory but may still rely on the reader's signal to power their transmission. They can initiate communication and operate over longer ranges (several meters), but are also more expensive and bulkier than passive tags. Passive tags, however, are also more popular and cheaper, making them more likely to be used within the broad range of applications stated earlier. Therefore, this chapter will only focus upon devices and protocols that have been designed for passive tags.

RFID tags are able to uniquely identify individual items of a product type, unlike barcodes, which only identify each product type. This is particularly useful when the transaction history of each item needs to be maintained or when individual items need to be tracked. Furthermore, RFID tags do not require line-of-sight reading like barcodes, increasing the scanning process of a tag significantly. Due to these and other advantages that RFID tags have over barcodes, RFID is increasingly becoming more popular and is expected to replace the current barcode technology in the near future. However, there is also a growing concern among people about consumer privacy protection and other security loopholes that make RFID tags an easy target for malicious attacks. Passive RFID tags in their current form are vulnerable to various types of attacks and thus there is a pressing need to make this technology more secure before it is viable for mass deployment. Therefore, privacy and authentication are the two main security issues that need to be addressed for the RFID technology.

The two primary concerns of privacy with RFID tags are clandestine tracking and inventorying [1]. Clandestine tracking deals with the issue of a nearby RFID reader being able to scan any RFID tag, because these tags respond to readers without discretion. Clandestine inventorying on the other hand is a method of gathering sensitive information from the tags, thus gaining knowledge about an organization's inventory. An organization called EPCGlobal [2] manages the development of the Electronic Product Code (EPC), a code in RFID tags that is equivalent to the code used to store information in a barcode. EPC-compliant RFID tags have fields to store the manufacturer code and the product code that makes it easy to follow the inventory patterns of a store [1] or the assignment of ID numbers to employees of a business, for example.

RFID privacy is already a concern in several areas of everyday life. Here are a few examples. Automated toll-payment transponders, small plaques positioned in windshield corners, are commonplace worldwide. In a recent judiciary, a court subpoenaed the data gathered from such a transponder for use in a divorce case, undercutting the alibi of the defendant [3]. Some libraries have even implemented RFID systems to facilitate book checkout and inventory control and to reduce repetitive stress injuries in librarians. Concerns about monitoring of book selections, stimulated in part by the USA Patriot Act, have fueled privacy concerns around RFID [4]. Lastly, an international organization known as the International Civil Aviation Organization (ICAO) has promulgated guidelines for RFID-enabled passports and other travel documents [5,6]. The United States has mandated the adoption of these standards by 27 "visa waiver" countries as a condition of entry for their citizens. The mandate has seen delays due to its technical challenges and changes in its technical parameters, partly in response to lobbying by privacy advocates. One may see how verification of the information stored upon the passport would also become an issue as well. This brings us to the other security threat in RFID, authentication.

Authentication is another major security issue for RFID tags. Privacy deals with authentic tags being tampered by attacking readers, while authentication deals with valid readers being misled by deceptive tags. One example, where authentication would play a useful role, is when scanning counterfeit tags. It has been shown that one can rewrite what a tag emits onto another tag, effectively making a clone [1]. Therefore, authentication is as much of a concern as privacy is.

The key challenge in providing security mechanisms to passive RFID tags is that such tags have extremely weak computational power because they are designed to be ubiquitous low cost (i.e., a few cents) devices [7]. Numerous solutions have been developed to solve both security threats for RFID tags. These solutions include separate devices that were developed to protect an RFID tag and low-computation algorithmic protocols developed within the tag itself. This chapter discusses the advantages and disadvantages of a few of these solutions from which the field of RFID security has contributed. Insight is then provided into the current state of RFID security and its future direction.

5.2 Premier RFID Authentication and Privacy Protocols

Upon realizing the need to provide consumer privacy to RFID tags, researchers first began to disable the tag upon scanning. In other words, once a product has been through the checkout procedure, a magnet-like device would disrupt the wiring of a tag, disabling it from use again. This procedure is currently used in many bookstores and other retail environments to prevent theft. However, this approach is not suitable for all items. Therefore, researchers began borrowing the concepts of well-known cryptography methods to not only establish privacy but also provide mutual authentication for tags that contain a higher security risk such as identification badges and electronic passports. This section describes the disadvantages of disabling a tag and looks into how cryptography methods will compare to RFID tags.

5.2.1   Tag "Killing" Protocols

The first approach to dealing with consumer privacy was developed by the company that will oversee the barcode to RFID transfer, EPCGlobal Inc. Their approach is to just "kill" the tag [2]. In other words, the tag will be made inoperable, allowing it not to be scanned by malicious readers. This process is done by the reader sending a special "kill" command to the tag (including a short 8-bit password). For example, after you roll your supermarket cart through an automated checkout kiosk and pay the resulting total, all of the associated RFID tags will be killed on the spot.

Though killing a tag may deal with consumer privacy, it eliminates all of the postpurchase benefits for the consumer. One example of these types of postpurchase benefits are items being able to interact with what are being called "smart" machines. For example, some refrigerators in the future will interact with the RFID tags on food items. This will allow the refrigerator to scan what items you normally buy, and once it notices that so many items have been removed over a period of time, it will inform what items are missing so you may purchase some more. Another example of a "smart" machine would be a microwave. The microwave would scan the RFID tag from the purchased item and automatically set the timer to the correct amount of time needed. From these examples, you can see that killing a tag would not be an appropriate approach to deal with consumer privacy.

5.2.2   Cryptography Protocols

Due to the invention of "smart" machines and other devices that will need to reuse a tag's information repeatedly, researchers began developing ideas that will insure one's privacy and mutually authenticate a tag and a reader to one another without actually disabling the tag upon its scanning. One of the first set of approaches developed to fit this criterion borrowed the concept of the public key cryptography approach [8–11]. The cryptography approach is a simple and well-defined algorithm to be implemented. An example of this type of protocol would be as follows. Consider a matching pair of encryption public and private keys for both a reader R and a tag T. Each tag would initially be embedded with its reader's public key, R_{pu}, and its own unique private key, T_{pr}. Upon query from a reader, both a tag and the reader themselves may mutually authenticate each other using the tag's embedded keys. Tag T would encrypt an arbitrary nonce n with its private key within an additional encryption of the reader's public key $R_{pu}(T_{pr}(n))$ and send it to the reader. Once the reader has decoded the outer encryption using its private key R_{pr}, it will search its back-end database and obtain the tag's matching public key T_{pu}, decrypting the inner lock to retrieve the nonce from the tag. The reader would then reencrypt this nonce with its own private key, and similar to the tag, it would establish a second encryption around the previous one with the tag's public key, $T_{pu}(R_{pr}(n))$. Before establishing this encryption, the reader would use its keys in combination with the current tag to form a temporary key T_k, which would be sent along with the nonce to the tag. Upon retrieval and decryption of the message sent from the reader, the tag would use this T_k to send any further information between itself and the reader, such as a person's job title, to enable them access to a certain location of a building. Figure 5.1 illustrates such a protocol. For a more time-based protocol, any important information from the tag would be embedded within the first message sent, such as a product's EPC [2] for a supply chain marketing system. This approach prevents the unwanted listening of the transmitted messages from a tag unless a hacker was to retrieve the private key of that tag, which is very unlikely in a given short period of time.

Despite the security strength of this approach, public key cryptography (especially for execution within RFID tags) requires strong computational power to establish encryption upon the transferring messages. This not only increases the size of the tag to become quite large, but it also significantly increases the cost of each tag, causing this approach to be used within very limited sectors of RFID security.

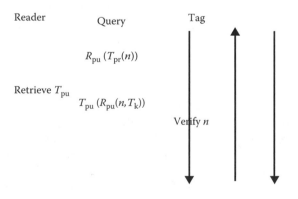

Figure 5.1   An example cryptography protocol.

5.3 RFID Privacy Devices

As explained earlier, the elongated broadcasting range of an RFID tag is susceptible to unwanted scanning (or reading) by a hacker or any malicious third party. This fact is more prevalent when dealing with the issue of privacy; whether it be consumer privacy (gathering information about previously purchased items) or something of a more secure nature (retrieving job or governmental information from a identification badge). To combat this issue, many of the earlier attempts suggested carrying or attaching an external device to protect a tag from unwanted access. The following three sections provide examples of such approaches.

5.3.1 Faraday's Cage

One of the first approaches in dealing with consumer privacy involves what is called a Faraday Cage [1]. A Faraday Cage is a container made of metal mesh or foil that is designed to block certain radio frequencies. In fact, this method of shielding an item with certain metal material is known to be used by thieves when trying to surpass shoplifting detection systems. Recently, the U.S. State Department has even indicated that U.S. passport covers will include metallic material to limit RF penetration, thus preventing long-range scanning of closed passports [5,6]. This approach, however, does come with its disadvantages. The main disadvantage is that the cage is not designed to fit around certain items, such as wrist-watches, containers, and bigger items such as televisions or computers. This disadvantage limits the use of this approach, restricting it from more commercial investments such as the supply chain market.

5.3.2 Active Jamming Device

Another approach dealing with consumer privacy is called the active jamming approach [1]. This approach will allow an individual to carry a device that would block nearby RFID readers by transmitting or broadcasting its own signals. However, this approach could be illegal if the broadcast signaling power of this device is too high. This might cause the jammer to interfere with surrounding legitimate RFID readers where privacy is not a concern, disrupting a company's business. Therefore, due to the legality of this approach, it is not a suitable solution for the privacy protection of RFID.

5.3.3 Blocker Tag

There have been multiple studies related to RFID security in the past few years and enumerating all of them is beyond the scope of this chapter. However, Juels has discussed a number of these techniques in detail and highlights the pros and cons of each method in Ref. [1]. In particular, we would like to discuss a technique of his that one of our protocols was inspired from—the privacy bit concept from Ref. [12].

The approach Juels uses in Ref. [12] is similar to that of the "jamming" approach described earlier; however, its effect is not as strong upon operation, cleverly interacting with the RFID "singulation" protocol to disrupt only certain operations. The singulation

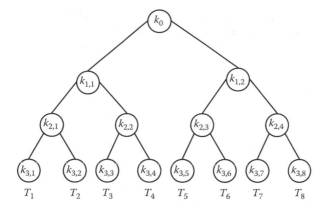

Figure 5.2   A singulation protocol tree.

protocol for RFID tags is a tree-based method upon which a reader may distinguish multiple tags scanned at the same time. This process works by repeatedly querying all present tags within the area, distinguishing each separate tag by keeping a count of the number of collisions a reader has received. For a more detailed description of this procedure, consider the example given in Figure 5.2. This figure represents a tree with a depth of 3 containing $2^3 = 8$ tag serial numbers represented at each of its leaves. Assume we want to distinguish between tags "001" and "011." A reader would first query for all tags with "0" prefix. This query would return a collision because both tags have this criterion. Upon querying for a "1" prefix, no signal would be returned; therefore, the reader would not continue to query for any tags beginning with that prefix. The singulation protocol would then follow the recursion upon the collision points of the tree as before to eventually reach the corresponding leaves "001" and "011," receiving only one response back denoting their presence.

To interact with this protocol, in Ref. [12], Juels uses a special bit in tags called a "privacy bit" that can take a value of 0 or 1 and can be easily toggled by a reader after authenticating with a unique pin for that tag. The tag bears a value of 0 when inside a store, indicating that it has public access and on checkout the tag is moved to a "private" zone by flipping the privacy bit to 1. But doing just this does not ensure security of the tag. An additional specialized tag called the "blocker" tag must accompany the tag in question to secure it [12]. The blocker tag confuses malicious readers into thinking that tags with all possible values are present with the bearer. The tag specifically achieves this by sending the corresponding value a reader is currently querying for within the singulation protocol. For example, if a reader is asking for a serial number beginning with "11," the blocker tag would send this value to collide with any tags that may actually contain this value, confusing the reader upon reaching the leaves of its tree. This "confusing act" can either be done to overwhelm the reader (full blocker) or in a "soft" or polite way (partial or selective blocker) [13]. A soft blocker, however, would only interact with leaves that are within a "privacy zone." For example, as described earlier, the privacy bit of a tag would change from 0 to 1 upon authentication of tag. The soft blocker would then only protect tags with a prefix of "1," allowing other items within the area to be scanned if necessary. Either way, the tag is secure only in the presence of the blocker tag.

This concept is specifically designed to promote consumer privacy; however, given the difference in signaling strength of each tag, they may hold a better place within different retail environments. A normal blocker tag may be used within cellular phones to disrupt malicious transmissions trying to attack or retain information from different calls or text messaging. A soft blocker would be more suited for the supply chain market, and thus would be embedded within a grocery or shopping bag. This temporary method of security provides the advantage of allowing "smart" machines as described earlier access to certain RFID-related items without any additional processing. However, similar to the previous two devices, this concept has its flaws as well. Even a well-positioned blocker tag has a chance of failing given the unreliable transmission of RFID tags [1]. Also, readers may eventually evolve in exploiting weaknesses to blocker tags and overtake their signal strength [14]. To fully understand the attacks and defenses upon this approach, research and evaluation will have to continue before any deployment is considered.

5.4 RFID Protocols Based on Hash Functions

Upon the failure of separate mechanisms devised to provide efficient authentication to RFID tags, researchers began developing schemes to provide such security within the tag themselves. A popular method among many protocols in achieving such security involved using an encryption-like method to secretly transfer messages between a reader and a tag, known as hash functions. A hash function is any mathematical function or well-defined method that rearranges any given data into a reasonably small integer, normally providing use as an index for an array [15]. Hash function algorithms such as SHA-1 [16] and MD5 [17] have been widely accepted as a secure form of protection for transferring data within a limited computational range. This section explores one of the original hash-based approaches developed upon the improvement of the searching time for a tag within a database.

5.4.1 Hash Lock: The Original Hash Function-Based Approach

One of the first protocols, which many were later developed from, is known as *hash lock* [18]. This hash function-based approach deals with unlocking a tag value through hash-based results to gather its secret information. The tag begins by starting from a "locked" state, where a reader sends a lock value to the tag, *lock = hash(key)*, where the key is a random value. This value is stored within the tag's reserved memory location (i.e., a Meta-ID value), wherefore the tag enters the locked state, not allowing any information about the tag other than what will need to be given during the authentication process. To unlock the tag, the reader must send the tag the original key that was used to make its

Meta-ID value. Upon receiving this value, the tag performs a hash function against the key and compares it to its Meta-ID value. If it matches, the tag unlocks itself, allowing its EPC [2] to be responded to readers upon forthcoming cycles of queries. This protocol is very simple and straightforward in providing security against the breach of secret information within the tag, that is, its EPC. Since authorized readers only know the original key value of each tag, only they are able to unlock the tag for its information, upon which they lock the tag again after reading the code.

Despite its simplicity, this protocol provides a heavy breach in security. It fails to provide mutual authentication between the tag and the reader, only authenticating the reader because it must provide the key value that should be unknown to the public. This breach can simply be exploited by a malicious third party scanning the tag for its Meta-ID value. This value in turn would be randomly broadcast to nearby readers where eventually the key value for that particular tag is returned. The party may then use this information to send to the original tag, obtaining its EPC and other sensitive information.

To combat the above-mentioned flaws, the same authors of the paper [18] devised a new approach, dealing with randomization. The emphasis here was to disguise the Meta-ID value with a random number for each query, thus the tag and its value could not easily be traced. Therefore, an additional pseudorandom number generator is embedded into the tag for this approach. Similar to the original hash lock protocol, a tag will prestore a key value (known as an ID) to place itself in a locked state. However, this approach will not store the hashed result of this ID, but rather upon each query, the tag will hash its ID with a random value given by its pseudorandom generator, resulting in hash(ID_k, r), where k represents the kth tag among a number of tags within the system, ID_1, ID_2, ..., ID_k, ..., ID_n.

Upon the reader's query of a tag, the reader will obtain two values. These values include the random number generated by the tag and the hash function result value generated by the hash against the ID value of the tag and its current random number at that time. To unlock the tag, a reader must send the tag its original ID value. Therefore, the reader will begin to search its back-end database containing all ID values for all tags, where it must repeatedly perform a hash function against each separate value with the random number given from the tag. This will allow the reader to compare each hashed result against the one sent from the tag, wherein if they match, the reader will obtain the matching kth ID value, returning it to the tag to unlock itself. Once the tag is in an unlocked state, any reader may perform a query to gain the tag's EPC information.

In addition to successfully achieving security on RFID tags, this approach also provides location privacy. In the previously developed protocol Hash Lock, each tag still reveals its Meta-ID. However, this approach only discloses a random number and a hashed value based on that number. Therefore, a malicious third party cannot trace a specific tag (i.e., a product of a store) based from its Meta-ID. In this case, the randomized hash lock protocol is able to provide location privacy.

Despite the vast improvements that this randomized protocol presents over the original hash lock scheme, this approach may not be suitable for all cases. Due to the fact that the reader must search through as many ID values as possible to find the matching hash result, its running time is that of $O(n)$, for n tags within the system. Therefore, this approach would have a vast scalability problem when the number of tags for a system increases enormously. The additional cost of producing a pseudorandom generator for a tag also presents another problem for a system of that scale.

5.4.2 Tree-Based Approaches

As explained earlier, the randomized hash-based approach in Ref. [18] fails to maintain a less than optimal running time and low-cost tags against a system whose number of tags increases expeditiously. However, as this type of system's popularity continues to expand over time, the search to find an efficient algorithm grows larger. In an effort to decrease the running time of a hash-based function while maintaining its security, authors of Refs. [19–21] developed an approach which improves the key search efficiency from linear complexity to logarithmic complexity. The key in achieving this is based upon the structure of the back-end database, which is set up as a tree-based graph. Every node will then require $O(\log N)$ search time, as proven in multiple mathematical theories. One of the first groups to use a tree-based approach was Molnar et al. [21], who in their approach used a challenge-response protocol. This procedure required multiple rounds to identify a tag, each round consisting of three messages between both parties. Due to the tree-based nature of the back-end database, however, this caused each round to have a $O(\log N)$ running time, incurring relatively large communication overhead between each message. Therefore, another paper [19] made an improvement upon this algorithm that shortened the length of a round to only one message from the tag, providing no further interaction between the tag and the reader. To further explain the concept of tree-based protocol, a detailed explanation of the procedure in Ref. [19] is provided below.

The back-end database for a reader in a tree-based approach consists of a set of keys formed in a binary tree manner. Consider the tree in Figure 5.3. Each node contains a distinct key and each tag denotes a leaf node. Therefore, there exists a unique path of keys from the root to each leaf node, whereas these set of keys are assigned to each leaf

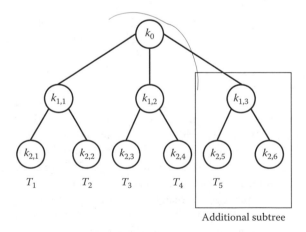

Additional subtree

Figure 5.3 **A static binary tree with eight tags.**

node, which is used for authentication. For example, tag T_8 contains keys k_0, $k_{1,2}$, $k_{2,4}$, and $k_{3,8}$. When the reader R authenticates T_8, it first sends a nonce n to the tag. The tag T_8 then encrypts its set of keys with n by executing a hash on both values and sends the result back to the reader. The reader then searches for the leaf node corresponding to the tag by repeatedly hashing the sent nonce n with its binary tree of keys and matching those results to the sent list of encrypted results from the reader. If such a path to a leaf node exists, T_8 is identified and reader R regards this tag as valid.

From the above procedure, we can tell that the path to a tag will end up sharing certain keys with other similar paths to different tags. For example, tags T_8 and T_7 share key $k_{2,4}$, and of course all tags will share the root key, k_0. An advantage in using this static architecture is that its running time is logarithmic. For example, in Figure 5.3, any identification of a tag only needs $\log_2(8) = 3$ search steps. This structure, however, also presents a security flaw in that if a tag is compromised, an adversary would obtain multiple paths from the root node to a leaf node, including the keys on those paths. Since the keys of the architecture are never updated, the captured keys will still be used by the uncompromised tags, allowing further knowledge of what key combinations an uncompromised tag may contain.

A practical solution to solving the above-mentioned problem is to update the keys after every authentication; however, the aforementioned static tree architecture cannot handle such a task due to its complexity. For example, if one were to update tag T_1, they would have to change keys k_0, $k_{1,1}$, $k_{2,1}$, and $k_{3,1}$ partially or totally. This in turn would single-handedly affect every path of tree, not allowing the other nonauthenticated tags to update themselves to authenticate correctly with a reader. Therefore, using this architecture, every tag would need to be updated periodically and simultaneously (along with static tree used by the reader) to continue the synchronization of the approach. Unfortunately, this solution is not practical in large-scale systems with hundreds or millions of tags. An alternative solution would be to collect only those tags affected by each authentication of another tag's path periodically; however, this idea would be more cumbersome than the first in trying to collect a number of tags only affected by one tag changing's its keys. Therefore, we have developed our own protocol called "HashTree," a dynamic key-updating algorithm for private authentication in RFID systems. Our protocol is explained in Section 5.4.3.

5.4.3 HashTree: A Dynamic Key-Updating Approach

Though tree-based approaches have an efficient search time, they lack in the long-term security guarantee. Due to the infrastructure of a tree-based approach, sets of keys in the path between two or more tags are commonly shared. Consequently, if one tag compromised, this could lead to the leaking of information about other tags within the system. To resolve this issue, the keys of both the reader and the tag must be simultaneously updated. To our knowledge, Ref. [22] is the first paper to discover this flaw and develop its own dynamic key-updating approach to this problem. In Ref. [22], the back-end database for a tree contains a set of keys k in a similar binary tree structure from that of Ref. [19]. To update the tree, however, each node contains an additional temporary key tk. Initially, each temporary key equals the value of the current key for each node, $tk = k$. The authentication procedure for a tag is essentially the same as a normal tree-based approach, where the tree is repeatedly searched for a path leading to a matching leaf node that represents a tag. Upon authentication of a tag, the reader performs a hash function upon k and sets its results as its current key value, $k = h(k)$. The temporary key is then always updated to the current key's value beforehand. This is to insure that every search for a key's path to its leaf node is correctly calculated, because the protocol will check both the current key value k and the temporary tk value for a node during authentication of a tag. Though this protocol prevents many of the problems with Ref. [19], it is only suitable for systems that plan on never scanning the same tag twice. This is due to the value of the temporary key only being capable of holding one key value at a time. If a tag were to be scanned multiple times, the temporary tk value of a node would only hold the ith -1 current key value of a node of i scans. This would not only allow paths sharing nodes with this tag's path to be wrongfully unauthorized, but would eventually misauthorize the repeatedly scanning tag as well. Therefore, we have developed the "HashTree" protocol, which provides a simple dynamic key-updating system for tree-based approaches. This approach simultaneously updates the keys of a reader and tag upon each authentication, while still allowing nonpreviously queried tags to continually authenticate themselves with ease.

Our HashTree protocol is comprised of three components: system initialization, tag identification, and system maintenance. The first two components are very similar to the static tree-based approaches presented in Ref. [19] and performs the basic identification functions. However, unlike the previously described dynamic-key approach [22], our protocol secures the RFID system from the Compromising attack without actually updating a reader's back-end database. Lastly, the third component is used to direct the joining and disjoining of tags to and from the system.

Our protocol begins by providing a similar arrangement for a reader and a tag of that from [19] paper. The back-end database of a reader contains the similar binary tree-like structure that both [19] and [22] provide. Assuming that there are N tags T_i (where $1 \leq i \leq N$) and a reader R in the RFID system, reader R will assign the N tags to N leaf nodes of the balanced binary tree S. Each node in the tree S will be assigned a preset key k. Upon the introduction of a tag T_i, the reader will distribute a path of k keys to a tag denoting an unassigned leaf node within the tree. However unlike previous protocols, these keys will not be stored in their natural state within the tag. Instead, an array of hashed results from a randomly generated nonce n by the reader against each k key from the original given set of keys will be stored in tag T_i along with the n value used to generate those results. Storing the key values this way is very important to the logic of the algorithm, which will be explained shortly.

As stated earlier, our authentication process for a tag is very similar to that of the protocol from Ref. [19]. However, the key difference is in how each tag has a precalculated set of hash results instead of storing the exact keys from the tree in a reader's back-end database. Upon query from a reader, the tag will send its array of hash results including the nonce n used to calculate them as described in the earlier paragraph. The reader will then take the n value and hash it against every k value of a node starting from the root node in a repetitive logarithmic manner until it has reached the matching leaf node corresponding to that tag. Assuming that a path of nodes containing a matching hash result for the array of results sent is traversed and found, then tag is authenticated. After authentication of the tag, each tag will be updated to prevent the Compromising attack described earlier. To achieve this, the reader will generate another random nonce n_2 different from the one given by the tag. This nonce will be then be used by the reader to perform another set of hash functions in accordance with the tag's T_i keys as recorded in the reader key tree. For a clearer view of what's being transmitted, refer to tag T_3 in Figure 5.3. Using the nonce value n_2, the sequence of hash function results against the original set of keys for tag T_3 would be $(h(k_0, n_2), h(k_{1,1}, n_2), h(k_{2,2}, n_2), h(k_{3,3}, n_2))$. This new list of hash function results and the nonce n_2 generated will then be sent to the tag, thereby updating its key values.

Before explaining the system maintenance portion of this section, we explain the theory behind only updating the tag's values and not the reader's. The key behind paper [22] is that it assumes an attacker may gain access to the secret values of a tag, therefore compromising a reader's back-end database key tree system. This is due to each tag's set of keys having a direct correlation with another tag's set of keys, as earlier described within the description of the static tree paper [19] in Section 5.4.2. Our algorithm handles this type of compromise attack in several ways. First, by updating a tag's set of key values, it can no longer be traced to other tags held within a secure holding area. For example, if this protocol were used for supply chain management within a retail store, all tags bought from the customer could not be traced to items held within the store. More importantly, the process in which we assign the tag initially and after authentication (i.e., using hash results instead of original key data) blocks the attacker from knowing the actual key values held within the key tree of the back-end database connected to a reader. Because of this, key locations from one tag to another (even tags that originally share a key path) are not associated with one another. This is one of the main reasons we do not have to update the key tree of the reader. To explain this further, consider tags T_1 and T_2 in Figure 5.3. Both tags in this tree structure share the key $k_{2,1}$. In a static tree, compromising one of these tags would compromise the secret of the other because each key is given to the tag in its normal form. However, our algorithm produces a different hash result of this value for both tags because each tag would have a different nonce (i.e., $h(n_1, k_{2,1}) \neq h(n_2, k_{2,1})$); therefore, the attacker would have no way of relating these two values together.

Lastly, in a real-time execution of this protocol, users may need to simultaneously remove and add additional tags to the key tree; therefore, we provide a series of small steps to perform such tasks. Assume an additional tag T_i needed to be added to the system. This task may be performed in one of two ways. The reader will first search through its list of leaf nodes; if an empty space is available, the tag's information will be placed there. However, if all the leaf nodes are currently pointing to a tag, a new subtree will have to be established. This subtree will be of length d corresponding to the depth of the tree that currently exist in the back-end database. Each node of this subtree will be assigned a randomly generated key and the tag T_i will be assigned these values with a random nonce as described in the system initialization portion of the protocol as before. An example of the addition process where an empty node could not be located is shown in Figure 5.4.

Consequently, one may need to withdraw tags from the system as well. This simple task is done by emptying the leaf node associated to tag T_i. This allows for an additional tag to be associated with that node as described in the previous paragraph.

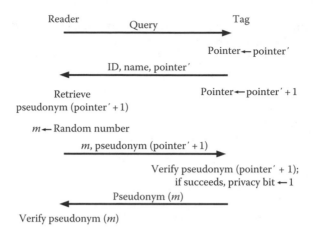

Figure 5.4   Example of an additional subtree.

A secure authentication protocol should meet the following security requirements [19]: privacy, untraceability, cloning resistance, forward security, and compromising resistance. Specifically, the untraceability requirement deals with the notion that a tag's output should be able to correlate itself with a tag; otherwise, this tag may be traced by attackers. From looking at our protocol, some may argue that even though each tag contains a different nonce *n* value, the output is the same upon multiple queries. However, the updating of each tag's values upon authentication prevents this attack from occurring. As earlier stated, this protocol is made general enough for a system to interact with tag such an identification badge, for example. Therefore, assuming this badge is used everyday to access a building, it would be difficult for an attacker to keep track of this type of tag unless the person possessing it is being physically followed throughout the day. Another issue one may bring is that of privacy. There is no outlining protection upon an attacker gaining the current values from the tag and performing a replay attack. If this presents itself as an issue for a system, then each tag may simply use a Faraday Cage as explained in Section 5.3 to protect it until it is ready for use, similar to many new electronic passports providing the same covering for extra security [5,6]. With this modification, our protocol successfully fulfills each security requirement for a tag, thus making it cost effective and secure enough for deployment within a real-time RFID passive tag authentication system.

5.5   Other RFID Authentication and Privacy Protocols

While many protocols were developed from earlier approaches such as most tree-based approaches (including our "HashTree" protocol), other researchers decided to take another route. The next protocol you will read about is developed by Juels [23] and involves the enablement of a passive tag to relabel itself, like many of its active counterparts. The idea behind this protocol combined with Juels' earlier soft-blocking [13] protocol makes another one of our protocols and the last one of the chapter, "RFID-Guard: An Authentication and Privacy Protocol Designed for Passive RFID Tags."

5.5.1 Minimalist Cryptography

Many active tags contain the ability to relabel themselves in a fashion that is indistinguishable to third-party malicious attackers, but may still be authorized by certified readers. Since passive tags to do not contain enough power to withhold this ability, many researchers began developing additional items to somehow block or distort the transmission of the tag from hackers, as described in Section 5.3. However, Juels has developed a protocol called the "minimalist" system [23], allowing the relabeling of a passive tag within its limited computational capabilities. Within this system, each tag contains a small set of pseudonyms; a different pseudonym is then given to a reader upon query because each tag will rotate through its list of pseudonyms per each scan. The security holds in that only an authorized reader will contain the entire set of pseudonyms. An unauthorized reader does not contain each pseudonym for a tag, and thus will not be able to gain any secure information from the tag given its different appearances. A more thorough example of this scheme is provided below.

As previously stated, each tag contains an array of pseudonyms α_i, where i denotes the current pseudonym of a tag for m pseudonyms, $1 \leq i \leq m$. However, the protocol would not be secure from these pseudonyms alone. If that were the case, the minimalist system would be vulnerable to the cloning attack, an attack of which many static designed protocols for RFID tags suffer from [24]. In this case, an attacker would query the tag, thus obtaining its current pseudonym α_i and replay this value to the reader, allowing itself to be recognized as the currently scanned tag. To combat this issue, a tag only verifies itself to a reader after the reader has authenticated itself to the tag. To accomplish this, for every pseudonym α_i, each tag contains two additional key values, β_i and γ_i, for which the tag and reader will authenticate themselves with. Upon query by a reader, a tag will respond with its current α_i pseudonym value. Assuming the value is valid, the reader will retrieve the tag's corresponding β_i and γ_i values from its back-end database and send the tag β_i. Upon authentication of the reader, the tag will respond with its corresponding γ_i value. As you can see, this protocol mimics that of a simple challenge-response protocol, but one that is designed upon pseudonym rotation.

To successfully achieve long-term security for a tag, one must continually update its α_i, β_i, and γ_i values. Logically, this update method would need to occur after each mutual authentication between a tag and a reader to maintain the low cost of not periodically updating a large amount of tags at once. However, updating the tags also presents a new problem: an attacker may still eavesdrop or tamper with the updating process. To address this issue, Juels proposed using one-time pads that have been used across multiple authentication protocols to update the current values of a tag and a reader. Using this method, a malicious third party who only eavesdrops periodically will be unlikely to gain the updated α_i, β_i, and γ_i values.

One-time pads [25] may be thought of as a simpler form of encryption than that used within cryptography; thus, they are able to be used within tags that have smaller computational capabilities (i.e., passive tags). One-time pads are essentially a random bit string of length l. If two parties then share a secret one-time pad δ, it has been proven that a message M may be sent secretly via cipher text $M \oplus \delta$ between both parties, where $\oplus$ denotes the XOR operation. Thus, after mutual authentication between both the tag and the reader, the reader uses these one-time pads to update the α_i, β_i, and γ_i values noted earlier and sends these pads to the tag so that it may update its values as well. Provided a malicious third party does not eavesdrop upon a reader transmitting the message and obtain these pads, they achieve no knowledge of the newly updated tag values. However, in case a third party were to obtain one of the pads, this scheme also has an additional spin on what is considered to be the normal use of a one-time pad. This involves using the one-time encryption across multiple authentication sessions. To achieve this, pads from two different authentication sessions are XORed with a given tag value w to update it, where $w \in \alpha_i \cup \beta_i \cup \gamma_i$. Therefore, even if a third party successfully obtains a pad used in a prior session, it is seen that they will still not be able to obtain any information about the updated values of w.

The minimalist approach offers resistance against spying upon corporate businesses, such as the clandestine scanning of product stocks in a supply chain market. Since our first developed protocol was made for retail environments, we borrowed this idea of pseudonym rotation along with Juels soft-blocking technique [13] to develop what is known as "RFIDGuard."

5.5.2 RFIDGuard: An Authentication and Privacy Protocol Designed for Passive RFID Tags

Large organizations such as Wal-Mart, Procter & Gamble, and the U.S. Department of Defense have generated a lot of attention for RFID technology in recent years due to their need of deploying RFID as a tool for automated oversight of their supply chains [1]. This is due to a significant advantage that RFID tags have over standard barcodes; they are able to uniquely identify individual items of a product type, unlike barcodes, which only identify each product type. However, the elongated reading range of an RFID tag enables third party users to eavesdrop upon transmissions between authenticated readers and tags, thus enabling a security breach. RFID security within supply chain management was our original intent for studying and creating different protocols for passive RFID tags, and thus led to the creation of our first developed protocol, "RFIDGuard: An Authentication and Privacy Protocol Designed for Passive RFID Tags."

The idea behind RFIDGuard is comprised of a modified version of two previously developed protocols developed by Juels, the "minimalist" system [23] and the soft-blocking technique [13] as discussed earlier in this chapter. The minimalist approach suggests each tag contains an array of pseudonyms, such that upon each query from a reader, the tag will rotate through its list of pseudonyms, providing protection from any malicious third parties because they will not have the entire list themselves. To establish mutual authentication between a reader and a tag, our protocol uses this pseudonym rotation concept; however, unlike the minimalist approach, it will not need to update the tag's pseudonyms or contain additional keys to perform this task. The original soft-blocking approach interacts with the singulation protocol used to distort multiple RFID tags scanned at the same time by blocking part of the tree using what's known as a "privacy bit." Our protocol establishes privacy to RFID tags using a privacy bit, but instead of interacting with the singulation protocol, the privacy bit toggles the tags between locked and unlocked states. During a locked state, tags will only provide enough information to an authenticated reader to unlock it where it will then supply its EPC [2] to the reader, whereas an unauthorized reader would not contain enough information to retrieve the secret data. The rest of this section provides a more detailed look at our protocol and its security measures.

The RFIDGuard protocol is comprised of four smaller protocols: in-store, checkout, out-store, and return. Each protocol represents a presumed location of a tag, thus providing it with different amounts of security. Each tag contains several different items: a list of pseudonyms, a pointer representing what pseudonym the tag is currently pointing to, a number representing a generic *name* of the product, a privacy bit contained as the first bit in a tag's EPC, and any secret information a tag may need to send to a reader upon authentication including the aforementioned EPC (referred to as ID in this protocol). Each tag is assumed to have no further interaction when it is made until it has reached the retail environment and therefore it will start in the in-store/checkout protocol. This is denoted by a tag's privacy bit starting with 0 to denote the tag in an "unlocked" state. The in-store protocol is simply the first two steps within the checkout protocol (as is the out-store protocol the first two steps of the return protocol), so we will only refer to them as such.

Upon query of a tag within the checkout protocol, a tag will make a copy of its current pointer sending it, its name, and its ID to the tag, while increasing its pointer an additional time after transferring the message. How the pointer is increased upon every query will explained in the following protocol. As stated earlier, the tag is assumed to have no further interaction until it has arrived to the store; therefore, any query beyond that point while still be in the unlocked state is assumed to be given by an authenticated reader, and therefore will not need additional security before giving out its *ID*. Upon receiving the tag's message, the reader will use the tag's generic name and retrieve the pseudonym after the location of the pointer given, pseudonym[pointer+1]. Only an authenticated reader will know that you should return the pseudonym after the pointer given and not its current one; thus upon verification by the tag, the tag will flip its privacy bit to 1, denoting it to be within the "locked" state and the tag will follow the return protocol until it has been unlocked again. Before this occurs, however, the reader also sends an additional random number m denoting a pointer location within the tag where $1 \leq m \leq n$ for n pseudonyms in a tag. After verification of the reader, the tag will also send pseudonym[m], the pseudonym according to the location of the tag. This allows the protocol to combat against cloned tags. Some cloned tags may contain valid pseudonyms of an authentic tag but do not contain all of them or have them placed in the incorrect order. If this is true, the query for a random pseudonym will have a great chance of catching this, thus if the pseudonym[m] is not valid or the tag does not send one in a designated timing period, the reader will alarm the store that the consumer is carrying a corrupt or cloned tag. A pictorial example of this tag is given within Figure 5.5.

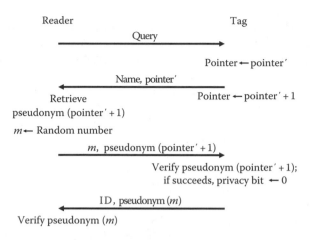

Figure 5.5   The checkout protocol of RFIDGuard.

As earlier stated, after the tag has completed the checkout protocol, it will enter the locked state by setting its privacy bit to 1 where it will the follow the return protocol procedures until it is unlocked. In this locked state, the tag is assumed to be present to malicious third-party readers and will not give information as easily as it has in the previous protocol. Therefore upon query, the tag will repeat the same process with its pointer but only its copy and its name, not its ID. How the pointer is increased upon every query is more prevalent in this subprotocol than in the previous one. Any hacker trying to repeatedly scan a tag for its information will trigger the tag to continue to switch its current pointer. Therefore, if the hacker were to retrieve one or two pseudonyms from the reader within the checkout protocol, it would confuse the hacker as to which position they should be in. This would enable the reader to better catch a cloned tag made from the stolen pseudonym and not verify the tag. Returning to the subject of the return protocol, a valid reader would take the information sent from the tag and repeat the process of sending the tag a random pointer m and the pseudonym[pointer+1] to verify itself. Upon verification of the reader, the tag will not only send pseudonym[m] but its ID as well. The purpose of sending pseudonym[m] here is not to check for clones, but to check for corrupted tags. A valid reader would only be able to pick up a tag within a store for this protocol if a consumer is returning an item; thus, one may corrupt the tag to increase its original value worth, enabling the consumer to receive a larger cash amount or exchange the item for a higher priced one. Thus this randomized pseudonym feature would try and catch corrupted tags before the consumer is able to steal money from the business. A pictorial example of this tag is given within Figure 5.6.

This protocol achieves many of the security requirements stated within [19] as listed earlier in the paper; however, it may not be clear as to why certain items within the tag are disclosed within certain parts of the protocol. By looking at checkout protocol, one would wonder why the *ID* is released before validating the reader or why multiple pseudonyms are given on both subprotocols. The reason for both of these questions are because we

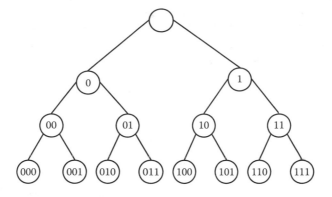

Figure 5.6   The return protocol of RFIDGuard.

make an assumption of both protocols (at least ones with authenticated readers) being executed within a physical building. Therefore, given the limited broadcasting range of a passive RFID tag, it would be very unlikely for someone to eavesdrop upon the transmission of messages during the execution of these protocols for an extended period of time. Furthermore, because our protocol is contained to a physical building, we assume that a retail store has some security mechanisms that prevent unauthorized readers from entering the store. This can be easily achieved by installing detection devices near the entrance of the store to detect unauthorized readers [26].

Another security issue that needs to be discussed within this protocol deals with the requirement of preventing a tracking attack. As stated earlier in discussing the HashTree protocol, tracking involves knowing the whereabouts of a tag by continually receiving a static value returned from it. This issue deals with the generic name used to search for tag's information within the protocol. No additional device method is needed to combat this issue, but an understanding of what the generic name value represents is needed. This *name* stands for a product type and not every individual product within the store. For example, two bottles of soda may have the same generic name as long they are made by the same company. If the protocol were not setup this way, the back-end database of the system would exponentially increase in cost, both physically and memory wise, using extra slots that may never need to be returned such every individual bottle of soda.

Upon explanation of the above concerned security requirements, RFIDGuard has the potential of becoming the standard retail environment protocol with further research.

5.6   Conclusion

RFID is a promising technology that can revolutionize the way we lead our lives. However, before this becomes a reality, certain security issues like consumer privacy protection, fraud prevention, and detection must be addressed. This chapter covers only a minimum amount of protocols used within RFID technology to establish the current security that protects this field. Before RFID officially overtakes the use of barcodes,

extensive research must continue to ensure an efficient speed and protection of the individual tags. However, it is still believed that not every system using passive RFID tags will encase the same approach, but rather a number of security protocols will become the standard by EPC regulations for each separate RFID system. There is an ample scope in the field of RFID security to improve and innovate to allow RFID technology to be incorporated into our daily lives.

References

1. A. Juels, RFID security and privacy: A research survey, *IEEE Journals on Selected Areas in Communications*, 24(2): 381–394, 2006.
2. EPCglobal. Epcglobal website. http://www.EPCglobalinc.org/, 2007.
3. S. Stern, Security trumps privacy, Christian Science Monitor, 2001.
4. D. Molnar and D. Wagner, Privacy and security in library RFID: Issues, practices, and architectures, in B. Pfitzmann and P. McDaniel, eds., *Proc. ACM Conf. Commun. Comput. Security*, pp. 210–219, Washington, DC, 2004.
5. International Civil Aviation Organization ICAO, Document 9303, Machine readable travel documents (MRTD), Part I, Machine readable passports, 2005.
6. A. Juels, D. Molnar, and D. Wagner, Security and privacy issues in e-passports, in *Proceedings of the First International Conference on Security and Privacy for Emerging Areas in Communications Networks (SecureComm)*, pp. 74–88, Athens, Greece, September 2005.
7. G. Barber, E. Tsibertzopoulos, and H. B.A., An analysis of using epcglobal class-1 generation-2 RFID technology for wireless asset management, in *Military Communications Conference*, vol. 1, pp. 245–251, Atlantic City, NJ, October 2005.
8. S.E. Sarma, S.A. Weis, and D.W. Engels, RFID systems and security and privacy implications, in *Workshop on Cryptographic Hardware and Embedded Systems (CHES) 2002, LNCS no. 2523*, pp. 454–469, Redwood Shores, CA, 2003.
9. M. Ohkubo, K. Suzuki, and S. Kinoshita, Cryptographic approach to privacy-friendly tags, in *RFID Privacy Workshop*, MIT, Cambridge, MA, November 2003.
10. S. Bono, M. Green, A. Stubblefield, A. Juels, A. Rubin, and M. Szydlo, Security analysis of a cryptographically-enabled RFID device, in *USENIX Security Symposium*, pp. 1–16, Baltimore, MD, July–August 2005, USENIX.
11. J. Wolkerstorfer, Is elliptic-curve cryptography suitable to secure RFID tags?, in *Hand-out of the Ecrypt Workshop on RFID and Lightweight Crypto*, Graz, Austria, July 2005.
12. A. Juels, R.L. Rivest, and M. Szydlo, The blocker tag: Selective blocking of RFID tags for consumer privacy, in *Proceedings of the 10th ACM Conference on Computer and Communication Security*, pp. 103–111, Washington, DC, 2003.
13. A. Juels and J. Brainard, Soft blocking: Flexible blocker tags on the cheap, *Proceedings of the 2004 ACM Workshop on Privacy in the Electronic Society*, pp. 1–7, Washington, DC, 2004.

14. M. Rieback, B. Crispo, and A. Tanenbaum, RFID Guardian: A battery-powered mobile device for RFID privacy management, in C. Boyd and J. M. Gonzlez Nieto, eds., *Proceedings of the Australasian Conference on Information Security and Privacy*, Springer-Verlag, New York, 2005, vol. 3574, *Lecture Notes in Computer Science*, Brisbane, Australia, pp. 184–194.

15. I. Mironov, Hash functions: Theory, attacks, and applications, Microsoft Research, Silicon Valley Campus, November 2005.

16. National Institute of Standards and Technology, Secure hash standard, Federal Information Processing Standards Publications (FIPS PUBS), April 1995.

17. R. Rivest, The MD5 message-digest algorithm, MIT Laboratory for Computer Science and RSA Data Security, Inc., April 1992.

18. D. Johnson, C. Perkins, and J. Arkko, Mobility support in IPv6, RFC 3775, IETF, June 2004.

19. T. Dimitriou, A secure and efficient rfid protocol that could make big brother (partially) obsolete, in *PERCOM 06: Proceedings of the Fourth Annual IEEE International Conference on Pervasive Computing and Communications*, pp. 269–275, Washington, DC, 2006. IEEE Computer Society.

20. D. Molnar, A. Soppera, and D. Wagner, A scalable, delegatable pseudonym protocol enabling ownership transfer of rfid tags, Cryptology ePrint Archive, Report 2005/315, 2005. http://eprint.iacr.org/.

21. D. Molnar and D. Wagner, Privacy and security in library rfid: issues, practices, and architectures, in *CCS 04: Proceedings of the 11th ACM Conference on Computer and Communications Security*, pp. 210–219, New York, 2004. ACM.

22. L. Lu, J. Han, L. Hu, Y. Liu, and L.M. Ni, Dynamic key-updating: Privacy-preserving authentication for RFID systems, in *PERCOM 07: Proceedings of the Fifth IEEE International Conference on Pervasive Computing and Communications*, pp. 13–22, Washington, DC, 2007. IEEE Computer Society.

23. A. Juels, Minimalist cryptography for low-cost rfid tags, in *Proceedings of the 4th International Conference on Security in Communication Networks*, vol. 3352, pp. 149–164, 2004.

24. S.E. Sarma, Towards the five-cent tag, Technical Report MIT-AUTOID-WH-006, Auto-ID Labs, 2001. http://www.autoidlabs.org/.

25. A.J. Menezes, P.C. van Oorschot, and S.A. Vanstone, *Handbook of Applied Cryptography*, CRC Press, Boca Raton, FL, 1996.

26. T. Li and R. Deng, Vulnerability analysis of emap-an efficient RFID mutual authentication protocol, in *International Conference on Availability, Reliability and Security*, Vienna, Austria, 2007.

Chapter 6

RFID Security

Denis Trček and Pekka Jäppinen

CONTENTS

We are witnessing a strong proliferation of ubiquitous and pervasive computing, where devices with weak computing resources are playing an increasingly important role. A large number of these devices are various sensors, most notably radio-frequency identification

(RFID) tags. Emerging situations stimulate the permanent need for lightweight protocols to enable and preserve security in such environments, which is certainly not an easy task. This chapter, therefore, provides an extensive overview of the field of RFID security, starting with basic definitions and a reference scenario, and continuing with appropriate metric that enables quantitative evaluation of RFID security protocols. Next, the main solutions in this field are given together with the identification of their weaknesses (some of them have not been addressed so far). Based on this, new nondeterministic (ND) cryptographic protocols are presented that are designed for provision of security in RFID environments. Finally, an outlook of this area is given with a description of open issues and expected trends in the near future.

6.1 Introduction

Current computing trends are shifting toward wireless communications that enable ubiquitous computing paradigms. Within such environments a significant number of devices will be devices with limited computing resources, be it processing power, available storage, or power supply. The main representative among them will be radio-frequency identification (RFID) tags (according to Gartner Group their expected market share is to reach 3 billion U.S. dollars by the year 2010 [1], especially due to their wide use in retail [2]).

RFID devices are soon expected to be among the most numerous communication devices in ubiquitous computing environments with application areas ranging from retail to health-care systems. Therefore security is becoming increasingly important, not only from the users' perspective and expectations, but also from a legislative viewpoint. Because RFIDs are weak with resources, this means a significant challenge for assurance of security in RFID-based computing environments—many stringent security requirements have to be met by solutions for devices that lack processing power, memory, and communication capabilities.

This chapter presents an extensive study of RFID security. In Section 6.2, the basic definitions and an appropriate reference scenario are given. In Section 6.3, the current status of the field is presented, where existing protocols with known weaknesses are described. In addition, some new weaknesses of existing protocols are described. Further, a metric for quantitative evaluation of those protocols that are intended for RFID systems is presented in this section. Based on these lessons, two new protocols are presented in Section 6.4. These new lightweight protocols are nondeterministic (ND) and well suited for provision of security in RFID environments (their brief analysis is given in the same section as well). Section 6.5 gives an outlook about the most likely future research scenarios in RFID security area, while conclusions are drawn in Section 6.6. This chapter ends with references.

6.2 Basic Definitions and the Reference Scenario

RFID systems consist of a front-end part, where RFID tags are the main component, and the back-end part that includes a reader, communication links to a database, and

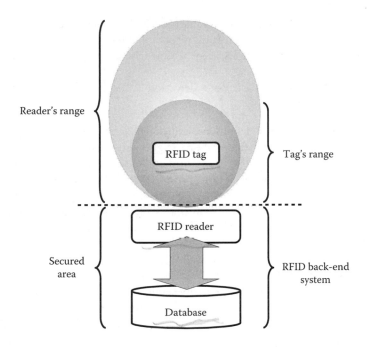

Figure 6.1   RFID reference environment scenario.

the database itself [3]. The borderline presents a reader, which is assumed to belong to the back-end part of the system. Further, it is often assumed that the back-end part is secure and that it is the front-end part that has to be treated from security point of view, which will also be the case in this chapter. The situation is shown in Figure 6.1.

An RFID tag consists of a microchip with encoded identification (ID), and an antenna. Communication between a tag and a reader takes place on radio frequencies by electromagnetic coupling. A reader induces a voltage in the tags circuitry that provides sufficient power for a tag to perform the necessary calculations and to respond. Such functioning is characteristic for so-called passive tags. However, tags can also have power autonomy (provided by a battery), and these tags are called active tags. Passive tags are cheaper; they have an operating perimeter up to 3 m, and a relatively high error rate. On the other hand, active tags are more expensive, but have an operating perimeter up to a few hundred meters with a lower error rate. Both kinds of tags can be read only; write once, read many; or rewritable. Because of cost constraints, the majority of tags on the market are passive tags and they will be the focus of this chapter.

Now according to Ref. [4], security means minimization of vulnerabilities of assets and resources. It is achieved by deployment of security mechanisms that include crypto-graphic primitives, and logical and physical mechanisms, which serve for implementation of the following security services:

- Authentication that ensures that the peer communicating entity is the one claimed
- Confidentiality that prevents unauthorized disclosure of data

- Integrity that ensures that any modification of data is detected
- Access control that prevents unauthorized use of resources
- Nonrepudiation that provides proof against false denying of the message content by its creator
- Logging and auditing that enables detection of suspicious activities, and analysis of successful breaches

With regard to the above definitions, additional clarifications have to be given. In case of a passive tag, authentication is always triggered by some outer source. This kind of authentication will be referred to as enforced authentication. Enforced authentication is very important, because it can lead to privacy threats—depending on a context, of course. For example, as long as an item is on a store shelf, the tag communicates freely its ID to nearby devices. However, as soon as a user buys the tagged item, the possibility of privacy breach emerges—all that has to be done is to link the identity of the user with the tag at the cashier. Afterwards, whenever enforced authentication by an unauthorized reader takes place, privacy can be broken.

Although this chapter does not cover privacy, the above explanation is necessary to understand the need for appropriate security services, because current use of security in the area of RFID systems is mainly oriented toward protection of privacy. However, use of RFID systems is expanding and new security issues are emerging that will also be discussed in the rest of this chapter.

6.3 Current Status of the Field

This section first gives an overview of cryptographic primitives-related issues (security mechanisms), followed by cryptographic protocols-related issues (security services), and security metrics issues.

6.3.1 Overview of Cryptographic Primitives Issues

The main barrier to security implementations is price. Recent reference RFID implementation was expected to have the following characteristics [5]. It was passively powered and had 96 bits of read-only memory that carried the tags identity (ID), which was unique for each tag. Chip operated at speeds providing 200 read operations per second. It was estimated that a maximum of 2000 gates could be allocated to security within the economically acceptable range. Taking into account Moore's law (and being a bit conservative), the upper limit is now approaching 4000–5000 gates.

This puts stringent requirements on security protocols, which have to be lightweight. Although many protocols in the literature are claimed to be lightweight, they are based on many hidden assumptions that do not take into account the additional gates that are needed for implementation. For example, they assume that use of one-way hash functions automatically qualifies a protocol to be lightweight. But this does not hold true for the majority of functions like the MD-x or SHA-x family [6]. Further, each additional step in a protocol often requires additional dedicated circuitry; steps in security protocols are semantically related and, as a consequence, each additional step results in a more complex algorithm that has to be implemented at the RFID tag.

Therefore only particular cryptoprimitives can be chosen, most notably lightweight AES [4] and lightweight DES (DESL) [7]. With lightweight AES, roughly 3400 gate equivalents are used and the circuit is optimized for low-power operation. With DESL, authors claim that a comparable strength to that of AES is achieved, with 45 percent less chip size, 86 percent fewer clock cycles and roughly 1800 gates. The latter can encrypt 64 bit plaintext in 144 clock cycles. DESL is particularly appropriate for our purposes. It will be the basis for producing 128 bit long hashed values (various principles of using symmetric block ciphers for one-way hash functions can be found in Ref. [8]).

6.3.2 Overview of Cryptoprotocols Issues

Because of the above constraints, protocols for security (with emphasis on authentication) should consist of as few steps as possible, where a simple "challenge–response" remains the most desirable architecture. If more rounds in a protocol are needed, then it is preferred that these additional messages are syntactically equivalent, which means that the same circuitry can be used (of course, with a different input). This is also in line with the requirement that serialized computations are preferred over concurrent ones. What matters with passive tags is power consumption per clock cycle (mean power consumption minimization). Therefore concurrent ("parallel") computations should be replaced by serialized ones [9].

The following common threats can be identified in the area of RFID protocols:

- Passive attack is an attack in which enough information can be obtained by simply monitoring the communication between a tag and a reader.
- Man-in-the-middle attack is done in a way where an adversary modifies challenges with its own data; an appropriately selected challenge may mislead a tag or a reader to believe that they communicate directly one with another, which is not the case, whereas the messages are handed over and modified by an adversary.
- Active attack is an attack where an adversary is actively involved in the communication and modifies messages (man-in-the-middle is a kind of active attack).
- Reply attack is an attack, where an adversary records exchanged messages and simply reuses them later without necessarily knowing what is contained in them, or how to actually calculate the content of these messages.
- Reply attack is an attack where relaying of messages is deployed between a tag and a reader to falsely convince the reader that the tag is in its close proximity so it can act accordingly.
- Malicious reader attack can be of many kinds, but we will concentrate on the unauthorized tracking attack, where a malicious system tracks a tag without necessarily knowing its true identity, but just recognizes it in various places on the basis of its responses to challenges.
- Physical attack is an attack, in which the tag's content is read directly from a circuitry instead of using wireless connection.

6.3.3 Some Important Cryptoprotocols for RFID Security

RFID cryptoprotocols can be divided into single tag protocols and multiple tag protocols. Further, each of these groups can be divided into single round protocols and multiple round protocols. The following multiple round single tag protocols should be mentioned (summarized from Ref. [10]):

1. *Protocol of Weis, Sarma, Rivest, and Engels [11]:* A reader and a tag share a secret x. After being triggered by the reader, the tag produces a random r and computes a string $(r, (ID \parallel H(ID) \oplus f_x(r))$, which is sent to the reader (here "$\parallel$" denotes concatenation of strings, "$\oplus$" bitwise XOR operation, and "f_x" a pseudorandom function that uses secret x as a parameter). After verification, the reader replies with the tag's ID. It is evident that exposure of the plain ID in the third step can be problematic, not to mention that replay attacks are trivial, because the first and the second messages are cryptographically independent.

2. *Protocol of Henrici and Muller [12]:* A reader sends a request, after which a tag calculates $H(ID)$ and $H(s \circ ID)$. Next, it sends $H(ID)$, $H(s \circ ID)$, and δs to the reader ("$\circ$" denotes some chosen operator, "s" the number of the step, and "δs" the difference between current and previous session number—this difference is equal to 1 when the previous transaction is valid). After receiving this message, the reader computes a new ID for the tag ($ID \leftarrow ID \circ r$), updates the database, and sends r and $H(r \circ s \circ ID)$ to the tag. After receipt, the tag is able to verify the integrity of r and is able to calculate the new $ID(ID \leftarrow ID \circ r)$. Therefore the tag and the reader are supposed to stay in synchronism. However, one problem with this protocol is that an attacker can achieve database desynchronization if XOR is used for "$\circ$." In this case, an attacker replaces r in the third step by a zero-bits string, and as a result $H(r \oplus s \oplus ID) = H(s \oplus ID)$ is obtained. This value is the same as the value from the second step, which the tag has sent to the reader and which can also be read by the attacker. Therefore, when the tag checks messages from the third step, it updates its new ID with a value that differs from that calculated by the reader. The result is database desynchronization.

3. *Protocol of Ohkubo, Suzuki, and Kinoshita [13]:* With this protocol, a tag and a reader share two hash functions G and H, and an initial secret s_i. A reader sends a request to the tag, and this triggers the tag to compute a new secret by calculating $H^1(s_i) = H(s_i)$ and storing this new value. At the same time, the tag computes $G^1(s_i) = G(s_i)$ and sends this value to the reader. The back-end database hashes each of the stored secret values and finds a matching pair $(ID, G^1(s_i))$. In the second run, $H^2(s_i) = H(H(s_i))$ and $G^2(s_i) = G(G(s_i))$ are calculated and used, etc. However, this version is vulnerable to replay attacks, because the second message that is sent from the tag to the reader is not linked to the first message. Therefore an adversary can send a request to the tag and record the reply to use it at some later time to respond to the reader. Avoine, Dysli, and Oechslin propose a solution, whereby the first message contains a fresh challenge r, and the second message is calculated as $G(s_i \oplus r)$ [14].

4. *Protocol of Molnar and Wagner [15]:* This is an example of a protocol that is supposed to be without flaws. A tag and a reader share a secret x. Initially,

the reader chooses random r_r and sends it to the tag. The tag chooses random r_t, computes $\sigma_1 = ID \oplus f_x(0, r_r, r_t)$, and sends it to the reader. The reader uses σ_1 to retrieve ID by calculating $ID = \sigma_1 \oplus f_x(0, r_r, r_t)$, and then replies with $\sigma_2 = ID \oplus f_x(1, r_r, r_t)$. After receiving this message, the tag checks if ID is okay.

However, some shortcomings still exist, even with the latter two protocols, which have not been previously described:

■ With regard to the modified Ohkubo, Suzuki, and Kinoshita protocol, if an adversary always sends the same challenge, the response from the tag will always be the same. This is a serious problem for enforced authentication, because the tag becomes traceable despite its unknown identity.

■ In case of the improved version of the Molnar and Wagner protocol, the short-coming is related to the third message. What happens if there is no match after the tag receives this message? If the tag is supposed to react, this case certainly means a more complex protocol with additional steps, but these steps are missing. And without these additional steps, using the protocol as given in Ref. [10], this means that an adversary can tweak some bits in the second message and a wrong ID is determined by the reader. So the protocol does not assure the integrity of the checked ID.

As described in Ref. [10], many protocols that deploy a XOR function can be successfully attacked by submitting a zero vector (all bits being zero) as an input to computation (see, for example, the protocol of Henrici and Mueller). The reason is straightforward—any bit sequence XORed with a zero vector produces the same bit sequence. In such cases it is a wise practice to check that the input is not a zero vector. Similar reasoning applies when a unit vector (all bits are set to 1) is used for XORing, which leads to negation of the input bit sequence.

The above described protocols are, so to say, deterministic protocols. A different approach has been taken by Hoper and Blum with the HB protocol [16] (its successors are HB+ [17] and HB++ [18]). All HB variants are based on the learning parity with noise (LPN) problem. This problem requires an attacker to calculate a k-bit secret x, shared between a reader and a tag, after being given several calculations of $b_i = a_i * x \oplus \nu_i$, where ν_i (also called noise) is equal to 1 with a probability that takes on values from the interval $[0, 1/2]$. Because the probability of noise being 1 is strictly less than 0.5, an adversary can challenge a tag with some chosen a several times successively. Once k equations with linearly independent a-s have been obtained, x can be recovered by Gaussian elimination. Further, exploitation of this principle is the basis for active attacks, to which the HB protocol family is not resistant. There are other weaknesses of this family like man-in-the-middle for HB+, which are described in Ref. [19].

Later, some other representatives of single tag protocols followed, and an extensive overview with the description of their vulnerabilities is given in Ref. [10]. In addition, this paper also covers multiple tag protocols, which are intended for scenarios where the simultaneous presence of two tags in a reader's field is required.

6.3.4 Measuring Lightweight Properties of Cryptographic Protocols

Based on facts given so far it clearly follows that there exists an important issue related to security in RFID environments. Actually, all of the above described protocols are assumed (or claimed) to be lightweight, because only such protocols are suitable for implementation in RFIDs. There is now an open room for discussion as to which properties qualify a protocol to be lightweight. To provide metrics for quantitative evaluation of such properties, appropriate methodology has been suggested recently [20]. Its background will be briefly described in the following text.

One of the most widely used theoretical models in computer science is Turing machine. It is aimed at studying what is theoretically computable. In addition, this model enables the measurement of computational complexity (in terms of time and space) of a certain problem, defined as a function of growing input. However, being faced with engineering problems when designing cryptographic protocols, the above model turns out to be inappropriate:

- First, the input size in case of cryptographic protocols is often small, having nothing to do with asymptotic behavior.
- Second, a concrete and real computing architecture has to be considered, because the newly designed protocol will run on this architecture. Therefore its architecture has to reflect the general properties of current computing devices.
- Third, it does not address the communication process, which is essential for RFIDs.

Further, current computing architectures are still best described by von Neumanns model. However, treating RFIDs with this model is inappropriate. First, these devices contain crippled processors, that is, the processing part does not contain the functionality that is usually supported by ordinary processors. Second, this architecture cannot be directly mapped to the problem area of lightweight protocols. And third, communications are not covered as well.

RFID devices consist of two basic circuitries. The central part is the processing part with memory (this one contains logical gates), while the second part is radiofrequency communication circuitry (RF part). Therefore a suitable model for measuring lightweight characteristics of protocols should cover the following: It should support any logical function, it should include communication costs, and it should take into account current production technology.

Based on the above requirements a model is given in Figure 6.2. As far as metric is concerned, the number of required NAND gates to implement a particular protocol is taken for this purpose. Taking only NAND gates as a basis for metric may seem odd at a first glance. But it is one of the basic facts of Boolean algebra that any logical function can be implemented with a logically complete set of Boolean functions. Further, this reflects the technological reality, because most logical circuits are implemented this way.

One complete set of Boolean functions consists of conjunction and negation, and can be implemented with NAND gates. This implementation requires one NAND gate for negation and two NAND gates for conjunction (NAND Boolean function is also logically complete over the Boolean algebra). Finally, it should be taken into account

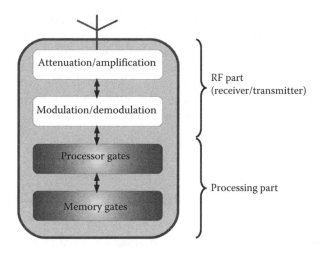

Figure 6.2   RFID model for derivation of lightweight protocols metric.

that RFIDs require clocked elements. These facts represent the basis for (lightweight) protocols metrics.

Further, every protocol implementation requires some storage, and for this purpose D storage cells are chosen for metrics, because they are clocked. Each D cell, that is, flip-flop, requires five NAND gates [21]. In addition to storage, typical logical functions that are needed for provision of security (cryptographic protocols) are bitwise XOR and addition mod 2^n. Bitwise XOR of variables x and y can be obtained as $(x\bar{y})$ AND $(\bar{x}y)$. After some optimization steps, the result are four NAND gates. Similarly, one-bit full adder can be implemented with 11 NAND gates. For performing mod 2^n addition, n-times 11 NAND gates have to be used. This explanation should be sufficient for the purpose of this chapter; more details about implementing Boolean functions with NAND gates can be found in Ref. [21].

A slightly harder issue is the introduction of metric for the communications part. This communications part is actually doing the coding of the messages and sending/receiving them over the distance through electromagnetic coupling. The more complex the protocol, the larger the number of bits that will be transferred. These bits can be thought of as being stored in some memory and transported from the transmitter to the receiver by means of this memory. Thus, D storage cells implemented with NAND gates can be introduced for communication metrics.

To summarize, the metric for the RF part will be the number of bits that needs to be transferred, and the cost of these bits will be measured with the number of D NAND storage cells required to store them. It is now possible to formally define the notion of a lightweight protocol. We introduce the cost N, which includes the total number of NAND gates required for implementation of a certain protocol. This means that N includes storage S, processing P and communications C gates, that is, $N = S + P + C$. This cost N is used as a metric that serves to evaluate lightweight protocols:

The cost N of a protocol is measured by the total number of NAND gates that is needed for its implementation, which includes storage, processing, and communications.

The above definition is useful for comparing costs of lightweight protocols. To state an exact limit on whether a protocol is lightweight or not is a matter of context and depends on the actual state of technology. Currently, a reasonable limit for lightweight protocols in RFIDs environments is the cost of 2500 NAND gates. One thousand five hundred gates *is on account of crypto* operations supporting gates in RFID, and 1000 on account of communications (the reader should note that the total number of gates in current RFIDs is approx. 5000).

6.4 New Nondeterministic Cryptographic Protocols

The protocols introduced in this section are suitable for single tag and multiple tag applications. They are ND (ND protocols), meaning that when a reader gets a response from the tag, the expected values of this response lie in a certain interval. The reader has to check all possible discrete values within this interval to find a match. Such protocols put the majority of the computational workload on the reader and back-end systems. In most cases this is acceptable, especially in the case of RFID architectures where significantly larger computational resources are available at the back-end side (such a principle is common nowadays, especially with digital signatures).

6.4.1 The First ND Protocol

The first protocol goes as follows (see Figure 6.3). A reader and a tag share a common secret x_i and both are able to compute the same strong one-way hash function H. The

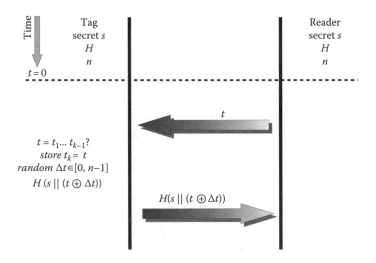

Figure 6.3 The first ND protocol.

tag and the reader also share n that determines the interval for calculation of random values Δt (this n does not need to be secret).

Now the authentication process takes place:

1. The reader challenges the tag with a time stamp.
2. After receiving the time stamp, the tag optionally checks it against past received values. It is unreasonable to assume that the RFID tag will have autonomous time circuitry and, to prevent replies, former values have to be stored. Due to limited resources, the memory for storing received challenges is FIFO, consisting of, for example, four locations; when the fifth value is received, the first value is overwritten. So if the time stamp value is fresh, the tag stores it and computes random Δt from the interval $[0, n - 1]$, meaning that it may have n different values. The tag concatenates secret s with $(t \oplus \Delta t)$ and hashes the string.
3. The tag sends the result from the previous step to the reader.
4. On receipt of the message from the second step, the reader starts calculations to find a match. For this match to be found, the reader calculates $H(s \parallel (t \oplus 0))), \ldots, H(s \parallel (t \oplus (n - 1)))$ with s-es being taken from pairs (ID, s) that are stored in the database. If a match is found, the tag is authenticated.

A random challenge is optionally checked for freshness in the second step to prevent enforced authentication. If a malicious reader constantly uses the same challenge, the responses of the tag will always be the same and the tag will be traceable. Of course, due to limited resources of a tag the list of all stored challenges can be currently relatively short, but available memory will grow in the future and this step can then become obligatory. However, the analysis at the end of this section shows that optional checking of freshness can already be mandatory with available technology, especially if 48 bits are allocated for challenge, which means that eight challenges can be stored in four 96-bit memory locations.

6.4.2 The Second ND Protocol

The second protocol goes as follows (see Figure 6.4). A reader and a tag are able to compute a strong one-way hash function H. The tag is given a secret s that is also known to the reader. Again, the tag and the reader share n that determines the interval for calculation of random values Δr.

Now the authentication process starts taking place:

1. The reader sends the challenge r to the tag.
2. On receipt, the tag optionally verifies if the received r is on the list of already used challenges. If not, it stores the received challenge, and calculates random Δr. Afterwards, the tag computes $H(s \parallel r)$ and further randomizes this result by XORing it with $H(\Delta r)$. It sends this result to the reader.
3. On receipt of the message from the previous step, the reader calculates $H(s \parallel r) \oplus H(\Delta r = 0), \ldots, H(s \parallel r) \oplus H(\Delta r = n - 1)$ until a match is found that authenticates the tag. Of course, the reader has access to a database with pairs (ID, s).

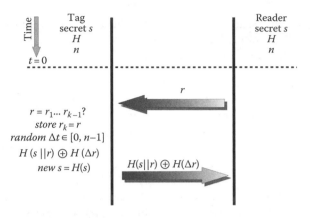

Figure 6.4   The second ND protocol.

This protocol has some important properties that have to be discussed. It is based on a random challenge that is optionally checked by the tag for freshness to prevent enforced authentication from malicious readers. Again, because available memory will grow in the future, this step can then become obligatory. Next, after the challenge is checked, it is concatenated with the secret s and only then hashed. Applying the hash function to two arguments (s and r) is often done by XORing s and r (such an example is the modified protocol of Okhubo, Suzuki, and Kinoshita [14]). This can cause problems if r consists only of zero bits, or bits that are all set to 1. In the former case, the XOR operation results in s itself, while in the latter case XOR results in negation of s. By concatenating arguments (instead of XORing them) this problem is avoided and there is no need for additional circuitry to check whether all bits in a challenge are 0 or 1. Finally, this result is XORed with hashed Δr. This supports multiple tag applications. By offsetting the initial value in circuitry that produces Δr, the two values for Δr will differ, and so will the final result. Thus an attacker will not know that tags are actually responding as twin-tags. The reason for hashing Δr is that the interval of these values is relatively small, for example, defined by 8 bits. Without hashing, only the last 8 bits of $H(s \parallel r)$ would be affected.

6.4.3   A Brief Analysis of ND Protocols

Before going into details it should be stated again that readers are assumed in our scenarios to belong to a secure environment, as well as the back-end part of a system.

Security of ND protocols: For the two ND protocols for RFID environments it can be concluded that all the messages are unique, look random to a third party, and are optionally checked for freshness as well as responses (of course, in engineering reality this uniqueness will certainly be limited because of the number of available FIFO memory locations and limited extent of Δt and Δr intervals).

Further, the first message in both protocols is tied cryptographically to the second message. Thus active and passive attacks are prevented. Reply attacks are also prevented. Malicious reader tracking is impossible because of constantly changing messages. However, preventing physical attack remains an open issue [22].

Further, relay attacks can be prevented by adding a distance-bounding protocol developed by Hancke and Kuhn [23]. This is possible, because the ND protocols are logically independent from the Hancke and Kuhn protocol that is aimed at distance-bounding, but the ND protocols are aimed at authentication and prevention of enforced authentication.

Consumption of resources: Let us provide quantitative estimates of the number of (NAND) gates needed for implementation of the above ND protocols:

- Storing 1 bit requires 5 gates (assuming D flip-flops).
- Due to the fact that block ciphers can be used for cryptographic hashing, our assumption for implementation is DESL that requires approx. 1800 gates [7]. Although using block ciphers for hashing is not efficient for ordinary implementations, in the case of RFIDs it makes sense (dedicated hash functions that are implemented in software are faster on ordinary computing devices, however if they are implemented in hardware they significantly exceed the number of the gates required for DESL).
- Values Δr and Δs are generated with an implementation that is suitable for lightweight purposes and that deploys a shift register [24]. With this implementation, the shift register has an XOR feedback loop, where one input is the output of the shift register, and the other input is the nth bit in the register. The output of the XOR gate is fed into the first storage cell of the shift register. A register with m bits can represent 2^m different values, however all zeros would get stuck in the circuit, so the actual number of different values is $2^m - 1$. By choosing an appropriate n, the resulting sequence is pseudorandom and of a maximal length, if m is such that $p(x) = x^m + x^n + 1$ is irreducible over GF. A shift register with 4 bits requires approx. 60 gates, with 8 bits approx. 120 gates, and so on. Therefore assuming 8 bits for our implementation, 120 gates are needed.
- Optionally $k = 4$ (optional) n-bit locations are needed for storage of used r and t values, and one location for secret s (ID). Therefore, assuming 96 bit values this means $(4 + 1) * 96 = 480$ bits and 2400 gates. Note that one n-bit location can store 3 challenges r and t, if these are 32 bits long, the total number of stored challenges in this case is 12.
- Bitwise XOR requires 4 gates, so XORing 128 bits requires 512 gates (this is the figure for the second protocol, while the first one requires only $8 * 4 = 32$ gates).

An estimate of the total cost, using the above values, is 4800 gates, where logic gates that are needed for comparisons of freshness of received challenges are not included. So the main cost is contributed by storage cells that are needed for freshness checks. If optional freshness checks are excluded, the number of gates is approx. 2400, which certainly qualifies the above protocols as lightweight. But even with the inclusion of optional steps, the two ND protocols stay within the limits for lightweight RFID implementations.

For demonstration purposes, let us calculate how lightweight the most typical RFID protocols that were discussed in Section 6.3.3 are. More precisely, let us focus on protocol of Molnar and Wagner, which contains structures that are typical also for other three protocols. So in the case of Molnar and Wagner protocol, the tag's response is as follows: $\sigma_1 = ID \oplus f_x(0, r_r, r_t)$. Storage of ID requires 480 gates, and storage of r_r requires $5 * 8 = 40$ gates, assuming that r_r is 8 bits long. The same holds true for secret x—if its length is assumed to be 8 bits, we need 40 gates to store of x. Further, calculation of 8 bits long r_t with a shift register requires 120 gates, and the implementation of f_x using DESL requires 1800 gates. Finally, XORing 96 bits requires $4 * 96 = 384$ gates, so the total number of gates needed for this protocol is approx. 2864. This also demonstrates how efficiently ND protocols provide security services, including prevention of malicious tracking.

6.5   Open Issues in RFID Security

A research conducted for IPTS in 2006 revealed that social acceptance and trust in RFIDs is quite low, which in turn was seen as an obstacle for their widespread deployment [25]. Further, a survey conducted by CapGemini showed that consumers see RFID security as a real problem, because they are more intrusive than several other privacy-invading technologies like loyalty cards [26]. As a consequence, RFIDs' security has attracted large attention during the recent years. According to one of the most representative Web pages in this area [27], there was only 1 paper published in this field in 2002, 11 in 2003, and almost 50 in 2007.

In 2006, Rieback et al. exposed that the main security issues that require feasible solutions are on-tag cryptography, key revocation, standardization, and legislation [28]. And indeed, during the last 2 years, as already mentioned, several cryptographic solutions have been developed and evaluated by the research community [27]. However, many solutions are merely of theoretical nature and have not actually been implemented (the reasons have been analyzed in the previous sections). In addition, even those solutions that have been implemented have not been tested and evaluated in the real world [29].

Although security (and consequently privacy) risks are known by the industry, so far there have not been enough incentives to implement security solutions in massive RFID tags production. It is quite realistic assumption that security features will be implemented in the tags when they are addressed adequately also in standards.

In general, security problems related to RFID systems vary quite a lot depending on the type of tags used. Consequently, the solutions differ as the capabilities of tags change. Passive tags with very little space for implementation of cryptographic solutions cannot use the traditional security solutions, and special lightweight solutions like those described in previous section have to be developed. The bigger and more powerful the tag is, the more powerful the security solutions are to be implemented. On the other hand, more capabilities bring up new challenges that have to be solved. On a simple passive tag it is enough to concentrate on protecting the reading process. In contrast, on a more powerful multipurpose active tag, the security measures have to also consider the access control from reading to writing and erasing.

In the following sections we will address existing problems in the five most important future areas: physical RFIDs security, cryptographic primitives and cryptographic protocols for tags, the application-specific issues with emphasis on back-end systems, legal issues, and general open issues.

6.5.1 Physical Security of RFIDs

An RFID tag can be attacked in several ways if an attacker has physical access to the tag. The tag's circuitry can be demolished with sharp objects, and connection to antennas can be severed (this has been suggested by Karjoth and Moskowitz as a privacy-protecting measure [30]). Further, tags can also be killed with a strong electromagnetic pulse. As an electromagnetic pulse does not leave any physical traces on the tag, it is not possible to visually notice if the tag is killed. Of course, users of tagged items would prefer a method where they could easily see whether the tag is active or not. With the help of a blocker mechanism the reading of tags can be blocked while the tag itself stays intact allowing more flexibility on tag use [31]. On the other hand, in some applications a possibility to kill or prevent tag reading is unacceptable and has to be prevented; therefore methods for appropriate protection of tags are required.

Tags can also be physically separated from the tagged item, which is a very trivial, but one of the most serious threats. Later on, a separated tag can be attached to a new item and thus the identification system is broken. This is a consequence of the fact that it is not always possible to embed an RFID tag into a tagged item in a way that prevents its physical removal. Again, on the other hand, for some application areas it is important that the tags can be removed. In addition, it might also be important to assure that it can be recognized when a tagged item contains an invalid tag.

A physical access to the tag opens up the possibility for side channel attacks such as timing and power analysis attacks. Oren and Shamir have even developed a power analysis attack where no physical contact is required. This attack can be conducted without attacker and tag transmitting any data, making it very hard to detect [32]. Therefore solutions against these types of side channel attacks are required. Last but not least, a well-known methodology that generally addresses tamper resistance, and which was demonstrated by Ross Anderson more than ten years ago, remains valid also for RFID tags [22].

6.5.2 Cryptographic Primitives and Cryptographic Protocols

A lot of research has been concentrated on developing secure protocols that prevent enforced authentication (i.e., protect a tag against unauthorized reading). Many of the suggested protocols rely on hash functions as a lightweight solution. Feldhofer and Rechberger pointed out though, that existing hash functions are not feasible for protecting the simple passive tags (this is due to the requirement for storing inner vectors which requires thousands of logical gates [6]). A symmetric cipher-like DESL can be used to create hash values in many cases. Still a full-fledged hash function that can be implemented in an RFID tag would be preferred due to good performance.

Another important assumption that requires additional research (and which is assumed to be resolved by many RFID protocol developers), is availability of quality random number generators. However, the research in this area has not been very active. The researchers from auto-ID labs have just published their solution, which is promising to resolve this issue [33]. But taking into account the fact that the solution has just been published, it remains open whether it will successfully resist other researchers' attempts to find security flaws in it.

As far as the majority of current (lightweight) security solutions are concerned, they have been concentrating mainly on protecting the tags identity, while other important issues, including digital signatures, remain open. This is largely due to the fact that asymmetric cryptography seems to be out of the scope of simple passive tags for the time being, but due to technological advances these issues may soon get on the agenda.

6.5.3 Back-End Systems

The back end of the RFID system includes the transmission of the information from the RFID reader to a database, as well as the use of this database. The back-end protection does not suffer from the limitations of RFID tags and thus traditional security measures for protecting the communication and databases can be used. It is good to realize that a possibility of reading a single tag is a relatively limited security threat, although access to a huge database containing data of several reads of RFID tags often has much larger dimensions (especially for privacy).

Thus protection for the back-end system has to be designed carefully. A reader in particular is an interesting target. The attacker may try to force malware inside the reader that would leak the read data to an intruder, or even completely destroy the back-end system. To better understand this kind of attack, it should be noted that RFID's response may contain not its ID, but rather some command. Knowing that tag readers are directly connected to databases, a simple SQL command can be used to cause various kinds of damages. For example, an SQL statement like "; `shutdown-`" would result in deny-of-service attack. Further, database integrity could be attacked by an SQL statement that would delete a relation, for example, "; `drop table <someAttackedTable>`." Last but not least, using the described code injection techniques that link RFIDs with the SQL world, even viruses can be designed (of course, when writable RFIDs are used) [34].

Combining tags' data with other data in the database opens up new additional problems. In a so-called breadcrumb threat a user is associated with an RFID tag in an item that is stored in a database [35]. When this user gives the item away, the association is not necessarily changed in the database. Thus later on the user may be associated wrongly with a location where the RFID-enabled item has been seen.

Because different applications have different requirements and different security threats, the focus is moving toward application-specific research. For example, tags that are used for passports require much higher tamper resistance than those that are used for keeping up product logistic. Last but not least, reader's memory and possible temporary files inside it can also be very attractive target for a malicious person.

As new areas for RFID tag use are invented, new requirements will arise like solutions for problems where transfer of ownership is required [36].

6.5.4   Legal Issues

Legal questions present one permanent challenge and have been discussed in many places in the literature (see, e.g., discussion by Flint [37]). Once the legal debates materialize into laws addressing also RFIDs' specifics, a secure solution will follow.

It is interesting to note that many states like California and the EU have debated legislation in the area of RFIDs. In case of the EU there have been discussions as to whether legislation is needed that specifically addresses RFIDs [38]. Although an official statement has been made that RFID-tailored legislation will not be introduced (because legislation has to be technology-neutral, and has to cover technology more implicitly), things seem to be changing again. Security (and consequently privacy) of RFID systems is so important and technologically specific that it seems to be impossible to completely abstract it in legislation. So the EU commission is now considering a compromise by providing some kind of "soft law guidelines" [39]. Despite all this, many laws already remain applicable to ubiquitous computing environments, the most notable one being the Data Privacy Directive [40].

6.5.5   General RFID Security Issues

When the basic security primitives and protocols will be lightweight enough, the emphasis on security development is expected to move toward application-specific issues. More complicated tags will open up new types of security threats, which, in turn, will require new solutions. An example for justification of this claim is the case with a possibility to write into a tag, which has enabled an insertion of malware that could affect a tag reader [41,42]. Another example that is anticipating the above-mentioned trends is suggested by Fishbin and Roy [43], where a tag should provide more information to readers that are closer to it than those that are farther away. The feasibility of the concept, and how it could resist different kind of specialized devices that attackers could create, is open.

Prevention of threats requires also appropriate threats models. Rao et al. have developed a simple model that can be used for analyzing the extent of a threat [44]. This threat assessment model, although superficial, can be used as a basis for further research to develop better threat analysis methods in different environments.

Besides technical, there are also nontechnical issues that will have to be addressed. Garfinkel et al. urge that customers should know more about RFID tags in products and how these tags work [35]. They suggest defining public policies for RFID tags behavior in different cases. Such policies would give the customers information that they can use to evaluate the risk a given tag will provide against their privacy. Ayode suggests that RFID readers should produce a sound when reading a tag, similar to the manner in which digital cameras make sound when a picture is taken [45]. Thus customers would know when the tag is read and could ease the acceptability of technology.

Finally, it is important to continually address relevant legal issues surrounding RFID tags. By deploying these tags it is easy to generate huge databases with sensitive personal information. RFIDs may cause the need for additional clear rules about such databases, and how the affected people will be properly informed. Responsibility issues should also be clarified [46]. For example, what would be the legal implications if a store does not kill RFID tag in an item after a customer has purchased it? If the killing fails, is it the responsibility of the tag manufacturer or the store? Legal problems will likely bring up research opportunities for engineers to develop new solutions that will improve secure use of RFID tags in the future.

6.6 Conclusions

The area of security of RFID solutions is becoming more and more important because of increasing and widespread use of such systems, which are becoming ubiquitous. To assure security in such environments, many issues have to be addressed: technological issues that deploy security mechanisms based on security services, users' awareness of importance of security (and consequently privacy), and legal issues.

As is often the case, issues in various areas of information systems security are first treated from the technological point of view, which also holds true for RFID systems. Therefore this chapter starts with the technological view of security and covers it extensively. It presents the most typical and important current protocols. These protocols are aimed at ensuring legally acceptable authentication, and related data integrity (e.g., database synchronization). We have described their weaknesses that have been found in the literature so far. In addition, we have described some new weaknesses.

Taking this into account, we have developed two new lightweight cryptographic protocols. These protocols are ND and require minimal resources on the tag's side. To achieve this, heavier computations are put on the reader/back-end side, which is acceptable in the majority of information systems. Because of these basic properties, the presented ND protocols effectively provide security by preventing enforced authentication. We have analyzed briefly their resistance to known attacks and we believe that they fulfill their intended use. Further, this chapter presents an apparatus that provides metric for evaluation of lightweight protocols. Such metric is essential for RFID environments, because these devices have very limited resources and security has to be provided by taking very stringent requirements into account. Last but not least, this chapter discusses those open issues that are likely to get on the agenda in the near future.

Acknowledgments

D. Trček would like to thank to ARRS (Slovenian Research Agency) for financial support of this research (grant number J2-9649). P. Jäppinen would like to thank Nokia Foundation for the Nokia Visiting Researcher grant that made possible the visit to Jožef Stefan Institute.

References

1. Gartner Group, RFID market $3 billion in 2010, *RFID Update*, Dec. 13, 2005, http://www.rfidupdate.com/articles/index.php?id=1014.
2. G. Roussos, Enabling RFID in retail, *Computer*, 39(3), 25–30, 2006.

3. D. Trček, Security and privacy in RFID based wireless networks, in *Handbook of Research on Wireless Security*, eds. Y. Zhang, J. Zheng, M. Ma, Vol. II, pp. 723–731, IGI Global, New York/Hershey, 2008.

4. International Standards Organization, Information processing systems: Open systems interconnection—Basic reference model, security architecture, part 2, ISO 7498-2, Geneva, 1989.

5. S. A. Weis, Security and privacy in radiofrequency identification devices, unpublished Masters thesis, MIT, Cambridge, MA, 2003.

6. M. Feldhofer and C. Rechberger, A case against currently used hash functions in RFID protocols, in *Workshop on RFID Security Security 06*, Graz, 2006, http://www.iaik.tugraz.at/aboutus/people/feldhofer/papers/RFIDSec06_slides.pdf.

7. A. Poschmann, G. Leander, K. Schramm, and C. Paar, New light-weight crypto algorithms for RFID, in *Proceedings of the IEEE International Symposium on Circuits and Systems-ISCAS 2007*, New Orleans, LA, 2007.

8. B. Schneier, *Applied Cryptography*, 2nd edn., John Wiley & Sons, New York, 1995.

9. M. Feldhofer, J. Wolkerstorfer, and V. Rijmen, AES implementation on a grain of sand, *Information Security, IEE Proceedings*, 152(1), 13–20, 2005.

10. S. Piramuthu, Protocols for RFID tag/reader authentication, *Decision Support Systems*, 43(3), 897–914, 2007.

11. S. A. Weis, S. E. Sarma, R. Rivest, and D. W. Engels, Security and privacy aspects of low-cost radio frequency identification systems, in *Proceedings of the 1st Security in Pervasive Computing, Lecture Notes in Computer Science*, 2802, 201–212, Boppard, Germany, 2004.

12. D. Henrici and P. Muller, Hash-based enhancement of location privacy for radio-frequency identification devices using varying identifiers, in *Proceedings of the 1st International Workshop on Pervasive Computing and Communication Security*, pp. 149–153, Orlando, FL, 2004.

13. M. Ohkubo, K. Suzuki, and S. Kinoshita, A cryptographic approach to a 'privacy-friendly' tags, in *RFID Privacy Workshop, MIT*, November 15, Cambridge, MA, 2003.

14. G. Avoine, F. Dysli, and P. Oechslin, Reducing time complexity in RFID systems, in *Proceedings of the 12th Annual Workshop on Selected Areas in Cryptography*, pp. 291–306, Kingston, Canada, 2005.

15. D. Molnar and D. Wagner, Privacy and security in library RFID: Issues, practices, and architectures, in *Proceedings of the 11th ACM Conference on Computer and Communications Security*, pp. 210–219, ACM Press, Washington, DC, 2004.

16. N. J. Hopper and M. M. Blum, Secure human identification protocols, in *Advances in Cryptology ASIACRYPT 01, Lecture Notes in Computer Science*, 2248, pp. 52–66, Gold Coast, Australia, 2001.

17. A. Juels and S. A. Weis, Authenticating pervasive devices with human protocols, in *Advanced in Cryptology—CRYPTO'05, Lecture Notes in Computer Science*, 3126, 293–308, Santa Barbara, CA, 2005.

18. J. Bringer, H. Chabanne, and E. Dottax, HB++: A lightweight authentication protocol secure against some attacks, in *IEEE International Conference on Pervasive Services, Workshop on Security, Privacy and Trust in Pervasive and Ubiquitous Computing*, Lyon, France, 2006.

19. H. Gilbert, M. Robshaw, and H. Sibert, An active attack against HB+—A provably secure lightweight protocol, *IEE Electronic Letters*, 41(21), 1169–1170, 2005.
20. D. Trček and D. Kovač, Formal apparatus for measurement of lightweight protocols, *Computer Standards and Interfaces*, 31(2), 305–308, 2008, http://dx.doi.org/10.1016/j.csi.2008.02.004.
21. L. Vodovnik and S. Rebersek, Digital circuits, Faculty of Electrical Engineering, Ljubljana, 1986.
22. R. Anderson and M. Kuhn, Tamper resistance—A cautionary note, in *Proceedings of the Second USENIX Workshop on Electronic Commerce*, pp. 1–11, Oakland, CA, 1996.
23. G. P. Hancke and M. G. Kuhn, An RFID distance bounding protocol, in *Proceedings of the IEEE/Create-Net SecureComm*, pp. 67–73, Athens, Greece, 2005.
24. P. Horowitz and W. Hill, *The Art of Electronics*, Cambridge University Press, New York, 1989.
25. M. van Lieshout, L. Grossi, G. Spinelli, S. Helmus, L. Kool, L. Pennings, R. Stap, T. Veugen, B. van der Waaij, and C. Borean, RFID technologies: Engineering issues, Challenges and policy options, in *IPTS*, Sevilla, 2006, http://ftp.jrc.es/eur22770en.pdf.
26. Capgemini, RFID and consumers—What European consumers think about radio frequency identifications and implications for businesses, Capgemini report, 2005, http://www.capgemini.com/news/2005/Capgemini_European_RFID_report.pdf.
27. G. Avoine, RFID security and privacy lounge, UCL, Louvain, 2008, http://www.avoine.net/rfid/.
28. M. Rieback, B. Crispo, and A. Tanenbaum, The evolution of RFID security, *IEEE Pervasive Computing*, 5(1), 62–69, 2006.
29. J. Ayoade, Roadmap to solving security and privacy concerns in RFID systems, *Computer Law and Security Report*, 23(6), 555–561, 2007.
30. G. Karjoth and P. Moskowitz, Disabling RFID tags with visible confirmation: Clipped tags are silenced, in *WPES '05: Proceedings of the 2005 ACM Workshop on Privacy in the Electronic Society*, pp. 27–30, Alexandria, VA, 2005.
31. A. Juels, R. L. Rivest, and M. Szydlo, The blocker tag: Selective blocking of RFID tags for consumer privacy, in *Proceedings of the 10th ACM Conference on Computer and Communications Security*, pp. 103–111, Washington, DC, 2003.
32. Y. Oren and A. Shamir, Remote password extraction from RFID tags, *IEEE Transactions on Computers*, 56(9), 1292–1296, September 2007.
33. W. Che, H. Deng, X. Tan, and J. Wang, A random number generator for application in RFID tags, *Networked RFID Systems and Lightweight Cryptography*, pp. 279–288, Springer, 2008.
34. M. R. Rieback, B. Crispo, and A. Tanenbaum, RFID malware: Truth vs. myth, *IEEE Security & Privacy Magazine*, 4(4), 70–72, 2006.
35. S. L. Garfinkel, A. Juels, and R. Pappu, RFID privacy: An overview of problems and proposed solutions, *IEEE Security & Privacy Magazine*, 3(3), 34–43, 2005.
36. K. Osaka, T. Takagi, K. Yamazaki, and O. Takahashi, An efficient and secure RFID security method with ownership transfer, in *International Conference on Computational Intelligence and Security*, pp. 1090–1095, 2006.

37. D. Flint, RFID tags, security and the individual, *Computer Law and Security Report*, 22(2), 165–168, 2006.
38. S. Pritchard, CeBIT 2007: Europe opts out of RFID regulation, PCPro, Dennis Publishing Ltd., March 15, 2007, http://www.pcpro.co.uk/news/107699/cebit-2007-europe-opts-out-of-rfid-regulation.html.
39. e-practice.eu, Commission launches consultation on radio frequency identification (RFID), March 3, 2008, European Communities, http://www.epractice.eu/document/4426.
40. European Commission, Privacy and electronic communications directive, 02/58/EC, *Official Journal of the European Communities*, L201, July 31, 2002, Brussels, 2002.
41. M. R. Rieback, P. N. D. Simpson, B. Crispo, and A. S. Tanenbaum, RFID malware: Design principles and examples, *Pervasive and Mobile Computing*, 2(4), 405–426, November 2006.
42. S. Ortiz Jr., How secure is RFID?, *Computer*, 39(7), 17–19, 2006.
43. K. P. Fishkin and S. Roy, Enhancing RFID privacy via antenna energy analysis, Technical Report Technical Memo IRS-TR-03-012, Intel Research, Seattle, WA, 2003.
44. S. Rao, N. Thantry, and R. Pendse, RFID security threats to consumers: Hype vs. Reality, in *The 41st Annual IEEE International Carnahan Conference on Security Technology*, pp. 59–63, Oct. 8–11, Ottawa, Ontario, Canada, 2007.
45. J. Ayoade, Roadmap to solving security and privacy concerns in RFID systems, *Computer Law and Security Report*, 23(6), 555–561, 2007.
46. E. P. Kelly and G. S. Erickson, RFID tags: Commercial applications vs. privacy rights, *Industrial Management and Data Systems*, 105(6), 703–713, 2005.

Chapter 7

RFID Deployment: Supply Chain Case Study

Eugene Lutton, Brian Regan, and Geoff Skinner

CONTENTS

The deployment of radio-frequency identification (RFID) systems requires careful analysis, planning, and control to enable the organization to obtain an optimal solution, otherwise the perceived benefits may never materialise. This chapter discusses a supply chain case study within the context of a RFID rationale and deployment methodology. The case study involves the tracking and identification of pallets and cartons amongst various members of a RFID trial. The methodology is divided into three phases: business, infrastructure, and deployment environments. Understanding the current factors affecting the discretionary behavior of the organization and the motivating aspects to the deployment is appraised in the first phase of the methodology. Once the organization has assessed a business case, the investigation of the technical needs of RFID is undertaken in the physical environment. If the motivation for deployment is still active, the instigation of a RFID testing and pilot phase is commissioned. The specification of phase transitional motivators and their outcomes will guide the organization throughout the deployment methodology.

7.1 Introduction

Radio-frequency identification (RFID) deployment requires a methodology for a successful initiation and implementation. This chapter will discuss a rationale deployment methodology using a National Supply Chain (NSC) case study as an example. What is different about this methodology is it looks at why the organization is contemplating or deploying RFID technology and how this then effects the decision on whether or not the deployment is completed. This methodology is flexible in that a single small business or a complete supply chain can utilize this methodology. Note that this methodology is focused on the collection of data from RFID architecture and not on the integration of data into corporate information systems.

Similar to other changes in business process and IT infrastructure, RFID deployment requires a business commitment, combined with a thorough analysis and planning. The whole process necessitates the production of documentation of the required or possible developments by the organization itself and the RFID consultants.

An organization may exist within an environment which may mandate the utilization of RFID technology. Alternatively the organization or business may initiate the examination of this type of identification system. The discretionary behavior of the organization will depend on the instigating environment. If the technology has been mandated then the scope for discretionary behavior may well be limited. For example, Marks & Spencer in Britain had mandated RFID implementation but had also supplied RFID equipment [1]. In the case study to be examined in this chapter, the participants had willingly entered into a consortium to study the possibilities of RFID data interacting in a supply chain. The rationale was to study the Electronic Product Code (EPC) network across a range of sites with the objective of increasing the real-time knowledge of an identified object while delivering business benefits to the participants. The scope included tracking an object and showing the reassignment of ownership and movement of the object between business sites.

This chapter discusses three phases of the RFID Rationale and Deployment Methodology; Section 7.2 *Phase 1*: Business Environment which examines the outlook the organization has in regard to business cases to produce the phase transition motivators (PTMs) of RFID deployment. Section 7.3 *Phase 2*: Infrastructure Environment investigates the physics and technical needs of RFID that must be understood along with the business case, to be enhanced or developed with a RFID enabled process. Section 7.4 *Phase 3*: Deployment Environment discusses the prototype testing and pilot steps; and finally Section 7.5 the conclusion. The Working Case Study will be continually explained in relation to each process in the methodology. Throughout the chapter data acquisition requirements, system integration or system development, and possible business enhancements for RFID deployment will be appraised to vary degrees.

7.2   Phase 1: Business Environment

The business environment examines the current status of the organization and how their discretionary behavior is modified by the instigating environment. If RFID is mandated, then the organization will be obliged to adopt the new technology and integrate this functionality into the required business case. There will be circumstances when the mandate will conflict with the survival of the organization; therefore, the business may not proceed with the mandate and have to choose to source other suppliers or customers. Thus in this case the cost to the organization is greater than the advantage. If the organization is analyzing nonmandated business cases that may be enhanced with RFID, it may conclude its research if the costs outweigh the benefits. In both of these previous examples, the parameters of the PTMs are not satisfactory for the organization to progress to the next phase of the methodology.

Figure 7.1 presents an overview of the first phase of the RFID rationale and deployment methodology. This phase examines the instigating environment and analyzes the

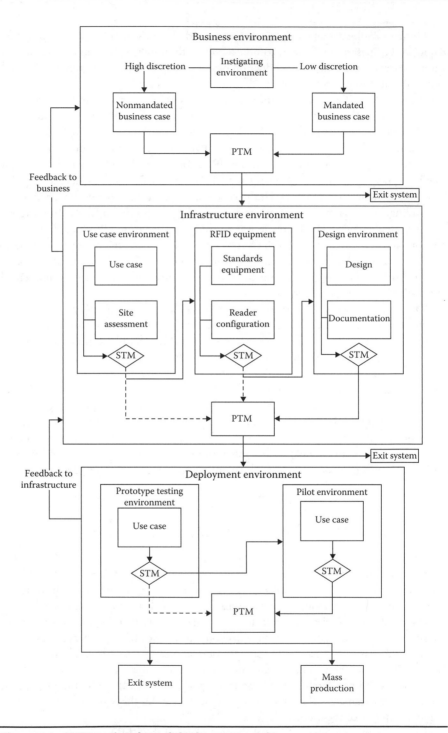

Figure 7.1 RFID rationale and deployment mode.

reasons why the organization is contemplating the deployment of RFID technology. The instigating environment will evaluate if the business case has come from a mandated directive. In turn the organization will exhibit varying degrees of discretionary behavior due to the nature of the instigating circumstances. The trigger to progress to the next phase is measured by the outcome from the PTMs step in this phase.

7.2.1 Business Environment: Instigating Environment

It is necessary to understand the motivation of the organization and the reasons for the examination and analysis in regard to the deployment of RFID. Why is the business organization looking at RFID and what is the strategy and planning model? This extends to the rationale, objective, and scope of the planned RFID investigations and any relationship to other identification technologies. In other words the business environment strategy may not be the final outcome as the planned scope may require it's boundary to be changed due to internal or external factors. Similarly the objectives may be continually refined due to feedback throughout the three phases of the methodology. If the deployment is a closed wall process, then the feedback loop will be mainly concerned with internal factors. In the case study of a supply chain deployment and integration, it is evident that different organizational interrelations will play a major factor in the discretionary behavior and collaborations of the organizations.

7.2.1.1 Examining Discretionary Behavior

Discretionary behavior in an organization postulated [2] that the amount of discretion impacts on the decision-making process within the business. Thus it is important to understand the reasons why the organization is investigating the use of RFID and what was the instigating circumstance. The following paragraphs will outline three scenarios (Firms 1, 2, 3) of how the starting point of an organization within the model will affect their discretionary behavior.

Firm 1 noted that RFID is being used by organizations in the supply chain and purchased tags for their vehicles to travel on tollways [3]. A directive was made to discover the values of RFID technology and report on the feasibility of an RFID-enabled identification system. In this scenario the organization has a high level of discretion in choosing those business cases to explore and how these will be enabled with RFID systems. The instigating environment is of an investigative nature and this then flows into the nonmandated business case step of this phase.

Firm 2 supplies products to a customer (a major retailer), who has mandated the use of RFID so the customer can identify objects as they are received. In this example, the firm's discretionary behavior has been reduced and they need to enable certain business cases with RFID technology. The organization still has the ability to source their infrastructure requirements, though the tag and spectrum specifications must be compatible with their customer systems.

Firm 3 is similar to firm 2 but in this case RFID technology is mandated and the necessary equipment to implement the system is supplied. In this case the organization has very little discretionary behavior except in how they physically implement the system.

This discretion may be further reduced if the supplier has to allow the customer to be aware of the transition of the object through the business. In other words the visibility of the object as it changes state may be a requirement of the business case. This can be likened to a manufacturer knowing the stage of a product item as it is being manufactured. For instance, Bayerische Motoren Werke (BMW) uses an active tag to identify the stages of a car in the production line. The tag also contains technical specifications for assembling the car [4].

The instigating environment for firms 2 and 3 has emerged from a mandated approach and this leads to the mandated business case step of this phase. In these three previous examples the organization still has the ability to decide not to implement RFID to identify objects. If the firm decides not to deploy RFID, in the case of firms 2 and 3 it may be necessary to source new customers for the survival of the business.

Table 7.1 displays the levels of discretionary behavior an organization has in enabling business processes with RFID technology. This is directly related to the instigating environment as the organization will have nil to high discretion depending whether or not the directive is mandated. It has been assumed that the RFID frequency spectrum and tag characteristics have been decided, in a mandated situation.

7.2.1.2 Working Case Study: National Supply Chain

Some initial questions about the working NSC case study:

1. *What is the instigating environment?*
2. *What is the level of discretion for each member?*
3. *Do they progress to the mandated or nonmandated business cases?*

The case study's instigating environment came from an investigative starting point. The members came from a nonmandated instigating environment and displayed a high level of discretionary behavior. The members of the supply chain and other interested parties had decided to test the deployment of RFID technology to investigate its benefits. They were interested in examining the business benefits from tracking objects through various stages of the supply chain from the manufacturer to the showroom.

It is of value to note that the initial entry within the business environment model was of an investigative nature and the business cases were not mandated. In this case

Table 7.1 Discretionary Behavior

Discretionary Behavior	Business Cases		Technology	
	Mandated	*Nonmandated*	*Mandated*	*Nonmandated*
High		×		×
Moderate				×
Low	×			
Nil	×		×	

study it is apparent that the members would have to select business cases, which allow the capture of RFID data and the sharing of that data between the organizations.

7.2.2 Business Environment: Business Cases

The business case is defined as the argument supporting a proposal to encompasses all the processes and factors necessary to fulfill the business objective of a particular RFID implementation proposal. This includes all resources, such as data systems, personnel, and hardware required. The model shows two paths leading from the instigating environment with identification tags of low and high discretion. Depending on the instigating environment the organization will be analyzing a mandated or nonmandated business case. Though similar questions and analysis are needed, there will be a difference in the imperative decision making in the situation. All the organizations involved in the examination, assisting or deploying the RFID system must investigate and consider a common collection of questions.

The first stage of questions is needed to comprehend the current business environment of the business case. The second stage analyzes the information systems and data strategy required for the development of the RFID-enabled business case. The format of the following indicative questions which are in no particular order of importance will be

- *Question proposed; additional questions leading from the initial query*
- *Suggestions or examples given*

Stage one questions: What is the current business environment?

Q1: Has the business come from a mandated or nonmandated instigating environment? In this respect it is important to know whether or not the business case has been mandated for the following reasons:

AQ: How much discretion does the business organization have in the business case or in defining the total RFID environment?
This affects decision making and the amount of emphasis and resources applied to the analysis of the business case. Whether or not the business case is mandated will have a flow on affect to the weighting given to the PTM indicators. Some PTM variables will not be as influential in the discretionary behavior demonstrated by the organization. For example, identification performance or visibility of an object will be more influential than cost factors (unless excessive) in the case of a mandated environment. In the case of a dominant partner mandate, [5] suggests incentives may flow from the dominant to the subordinative partner. General Electric subsidized the cost of the RFID tags for its trading partners [6]. These two examples are important when it comes to the descriptions and analysis of the PTMs.

Q2: What are the likely changes to human resources?
In the case of RFID installed in hospitals, staff complained of the additional monitoring activities created [7]. It is important to have the support and compliance of personnel to increase the possibilities of a successful deployment.

Q3: To what extent does the economic and business environment affect the discretionary behavior of the organization?

Q4: Is the business case linked to strategic business outcomes?

Q5: Are there time constraints for implementation?

Business case mapping relates to processes, interactions with employees, machinery, data systems, suppliers, customers both internal and external, and strategic business objectives. Within the business case there is a system that interacts with data systems and these systems may be internal or external. These interactions need to be modeled so the communication (exchange of data) is documented and captured by the appropriate devices. In this case, a Unified Modelling Language (UML) use case is utilized to document the interaction with data systems and so form an important component of the business case by showing among other things the participants (UML actors) and the system boundary. It is necessary to model that part of the business case which interacts with the data system. Not all processes in the business case will require data capture or have the need for output functionality. For example, the act of placing an identification object on a pallet or the tagging of a patient may not be recorded.

The following stage two questions examine the information system and data strategy. This methodology does not cover the integration of information systems or the interoperability of said systems between different links in an intra- or inter-organizational system. It is still a crucial exercise to investigate what data is to be captured and how it supports the business case.

Stage two questions: What is the information and data strategy? Information integration and the collection of data need to be investigated and documented.

Q1: Is this business case deployment part of a supply chain?

If the organization is a link in a supply chain, it is imperative they understand the interrelationships between their business case and proceeding and forward business cases of their customer or supplier. There will be circumstances when the organization in a supply chain will have attached tags to objects that have come from a supplier, who does not have RFID enabled identification systems.

Q2: What protocols are in place to handle incorrect RFID data?

It has been seen that RFID is prone to data collection errors [8–10]. It is necessary to reduce the likelihood of errors at the source, instead of handling them once they have been collected. The business organization needs to be aware that RFID is prone to errors and determine what detection rate is an acceptable percentage. This may mean trialling the detection and changing the way an item/s is tagged.

Q3: What is the current information system?

RFID systems create a large amount of data and this data will need to be stored and used in a cost effective manner. While [11,12] discusses the amount of increased data flow, which requires transformation into information, although [13] suggests that changes to the data models may be required for the increased data and how this will be stored in a data warehouse.

Q4: What data will be available for all users of the supply chain?

Q5: What is the object to be identified?

The business case has examined a set of questions that need to be investigated from a business case environment and an information system perspective. It must be noted that the previous section did not produce an exhaustive list of questions. There will be organizational specific questions that will be raised during the examination of the business case, which was not covered.

7.2.2.1 Working Case Study: National Supply Chain

Briefly the participants of the NSC included a pallet supplier, packaging company, product manufacturers, product retailers, transport suppliers, and software, hardware, and service providers. The consortium was lead by representatives from all the participants. The following sample of questions from the previous section will be discussed in relation to the working case study.

Stage 1 questions: What is the current business environment?

Q1: Has the business come from a mandated or nonmandated instigating environment?

The members of the NSC had come from a nonmandated background. To allow the interaction between the parties there would be business cases that require RFID-enabled processes. Some members also looked at areas that were not crucial to a supply chain scenario. Thus the organizations had a high and low discretion in regard to business cases.

Stage 2 questions: What is the information and data strategy?

Q3-4: What data will be available for all users of the supply chain and what is the object to be identified?

Data to be shared was change of ownership of a pallet and carton in transit among the members. The objects to be identified were pallets and products at the carton level.

The next section examines the purpose and development of PTMs in the context of the business environment. It also discusses them in the context of the NSC.

7.2.3 Business Environment: Phase Transitional Motivators

PTMs guide the organization in their decision making in regards to progressing to the next phase of the methodology. It is important to have some form of metric to clarify an outcome from analyzing, developing, implementing, or maintaining the requirements of an organizational business case. Myerson [1] uses a three-parameter performance index to measure the success for each stage in the supply chain management organizational maturity model. This methodology uses the PTMs in the formation of a decision-making template. This extends the [1] model by including specific metrics for defining PTMs and outcomes for transitions from one phase or step to another.

The development of PTMs will be guided by analyzing the questions suggested for a particular phase or step in this methodology. In this instance, the organization is investigating the business environment and will reflect upon the reasons why they should deploy RFID technology? If the reasons to implement RFID are conducive, then the organization may continue the assessment. So the organization will explore and analyze parameters that must be fulfilled to satisfy the need to deploy a RFID system. The following paragraphs discuss examples of relevant PTMs.

Internal and external influences are important considerations for the development of PTMs. For example, the mandate from a customer may give the organization a set time frame to have a fully functional RFID system implemented. In this case, the ability to fulfill the time constraint may be a crucial factor in the decision whether to progress to the next phase. This mandate will be a fundamental issue to the organization if it has come from a major customer. In the context of this methodology, the question asked is, "will the organization start analyzing the Infrastructure Environment?"

Financial considerations are a major indicator in the economic survival of any organization. Depending on the economic position of the organization, it may be crucial that the costs are below a certain figure or the return on investment (ROI) target is met. In this case, financial metrics must be met to enable the organization to progress to the next phase. PTMs may have links to other PTMs, such as financial indicators, with a subgrouping of ROI, net present value (NPV), and a specific budget. Each of these factors may have to be met for the PTMs conditions to indicate progression to the next phase.

Another important consideration is the impact of legal implications when deciding to implement RFID technology. The organization may be mandated to implement this type of system due to changes in government regulations. In the business world, maintaining competitive advantage may be another reason to investigate this technology. Other possible PTMs may also include human or technical constraints; operational constraints; and information sharing between partners in a supply chain or integration with existing organizational processes and information systems.

Another point to consider is how the effect of a mandated or nonmandated business case influences the relationship and development of PTMs. If the business has come from a mandated business case, they may display low coupling to the outcome of the PTMs. This is pertinent to an organization in the supply chain if the mandate has come from an influential customer in respect of sale volumes. The organization may be required to implement RFID-enabled business cases if it wishes to continue in their present business environment. Therefore the outcome of the PTMs step may not be strongly linked to financial benefits. For example the NPV or ROI may not be as influential as the need for the organization to implement this mandate. On the other hand, the nonmandated business case may have a high coupling with the output from the PTMs. For instance, the ROI will have a strong correlation with continuing RFID deployment.

In the mandated business case, discretionary behavior will be reduced and this may impact on setting the parameters for the PTMs. In this scenario, it is necessary to link the goals and objectives of the business environment to the PTMs. If the goals and objectives are not met, then a further revaluation of the business case and RFID deployment will be necessary. Once the PTMs are developed, categorized and given a desired outcome, such as a monetary or a time-based objective, it is then entered into a PTMs template. Table 7.2 outlines the PTMs rating system for each phase or step of the methodology.

Table 7.2   PTM Template

PTM			
Weighting	*Feasibility*	*Outcome*	*Action*
Imperative	Viable	Successful	Continue
Recommended	Challenging	Reevaluation	Feedback
Optimal	Unworkable	Failure	Exit

From Table 7.2 it can be seen that a weighting is assigned to a PTM or step transitional motivators (STMs). The PTM relates to a phase and the STM relates to an internal step in a phase. For ease of reading the term PTM will be used for this section. Each PTM is then specified a feasibility scale, which relates to the ability of the organization to complete the PTM. Once analysis of the PTM is undertaken an outcome rating is assigned to the particular PTM. This occurs after a step or phase is completed. From this outcome an action is decided upon by the organization. This action may be to continue the process of evaluation, produce feedback to a previous phase, or exit the methodology. In the last case this relates to the cessation of examining RFID deployment.

In Table 7.3 the financial consideration is given an imperative weighting. In other words the financial metrics must be met otherwise the organization will not continue with the RFID deployment. In this case the business case was not mandated and financial outcomes were crucial to the decision-making process. All PTMs go through an iteration cycle in the methodology. Table 7.3 illustrates that if the first cycle outcome is failure then feedback is given to the previous phase or the existing phase. Then if the next iteration outcome is failure then the action is exit and the organization will cease the deployment. This will certainly be the case if the motivator weighting is imperative. Another example is evident in Table 7.3, where the organization has decided it wants to source the RFID equipment locally. The weighting is recommended and action is

Table 7.3   Financial and RFID Suppliers PTMs

PTM					
Iteration	*Motivator*	*Weighting*	*Feasibility*	*Outcome*	*Action*
1	Financial	Imperative	Viable	Failure	Feedback
2	Financial	Imperative	Challenging	Failure	Exit
1	Local RFID suppliers	Recommended	Challenging	Failure	Feedback
2	Local RFID suppliers	Recommended	Challenging	Failure	Continue

Table 7.4   Legal Considerations PTM

PTM					
Iteration	*Motivator*	*Weighting*	*Feasibility*	*Outcome*	*Action*
1	Legal	Imperative	Viable	Successful	Continue

to continue even though the outcome was a failure. Because the organization is unable to obtain the equipment locally this PTM's result will not stop the progression to the infrastructure phase.

This stage of the methodology will be a determinant of the number of steps the organization will follow. In other words if the PTMs are not congruent for the organization, then the examination of RFID deployment may well be canceled.

7.2.3.1   Working Case Study: National Supply Chain

Due to the lack of space the discussion of the PTMs or STMs in relation to the working case study will only give one example. In this case the organization is investigating the decision to continue to the infrastructure environment phase. Table 7.4 shows that the PTM was related to legal considerations for the business environment consideration. It was given an imperative weighting because it is crucial to conform to all regulations. The feasibility was determined to be viable at this stage of the deployment. The outcome was successful and the action for this PTM was to continue to the next phase.

The business environment determines the objective and scope of a business case. The next step is to determine the PTMs and analyze their outcomes and decide on an action. If the outcome is to continue then they will progress to the next phase. This phase examines the infrastructure environment and determines the feasibility of the business case in the context of an operational setting.

7.3   Phase 2: Infrastructure Environment: Manufacturer to Retailer

This phase of the RFID Rationale and Deployment Methodology investigates the infrastructure environment, with a key ingredient being the analysis of the physical characteristics of RFID. This understanding will be mapped to the business case and then to the use case, which will be enhanced or developed with a RFID-enabled process. Radio-frequency signals are vulnerable to impairment by interference from electrical or magnetic fields. In this phase of the deployment as shown in Figure 7.1, the organization is investigating where and how the use case is to be completed. The site assessment creates a picture of the likely interference where the use case is being performed and inputs information into the RFID equipment step. The STMs are an evaluation checkpoint for each step in this phase namely, the use case environment, RFID equipment, and design segments. Once the STM metrics for the use case environment are calculated and are consistent with the organizational requirements the next step in this phase is instigated. The second step is the RFID equipment and involves the configurations of the RFID readers and tags to optimize the identification of an object with minimal interference from other activities in the use case environment. The design step is further developed from inputs from the RFID equipment environment. This step is completed for each use case and all activities in the environment that need to interact with the object. Once examination of the PTM for this phase is completed, the organization may progress to the deployment phase.

7.3.1   Use Case Environment

The use case environment consists of a use case that is a component of the business case and the site assessment of each use case. The use case primarily relates to the identification of an object in an organizational setting. This will be further explained in the following sections.

7.3.1.1   Use Case

The details of each business case from the business environment are the basis for the foundation for this stage. Analysis and documentation completed during the preceding phase will set the direction in this step. The use case environment is related to the capturing of data from the interaction of a tagged object and a reader in a particular business setting. This is why it is necessary to have clear documented knowledge of the business case and its objectives and scope. This knowledge gives additional insight into how the use case will be enabled with RFID technology. The primary objective is to document how and where the data communications will take place so the site assessment will have a draft location blueprint. Further assessment of the use case environment can be gained from careful observation, interviews with employees, documentation, and prior experience of the system integrators.

This step requires questions to be defined and results analyzed to assist in the understanding of the use case.

Typical questions that need to be answered involve the following examples:

Q1: What business case does the use case belong to?

AQ: What is the primary objective and scope of the use case?
Knowing the business case and how the use case maps to it is an important consideration. This use case may be a subset of use cases that are required to fulfill the functionality of a particular business case. Knowing the links between use cases gives a greater understanding of the scope and objective of each data interaction. This is why it is important to know the context of a particular use case.

An example is the packing of an order onto a pallet:
First the empty pallet is retrieved (use case 1); the products are packed within a carton (use case 2); the carton is placed on a pallet; the pallet is wrapped (use case 3), and these three may circle till the order is complete.

The pallet/s are placed in a loading zone if the transportation is ready (use case 4); or they are placed in the competed order zone (use case 5).

This is the scope of the order processing business case. Each use case is documented and tested in a conceptual model and also in regard to the interaction between a reader and a tagged object in the infrastructure environment. Some useful modelling templates are the use case description and diagram in conjunction with the activity diagram. The following Figure 7.2 is a use case diagram showing the actors interacting with the system. This includes a wrapping station, an employee, and a forklift with a RFID reader. When a forklift picks up an empty pallet the system is notified (the reduction in pallet numbers), as the employee finishes packing a carton it is tagged and recorded (tagID). The pallet is then wrapped with plastic and is ready for delivery.

Q2: What are the data requirements for the use case?

AQ: What data is to be transmitted from or written to the tag?

AQ: What data parameters are required by the information system?

As shown in Figure 7.2 the data needed for the pick up empty pallet is the tagID of the pallet. In this case there is no data written to the tag. At the pallet wrapping station, the pallet tag is recorded, and the tagID for each carton is recorded, with a time stamp for each use case. The location of each read may also be recorded, depending on the organizational use case scenario.

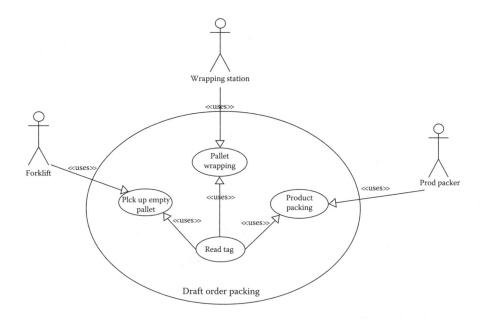

Figure 7.2   Use case of product order.

Q3: What is the location of the use case and interrogation zones (IZ)?

AQ: What RFID frequency spectrum and tag type is appropriate for the use case?

AQ: What is the ambient electromagnetic noise (AEN)?

AQ: What systems are in place if object is not identified?

AQ: Is this use case a link in a supply chain?

Documentation of the results for this step needs to be formulated into a format that is understood by the organization and the deployment team. The Unified Modelling Language (RFID) model called an activity diagram in conjunction with the use case description is a useful tool to document the processes that will require the transfer of RFID data. These models will also help in determining what activities occur before and after each data exchange. This knowledge can assist in optimizing the location of each interrogation zone (RFID) and highlight the data required for each RFID and how the state of the identified object may be changed due to this interaction. Information is also needed for the information systems and how and where the data will be stored, what systems require data from this process and where is this data stored.

The basic requirement of a use case is the identification of an object in an RFID. So from this step an organization may be able to build a template of generic RFID processes. Some generic examples of identification in a supply chain include: Identify object at door; identify object at wrapping station; identify object at transportation stage; identify object at inventory location; and identify object at assembly stages.

In the case of Ref. [14] survey participants identified shipping and transportation as the dominate area of utilization of RFID technology in the Department of Defence (DOD) supply chain. The use of item level tracking was still an area to be expanded and utilized. They also observed that passive tags with identification data were the primary RFID tag. The development of a generic template will assist in creating a standard procedure for use case analysis and documentation. For example, an organization may be investigating RFID at one location and these templates will assist in the scalability to another geographical setting with similar requirements. This template is shown in Table 7.5 and includes information from the use case environment, RFID equipment, and is used as a foundation of the design environment. With all templates, change may be required to fulfill the requirements of a specific organization deployment. The template is based on a use case description, though it is enhanced with information relating to the type and placement of a RFID reader, antenna and tag, etc. The template is updated, as more information is available during the completion of the infrastructure environment steps.

7.3.1.1.1 Working Case Study: National Supply Chain

The following questions from the previous section are discussed in relation to the working case study.

Q1: To What business case does the use case belong to?

AQ: What use cases were examined?

This question was entirely dependent on each member. One member investigated counting the number of pallets coming into and out of the business. Others examined the packaging and transportation of an order. In general, the use case is related to the identification of an object in the context of the organization. The use case varied between each entity in the supply chain. The work environment varied from a pallet supplier, to a manufacturer, a transportation company, product to a warehouse, and then the product to an individual store.

Table 7.5 Use Case Template

Use case	Wrap pallet			
Organization	Firm X			
Location	Wrapping station 3 Building B (see plan B)			
Operating conditions	Indoors, low humidity. The wrapping station is near the main entry point for the building			
AEN	Results from site assessment			
Object type/s	Pallet	Carton	Product	Item
Object tag	EPC passive RO	EPC passive RO	No tag	
System data	Time, location	Time, location	No tag	
Tag placement				
Successful read (per)				
Operating range	Tag to reader			
Frequency	UHF 918–926 MHz			
Reader	Model X (IP config) handheld/4 ports			
Reader placement	To be decided			
Antenna	A1	A2	A3	A4
Antenna placement	To be decided			
Preconditions	An order must be created and the pallet has been packed. Now wrapping the pallet is commenced			
Postconditions	The pallet is wrapped and identified and moved to the loading zone			
Exceptions	Items missing from pallet			
Notes	The tags have a preset EPC ID number			
	The state of the objects is waiting for shipment			
Additional requirements	Bollard needed for reader portal			

Q2: What are the data requirements for the use case?

All participants in the supply chain wanted to see the movement of an object from the manufacture to the showroom floor. This meant capturing and sharing data such as the time and location of an object. This knowledge is needed to supply the correct data to members up and down the chain. For instance, the pallet supplier wanted to know the movement of a pallet through the chain. Thus it was important that the receiver of a pallet included the pallet identification in a read and was viewable on the web-sharing portal.

Once the use case is documented the site assessment is commenced to analyze the use case environment for the organization. This is further discussed in the following section.

7.3.1.2 Site Assessment

The assessment of the site will involve a Full Faraday Cycle Analysis (FFCA) over a complete business cycle of a given use case. It is important to understand the likely objects that may or could cause interference in communications between a tag and reader. This interference may be due to some of the following examples: RF communication equipment such as alarm systems; cordless phones; and materials used in the construction of the building such as metal frames.

One important question for the entity and the deployment entity is, "are there activities undertaken in a nonpredicable or infrequent time cycle?" For example couriers who use RF equipment that collect specific items infrequently. Such questions affirm that the site assessment must be executed over a complete business cycle. This entails mapping the environment for ambient AEN during all business procedures. In the former case the assessing team must be aware of exception circumstances in the overall business cycle, to be able to capture all instances of possible reader to tag interference.

The primary reason to undertake a site assessment is to examine the environment where the network is to be deployed and the items that are to be tagged. During this process it is necessary to map the business process, so locations that entail the transfer or possible transfer of RFID data may be documented. This necessitates the determination of the interrogation zone for each location and the creation of a blueprint of the deployment zone. The client and system integrator requires specific knowledge of the placement of reading zones, thus the blueprint is crucial for installation success. This knowledge in turn allows the client to be aware of possible changes to business process to maximize the identification of tagged objects.

The following paragraphs will give an overview of the process to undertake a site assessment. The topics that will be discussed are: equipment required for a FFCA; how is the FFCA organized; what are the standard configuration parameters; and what are the outcomes from this examination?

The equipment necessary for FFCA includes the following: Spectrum analyzer; 1/4 wave dipole 922 MHz antenna with a ground plane plate, that captures electromagnetic waves over 360°, which is connected to the spectrum analyser; recording device connected to spectrum analyzer.

The equipment will be set up for a detection zone and readings will be recorded over the required use case time cycle. The spectrum analyzer is set to the middle of the UHF frequency for your country. The antenna is placed in the centre of the IZ, which will be dependent on each site scenario. The recording device is set to log data from the spectrum analyzer at a set interval or for specific business operations during the use case time cycle. From this one measurement process it maybe necessary to repeat this test for a number of reading zones to track potential interference to reader and tag communications. This is dependent on the noise data collected during the IZ analysis and the IZ scenario. It is important to have all machinery that may cause AEN to be passed through the interrogation zone so data may be collected on its operational noise. This may not be possible if only a conveyor belt assembly line is measured. Thus it is likely that a number of scenarios may need to be examined to determine the production of AEN.

While this operation is undertaken, all sources of noise need to be identified on the blueprints of the deployment zone and details recorded into the use case template. This blueprint is the primary output from this activity, with suggestions on the placement of IZ locations. From the site assessment it will be clear what the performance of the RFID will be and guidance on equipment selection, configuration and installation placement. It must be pointed out that each site will require analysis and no two sites will be the same.

7.3.1.2.1 Working Case Study: National Supply Chain

What were the outcomes from the site assessment?

It was observed that each organization had site specific issues. It was not possible to draw generic conclusions from a particular site assessment result to apply to another. One member had an issue with interference from an existing radio frequency system. This was overcome by creating a time frame for the identification of the pallet orders.

Sites had existing systems operating at the 950 MHz frequency and concerns were raised that there may be interference from the new RFID system. Analysis was completed and the resulting report suggested that there should be no problems.

7.3.1.3   Use Case Environment: Step Transitional Motivators

The setting of STMs for the use case environment is undertaken to gauge the effectiveness of this step. The following list of suggested motivators may be a starting point for an organization. It is necessary to know if the next step in this phase will be commenced.

STM 1: What is the level of AEN?
This will have a weighting of imperative. This motivator is crucial to the operation of a RFID system. If the AEN is too high, the communication between a reader and a tag in an IZ may not be workable.

STM 2: Use case re-engineering
Initially this would have an optional weighting but it may become necessary if the location of the IZ requires moving because of high level of AEN. There may be a need to make structural adaptations to the site environment, to optimize the communications between readers and tags. Further information will be available, once the RFID equipment step of this phase is completed.

If the business environment PTMs included an imperative time constraint or budget constraint, then these will need to be taken into consideration for this step. If the results from this use case environment are beneficial to the entity then the next step of the infrastructure environment is initialized, which is the RFID Equipment Environment. If all imperative STMs have a continue action then the organization will commence the RFID equipment step. If the action is to exit, then this step will progress to the PTM process of this phase with the results.

7.3.1.3.1   Working Case Study: National Supply Chain

Due to space constraints one STM or PTM for this phase will look at the NSC example in relation to the methodology.

This STM (Table 7.6) relates to the speculation of the RFID deployment causing interference to the existing system frequency. After evaluation of the site it was decided that there will be no cause for concern.

7.3.2   RFID Equipment Environment

The RFID equipment environment examines the hardware and software required to implement a RFID deployment. The testing and configuration of reader and tag communication is undertaken to obtain an optimal identification rate. The standards equipment and reader configuration will be discussed in the following sections.

7.3.2.1   Standards Equipment

Quality assurance dictates that the RFID equipment needs to be inspected before deployment. This inspection of the equipment checks that the readers and antennae operate within national guidelines for power, spectrum coverage and are operational. Testing of tags can be done in a controlled testing laboratory. This may be necessary to determine the minimum power, needed to activate the tag and to obtain an average tag metric. It is also important that optimal equipment is purchased for the required use case/s. Within a supply chain, an organization RFID system must be able to communicate with a customer or supplier systems. As mentioned before the radio-frequency spectrum is usually application-specific. It is imperative that before equipment is purchased, information about interoperability issues is avoided. Different standards between

Table 7.6   AEN Level STM

STM					
Iteration	Motivator	Weighting	Feasibility	Outcome	Action
1	AEN level	Imperative	Challenging	Reevaluation	Feedback
2	AEN level	Imperative	Challenging	Success	Continue

RFID equipment and country spectrum parameters may cause problems in a RFID deployment. There is a requirement to be aware of propriety systems as compatibility with an organization's future expansion or interaction with other organizations may be impaired. The development of the Low-Level Reader Protocol (LLRP) standard for networking readers, will address some of the interoperability issues in RFID technology. The use case template assists in the selection of RFID equipment. This particular template ensures that all necessary components are available for each reading area. This includes readers, antennas and uninterruptible power supply (UPS), etc.

The use of standards will assist the organization in reproducing a generic use case so it can be deployed into another organization setting. Standards are needed for: testing procedures; documentation; equipment specifications; spectrum parameters; procedures to procure equipment; and data transfer and format.

These protocols will guide all relevant members in a RFID deployment, whether they are part of a supply chain or a closed loop system.

7.3.2.1.1   Working Case Study: National Supply Chain

UHF standard of Australia was the standard used in the trial. This has the spectrum of 918–928 MHz and antenna power of 1 W. The tags used were passive of type class 1 generation. The readers were fixed or handheld devices. The fixed readers were connected to the network by an IP address and had the capability to communicate with up to four antennas. The consortium was supplied with readers and tags by equipment supplier members of the working case study. This ensured all members were sourcing similar equipment from the same suppliers.

7.3.2.2   Reader Configuration

The configuration of the readers and their antennae is critical to the identification of the tags. The conducting of an RF path loss contouring mapping survey is undertaken to optimize this formation of reader to tag communication. The signal of the reader and its antenna has to be investigated, to identify if the tag will be identified. Knowing the boundary of each IZ will clarify the distance needed for the antenna coverage and tag identification. To overcome cross reads, the reduction in the power of an antenna may be necessary. Alternatively if interference on one side of an IZ is affecting reads then the antenna/s on that side may have to have their power increased.

During the configuration it is important to note the position of the antennas and readers, which may change due to site constraints and interference. The actual position of the antenna will include their angle to the tag and IZ. Configuration of the reader includes some of the following: IP address; antenna ports; read cycle; tag sample rate; and connection to a controller and middle ware. This chapter does not consider the interaction and integration of the reader and organizational systems.

7.3.2.2.1   Working Case Study: National Supply Chain

There were problems with reading the tags on wet or freshly painted pallets. This identification issue was reduced by using foam and aluminum foil behind the tag as it was attached to the pallet.

7.3.2.3 RFID Equipment: Step Transitional Motivators

The development of STMs for the RFID equipment is undertaken to analyze the outcomes from this step. The following list of suggested motivators is a sample of what an organization may use as STMs.

STM 1: Successful object identification percentage
The weighting applied to this STM is imperative, though a secondary identification system may be needed.

STM 2: Optimize tag placement and orientation
This is related to STM 1 and will have an important impact on successful reads. Another link to this motivator is the configuration of the tag to reader communication.

7.3.2.3.1 Working Case Study: National Supply Chain

During the configuration of readers it was found that the physical port numbers did not coincide with one of the reader's software port numbers. The reader was set to read from 0 to 3 but it was trying to read from ports 1 to 4. This was a standards issue and the decision was to continue with the deployment. This problem highlights the need to test equipment before it is deployed (Table 7.7).

7.3.3 Design Environment

This step in the infrastructure environment develops the required design to fulfill the objectives and scope of a use case. Inputs from the preceding steps of this phase are used as the foundation for this step. The output is a collection of documents pertinent to the deployment of the RFID system.

7.3.3.1 Design

The design is looking at developing the infrastructure blueprints to enable RFID technology to be used within a use case. The design step is the implementation plan and operational document of a RFID deployment. Designing a RFID deployment requires knowledge of the actual positioning of readers, tags, antennae, and data requirements. This includes creating a plan for the positioning and location of electrical cables,

Table 7.7   Reader Standards STM

STM					
Iteration	*Motivator*	*Weighting*	*Feasibility*	*Outcome*	*Action*
1	Reader standards	Recommended	Viable	Reevaluation	Feedback
2	Reader standards	Recommended	Challenging	Reevaluation	Continue

uninterrupted power supply, earth locations for grounding the equipment, and local area network connection points. Additional settings are created for preinstalled baseline configurations of site scenarios such as conveyor situation, pallet bay, and dock door for readers and antennas that are analyzed in the site assessment. The design step also produces documents relating to information systems and servers requirements, such as what site infrastructure is needed to contain the readers and antennas and placement of tags on objects.

Documents from the use case environment, the RFID equipment, and the business environment are utilized for this stage of the methodology. One necessary input is the use case template which may be documented for identification of an object at a wrapping station, product shelf, or antitheft prevention (entry points to retail stores). The template parameters may need tweaking for each implementation, as a particular site may have inference that is not occurring at another site.

When designing the installation procedures and diagrams for equipment in a business process, it is important to minimize the disruption to the work flow. For example the positioning of readers and antennas may require a change in direction for vehicles due to the placing of RFID infrastructure. Generally it will be necessary to install protective guards to the reader station as it may be hit by machinery. Once the design is finalized, the documentation step is commenced.

7.3.3.1.1 Working Case Study: National Supply Chain

It was noted that the thorough understanding of the use case and how its processes meld with the RFID design plan. The development of user interfaces and the new processes must take into consideration the personnel involved.

7.3.3.2 Documentation

The documentation contains the instructions and protocols for the deployment of one to many use cases. It is also the central repository for all documents produced in this phase. It is efficient to divide the documentation into operational, design, and installation partitions, because not all documents will be used by participants in the implementation of the system. Example of documentation include: training materials for employees; a complete bill of materials; testing procedures for the deployment; risk assessment; human resource consideration in relation to deployment, and user acceptance testing protocols.

Quality assurance of documents and the mapping of design parameters back to the use case is an important procedure in this stage. This mapping will help discover information that may be missing or incorrectly understood. This is particularly important when planning a supply chain deployment.

7.3.3.2.1 Working Case Study: National Supply Chain

It was crucial to have accurate documentation of use cases and their interactions with the organization. Since this was a supply chain, it was necessary to have a clear understanding of the scope and data requirements for all participants.

7.3.3.3 Design: Step Transitional Motivators

The design step motivators if fulfilled progress this stage to the PTMs step of infrastructure environment. Possible motivators include:

STM 1: Regulations compliance
Imperative because you do not want noncompliance, for example, with electrical cabling or deployment protocols occurring.

STM 2: Accurate documentation
This is an imperative STM and must map with organizational business case requirements. It is important to have clear documentation of the use case and its deployment and integration with existing or new systems. Depending on the scenario a recommended weighting is applied.

7.3.3.3.1 Working Case Study: National Supply Chain

An example of a STM for the design step was the production of accurate documentation (Table 7.8). It was clear that adequate information had to be shared between the different businesses.

7.3.4 Infrastructure Environment: Phase Transitional Motivators

The PTM in this phase is the gateway to the deployment environment, only if the action from the motivators is to continue. If the action is feedback then this may be internal to the business environment. If it is a second iteration and the action is exit then the organization may cease examination. Examples of possible PTMs are:

PTM 1: System compliance with regulations
This relates to government, work place safety regulations, and RFID equipment. Given an imperative weighting the system must comply otherwise financial penalties may occur.

PTM 2: Mapping of use case to the business case
This phase analyses the scope and objectives of the business case in relation to a use case.

7.3.4.1 Working Case Study: National Supply Chain

A motivator that may be applied to the NSC is the location of IZ (Table 7.9). This was important as it showed state changes to the object. In other words the identification

Table 7.8 Accurate Documentation STM

STM					
Iteration	*Motivator*	*Weighting*	*Feasibility*	*Outcome*	*Action*
1	Accurate documentation	Recommended	Viable	Successful	Continue

Table 7.9 Interrogation Zone

PTM					
Iteration	*Motivator*	*Weighting*	*Feasibility*	*Outcome*	*Action*
1	Location of IZ	Imperative	Challenging	Reevaluation	Feedback
2	Location of IZ	Imperative	Challenging	Successful	Continue

of an object at certain interrogation zones produced a change in ownership of the object.

Now that the infrastructure environment has been completed the organization can progress to the deployment environment phase of this methodology.

7.4   Phase 3: Deployment Environment: Factory to Showroom

The deployment environment examines the outcomes from operating a use case process once it is enabled with RFID technology. As seen in Figure 7.1 there are two different stages in this phase namely, the prototype testing and pilot environments. The prototype testing is completed first and if successful the pilot is commissioned.

7.4.1   Prototyping Testing Environment

The prototyping test environment examines the use case/s that is part of a business case. If the outcomes of the STMs have an action of continue then the organization will commence a pilot study. Otherwise the methodology progresses to the PTM step.

7.4.1.1   Use Case

As mentioned in the business case section and shown in the use case of product order diagram (Figure 7.2), a business case may have one to many use cases. The prototype test looks at the integration of the uses case required to fulfill the scope and objectives of a business case. It is necessary to test these pieces (use cases), so they are able to complete one puzzle (business case). The individual use case IZ has been tested and designed during the infrastructure environment phase. Deployment and installation procedures developed during that phase are utilized for this use case, with refinements as needed. One objective of this phase is to understand how a use case operates during an operational situation.

The configuration of readers, antennas, and tags has already been documented. It is appropriate to confirm that the scope of the use case is operating at its required level. For example in the use case diagram (Figure 7.2) the contents of a pallet is to be identified at a wrapping station. This scenario is enabled with RFID technology and the results are compared with the use case operational metrics. These metrics can include: rate of identification; correct data elements captured; uptime, operational cost, and organizational information integration. Once this use case is validated then the next use case is tested. This continues till the performance measures of each use case has been tested and successful in regard to the business case.

In the case of a supply chain, each organization will test one-scope business case that has been agreed upon. They would test one process that allowed the designated objective to be identified whilst moving amongst the links in the chain. An example is the tracking of objects at a pallet and carton level. Only some members would have to identify all objects, such as pallet and cartons during the supply chain interactions. So to enable this object visibility to occur, each organization would have to test the required use case to perform this activity. Once the required use cases are operational within each organization then a pilot can be initiated.

This modular approach to the testing prototype will allow the organization to see if it is possible to implement a pilot for the selected business case. The organization may gain a picture of the possible outcomes for each use case, without major disruption and stretching of organizational resources. This includes the change management program, as alterations are slowly integrated into the organizational operations. Evaluation of results and lessons learnt can be recorded into the next piece of the puzzle. This creates a detailed assessment of the possible benefits of RFID technology before the pilot is initiated. Once the selected business case has been approved, it is now ready to be used for a pilot deployment.

7.4.1.1.1 Working Case Study: National Supply Chain

Though the NSC did not have a designated prototyping test environment step, it was important to continually test before deploying a pilot RFID system to overcome issues that had not been documented. For example, when the new RFID frequency was deployed it became apparent that it interfered with an existing forklift systems. This was previously tested and the results showed there should not be problems. The lesson from this is to continually test before deployment.

7.4.1.2 Prototyping Testing Environment: Step Transitional Motivators

The setting of STMs for the prototyping test environment will be undertaken to gauge the effectiveness of this step. The following are indicative list of suggested motivators:

STM 1: Use case integration
This will have a weighting of imperative. This motivator is crucial to the operation of a RFID system.

STM 2: Staff training
This STM has an imperative weighting and will improve the acceptance and success of a pilot system.

Table 7.10   Use Case Testing STM

PTM					
Iteration	*Motivator*	*Weighting*	*Feasibility*	*Outcome*	*Action*
1	Use case testing	Imperative	Challenging	Reevaluation	Feedback
2	Use case testing	Imperative	Challenging	Successful	Continue

7.4.1.2.1   Working Case Study: National Supply Chain

Applying a STM to the NSC in light of the methodology (Table 7.10), the continual testing of a use case before the pilot would overcome any unforseen problem, especially the forklift system interference.

7.4.2   Pilot Environment

A well-planned, scalable and scoped pilot study will be initiated, once the selected scenario is approved by all parties. The pilot will allow further examination and determination of the capabilities of the RFID system, before a complete integration is completed or contemplated.

7.4.2.1   Use Case

The identification of the use case/s will come from the output of the prototype testing environment. This may be a mandated for the most promising fully scoped business case that has been decided upon. The purpose of the pilot is to confirm the results of the previous step with a longer time frame of operation.

Contingencies must be in place to overcome difficulties that may arise, especially system downtime and missing or incorrect data capture. Tag placement and operational life expectancy will become more apparent as the trial proceeds. The tags may need to be moved or replaced because of how the objects are handled. For example, the tags adhered to a particular object may become dislodged over time, so an alternative tag type may need to be sourced.

The overall increase in data flow and the need to store and transform it into meaningful information will become more apparent. This is especially the case in a supply chain. Incorrect data may be uploaded from a particular partner, due to crossed reads in one IZ. This will necessitate the identification of the source of data and then the reconfiguration of that interrogation zone.

Metrics testing is similar to user acceptance testing of a new software product. The user organization has a sign-off agreement, that if the conditions are satisfied then that part of the deployment is successful. This may be an in-house installation or organized by a third party. One example is parameters are determined for accuracy rates, the number of successful reads in a given area and how the tests will be performed. For instance, the client wants to read 30 cartons on a pallet, which is being moved by a pallet jack through a dock door. The accuracy rate required is 99.5 percent and the test is to be carried out 10 times. It is important to have a measurable parameter, that gives both parties a sign-off milestone for a particular scenario. Once this has occurred then a possible pilot study may be initiated.

7.4.2.1.1 Working Case Study: National Supply Chain

The NSC pilot was a successful demonstration of the capabilities of tracking desired objects through a supply chain. The participants were able to increase the visibility and identification of pallets and cartons more efficiently than using conventional bar codes. Shared information was available in real time and this enhanced decision making for the organizations. Some important lessons included: full scope and documented use cases; internal and external organizational support for the pilot; the use of experienced RFID partners; continual testing and refinement of use cases; slowly expanding RFID implementation; and development of base line metrics before RFID is installed. Interesting hurdles included: lack of existing standards causing interoperability constraints with equipment; interference from existing systems; and the amount of incorrect data.

7.4.2.2   *Pilot Environment: Step Transitional Motivators*

The setting of STMs for the pilot environment will be undertaken to gauge the effectiveness of this step. The following list of suggested motivators may be a starting point for an organization.

STM 1: Milestone determination
For example, the sign-off agreement will be conditional on the metrics assigned for that use case. Clear objectives and performance measurements are integral for agreements between organizations.

STM 2: Change management performance
If the result from this STM is not positive, then the likelihood of the pilot being a success will be undermined.

7.4.2.2.1   Working Case Study: National Supply Chain

An example of a STM for the pilot environment for the NSC is the importance of successful training of the personnel (Table 7.11). Without the understanding of changes in work practices the case study would not have been able to perform the pilot successfully.

Table 7.11   Employee Training STM

PTM					
Iteration	*Motivator*	*Weighting*	*Feasibility*	*Outcome*	*Action*
1	Employee training	Imperative	Challenging	Reevaluation	Continue

Table 7.12   Mandated Business Case STM

PTM					
Iteration	*Motivator*	*Weighting*	*Feasibility*	*Outcome*	*Action*
1	Mandate	Imperative	Challenging	Reevaluation	Feedback
2	Mandate	Imperative	Challenging	Success	Continue

7.4.3   Deployment Environment: Phase Transitional Motivators

What are the motivators that produce the transitional motivation to progress to a mass production or exit from the methodology? The PTMs developed in the business environment may still have a bearing in this and the previous phase. This is valid with time constraints; financial considerations; regulation compliance; operational considerations.

PTM 1: Fulfillment of business case scope and objectives
If this PTM is successful then this business scope will become a production unit.

PTM 2: Supply chain agreement
If all the parties agree that this particular pilot was successful then it may well become a production unit. This would be given a weighting of optional, depending if it was a nonmandated approach.

7.4.3.1   Working Case Study: National Supply Chain

A likely example of a PTM for the deployment environment is the agreement between all parties or if it becomes mandated business case (Table 7.12). The NSC was a pilot study that had a defined lifetime to demonstrate the capabilities of a RFID supply chain.

7.5   Conclusion

The need to analyze, plan, coordinate, test, and document the rationale and deployment of RFID technology is crucial for a successful integration into an organization. This chapter outlined a methodology of RFID deployment using a NSC case study as a pointer to the methodology. The instigating environment will affect the discretionary behavior of an entity and how it chooses to follow the deployment methodology. It is important to analyze and scope the business cases, so a clear objective and performance metric is defined. The infrastructure environment examined the physics and technical needs of RFID that must be understood and the use case, which was enhanced or developed with a RFID process. An important key to possible success is the consultation and training of all participants and this requires a change management process to be initiated. A thorough understanding of the implementation environment and a pilot study will enable the business to gauge the effectiveness of a RFID deployment.

References

1. Myerson, J. 2007. *RFID Business Processes RFID in the Supply Chain: A Guide to Selection and Implementation.* Boca Raton, FL: Auerbach Publications.
2. Williamson, O. E. 1964. *The Economics of Discretionary Behavior: Managerial Objectives in a Theory of the Firm.* Englewood Cliffs, NJ: Prentice-Hall.
3. Landt, J. 2001. Shrouds of time—The history of RFID. Tech. rep., AIM Inc.
4. Poirier, C. and McCollum, D. 2006. *Chapter 9—Building the RFID Business Case and Roadmap for Execution RFID Strategic Implementation and ROI: A Practical Roadmap to Success.* J. Fort Lauderdale: Ross Publishing.
5. Whitaker, J., Mithas, S., and Krishnan, M. 2007. A field study of RFID deployment and return expectations, *Production and Operations Management,* 16(5): 599–612.
6. Lucas, H. 2005. *Information Technology: Strategic Decision Making for Managers.* Hoboken, NJ: Wiley.
7. Fisher, J. and Monahan, T. 2008. Tracking the social dimensions of RFID systems in hospitals. *International Journal of Medical Informatics* 77(3): 176–183.
8. Wang, F. and Liu, P. 2005. Temporal management of RFID data, in *VLDB 05: Proceedings of the 31st International Conference on Very Large Data Bases* (Trondheim, Norway, August 30–September 02, 2005). Very Large Data Bases. VLDB Endowment, pp. 1128–1139.
9. Jeffery, S., Alonso, G., Franklin, M., Hong, W., and Widom, J. 2006. A pipelined framework for online cleaning of sensor data streams, in *ICDE 06: Proceedings of the 22nd International Conference on Data Engineering (ICDE06).* Washington, DC: IEEE Computer Society, pp. 140.
10. Jain, J. and Das, S. 2006. Collision avoidance in a dense RFID network, in *WiNTECH 06: Proceedings of the 1st International Workshop on Wireless Network Testbeds, Experimental Evaluation and Charaterization.* New York: ACM Press, pp. 49–56.
11. Levinson, M. 2003. The RFID imperative, CIO Magazine, December 1, 2003. http://www.cio.com/article/32004/SuccessfulUseofRFIDRequirestheRight Infrastructure (accessed October 10, 2007).
12. Niederman, F., Mathieu, R., Morley, R., and Kwon, IK-Whan. July 2007. Examining RFID applications in supply chain management, *Communications of the ACM,* 50(7), 92–101.
13. Owens, J., Chalasani, S., and Sounderpandian, J. 2005. Use of RFID in supply chain data processing, in *Encyclopedia of Data Warehousing and Mining,* John Wang (Ed.). Idea Group, Information Science Publishing, pp. 1160–1165.
14. Wagner, M., Clark, J., and Thomas, C. RFID industry survey: Measuring RFID use and performance in the DoD supply chain, March 2008, www.xiostrategies.com (accessed April 20, 2008).

WIRELESS SENSOR NETWORKS

Chapter 8

Geographic Routing in Wireless Sensor Networks

Juan A. Sánchez and Pedro M. Ruiz

CONTENTS

8.1   Introduction

A wireless sensor network (WSN) consists of a set of autonomous devices called sensor nodes, equipped with short-range wireless interfaces and hardware for monitoring environmental parameters, such as humidity, pressure, and temperature. Sensor nodes are also equipped with a small microprocessor, and they are usually powered by batteries. Sensors use other nodes as relays to send messages to other sensors or data sinks, which are not within their coverage area. Thus, using efficient routing protocols is of paramount importance to support the distributed operation of WSNs.

WSNs are, in a way, similar to the traditional wireless ad hoc networks because both are unstructured networks consisting of an undetermined and variable set of nodes, and both use wireless interfaces to communicate among themselves. Thus, the deployment of WSN's applications might benefit from the use of the same routing protocols already defined and tested in wireless ad hoc networks. Nevertheless, the density of node deployment; the frequent topology changes, which is one of the characteristics of WSNs; the unreliability; and the severe power, computation and memory limitations of sensor nodes, prevent traditional Wireless Ad Hoc routing protocols from working properly in WSNs.

Geographic routing is a routing technique that relies on the geographic position information. First introduced by Finn [1], and Takagi and Kleinrock [2] in the 1980s, it was proposed mainly for wireless networks. Geographic Routing (also called georouting or position-based routing) is based on the idea that the source sends a message to the geographic location of the destination instead of using its network address. In Geographic Routing, nodes take routing decisions based solely on local information. Nodes need to know only their own positions, the position of the destination, and the positions of their neighbors. With this information, a message can be routed to the destination without the knowledge of the complete network topology or a prior route discovery mechanism.

The lack of global knowledge means that Geographic Routing cannot provide globally optimal solutions. Therefore, approximations are widely used in the literature. In fact, the localized operation of these protocols creates two common problems: routing loops and dead ends. There are several different approaches to tackle these problems, especially for the unicast case. Most of them are based on, or are extensions of, two well-known techniques: greedy forwarding and face routing. Greedy forwarding tries to avoid routing loops by bringing the message closer to the destination hop by hop. Thus, each node forwards the message to the neighbor that is most suitable (always from a local point of view).

However, greedy forwarding can lead to a dead end, that is, a node having no neighbors more suitable than itself. Face routing helps to recover from such kind of situations by finding a path to another node, where greedy forwarding can be resumed. A dead end is typically placed on the border of an area not covered by any sensor.

These empty areas are also called void areas. Basically, face routing (also called perimeter routing) is used to surround void areas.

The multicast case is even more complex than the unicast one, and only a small number of geographic multicast protocols have been proposed to date. Here, the challenge is reaching a set of nodes by building an efficient communication tree. The high number of nodes typical of WSN requires that each node in the network has several different one-hop neighbors, to whom the message can be forwarded. Thus, the number of forwarding alternatives grows exponentially. Additionally, the shape of a multicast tree directly determines its efficiency and, therefore, the good or bad usage of the scarce resources of sensor nodes.

Besides, some recent works propose a new approach for Geographic Routing protocols called Beacon-less Geographic Routing. The idea is to eliminate one of the common assumptions made by most works in the literature; concretely, the use of periodic short HELLO messages (also called beacons) among one-hop neighbors, as a way of neighborhood discovery mechanism. Recent results confirm that these beacons can cause several problems to routing protocols, such as interferences and collisions. In addition to that, they can be considered as the origin of a waste of bandwidth and energy, especially in those nodes not taking part in any routing task.

Finally, it is important to point out that Geographic Routing protocols are not completely free of problems. Some aspects are not clearly solved yet, such as the way nodes determine the coordinates of their destinations (localization problems). There are also some scalability issues of multicast protocols related to the limited size of messages that sensor nodes can manage. Moreover, the use of the unit disk graph (UDG) to model WSNs can lead to design protocols not well adapted to work in real scenarios. That is, the coverage range cannot be assumed to be constant because it hardly depends on the environmental conditions. These are only some examples of the problems that the protocols designed to date can experiment, when used in real scenarios.

This chapter studies the different approaches proposed to date in the field of Geographic Routing for WSNs. Starting with the unicast scenario, we overview the most important protocols in the literature, as well as their possible problems. We also discuss the multicast case, and then, we present the most recent beacon-less geographic works in the literature. Finally, we conclude this chapter with an overview of the major problems that the Geographic Routing protocol designers must face in the next years.

8.2 Geographic Routing Fundamentals

8.2.1 Preliminary

Motivated by the performance and scalability problems of table-driven routing protocols, Finn [1] defined the basis of what we currently call Geographic Routing. Finn was looking for a way to reduce the unacceptable overhead introduced by routing protocols to cope with large networks. Besides, he noticed the inability of routing protocols to deal with mobile nodes, and he additionally wanted to reduce the considerable amount of memory and computation power needed by routers to manage routing tables.

Finn's idea was simple but effective. Instead of using addresses to identify nodes in the network, he proposed to associate a unique location in a Cartesian coordinate system to each node in the network. With this scenario in mind, the routing task is reduced to selecting the next router, to which the message has to be forwarded. This creates a multihop path toward the destination. Given a destination node and its associated location, each router can determine, among the routers it is connected to, the one located closest to the destination. Then, forwarding the message to that selected router takes it closer and closer to the destination.

Nowadays, most routing challenges come from the world of wireless networks, in which totally decentralized and unstructured ad hoc topologies have to deal with heterogeneous mobile nodes. Among them, WSNs are a special case. A WSN can be formed by an enormous number of nodes. Additionally, sensor nodes are characterized by their constrained resources: memory, energy, and computation power.

Most routing protocols proposed for multihop networks are based on the work done by Bellman and Ford [3,4]. Some well-known examples are the Ad hoc On-Demand Distance Vector [5] (AODV) and the Optimized Link State Routing [6] (OLSR). These protocols are representatives of two different routing schemes. AODV is the representative of the set of reactive routing protocols (also called on-demand protocols). These protocols postpone the finding of routes to the moment at which they are needed. On the other hand, OLSR represents the set of proactive or table-driven protocols, which try to keep updated the routing information along time. The updating process is done independent of the necessity of that information.

In the last years, several studies have shown that reactive- and table-driven-based routing protocols, as well as the plethora of existing mixed approaches, cannot cope with the high requirements imposed by wireless networks, especially those of the WSN. Thus, with the adaptation of Finn's work to wireless networks, geographic routing solutions have gained a lot of momentum.

Unlike the traditional distance vector, or link-state-based routing protocols, Geographic Routing algorithms do not require the interchange of routing tables among 1-hop neighbors. Nodes take routing decisions solely based on their positions, the position of the destination, and the positions of their 1-hop neighbors. These decisions are taken using geometric criteria, so that the node's positions can be defined in a common coordinate system. Usually, the coordinates used are defined in the standard $\mathbb{R}^2$ Cartesian plane, but it is possible to use any coordinate system in which a metric distance can be computed.

The advent of wireless communications changed the way of designing networks and routing protocols. Unlike in wired networks, routers using wireless interfaces are able to communicate with every other device located close enough to be reachable. That depends on the relation between the coverage range of the wireless interfaces and the distance among routers. Obviously, the environmental conditions can be also a decisive factor. Thus, Finn's idea of using the position to identify network nodes is even more appropriate in the case of wireless networks because here, the location, and therefore, the distance among nodes is crucial to determine the connectivity.

Moreover, taking routing decisions in such a way makes it unnecessary to keep routing tables. Thus, this intrinsic, stateless property is ideal for building routing protocols that scale with the number of nodes in the network. Besides, as the routing decisions are taken using geometric operations, the computation power requirements are very low. These two properties of position-based algorithms make them very suitable for being run by sensor nodes.

Finally, the fact that wireless communications are held in a shared medium can help reducing the bandwidth consumption, especially when more than one nodes have to receive the same message. This effect is called wireless multicast advantage and can be seen as a free form of broadcast. All the devices located in the coverage area of a wireless sender receive the message; thus, a single transmission is enough to reach more than one node. This is especially useful in the case of multicast communications, as we will see below.

8.2.2 Geographic Routing Operation

Since the adoption of Finn's idea of cartesian routing, there have been several geographic routing proposals. Every proposal presents a new concept, an idea, or a solution to a concrete problem, but almost all algorithms share the same basic structure. Most aspects of this common structure have a direct influence on the performance and scalability of Geographic Routing protocols. Moreover, the differences between unicast and multicast approaches can be better understood keeping in mind the way these algorithms work.

Usually, all nodes deployed in a WSN consist of the same routing protocol. Source nodes can send messages at regular intervals, or when external events occur, but intermediate nodes send messages only after receiving and processing them. Those nodes taking part in a routing task without being the source or the destination are considered intermediate nodes.

We can divide the operation of a standard Geographic Routing algorithm in the following four phases:

1. *Determining the destination's coordinates.* The destination's coordinates are determined by the source of the message. These coordinates are included in the header of the message for the information of intermediate relays. Coordinates play the role of addresses in Geographic Routing protocols. We will see later that the determination of the destination's coordinates is an additional problem. There are some proposals to solve this problem but, in general, none of them is completely satisfactory.
2. *Determining 1-hop neighbor's coordinates.* The 1-hop neighbor's coordinates are usually obtained through the periodic interchange of short messages called beacons. Beacons contain the identifier of the sender and its coordinates. Beacons allow nodes to have an idea of their neighborhood because WSN topologies are unstable. The reason for this is that although sensor nodes do not normally change their positions, to save battery they normally work in a duty cycle. That is, some periods of time sensors are fully operational and the others are in the sleep mode.
3. *Determining the next relay.* The next relay is determined using the information about the position of the current node, the destination's coordinates, and the

coordinates of the 1-hop neighbors. The function in charge of taking this important decision is usually called the next-relay selection function. Most Geographic Routing protocols differ only in the definition of this function.

4. *Message delivery.* After choosing which 1-hop neighbor will become the next relay, the last step consist of delivering the message to it. Usually, the message header includes the routing assignment, that is, the association between the destination and the identifier of the node selected as the next relay. All 1-hop neighbors receive the same message but only the one indicated in the routing assignment must continue routing it. Obviously, the multicast case is slightly different because some nodes, the ones having tree branches, must deliver the same message to more than one neighbor. In this case, the header of the message can include several routing assignments.

These four phases are general enough to cover the different schemes followed by the Geographic Routing protocols proposed in the literature. Additionally, we can analyze some implicit assumptions made by most authors when designing Geographic Routing protocols. The most important is the assumption about the ability of the source node to determine its own location and the location of the destination (or set of destination nodes). These are in fact bigger problems than the routing itself.

Geographic Routing protocols are based on the fact that nodes have coordinates, but most proposals do not deal with the problem of how nodes determine their own location. It is normally assumed that sensor nodes incorporate some sort of GPS-like device, but this is not possible in all the scenarios. Until very recently, GPS devices used to consume too much energy and were too big for being equipped in a typical sensor node operated by batteries. Nowadays, there exist some GPS devices with a very reduced energy consumption and size, but GPS technology does not work properly in indoor scenarios.

On the other hand, the proposals to discover the location of a destination node are diverse. We can find solutions ranging from classical centralized schemes, similar to the domain name system (DNS) on the Internet, to very complex schemes based on distributed hash tables in charge of registering the location of the nodes located in certain areas. Each solution presents its own set of advantages and drawbacks, but all of them imply some level of undesired overhead. This chapter is devoted to routing, hence, we do not cover all these issues. Thus, we also assume along the text that the destination's coordinates are available.

Besides, the process of determining the coordinates of 1-hop neighbors is also a key factor, normally ignored by most protocols. That is, authors assume that each node in the network has a permanently updated table, including the identifiers and locations of all its 1-hop neighbors. This is usually achieved by the periodic interchange of short messages called beacons, which include the necessary information. Nevertheless, recent studies [38] have shown that, even having a very short length, these beacons can influence negatively on the overall performance of the routing protocols. In fact, interferences and collisions produced by beacons can degrade the throughput of the network significantly. These problems are not normally reported in the literature because beacons are not normally taken into account during the simulation phase. However, the effect is appreciable when testing real deployments.

To solve these issues a few proposals, usually called Beacon-less Geographic Routing protocols, have been suggested. In these protocols, the determination of 1-hop neighbor's coordinates is reactive. Instead of proactively interchanging beacons, nodes routing a message search for their current neighbors, using a 1 hop limited kind of broadcast. The results obtained to date are very promising. Some tests performed in real scenarios have shown that Beacon-less protocols adapt better than standard Geographic Routing protocols to interferences and collisions, which are inherent to wireless communications.

8.3   Geographic Unicast Routing

Working with coordinates and having a single target (the unicast destination) means that from every possible node location does exist a clear notion of the direction the message should follow. Therefore, every node can easily determine what is the best possible multihop path from itself to the destination. Let d be the distance between the node and the destination, and r the maximum radio range. The best path is a straight line to the destination containing d/r hops. It is very unlikely for a node to have a neighbor just where the best possible path would start, but knowing this ideal position it is possible to determine which neighbor could be the best next-forwarder.

Nevertheless, the next-relay selection function uses only local information, so that, in the best case, the decision can be locally optimal, but no guarantee of global optimality can be offered. That is the reason why the different protocols in the literature use different functions; each one uses a different heuristic to estimate the goodness of every possible next-relay candidate.

Additionally, there are situations where it is not possible to follow the ideal path. In fact, the path might reach a node with no neighbors closer to the destination than itself. These situations, called local minima, must be managed in a different way. Thus, Geographic Routing algorithms use two different operation modes: greedy and perimeter. The greedy mode is used whenever possible, and the perimeter mode is strictly used to scape from local minima.

8.3.1   Greedy Scheme

One of the first Geographic Routing protocols is Most Forward within Radius (MFR) by Takagi and Kleinrock [2]. MFR defines the concept of progress to measure the goodness of each candidate to become the next-relay forwarder. The progress of a node is defined as follows. Given s, the node currently holding the message, d, the destination node, and n, a 1-hop neighbor, the progress of n is defined as the difference between the length of the segment, $\overline{sd}$ and $\overline{nd}$ (see Figure 8.1). The next-relay selection function of MFR works on computing the progress of each neighbor and selecting as the next relay the one whose progress is the largest.

Later, the Cartesian Routing by Finn [1], which is an improved version of the Random Protocol [9], proposed a simpler next-relay selection function. A node holding a message selects the neighbor closest to the destination as its next relay. This

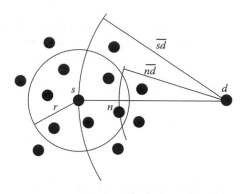

Figure 8.1 In MFR, the 1-hop neighbor selected as the next relay is that providing the most progress. Neighbor *n* has a progress of $\overline{sd} - \overline{nd}$.

algorithm was designed for wired networks consisting of network nodes located in a more or less regular geometric structure. However, when trying to use this protocol in a WSN we can easily find examples where the protocol enters into a cycle. For example, two nodes located at the same distance from the destination will interchange indefinitely the message, because each node is the other's neighbor closest to the destination.

A different criteria is used in the Compass Routing by Kranakis et al. [7]. Here, nodes consider the slope of the segments between itself and each neighbor. The 1-hop selected as the next relay is the one whose slope is closer to the slope of the line passing through the current node and the destination. This protocol can also lead to failure, as shown by Stojmenovic and Lin in Ref. [8].

8.3.2 Perimeter Scheme

As mentioned before, Geographic Routing protocols need mechanisms to recover from those situations that are usually called local minima or void areas. Some protocols, such as the Cartesian Routing, propose to use restricted flooding when the message being routed arrives at a dead end. However, this cannot be considered as a pure Geographic Routing technique. Besides, the probability of generating duplicate messages, and the overhead introduced, make these solutions unappropriated for WSN.

Most authors model WSNs as graphs whose vertices represent sensor nodes, and edges represent the existence of direct communication between nodes. This is very convenient because there are some well-known algorithms from the graph theory that are useful to solve certain routing problems, for example, Dijkstra's algorithm to find the shortest path in a graph.

The right-hand rule [13] is an algorithm that can be used to find out the exit of a maze. The algorithm works as follows. The player walks keeping one hand in contact with one wall of the maze, for example his right hand. If the maze does not contain disconnected sections, the player is guaranteed to reach an exit if there is one. In a WSN,

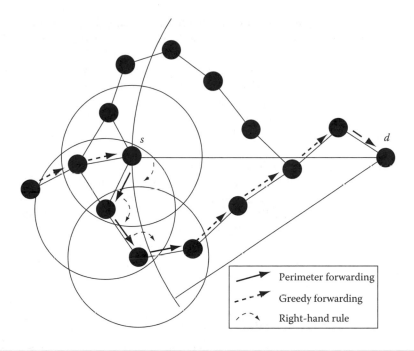

→	Perimeter forwarding
- - →	Greedy forwarding
⌒ ↖	Right-hand rule

Figure 8.2 Example of applying the right-hand rule during perimeter forwarding to surround a void area. Forwarding toward *d*, the node *s* is a local minimum; therefore, *s* initiates perimeter forwarding until it reaches a node closer to *d* than *s*, where greedy forwarding can be resumed.

a void area can be seen as a chamber in a maze where the walls are defined by the edges of the graph modeling the network. Starting in one of the vertices of the void, it can be surrounded by applying either the left- or the right-hand rule (see Figure 8.2).

The only requirement for this algorithm to work is the nonexistence of crossing edges, that is, the graph must be planar. As one of the keys of a Geographic Routing algorithm is its localized operation, we cannot assume nodes to know the whole network topology. Each node knows only its local vision of the whole graph, that is, its local subgraph.

Fortunately, there are some topology control algorithms that can be used to obtain a planar view of the portion of the graph that each node can locally know. These planarization algorithms can be applied locally [14]. Concretely, these algorithms work only with the subgraph in which the current node and its 1-hop neighbors are embedded.

The most commonly used planarization algorithms are gabriel graph [12] and the relative neighborhood graph [11]. They work by locally removing the edges connecting the current node with its neighbors. There are other algorithms whose planarized subgraphs have subtle differences but these are less employed because of their higher computational cost, for example, the localized delauney triangulation [15–17], and the Morelia test [18,19].

There are different proposals for dealing with voids; most of them are based on the idea of applying the right-hand rule over a locally planarized subgraph. Among them we remark FACE-2 proposed by Bose et al. [10], along with FACE-1 (a simpler and less efficient initial version). In FACE-2, a message is routed from the source to the destination through the interior of the adjacent faces (closed polygonal areas delimited by edges between nodes) intersected by an imaginary line between the source and the destination. For example, starting by following the right-hand rule, the first time the message crosses an edge intersected by the imaginary line; the message continues applying the contrary rule, in this case the left-hand rule. Thus, each time that the message changes from one face to the next, the rule is also changed.

Both FACE-1 and FACE-2 give a very poor performance because the paths they create can be too long. Therefore, the same authors proposed a third algorithm combining a greedy routing algorithm with a FACE-2 perimeter routing protocol. The new algorithm called greedy-face-greedy [10,14] (GFG) is thus a greedy approach, which is able to get over void areas through the use of the FACE-2 algorithm. As the perimeter routing (FACE-2) is only applied when necessary, the paths created with GFG are shorter than those created with FACE-2.

8.3.3 Dealing with Real Scenarios

Most Geographic Routing protocols, proposed in the literature, are designed to work in some ideal conditions. For example, assuming an ideal media access control (MAC) layer or considering the network as an UDG are two widely extended practices. Assuming an ideal MAC layer means that the protocol is not designed to handle the losses and transmission errors inherent in wireless links. Besides, real wireless links do not have a constant radio range, as it is supposed when the network is modeled using the UDG.

Several studies, such as the ones by Zhao and Govindan [20], and the work by Woo [21], confirm that wireless links are far from behaving like an ideal MAC. Therefore, considering that messages can be lost or corrupted is a must for achieving practical solutions for routing in real scenarios.

Besides, A UDG is a graph where two nodes are linked by an edge if and only if their Euclidean distance is less than or equal to the constant, typically associated with the maximum, radio range of a node. Thus, direct communication between two nodes is only determined by their locations, and the alterations caused by the environment (walls, reflections, etc.) in wireless signals are not considered.

Additionally, recent studies show that not only distance and the environment, but also the size of messages has an important effect on the message delivery ratio. Concretely, the work by Sánchez et al. in Ref. [31] clearly shows that there is a correlation between the packet size and the packet reception ratio (PRR), in a way that the higher the packet length the lower the PRR.

Obviously, making these assumptions reduces the effort in designing a protocol, and can even reduce the time needed to run simulations. However, the protocols designed under these idealized conditions cannot be expected to work correctly in real scenarios.

Energy consumption is another important factor to be considered when designing routing protocols for WSNs. Sensor nodes are usually operated by batteries, which have a limited duration, and in most real scenarios it is easier to deploy new sensors than to replace the exhausted batteries. This is particularly critical for WSNs. So, a lot of effort has been made in reducing the energy consumption.

The Locally Optimal Source Routing for Energy-Efficient Geographic Routing [22] (LOSR) is one of the most recent proposals in this field. LOSR can be used to reduce the energy consumption of every Geographic Routing protocol because it only affects the delivery phase of the protocol. The idea is to follow the shortest path, in terms of energy, between the current node and the one originally selected as the next-hop relay by the Geographic Routing algorithm where LOSR is being applied.

The shortest path between two nodes in a graph can be computed using the well-known Dijkstra's algorithm, but it requires to know the complete graph. LOSR uses Dijkstra's algorithm only on the local subgraph of the node currently taking the routing decisions, that is, the subgraph consisting only of that node and its 1-hop neighbors. The edges of the local subgraph are labeled with the estimated energy consumption needed to successfully transmit a message to the other end of the link. Then, by applying Dijkstra's algorithm we can compute the lowest energy path. Additionally, LOSR accounts for the energy due to possible retransmissions of messages to deal with the error-prone nature of wireless links.

Once the shortest energy path has been computed, the message is sent through that path. To do that, LOSR uses a source routing header (SRH), including the list of nodes that must be traversed. This makes the message follow the best energy-efficient path (eventually going through nodes that do not provide an advance but reduce energy consumption), according to the local knowledge of the current node. In addition, routing loops are avoided because the SRH always leads to a node that provides an advance, the node initially selected as the next relay.

8.4 Geographic Multicast Routing

The applications of WSNs are endless, including monitoring, distributed control, etc. In some cases, sensor networks also interact with the environment. For instance, a monitoring station can send commands to different sensor nodes, which control water sprinklers to open and close water valves. In many of these scenarios, one-to-many and many-to-one communications become a basic building block of the overall system. In this context, efficient multicast routing protocols, being able to transport data or commands to the intended destinations (a subset of the sensor nodes), can provide extensive bandwidth and energy savings compared to broadcast, or multiple unicast communications.

Moreover, imagine the members of a firefighter team entering a building with a fire declared. Every firefighter is interested in receiving the alarm signal produced by sensors inside the building. The sensor detecting the fire can be farther away from the team members, and the team can also be distributed over the building. Therefore, it is important to count on a reliable mechanism for transmitting the alarm message toward all the firefighters.

Building efficient multicast trees is highly difficult even when the whole network topology is known. For fixed networks, the minimal multicast tree is called the Steiner minimum tree (SMT), and its computation is an NP-complete [32] problem. For wireless multihop networks, the SMT is not optimal, and computing the optimal tree is also NP-complete [33]. Thus, most of the solutions in the literature aiming at computing efficient multicast trees resort to heuristic solutions. Although they do not provide optimal solutions, they can show a very good performance with a much reduced cost, in terms of message overhead and computation cost.

The multicast problem has been widely covered in the literature on mobile ad hoc networks (MANET). Among the plethora of existing solutions, we can cite some very well-known ones, such as MAODV [34], ODMRP [35], and MMARP [36]. But, it has been shown that these solutions do not work properly for WSNs [37] because WSNs are different from traditional MANETs. Though there are some similitudes, the differences are more than noticeable. For example, the number of nodes composing a WSN is normally several times bigger than in the case of MANETs. Moreover, in terms of energy, memory, and computational power, sensor nodes usually have much less resources than MANET's nodes, which can be laptops, or even PDA-like devices.

For the particular case of Geographic Multicast Routing (GMR), most protocols in the literature are based on extensions of GFG [10], because it is one of the most well-known geographic unicast routing algorithms. Nevertheless, adapting a unicast Geographic Routing protocol to the multicast case is not a trivial task. The next section depicts how this adaptation can be performed.

8.4.1 From Unicast to Multicast

Given a unicast Geographic Routing protocol, the easier way to adapt it to deal with multiple destinations is to transmit the same message toward each different destination separately. This adaptation (also called multiunicast) is not a good solution from the point of view of performance. Obviously, the message will be transmitted several times by the same nodes, and concretely, by those nodes that are shared by two or more paths between the source and the destinations. The overall poor performance of this initial solution justifies the efforts made by some authors to fully adapt a Geographic Routing unicast algorithm to the multicast case.

Adapting a geographic unicast routing algorithm to the multicast case requires modification of different aspects of the routing process. We have to consider that, unlike in the unicast case where there is only a destination, and we have to find a single path, in the multicast case, the goal is to build a tree connecting the source with each of the destinations. The resulting tree will probably have nodes where different branches start and nodes where the same branch continues. Each branch spans a subset of the destinations. Thus, decisions about where to branch determine the global shape of the tree. This has also a direct influence on the efficiency of the delivery, concretely in the bandwidth, and on the energy required to send a message to all the destinations.

Although the basic operation of unicast Geographic Routing algorithms is still valid, in the case of GMR, the problem becomes more challenging for a number of reasons. To begin with, nodes need to know the positions of all the destinations of every message.

Therefore, messages must include the position of every destination, and the source node must determine them.

As in the unicast case, in the multicast one, each message must include the next relay selected, and the set of receivers it is in charge of. Thus, upon receiving a message each node checks whether it is now a relay or not, then it checks if it is one of the destinations, and lastly, it runs the routing protocol to decide what should be done with the message. In this case, there are two options: passing the message to only one relay or to more than one, thus creating a branching point. The decision about branching or not, and how to do it, is one of the most challenging ones.

Then, when the node decides to split the path creating two or more branches, the selection of relays consists of deciding which next hops shall take care of routing toward which subset of destinations. As the number of possible options grows exponentially with the number of neighbors and receivers, some heuristics are needed to allow sensor nodes to run the protocol.

Figure 8.3 shows the different possibilities that a node S with three neighbors has, when routing a packet whose destination's set contains five nodes. In addition, the decision of selecting one or multiple next hops (i.e., whether to create a branch or not) has a crucial influence on the overall cost of the multicast tree.

The differences between the unicast and the multicast case are also noticeable in the delivery phase. After selecting the best next-relay candidate, the data packet received by the node currently acting as a relay must be transmitted. In general, most existing solutions have assumed ideal communication channels. However, packets can be lost or corrupted in real networks. The protocol should be able to reliably deliver messages to all the selected next hops. Otherwise, a single packet loss may cause multiple destinations to lose the data packet.

Additionally, as a node can decide to start a branch in itself, the message header must contain not only a single next-relay and a destination pair, but also as many as the new branches created. Finally, extending the perimeter mode to route toward multiple destinations requires a careful design to avoid routing loops, and to handle situations in which part of the destinations can be routed in the greedy mode, and the rest in the perimeter mode.

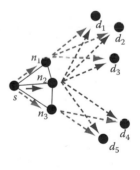

Figure 8.3 The number of alternatives in multicast routing grow exponentially with the number of neighbors and destinations. The example shows two possible options for node s routing toward five destinations and having three 1-hop neighbors.

8.4.2 *Multicast Greedy Routing*

The Position-Based Multicast (PBM) routing protocol, proposed by Mauve et al. in Ref. [23], was the first in its class. This protocol, although not initially thought for sensor networks, fulfills the criteria of locality and limited network overhead desirable for routing protocols operating in WSNs. PBM is a generalization of GFG [10], routing to operate over multiple destinations. It builds a multicast tree whose shape can vary from the shortest path tree to an approximation of a minimum cost multicast tree, depending on a parameter denoted λ.

As we mentioned before, the most important aspect of a GMR protocol is the selection of forwarding neighbors. This step of the operation of the protocol is controlled through the λ parameter. Authors try to find a good trade-off between the total number of nodes forwarding the message and the optimality of individual paths toward the destinations.

To select the forwarding neighbors, each node evaluates all possible subsets of the neighbors (W) using the following function to estimate the optimality of the alternative:

$$f(W) = \lambda N + (1 - \lambda)D, 0 \leq \lambda \leq 1$$

where

N is the number of selected neighbors ($|W|$) divided by the total number of neighbors

D is the summation of the minimal distances from nodes in W to destinations, normalized by the summation of distances from the current node to all destinations

For $\lambda = 1$, the number of neighbors selected is minimized. Thus, multicast tree branches are created as late as possible to approximate a Steiner tree. When $\lambda = 0$, the objective function rewards the alternatives in which the selected nodes provide the shortest routes for each of the destinations. In this case, the resulting multicast trees resemble the shortest path tree.

From all possible subsets of neighbors (W), the current node selects the one with optimal $f(W)$. If the best subset of neighbors is a single node, then that node will be the only relay for all the destinations. If the subset contains more than one neighbor, each of the nodes in the subset will take care of routing the data messages to some destinations. If at some node there is no node providing an advance toward one or more destinations, a different branch is initiated, individually, applying GFG for those destinations.

The main problem with this approach is that determining the optimal value for λ is not a trivial task. In fact, authors evaluated different values of λ, but they never determined an optimal value. An additional issue is the fact that the algorithm is computationally expensive. At each hop each node routing a message must test all possible subsets of neighbors to select the best among them. Therefore, given n neighbors, the number of possible subsets is $\sum_{k=0}^{n} \binom{n}{k} = 2^n$. This leads to an exponential time complexity, too high to be feasible for a sensor node.

For networks with a very large number of receivers, PBM may not scale well due to the need of including all the destinations in multicast data packets. Scalable Position-Based Multicast for Mobile Ad-Hoc Networks (SPBM) [24] was designed to improve scalability. It uses the geographic position of nodes to provide a scalable group membership scheme and to forward data packets. SPBM is mainly focused on the task of managing multicast groups in an scalable way.

However, SPBM fails to provide an efficient multicast forwarding, because it uses one separate unicast geographic routing for each destination, that is, a multiunicast approach. In addition, the proactive interchange of routing tables between neighbors makes the protocol not so scalable to the number of multicast groups, as much as PBM. Additionally, keeping membership tables implies that nodes have to store a large amount of information in memory, which is not in line with the philosophy of geographic routing principles.

The GMR [25] incorporates solutions for some scalability problems of PBM, while achieving better results, in terms of bandwidth consumption. In fact, the efficiency of the multicast trees computed by GMR is as good as the best of centralized heuristics when the network has enough density. GMR is a parameter-free algorithm based on the cost-over-progress framework. Concretely, to determine the best partition of destination nodes and then select which neighbor must be the next relay for each subset, in GMR, nodes evaluate the different alternatives using the ratio between the cost of the alternative and the progress it generates.

Considering the overall shape of a multicast tree, branching too early is as bad as branching too late. The cost-over-progress metric tries to find a trade-off between these two extreme cases, so that the cost is represented by the number of neighbors selected exceeding one, penalizing the ramification, and the progress is computed as the distance advanced toward the destinations, when using those selected neighbors as the next relays.

Results show that GMR outperforms PBM regardless of the selected value of λ over a variety of network scenarios. The main reason is that GMR's neighbor selection function manages to achieve a very good approximation of the minimum bandwidth consumption in a multicast tree. In addition, GMR is more efficient in computation time thanks to a greedy set merging algorithm proposed in the same work. This algorithm is used to achieve a polynomial of time approximation for the best branching option.

The GMR is a multicast Geographic Routing protocol optimized for reducing the bandwidth consumption. Its main goal is to build multicast trees using the least possible number of nodes, so that the number of transmissions needed to deliver a message to all the destinations is low. Reducing the number of transmissions can be seen also as a means of reducing the energy needed by the network to deliver multicast messages. Nevertheless, the same authors proposed a new protocol called the Localized Energy-Efficient Multicast Algorithm (LEMA), described in Ref. [42], which is exclusively designed to build multicast trees with the least possible energy consumption.

The LEMA and a recent proposal from Frey et al. called minimum spanning-tree-based Energy Aware Multicast routing (MSTEAM) described in Ref. [43] are based on the use of a multicast backbone, which is a totally different approach compared to GMR and PBM. Specifically, nodes take two kinds of decisions: decisions about branching, that is, when and how to split a message in a number of branches spanning subsets of destinations, and decisions about the next relays for each branch.

A node routing a message computes a minimum spanning tree (MST) of the graph made only by itself and the destinations. This tree is then used as a guide to determine when to initiate branches and how. A message split occurs when the MST over the current node and the set of destinations has multiple edges originated at the current node. Destinations spanned by each of these edges are grouped together. Destination nodes directly connected to the current node in the MST are also the closest to the current node, and are called first level destinations because, following the MST computed, the message should pass through them before reaching the rest of the destinations.

Then, the current node selects as the next relay the 1-hop neighbor providing a bigger advance toward each first level receiver. That next relay will be in charge of routing the message to the first level receiver, which is one of the destinations, and the rest of the destinations spanned by it in the MST. Obviously, the same 1-hop neighbor might be selected for two or more branches.

After that, the node currently holding the message computes the best possible way of delivering the message to the subset of 1-hop neighbors selected as the next relays. To do this, it applies the Dijkstra's algorithm to find the shortest energy path tree over the local graph consisting only of itself and its neighbors. Dijkstra's algorithm finds shortest paths, so that the problem can be solved just by labeling the edges of this local graph with the level of energy needed to transmit a message over that edge.

Finally, each routing assignment in the header of the message includes also the full shortest path computed, so that 1-hop neighbors selected as the next relays are reached using the optimum local path. This can be seen as a kind of local source routing, but normally there are only a few intermediate hops. The path can include nodes not providing an advance toward the first level receiver, but that is not a problem thanks to the source routing header used. That is, it is not possible to enter cycles during this phase.

The results obtained by LEMA are very good because in some scenarios they are close to the ones obtained using some centralized algorithms, such as MIP [41]. Additionally, comparing the number of messages transmitted, that is, the bandwidth usage, LEMA results are only a little bit worse than those of GMR.

8.4.3 Multicast Perimeter Routing

In the unicast case, the problem of finding a node where the greedy routing cannot continue due to a lack of neighbors providing an advance toward the destination is widely accepted to be solved using different localized protocols based on GFG. Nevertheless, the same problem appears in the multicast case. Some protocols resort to GFG, but in many cases, at the same node, more than one destinations may become unreachable in a greedy routing way. These cases, where multiple destinations need to be handled in a perimeter routing way are not well covered by previous proposals. In general, multiple single-perimeter routing tasks (one for every destination) are employed. This is a waste of network resources.

Although face routing was developed for the unicast case, it has recently been used in combination with multicasting protocols to handle multiple destinations. Some of the proposed solutions handle each destination separately, and thus lead to an increased energy consumption. Extensions of face recovery to the multicast case described so far are either limited to certain planar graphs or do not provide delivery guarantees.

Fortunately, the authors of MSTEAM have extended the idea of employing a multicast backbone to design a new multicast face recovery scheme. As we have already commented, a multicast backbone is a Euclidean spanning tree that contains at least the node currently holding the message and the destination nodes it is in charge of. The idea is to use the multicast backbone as a guide for taking routing decisions trying to follow the shape of the edges of the backbone to deliver a multicast message to all spanned destinations nodes. At the same time, as the message approaches the destinations, the shape of the multicast backbone adapts to the real topology of the network.

The multicast perimeter adaptation is called MFACE [44], and it uses the same idea of the multicast backbone for the sets of unreachable destinations found while routing in the greedy mode. Then, any edge of the backbone originated at the node currently holding the message will generate a new copy of the message. Next, each copy is routed toward the set of destination nodes spanned by the corresponding edge. Whenever the message arrives at a face edge intersected by a backbone edge different from the initial one, the message is split into two copies, both handling a disjoint subset of the multicast destinations defined by splitting the multicast backbone at that intersection point. MFACE behavior is similar to a deep first-tree exploration method, but with parallel exploration of subtrees at the same level.

8.5 Beacon-Less Geographic Routing

8.5.1 *Motivation*

The localized operation, reduced computation and storage requirements, and, above all, the scalability with the required number of nodes are the most important characteristics of Geographic Routing algorithms. Thus, routing protocols employing this technique have become very popular in the field of WSNs [39,40]. As we have already commented, one of the assumptions made by most authors is that nodes know the location of their 1-hop neighbors. In fact, this is easily achieved using the well-known beaconing mechanism. Beacons are short messages including the identifier of the issuer and its coordinates. Therefore, assuming that nodes know their own coordinates, and that nodes send this information included in the beacons with certain periodicity, every node can learn the location of its 1-hop neighbors.

By increasing the frequencies of transmission of the beacons, the mechanism can be more robust to network changes. Topological changes are usually induced by mobility of nodes, the incorporation of new nodes, the end of life of some others, or even due to the duty cycle that most sensor nodes employ to reduce energy consumption. However, the higher the frequency the higher the number of transmissions in the whole network. We must point out that all the nodes in the network are doing the same process; thus, every beaconing cycle represents n messages, n being the number of nodes, that is, it can be seen as a kind of flooding.

From the point of view of saving network resources, flooding the network at a high rate is not among the best practices. On the other hand, reducing the frequency of beacons can lead to a situation where nodes have a wrong view of their neighborhood. The periodicity of the beaconing mechanism is a hard to tune parameter, and it can even depend on the scenario.

Additionally, beacons not only can generate a big amount of network traffic, beacons can also interfere with regular data traffic causing data loss. Obviously, beacons also waste energy from sensor nodes, but the most important issue is that all undesired network traffic, interferences, and energy depletion affect all nodes in the network. In other words, sensor nodes not taking part in any routing task suffer from these problems, and unlike the nodes routing messages, they do not profit from having an accurate 1-hop neighbor table. Moreover, even adopting a high refresh rate for the beaconing mechanism, recent studies have shown the limitations that Geographic Routing algorithms have when mobility is taken into account [38].

To avoid these periodic transmissions, beacon-less protocols have been proposed in the literature. The general idea is to reactively discover the information about the neighbor's positions, which is needed to select the next forwarder when routing data packets. Instead of keeping neighbor tables proactively updated, Beacon-Less Routing (BLR) protocols try to save network resources by postponing the discovery of 1-hop neighbors to the last moment.

8.5.2 Noncollaborative Approach

In the first design of a beacon-less Geographic Routing protocol, the proactive identification of neighbors can be done using a simple query–answer–selection mechanism, that is, a node holding a data message transmits a control message, indicating its interest in discovering neighbors (Figure 8.4). Thanks to the shared medium, the message is received by all its 1-hop neighbors, so that everyone can answer with a control message similar to a beacon. The response includes the identifier and the location of the neighbor. The node receiving all the answers can then decide which neighbor to select as its next forwarder. Finally, the data message is transmitted, including the identifier of the selected next-relay in the header.

This first approach is simple and allows us to transform any standard Geographic Routing algorithm into a beacon-less one. However, there are some problems making impractical this initial design. First of all, we have to point out that this solution can be classified as noncollaborative because one single node, the one currently holding the

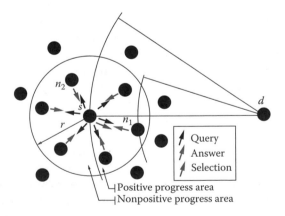

Figure 8.4 The general operation of the centralized approach in beacon-less routing has three phases: query, answers, and selection. In this example, *s* broadcasts a query, then 1-hop neighbors answer indicating their identification and location, and finally, *s* selects the best next-relay among those located in the positive progress area.

message, is responsible of taking the routing decision. This is not essentially bad, but we need at least three messages for every hop in the path toward the destination. Assuming a high node density typical from WSN, every node will have much more than one neighbor; therefore, the number of messages per hop can be very high.

Second of all, letting every neighbor answer just when it receives the query may not be the best option because the probability of losing messages due to collisions is very high. Additionally, responses from some neighbors are not necessary. Following a greedy routing approach, all neighbors located farther than the node currently holding the message cannot be selected as the next relays; therefore, their answers are unnecessary.

Contention-based forwarding (CBF) [28] and implicit geographic forwarding (IGF) [27] are good examples of the first design alternative. Both use a similar three-way (RTS/CTS/ACK) handshaking procedure to select the next forwarder, as we have detailed. In CBF, the node currently holding the message transmits a Ready To Send (RTS) control message, then candidate neighbors answer with a Clear To Send (CTS) message, and finally, the selection of the next forwarder is performed using an acknowledgment message (ACK). The operation of IGF is very similar, but it includes an additional control message. The reason is that the discovery of neighbors (first two messages) is done at the physical layer while the selection of the next relay is performed at the network layer using two messages. The first one identifies the next relay selected, and the second one is an answer from the selected neighbor confirming the reception of the data message.

Both protocols incorporate the same mechanism to avoid collisions among the responses issued by 1-hop neighbors. They force neighbors to wait for a certain time before answering. The determination of this delay is based on the relation between the location of the neighbor, the location of the node currently holding the message, and the location of the destination. Thus, neighbors whose position provides more advance toward the destination wait for a short time before answering. The advantage is twofold: it reduces the number of collisions due to the fact that answers are not generated at the same time, and secondly, it forces the most suitable neighbor to become the next relay to answer first. This allows the rest of the neighbors to cancel their answers, and reduces the overall delay (i.e., forwarding can start as soon as the first answer is received).

Concretely, the CBF incorporates an additional feature to try to guarantee that only one response is transmitted at each step. The idea is to assure that all candidate neighbors to the next relay are able to hear the transmission of the response of the first neighbor. To do that, authors use the idea of forwarding area, that is, only neighbors located inside a predefined area can be considered as candidates to become the next relays. Thus, authors define this area in such a way that, given a predefined coverage radius, all nodes inside the area can reach each other with their transmissions.

8.5.3 Collaborative Approach

As we have already seen, the noncollaborative approach has some drawbacks, especially the fact that the minimum number of messages per hop is three. This reduces the overall performance of the protocol by increasing the control overhead. Therefore, a number of authors have tried to overcome these problems by using a totally different approach, that is, a distributed scheme where neighbors collaboratively decide which one must be the next forwarder.

Obviously, the distributed selection of the next forwarder cannot be done using a heavy protocol to guarantee that a single node is selected. In that case, this solution would be worse than the centralized one because it usually implies several phases and multiple control messages. Instead, neighbors compete to become the next relay. In fact, using a mechanism to delay responses, as we have depicted above, the first neighbor answering should win the competition, and therefore, become the next forwarder.

The BLR protocol [26] exemplifies this distributed operation. BLR simplifies the process of selecting the next forwarder to reduce the number of messages needed at each hop. Instead of forcing neighbors to answer with CTS messages, and then deciding the next forwarder in a noncollaborative way, in BLR, neighbors are free to decide by themselves whether or not they are good next-forwarders. When a neighbor decides it is a good candidate, it directly forwards the message. Using timers in the same way as the CBF or the IGF do, the most promising neighbor should answer first; therefore, it transmits the message acting as the de facto next-relay. The rest of the neighbors are expecting to hear that forwarding, and consequently, cancel their own timers.

As in the case of CBF, neighbors placed farther away from the coverage radius of the one whose timer expires first, will not hear its transmissions. Thus, their own timers will finally expire making them do a second, and an unnecessary, transmission. To solve this problem, the authors propose also to limit the area in which a neighbor can be self-considered a candidate relay to avoid duplicates in the same way as CBF does it. This forwarding area is defined, so that all nodes covered can reach each other (Figure 8.5).

8.5.4 Dealing with Voids

The two different beacon-less schemes depicted so far can cope with the denominated greedy forwarding, but these solutions do not work properly when there are no nodes providing a positive advance toward the destination. Using the noncollaborative approach,

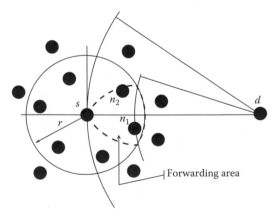

Figure 8.5 The forwarding area delimits the subset of neighbors which can hear each other transmissions.

no neighbor will answer the initial control message because none of them is in the appropriate area. The distributed scheme suffers from the same problem, but additionally, the node currently holding the message does not detect it, because its mission is to forward the message and let its neighbors decide themselves the next relay.

Some proposals, like CBF, deal with these situations resorting to some well-known perimeter-forwarding mechanisms, such as the one defined in Greedy-Face-Greedy Routing [10]. As these mechanisms work applying the right-hand rule over a planar version of the local subgraph, it is necessary to know the whole list of neighbors. Therefore, the proposed alternative transmits a special control message indicating a full neighborhood discovery request. That is, all neighbors must answer independent of their location. This represents an additional control message overhead as well as another opportunity to lose messages due to collisions.

To cope with these problems, Chawla et al. [29] improved the algorithm by reducing the number of responses during the recovery phase. They proposed a new timer-assignment function to avoid neighbors from answering if a crossing link appears after their addition to the planar graph. That is, farther neighbors answer first. Acting in that way, a neighbor can determine whether or not it belongs to the planar graph.

Besides, there are also solutions not using the perimeter forwarding approach. For example, in Blind Geographic Routing (BGR) [30], we can find an original alternative. BGR is a beacon-less Geographic Routing algorithm very similar to CBF; its major contribution is its recovering scheme. Unlike CBF, here the forwarding node can try up to three different forwarding areas looking for a neighbor's answer. The first area the protocol tries is the one covering possible neighbors providing advance toward the destination. When no response is received, a new forwarding area is used. This second area results from turning the initial one 60° to the left or to the right. If still no responses are received, a last area is used, the one obtained by turning the initial one in the opposite direction.

8.5.5 Dealing with Real Scenarios

The related protocols are designed assuming that messages cannot be lost. However, in real scenarios it is quite common that some messages get lost. Therefore, solutions such as the forwarding area do not guarantee a single response coming from the most promising neighbor. Its response cannot reach all the neighbors, so they will respond upon the expiration of their timers. In protocols like CBF or IGF, the number of responses rises when they are tested in real scenarios, but the worst problem appears in protocols following the collaborative scheme. For example, in BLR, the number of duplicate paths generated at each hop increases dramatically. This represents an increase in the control overhead as well as the number of duplicate messages arriving at the destination.

Moreover, the precision of location systems used by nodes might not be as accurate as required. For example, several nodes can determine they are in the same location, thus answering at the same time, because the delay is computed using only coordinates and distances.

Additionally, the probability of reception of a message is affected not only by the distance but also by the size of the packet. Therefore, a neighbor can be selected because it has previously received a control message, such as the discovery request mentioned above, and its response was the first received. Nevertheless, when the data message is transmitted, as this message is usually larger than those used as control messages, there exists a nonzero probability of not correctly receiving it.

The BOSS (Beacon-less On-demand Strategy for Sensor networks) algorithm, described in Ref. [31], has been designed to take into account losses and collisions typical of radio communications. BOSS is a protocol similar to CBF because it also uses a three-way handshaking, but differs in the way it handles real scenarios. For example, the discovery request used in BOSS includes the full data payload. Thus, only neighbors able to receive the data packet take part in the next phase. This also guarantees that the selected next relay has received the data, so that it can effectively act as a relay.

Moreover, trying to overcome the inaccuracies in the node's coordinates, BOSS defines a new timer-assignment function called the discrete dynamic forwarding delay (DDFD). DDFD divides the neighborhood area into subareas according to the progress toward the destination. Neighbors placed in the same subarea share the same base delay. The final delay is computed adding a random number of milliseconds. Thus, neighbors in a subarea with a high progress can answer before the neighbors placed in farther subareas, and neighbors in the same area do not transmit messages at the same time due to the random part of the delay. The goal of DDFD is to reduce the collisions of the answers during the selection phase.

Finally, BOSS includes another feature to reduce the probability of generating duplicates. Neighbors hearing a response from a different node cancel their own timer, but the timer is also canceled when the selection message is received. That is, the node currently holding the message transmits the selection message just when it receives the first response, so that neighbors that did not cancel their timers because they did not receive any response, can do so immediately.

8.6   Conclusions and Open Issues

Geographic Routing algorithms represent a fundamental building block for developing WSN applications. It is clear that the performance and scalability of these protocols is superior to the previous schemes. However, we must consider that there is still much work to be done. There are some issues to be better solved, and some problems not yet addressed.

As we have seen, the determination of the location of a concrete node is a prior phase to the routing protocol, and this is a problem even bigger than the routing itself. The solutions proposed to date work acceptably, but the overhead introduced is still high. In relation to routing itself, the most commonly used solutions to the problem of voids (i.e., GFG and its variants) are still not fully tested in real environments, but it is easy to see that the solutions depending too much on highly accurate location information can fail when tested in real scenarios.

Current hardware used in sensor nodes allows routing protocols to control transmission power. This can be used to reduce energy consumption, but power adjustment can lead to variations in the coverage range, and even most important, variations in the packet reception ratio. Most protocols designed to reduce energy consumption are based on energy models, in which the energy consumption is related to distance. Nevertheless, adjusting the transmission power to reach only the desired node located at a certain distance might change the probability of reception of a message for the same node. Thus, new energy models taking into account, at least, factors such as environment, distance, transmission power, and packet size should be developed.

Multicast protocols have additional problems. Concretely, we must highlight the scalability issues related to the number of destinations. As the message header contains the location of each destination, it is clear the this number is limited, especially in WSN, where messages have a very limited size.

Finally, although beacon-less protocols solve some problems related to the network overhead introduced by beacons, there are also some open issues affecting them. From our point of view, the most important one is that most beacon-less techniques have a strong dependency on knowing a fixing maximum coverage range. Most timer and forwarding area decisions are taken based on the distances related to the coverage range. Thus, in a real test bed, we can find some strange behaviors and wrong decisions if a node is reachable farther away from the preconfigured coverage range, or on the contrary, a node located inside the coverage range is not reachable due to propagation errors.

References

1. G.G. Finn, Routing and addressing problems in large metropolitan-scale internet-works, Tech report ISI/RR-87-180, University of Southern California, 1987.
2. H. Takagi and L. Kleinrock, Optimal transmission ranges for randomly distributed packet radio terminals, *IEEE Transactions on Communications*, 32(3):246–257, March 1984.
3. R. Bellman, On a Routing Problem, *Quarterly of Applied Mathematics*, 1(16):87–90, 1958.
4. L.R. Ford Jr., Network flow theory, The RAND Cooperation, Santa Monica, California, Technical report P-923, August 1956.
5. C. Perkins, E. Belding-Royer, and S. Das, Ad hoc on-demand distance vector (AODV) routing, RFC 3561, IETF, July 2003.
6. T. Clausen and P. Jacquet, Optimized link state routing protocol (OLSR), RFC 3626, IETF, October 2003.
7. E. Kranakis, H. Singh, and J. Urrutia, Compass routing on geometric networks, in *Proceedings of the 11th Canadian Conference on Computational Geometry (CCCG '99)*, Vancouver, Canada, August 1999, pp. 51–54.
8. I. Stojmenovic and X. Lin, Loop-free hybrid single-path/flooding routing algorithms with guaranteed delivery for wireless networks, *IEEE Transactions on Parallel and Distributed Systems*, 12(10):1023–1032, 2001.
9. R. Nelson and L. Kleinrock, The spatial capacity of a slotted ALOHA multihop packet radio network with capture, *IEEE Transactions on Communications [legacy, pre-1988]*, 32(6):684–694, 1984.

10. P. Bose, P. Morin, I. Stojmenovic, and J. Urrutia, Routing with guaranteed delivery in ad hoc wireless networks, *Wireless Networks*, 7(6):609–616, 2001.
11. G.T. Toussaint, The relative neighborhood graph of a finite planar set, *Pattern Recognition*, 12:261–268, 1980.
12. K. Gabriel and R. Sokal, A new statistical approach to geographic variation analysis, *Systematic Zoology*, 18:259–278, 1969.
13. J.A. Bondy and U.S.R. Murty, *Graph Theory with Applications*, Macmillan, London, U.K., Elsevier, North-Holland, 1976.
14. H. Frey and I. Stojmenovic, On delivery guarantees of face and combined greedy-face routing algorithms in ad hoc and sensor networks, in *Proceedings of the 12th annual ACM/IEEE International Conference on Mobile Computing and Networking (MobiCom '06)*, Los Angeles, CA, September 2006, pp. 390–401.
15. J. Gao, L.J. Guibas, J. Hershberger, L. Zhang, and A. Zhu, Geometric spanner for routing in mobile networks, in *Proceedings of the 2nd ACM International Symposium on Mobile Ad Hoc Networking and Computing (MobiHoc '01)*, Long Beach, CA, 2001, pp. 45–55.
16. X.Y. Li, G. Calinescu, and P.J. Wan, Distributed construction of a planar spanner and routing for ad hoc wireless networks, in *Proceedings of the 21th Annual Joint Conference of the IEEE Computer and Communications Societies (INFOCOM '02)*, 2002, pp. 1268–1277.
17. X.-Y. Li, I. Stojmenovic, and Y. Wang, Partial delaunay triangulation and degree limited localized bluetooth scatternet formation, *IEEE Transactions on Parallel Distributed Systems*, 15(4):350–361, 2004.
18. P. Boone, E. Chavez, L. Gleitzky, E. Kranakis, J. Opatrny, G. Salazar, and J. Urrutia, Morelia test: Improving the efficiency of the gabriel test and face routing in ad-hoc networks, *Lecture Notes in Computer Science*, 3104:23–34, 2004.
19. S. Datta, I. Stojmenovic, and J. Wu, Internal node and shortcut based routing with guaranteed delivery in wireless networks, *Cluster Computing*, 5(2):169–178, April 2002.
20. J. Zhao and R. Govindan, Understanding packet delivery performance in dense wireless sensor networks, in *Proceedings of the First International Conference on Embedded Networked Sensor Systems (SenSys '03)*, Los Angeles, CA, 2003, pp. 1–13.
21. A. Woo, T. Tong, and D. Culler, Taming the underlying challenges of reliable multihop routing in sensor networks, in *Proceedings of the First International Conference on Embedded Networked Sensor Systems (SenSys '03)*, 2003, pp. 14–27.
22. J.A. Sánchez and P.M. Ruiz, Locally optimal source routing for energy-efficient geographic routing, *Wireless Network (WINET) Journal*, November 2007.
23. M. Mauve, H. Füßler, J. Widmer, and T. Lang, Position-based multicast routing for mobile ad-hoc networks, Department of Computer Science, University of Mannheim, Technical report TR-03-004, March, 2003.
24. M. Transier, H. Füßler, J. Widmer, M. Mauve, and W. Effelsberg, Scalable position-based multicast for mobile ad-hoc networks, *First International Workshop on Broadband Wireless Multimedia: Algorithms, Architectures and Applications (BroadWim 2004)*, San Jose, CA, 2004.

25. J.A. Sánchez, P.M. Ruiz, J. Liu, and I. Stojmenovic, Bandwidth-efficient geographic multicast routing protocol for wireless sensor networks, *IEEE Sensors Journal*, 7:627–636, September 2007.
26. M. Heissenbüttel, T. Braun, T. Bernoulli, and M. Wächli. BLR: Beacon-less routing algorithm for mobile ad-hoc networks *Elsevier Journal of Computer Communications*, 27(11):1076–1086, July 2004.
27. B. Blum, T. He, S. Son, and J. Stankovic, IGF: A state-free robust communication protocol for wireless sensor networks, Department of Computer Science, University of Virginia, Technical Report, 2003.
28. H. Füßler, J. Widmer, M. Käsemann, M. Mauve, and H. Hartenstein. Contention-based forwarding for mobile ad hoc networks, *Ad Hoc Networks*, 1(4):351–369, 2003.
29. M. Chawla, N. Goel, K. Kalaichelvan, A. Nayak, and I. Stojmenovic, Beacon less position based routing with guaranteed delivery for wireless ad-hoc and sensor networks, in *Proceedings of 19th IFIP World Computer Congress (WCC '06)*, Santiago de Chile, Chile, August 2006.
30. M. Witt and V. Turau. BGR: Blind geographic routing for sensor networks, in *Proceedings of 3rd Workshop on Intelligent Solutions in Embedded Systems (WISES '05)*, Hamburg, Germany, May 2005, pp. 51–61.
31. J.A. Sánchez, R. Marin-Perez, and P.M. Ruiz, BOSS: Beacon-less on demand strategy for geographic routing in wireless sensor networks, in *Proceedings of 4th IEEE International Conference on Mobile Ad-hoc and Sensor Systems (MASS '07)*, Pisa, Italy, October 2007.
32. R.M. Karp, Reducibility among combinatorial problems, *Complexity of Computer Computations*, 43:85–103, 1972.
33. P.M. Ruiz and A.F. Gomez-Skarmeta, Approximating optimal multicast trees in wireless multihop networks, in *Proceedings of 10th IEEE Symp. on Computers and Comms. (ISCC'05)*, La Manga del Mar Menor, Spain, June 2005, pp. 686–691,
34. E.M. Royer and C.E. Perkins, Multicast operation of the ad-hoc on-demand distance vector routing protocol, in *Proceedings of 5th ACM/IEEE International Conference on Mobile Computing and Networking (MobiCom '99)*, Seattle, WA, August 1999, pp. 207–218.
35. S. Ju Lee, W. Su, and M. Gerla, On-demand multicast routing protocol in multihop wireless mobile networks, *Mobile Networks and Applications*, 7(6):441–453, 2002.
36. P.M. Ruiz, A. Gomez-Skarmeta, and I. Groves, The MMARP protocol for efficient support of standard IP multicast communications in mobile ad hoc access networks, in *Proceedings of International Mobile IP-based Network Developments (MIND) Workshop*, London, U.K., October 2002.
37. J.G. Jetcheva and D.B. Johnson, A performance comparison of on-demand multicast routing protocols for ad hoc networks, School of Computer Science, Carnegie Mellon University, 2004.
38. M. Witt and V. Turau, The impact of location errors on geographic routing in sensor networks, in *Proceedings of the 2nd International Conference on Wireless and Mobile Communications (ICWMC'06)*, Bucharest, Romania, July 2006.

39. S. Giordano, I. Stojmenovic, and L. Blazevie, Position based routing algorithms for ad hoc networks: A taxonomy, *Ad Hoc Wireless Networking*, 103–136, 2004.

40. J. Li, J. Jannotti, D.S.J. De Couto, D.R. Karger, and R. Morris, A scalable location service for geographic ad hoc routing, in *Proceedings of the 6th annual ACM/IEEE International Conference on Mobile Computing and Networking (MobiCom '00)*, Boston, MA, 2000, pp. 120–130.

41. J.E. Wieselthier, G.D. Nguyen, and A. Ephremides, Energy-efficient broadcast and multicast trees in wireless networks, *Mobile Networks and Applications*, 7:481–492, 2002.

42. J.A. Sánchez and P.M. Ruiz, LEMA: Localized energy-efficient multicast algorithm based on geographic routing, in *Proceedings of the 31st IEEE Conference on Local Computer Networks (LCN '06)*, Tampa, FL, November 2006, pp. 3–12.

43. H. Frey, F. Ingelrest, and D. Simplot-Ryl, Localized minimum spanning tree based multicast routing with energy-efficient guaranteed delivery in ad hoc and sensor networks, Institut National de Recherche en Informatique et en Automatique (INRIA), France, Technique Report RT-0337, June 2007.

44. H. Frey and F. Ingelrest, MFACE: A multicast backbone-assisted face traversal algorithm for arbitrary planar ad hoc and sensor network topologies, in *Proceedings of International Workshop on Theoretical and Algorithmic Aspects of Sensor and Ad-hoc Networks*, Miami, FL, 29 June 2007.

Chapter 9

Medium Access Control in Wireless Sensor Networks

Bashir Yahya, Jalel Benothman, and Ebtisam Amar

CONTENTS

A Medium Access Control (MAC) protocol defines rules to access and control the shared medium and plays a critical role in the efficient and fair sharing of wireless bandwidth. The nature of the wireless channel brings new issues like location-dependent carrier sensing, time-varying channel, and burst errors. Low power requirements add new challenges. Wireless MAC protocols have been heavily investigated by the research community and several protocols have been proposed. Protocols have been devised for different types of architectures, different applications, and different media.

Wireless sensor networks (WSNs) are usually battery powered, and to make applications economically viable, such networks have to operate for a long period of time (e.g., number of years) without recharging or replacing batteries. So, it is extremely important to develop techniques that prolong battery lifetime as much as possible. As a result, energy efficiency probably becomes the most important issue in WSNs.

The need for energy-efficient operation of a WSN has prompted the development of new protocols in all layers of the communication stack. Provided that the radio transceiver is the most power consuming component of a typical sensor node, large gains can be achieved at the link layer where the MAC protocol is controlling the usage of the radio transceiver unit.

Sensor network's MAC protocols differ greatly from traditional wireless network's MAC protocols in many issues. MAC protocols for WSNs must have built-in power conservation, mobility management, and failure recovery strategies. Furthermore, sensor MAC protocols should make performance trade-off between latency and throughput for a reduction in energy consumption to maximize the lifetime of the network. This is, in general, achieved by duty cycling the radio.

This chapter presents the fundamentals and concepts of wireless MAC protocols and explains the specific requirements and design constraints of a specific WSN MAC protocol. As well, we present, classify, and discuss a typical set of MAC protocols that are specifically designed for WSNs. Finally, we present research directions and identify open issues for future medium access research followed by a final conclusion.

9.1 Introduction

Recent advances in microelectromechanical systems, low-power highly integrated digital electronics, tiny microprocessors, and low-power radio technologies have created low-cost, low-power, and multi-functional sensor devices, which can observe and react to changes in physical phenomena of their surrounding environments. These sensor devices are equipped with a small battery, a radio transceiver, and a set of transducers that are used to acquire information about the surrounding environment. This emergence of such sensors has led engineers to envision networking of a large set of sensors scattered over a wide area of interest [1–6]. A typical WSN consists of a number of sensor devices that collaborate to accomplish a common task such as environment monitoring and reporting the collected data using the radio to a center node (sink node). WSNs can serve many civil and military applications that include target tracking in battlefields [7], habitat monitoring [8,9], civil structure monitoring [10], and factory maintenance [11]. In many applications sensor nodes should be deployed in an ad hoc fashion without careful planning [12]. They must organize themselves to form a multi-hop, wireless communication network to communicate with each other and with one or more sink nodes [13]. Because large number of sensor nodes are densely deployed, neighbor nodes may be very close to each other. Hence, multi-hop communication in sensor network is expected to consume less power than traditional single-hop communication. In addition, the transmission power levels can be kept low, which is highly desired in covert operations. Multi-hop communication can also effectively overcome some of the signal propagation effects experienced in long-distance wireless communication. Although one of the advantages of the multi-hop communication mode, the multi-hop solution is not always the best mode as discussed in [107]. To control the operation of the sensor network, a remote user can issue commands to the sensor network through a control center (sink) to assign data collection, processing, and transfer tasks to the sensors, and it can later receive the sensed data through the sink.

Provided that sensor nodes carry limited, generally irreplaceable, power source, WSNs must have built-in trade-off mechanisms that enable the sensor network to conserve power and give the end user the ability of prolonging network lifetime at the cost of lower throughput or higher latencies [1].

The energy constraints of sensor nodes and the need for energy-efficient operation of a WSN have motivated a lot of research on sensor networks which led to the development of novel communication protocols in all layers of the Open System Interconnection (OSI) communication stack. Given that the radio is the most power-consuming component of a typical sensor node, large gains can be achieved at the link layer where the MAC protocol is controlling the usage of the radio unit.

MAC protocols have been extensively studied in traditional wireless networks. Time division multiple access (TDMA), frequency division multiple access (FDMA), and code division multiple access (CDMA) are MAC protocols that are widely used in modern cellular communication systems. Their principle idea is to avoid interference by scheduling nodes onto different subchannels that are divided either by time, frequency, or orthogonal codes respectively. Since these subchannels do not interfere with each other, MAC protocols in this group are largely collision free.

Another class of MAC protocols is based on contention. Rather than preallocated transmission, nodes share the same channel. Collision happens during the contention procedure in such systems. Classical examples of contention-based MAC protocols include ALOHA [14] and CSMA [15]. In ALOHA, a node simply transmits a packet when it is generated (pure ALOHA) or at the next available slot (slotted ALOHA) [16]. Collided packets are discarded and retransmitted. In CSMA, a node listens to the channel before transmitting. If it detects a busy channel, it delays access and retries later.

WSNs differ greatly from traditional wireless networks in many characteristics. These special characteristics make traditional MAC protocols unsuitable for WSNs and have motivated a lot of research in the field of designing MAC protocols.

Various aspects of MAC protocols for WSNs are discussed in several surveys [1,3,17–31,80,81]. However, having the special characteristics of WSNs in mind, this chapter gives a comprehensive review on most recent developments and challenging issues that sensor network's MAC protocol should overcome and discusses a typical set of solutions proposed in the literature. Furthermore, we give a future view of the research directions for open problems that have not been studied or need deeper investigations.

9.2   Wireless Sensor Networks

WSNs are autonomous ad hoc networks designed for some potential applications in environmental monitoring, surveillance, military, health, security, and so on. A typical WSN is composed of a large number of sensor nodes, which are densely deployed either inside the phenomenon or very close to it. A sensor node is made up of four basic components as shown in Figure 9.1 a sensing unit, a processing unit, a transceiver unit, and a power unit. They may also have application-dependent additional components such as location management unit [1], a mobility management unit, and a power generator unit. Once the nodes are deployed into the target area, they collect data from the environment automatically and establish an ad hoc network to transfer their data to the base station (sink). The base station aggregates and analyzes the collected sensed data and decides whether there is an unusual or concerned event occurrence in the deployed area.

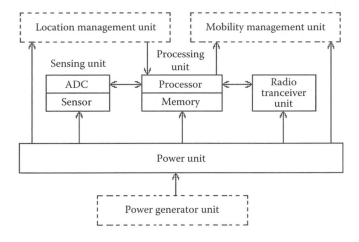

Figure 9.1   Components of a sensor node.

Sensor nodes in sensor networks, usually, are battery powered, and should have lifetime of months or years. Replacing or renewing energy sources of hundreds or thousands of sensors batteries after network deployment becomes infeasible or too costly. Therefore, energy efficiency has become the crucial design challenge in WSNs. This section presents and discusses the special characteristics that make WSNs differ from the traditional wireless networks.

9.2.1   *Wireless Sensor Network Characteristics*

WSNs are in many aspects quite similar to mobile ad hoc networks, but wireless sensor networks have several distinct characteristics that are different from other networks and create challenging problems. Here we limit the discussion to three characteristics that are of particular interest and affect the design of sensor network's protocols.

Limited resources. Sensor nodes have limited power, memory, and computational power. Therefore, any algorithm developed for WSNs must not rely on the assumption of unlimited resources, and must sparingly use the limited resources that do exist.

Sensor nodes are prone to failures. Due to the inherent instability and energy constraints of sensors, sensor nodes are prone to failures. It would thus be useful to determine which set of nodes or which areas within the network are experiencing high loss rates. Such information is potentially valuable to the design of fault-tolerant protocols or monitoring mechanisms, so that the problem areas may be redeployed, and critical data may be rerouted to avoid these areas from suffering high loss rates. These are just a few of the many possible applications of per node loss rate information to streamline data flow or enhance the reliability of large-scale sensor networks.

Wireless bandwidth constraints. Therefore, one cannot rely on the use of active acknowledgments, which are not scalable or bandwidth efficient, in the design of sensor network protocols. This renders the direct collection of loss rate data impossible in sensor networks. Furthermore, it would also be infeasible, due to limited bandwidth, for individual sensor nodes to collect and transmit loss rate data to a centralized location for processing.

9.2.2 Power Consumption of a Sensor Node

As mentioned above, power supply for a sensor node is at a premium: batteries have small capacity, and recharging by energy scavenging is complicated and volatile. Hence, the energy consumption of a sensor node must be tightly controlled. The main consumers of energy are the sensing unit, the digital processing unit, and the radio transceiver unit.

One important contribution to reduce power consumption of these components comes from chip-level and power technologies. Designing low-power chips is the best starting point for an energy-efficient sensor node. But this is only one-half of the picture, as any advantages gained by such designers can easily be squandered when the components are improperly operated. The crucial observation for proper operation is that most of the time a wireless sensor node has nothing to do. Hence, it is best to turn it off. Naturally, it should be able to wake up again, on the basis of external stimulation event or on time basis. Therefore, completely turning off a node is not possible, but rather, its operational state can be switched to a low-power state. Introducing and using multiple states of operation with reduced energy consumption in return for reduced functionality is the core technique for energy-efficient wireless sensor node [82]. Following, we give a more insight into the energy consumption of the main consumers of the sensor node.

Sensing unit: The sensing unit, which is composed of environmental sensors and analog-to-digital (A/D) converters, translates physical phenomena to electrical signals. There are several sources of energy consumption in sensor unit: signal sampling and conversion of physical signals to electrical ones, signal conditioning, and A/D conversion. The energy consumed by this unit is relatively constant, and improvements to their energy efficiency depend on increasing integration and skilled analog circuit design. And it also has been tested that passive sensors such as temperature, seismic, etc., consume negligible energy compared to other components in a sensor node [86].

Digital processing unit: The majority of digital circuits employed in sensor nodes are typically used in command and control functions, base-band signal processing unit, and execution of the protocol stack. The energy consumed in a digital circuit is determined by the sum of dynamic and static power dissipation. The dynamic power dissipation is described as the product of the switched capacitance and the gate supply voltage while static power dissipation originates from the undesirable leakage of current from power to ground at all times. Compared to dynamic power, static power dissipation dominates the total energy consumed in the digital circuitry unit. The easiest way to reduce static power dissipation is to shut down the power supply to idle state, which is called power gating. But one should note that shutting down complicated circuits may cause time and energy overhead. Methods that solve those problems and conserve energy is out the scope of this chapter. For more details refer to Ref. [83].

Radio transceiver unit: Wireless communication is the major energy consumer during the system operation. It is difficult to generalize energy consumption by communication system, as many variables influence the performance. Generally, the energy consumption of the radio consists of two components: (1) an RF component that depends on the transmission distance and modulation parameters and (2) an electronic component that accounts for the energy consumed by the circuitry that performs frequency synthesis, filtering, up-converting, etc. The average energy consumed by a complete radio transmission can be described by the following equation [38,85]:

$$E = P_{tx}(T_{\text{transmit}} + T_{\text{start-up}}) + P_{\text{out}} T_{\text{transmit}} \qquad (9.1)$$

where
P_{tx} represents the power consumed by the transmitter
T_{transmit} is the actual transmission duration
$T_{\text{start-up}}$ is the start-up time of the transceiver
P_{out} is the output transmission power that drives the antenna

The packets transmitted by the radio are likely to be small during idle periods because of the low data rates of the sensor network. The start-up power will dominate the power of active transmission. While the radio's high bias currents require acceptance of shutdown cost, the node should amortize the start-up power over more transmitted bits to reduce the power cost per transmitted bit.

Besides the factor pointed above, there are other reasons leading to energy dissipation in radios. Generally, radios can operate in four distinct modes of operation: transmit, receive, idle, and sleep. An important influencing factor is that a significant amount of energy is dissipated as the radio's operating mode changes. Table 9.1 shows that most radios operating in idle mode result in significantly high energy consumption [58], almost equal to the energy consumed in the receive mode. Obviously, it is important to completely shut down the radio rather than switch it to idle mode when not transmitting or receiving data. However, switching a radio on and off very frequently can sometimes result in even more energy consumption than leaving the transceiver unit in idle mode because of the start-up power. Moreover, as the transmission packet size gets smaller, the transition energy becomes dominant than the energy consumed during receiving and transmitting of packets [38,85]. Therefore, it is important to take this issue into account when designing energy-efficient MAC protocols.

Table 9.1 Power Consumption of a Typical Radio

Radio Mode	Power Consumption (MW)
Transmit	14.88
Receive	12.50
Idle	12.36
Sleep	0.016

9.2.3 Communication Patterns

Generally, most WSN applications have some remarkable characteristics that sets it apart from the traditional wireless networks. A common characteristic is that sensor nodes are deployed to just monitor the environment and report data to a processing center for further data processing. Monitoring applications do not need to transfer large amounts of traffic or data processing. Communication patterns that are responsible for generating traffic in WSNs are often very asymmetric with significantly more traffic appearing near the sink. However, three communication patterns can be defined for WSNs: broadcast, converge-cast, and local gossip [87]. *Broadcast* type of communication pattern is generally used by a base station (sink) to transmit some information to all sensor nodes of the network. Broadcasted information may include queries of sensor query-processing architectures, program updates for sensor nodes, and control packets for the whole system. The broadcast-type communication pattern should not be confused with broadcast-type packet. For the former, all nodes of the network are intended receivers whereas for the latter the intended receivers are the nodes within the communication range of the transmitting node.

In some scenarios, the sensors that detect an intruder communicate with each other locally. This kind of communication pattern is called *local gossip*, where a sensor sends a message to its neighboring nodes within a range. Sensors that detect an event, then, need to send what they perceive to the information center. That communication pattern is called *converge-cast*, where a group of sensors communicate to a specific sensor. The destination node could be a cluster head, data fusion center, or a base station.

In Clustering-based protocols, cluster heads communicate with their members and thus the intended receivers may not be all neighbors of the cluster head, but just a subset of the neighbors. To serve for such scenarios, a fourth type of communication multi-cast pattern is defined, where a sensor sends a message to a specific subset of sensors.

9.3 Concepts and Fundamentals of Wireless MAC Protocols

In this section, we discuss briefly some basic concepts and aspects of wireless MAC protocols, because the protocols used in WSNs inherit many of the problems and approaches already existing for this more general field.

Wireless MAC protocols have received a huge attention from researchers and commercial developers during recent decades, and there exists a huge body of literature [88].

Before the invention of WSNs, energy aspects were not one of the top priorities of earlier research on MAC protocols. But nowadays, energy has been established as one of the primary design keys especially in WSNs.

9.3.1 Requirements and Design Constraints for Wireless MAC Protocols

The most important performance requirements for traditional MAC protocols are throughput efficiency, high channel utilization, reliability, stability, fairness, and low latency, as well as low overhead. The overhead in MAC protocols comes from perpacket overhead, collisions, or exchange of extra control packets. Collisions can happen if the MAC protocol allows two or more nodes to send packets at the same time, which leads to the inability of a receiver to decode a packet correctly, causing upper layers to do a packet retransmission. For real-time applications, it is very important to provide deterministic or stochastic guarantees on delivery time or minimum available data rate. Sometimes, using some level of service differentiation by means of priorities concept is preferred to give a chance for serving the important packets over unimportant packets. The operation and performance of MAC protocols is heavily influenced by the properties of the under-lying physical layer. As WSNs use the wireless medium, they inherit all the well-known problems of wireless transmission. Common problems are time-variable bit rates, and high error rates, which are caused by physical phenomena like slow and fast fading, path loss, attenuation, or thermal noise. Furthermore, modulation schemes used, frequencies, distance between transmitter and receiver, and propagation environment have a large impact on bit error rates. Finally, the design of MAC protocols highly depends on the expected traffic load patterns [87].

9.3.2 Classification of Wireless MAC Protocols

Wireless MAC protocols generally could be classified into three different categories: fixed assignment protocols, random assignment protocols, and demand assignment protocols [88].

- *Fixed-assignment protocols*, which divide the channel bandwidth in a rigid static manner, independent of the channel activity. FDMA, TDMA, CDMA, and space division multiple access (SDMA) are common forms of this class.
- *Random-assignment protocols*, in which the entire bandwidth is provided to the users as a single entity to be accessed randomly. ALOHA and CSMA are examples of this class.
- *Demand-assignment protocols*, which require that explicit control information to be exchanged among the users. These protocols can be further classified as centrally controlled, such as polling or probing protocols, and as distributed controlled, such as mini-slotted alternating priorities (MSAP).

9.4 Medium Access for Wireless Sensor Networks

In this section, we narrow down our focus toward the specific requirements and design consideration for MAC protocols in WSNs. Sources that cause energy consumption in WSNs are also investigated in this section.

9.4.1 Sources of Energy Consumption in a Wireless Sensor Network

Prolonging the sensor node lifetime and keeping network operation viable as long as possible are the most important issues in sensor networks. Energy-efficient MAC protocol should consider a set of reasons that cause energy wastage and make sensor's battery drain quickly [33–36]. In this section we list and discuss a number of sources that should be taken into account when designing MAC protocols.

Packets collisions is the most dominant source of energy waste. When two packets are transmitted at the same time and collide, they become corrupted and must be discarded, and the retransmissions of these packets are required which increase the energy consumption. Another important source that causes energy waste in wireless domain is the *overhearing*. Overhearing means that a node receives packets that are destined to other nodes. Overhearing unnecessary traffic can be a dominant factor of energy waste especially in heavy traffic load environments and dense networks. Dense sensor network deployments are common because the sensing range of many physical parameters is much smaller than the communication range.

A typically radio unit can operate in four distinct modes; idle, receive, transmit, and sleep. As it is expected that the radio consumes more energy in transmit and receive modes, running in the idle mode is also costly especially when there is no data to send during the period when nothing is sensed, this is commonly named as *idle-listening*. It is thus desirable to completely shut down the radio rather than switching into the idle mode. However, frequent switching between modes, especially switching from sleep mode to an active mode, leads to more energy consumption, than leaving the radio transceiver unit in idle mode because of the start-up power [37].

Control packet overhead is also a major source of energy consumption that we consider here. Sending, receiving, and listening for control packets consume energy. Since control packets do not directly convey useful application data; they also reduce the effective throughput. Minimal number of control packets should be used to make a data transmission. Avoiding *Overemitting* in the sensor network improves the energy efficiency. Overemitting is caused by the transmission of a message when the destination node is sleep or not ready to receive. This again results in a waste of system's energy resources and ought to be avoided.

Traffic fluctuations, in some applications of WSNs generate traffic that fluctuates in place and time, which results in peak loads that may drive the sensor network into congestion which consequently raises the collisions probability, hence, much time and energy are wasted on waiting in the random back-off procedure [23].

Choosing the appropriate *packet size* is also an important issue from the energy point of view. As the packet size gets smaller, the transition energy becomes dominant than the energy consumed during receiving and transmitting of packets [38,85].

Most of these overheads are incurred by MAC protocol that is based on contention technique. When turning to MAC protocols that are based on scheduled techniques such as TDMA, may seem attractive at the first glance because idle-listening, overhearing, and collision simply do not occur, as sensor nodes are prescheduled, and each node knows clearly which slots should transmit and receive, before any data transmissions. But, these advantages come at the cost of protocol complexity which leads to reduced flexibility to handle traffic fluctuations and network topology changes, as well a significant increase in protocol overhead. One solution for these problems is to apply some kind of overprovisioning and use a frame size that is large enough to handle peak loads. Another approach is dynamically adapting the frame size but this largely increases the complexity of the protocol and, hence, is considered to be an unsuitable option for resource-limited devices such as sensor nodes.

WSN hardware as well as communication protocols should be designed to achieve their goals with a minimum of energy consumption through avoiding or reducing the energy waste due to sources listed above. A complete energy management scheme must consider not only the radio but also all hardware units of the sensor node that consume energy.

However, at MAC layer level, energy efficiency can be improved through avoiding or minimizing idle listening, retransmissions, unwanted overhearing, and overemitting. Turning off the radio when it is not needed is an important strategy for energy conservation.

9.4.2 Wireless Sensor MAC Design Requirements and Trade-Offs

Sensor nodes are battery powered and the use of large batteries is impossible because of the space and cost constraints. Additionally, it is often not feasible to change batteries on a regular basis. It is therefore essential to make sensor nodes save as much energy as possible and, hence, prolong the network lifetime. Given that the radio unit is the most power consumer within the sensor node, a significant amount of energy could be saved through controlling the radio operation. An energy-efficient MAC protocol possesses the greatest capability to decrease the energy consumption of the radio unit because it directly controls the radio unit operation.

MAC protocols are influenced by a number of constraints. A well-designed MAC protocol should consider a set of performance attributes and make trade-offs among them. The most important performance attributes that are required for wireless sensor MAC protocols are [17,32]:

Collision avoidance is the principal task of all MAC protocols. It determines when and how a node can access the medium and sends its data. Collisions are not always completely avoided in regular operation; contention-based MAC protocols accept some level of collisions. But all MAC protocols avoid frequent collisions.

Energy-efficiency: As explained earlier, energy is a scarce resource for sensor networks, and as the radio is the major consumer of sensor node's battery, especially for long-range transmission and when the radio is kept on all the time. Therefore energy-aware MAC protocol can save transmission and reception energy by limiting the potential for collisions, minimizing the use of control messages, utilizing most of the available frequency band to shorten the transmission time, turning the radio into low power sleep state when it is idle and finally, avoiding the excessive transitions among active and sleep states.

Scalability and adaptability are closely related attributes of MAC protocol that accommodate changes in network size, node density, and topology. Some of the reasons behind these issues are limited node lifetime, addition of new nodes to the network, and varying interference which may alter the connectivity and hence the network topology. A good MAC protocol should deal with and accommodate such network changes.

Latency is the time required to send a packet by the sender until the packet is successfully received by the receiver. In sensor networks, the importance of latency is application dependent. In certain WSN applications, the sensed object places a bound on how rapidly the network must react [84].

Reliability: Reliable delivery of data is a classical design goal for all network infrastructures. The problem of reliability is of a grand importance in WSNs, where there are WSN applications where guaranteed packet delivery should be ensured. In wireless networks, packet drops are mainly caused by buffer overflow and signal interference. Buffer overflow could be avoided through employing a buffer management strategy at MAC protocol to stop the number of backlogged packets from exceeding the maximum buffer size. Such buffer control could be achieved using traffic prioritization or filtering and aggregation. Knowing that data readings of neighboring sensors are highly correlated, employing a data filtering and aggregation mechanism may yield reduction in data traffic reducing communication, avoiding buffer overflow, and saving energy. Packet drops due to signal interference can be minimized through the use of sufficiently high transmission power and the prevention of contention for medium access among nodes.

Mobility in WSNs poses a challenge to the MAC protocol design. MAC protocol should adapt itself to changes in mobility patterns, making it suitable for sensor environments with both high and low mobility.

Channel utilization refers to how the entire bandwidth of the channel is utilized in communications. Channel utilization is normally a secondary goal in sensor networks.

Throughput refers to the amount of data successfully transferred from a sender to a receiver in a given time. Many factors affect the throughput in sensor network including efficiency of collision avoidance, latency, channel utilization, and control overhead. As with latency, the importance of throughput is application-dependent.

Fairness reflects the ability of different users, nodes, or applications to share the channel equally. It is an important attribute in traditional voice and data networks. However, in sensor networks, all nodes cooperate for a single common task. At a particular time, one node may have more data to send than some other nodes. Thus, rather than treating each node equally, success is measured by the performance of the application as a whole, and per-node or per-user fairness becomes less important.

In short, the above attributes reflect the characteristics of a MAC protocol. For WSNs, the most important factors are effective collision avoidance, energy efficiency, mobility, scalability, and adaptability to densities and numbers of nodes. Other attributes are normally of secondary importance.

9.5   Classification of Wireless Sensor Network MAC Protocols

According to the underplaying mechanism used to access the shared wireless channel, MAC protocols for WSNs can be categorized into three general groups: scheduled, unscheduled (or random), and hybrid protocols. Scheduled MAC protocols attempt to organize the communication between sensor nodes in an ordered way. The most common scheduling method organizes sensor nodes using TDMA where a single sensor node utilizes a time slot. Organizing the sensor nodes provides the capability to reduce collisions and message retransmissions at the cost of synchronization and state distribution. Unscheduled protocols attempt to conserve energy by allowing sensor nodes to operate independently with minimum of complexity. Although collisions and idle listening may occur and cause energy loss, the unscheduled MAC protocols typically do not share information or maintain state. Hybrid MAC protocols combine the strengths of scheduled and unscheduled MAC protocols while compensating their weakness in an attempt to improve the protocol efficiency. The greatest advantage of the hybrid MAC protocols comes from its easy and rapid adaptability to traffic conditions which can save a large amount of energy, but this advantage comes at the cost of the protocol complexity which limits its range of applications. Some proposed MAC protocols do not easily fit into this classification scheme and many other classifications exist. However, based on the idea behind the design of the MAC protocol, more specific classification of each category of MAC protocols is provided below.

In this chapter, we cannot cover all MAC protocols proposed in the literature because of the space constraints. Instead of that a set of typical protocols are included and discussed in this section.

9.5.1 Unscheduled MAC Protocols

A common MAC paradigm in wireless networks is CSMA [15]. It is popular because of its simplicity, flexibility, and robustness. It does not require much infrastructure support: no clock synchronization and global topology information are required, and dynamic node joining and leaving are well handled without extra operation. These advantages, however, come at the cost of trial and error—a trial may cost access collision where more than two "conflicting" nodes transmit at the same time causing signal fidelity degradation at destinations. Collision can happen in any two hop neighborhood of a node. Although collision among one-hop neighbors can be greatly reduced by carrier sensing before transmission, carrier sensing does not work beyond one hop. This gives rise to the well-known hidden-terminal/exposed-terminal problems, which cause serious throughput degradation especially in high data rate sensor network applications.

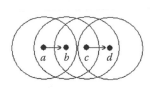

Figure 9.2 Hidden node problem (circles indicate transmission and interference range).

With reference to Figure 9.2, hidden-terminal problem could be explained as follows: Assume A, B, C, and D are sensor nodes with transmission range represented by the circles. Node A starts its transmission to node B. Node C does not catch the transmission of node A and starts its transmission to node D. The two transmissions collide at node B. In the exposed-terminal problem, consider the example in Figure 9.3; here, node B defers its

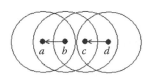

Figure 9.3 Exposed node Problem (circles indicate transmission and interference range).

transmission to node A because it hears transmission of node C to node D, even if there would be no collision at node A.

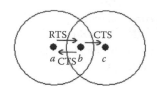

To overcome these problems, a collision avoidance technique like RTS/CTS (Request to Send/Clear to Send) handshaking is used. According to Figure 9.4, this RTS/CTS handshaking works as follows: Node A sends RTS, which blocks any possible transmission of all nodes within its radio range. Node B catches RTS of node A and it responds with CTS. With CTS, node B

Figure 9.4 RTS/CTS handshaking for collision avoidance.

blocks its neighbors and it lets node A to transmit. If the data packet is correctly received, node B sends an acknowledge packet to node A.

Unscheduled MAC protocols have several advantages. Unscheduled protocols allocate resources on demand; they can scale more easily and flexibly across changes in node density or network topology because they do not have to obtain the current schedule or join another sensor node group. Furthermore, unscheduled MAC protocols also allow sensor nodes to adapt more easily to changing traffic conditions because channel reservation can occur with finer granularity and sensor nodes can adaptively contend for the channel. Additionally, unscheduled protocols do not require fine-grained time synchronization as in TDMA protocols. However, unscheduled MAC protocols experience, in general, many drawbacks as it has all sources of energy waste, a higher rate of collisions, idle listening, and overhearing because transmission is not coordinated. In addition, fairness becomes an issue in unscheduled MAC protocols because no mechanism implicitly exists that equalizes the channel usage, unlike in a scheduled MAC protocol.

Based on the idea behind the design of the MAC protocol, more specific classification of unscheduled protocols is provided in this section.

9.5.1.1 Multichannel MAC Protocols

Using multiple radio transceivers in a single sensor node may seem a bad choice to conserve the energy of a sensor node, but several design approaches based on this technique could yield a significant energy reduction for the sensor node. Using multiple radio channels enables the sensor node to communicate simultaneously on separate channels if needed to increase bandwidth or response time. These benefits come at the cost of additional hardware requirements. First, radio transceivers constantly consume energy, even while asleep, so adding radio transceivers increases the energy consumption which lowers the overall energy consumption of the node. Second, a multiple radio transceiver system must possess the computational capability to receive and process data from multiple channels. Then, multiple radio transceivers system requires higher performance communication mechanisms and processor capabilities than single radio transceiver system. A typical protocol of this type is the PAMAS protocol.

9.5.1.1.1 PAMAS: The Power-Aware Medium Access Protocol and Signaling

PAMAS is a CSMA-based protocol in which the nodes that are not actively transmitting or receiving should power themselves off [89]. The approach requires the nodes to have two sperate radio channels for control and data. The control channel is used for handshaking and the data channel for regular traffic. Using two channels minimizes the potential for collisions. Message transfer in PAMAS starts with the source sending an RTS message to the destination on the control channel. The destination then decides if it should transmit a CTS by examining the data and control channels. If the destination does not detect any activity on the data channel and has not heard an RTS or CTS message recently it responds with a CTS message. A source that does not receive a CTS in time will back off using a binary exponential algorithm. Once the source receives a CTS message it transmits the data message over the data channel. The destination starts transmitting a busy tone over the control channel once it starts receiving the data message so that nearby nodes realize that they may not use the data channel. PAMAS implements a busy tone as a message twice the length of an RTS or CTS message. Furthermore, during the data reception the destination will transmit a busy tone any time it receives an RTS message or detects noise on the control channel to corrupt possible CTS message replies and prevent further data transmissions. Figure 9.5 illustrates the message transfer in PAMAS protocol.

Senders that cannot establish a connection switch to sleep mode and retry later. The duration of a node stay in the sleep mode is determined based on the exchange of special probe messages on the control channel among nodes in close proximity. Switching nodes that are not participating in communication to a sleep mode has been shown to result in energy savings of up to 70 percent. However the protocol requires the nodes to sense the medium to transmit and does not eliminate collisions completely. Furthermore, the protocol requires the nodes to have radio units for the two sperate channels (control and data), which increases the cost, size, and complexity of the sensor design. Additionally, controlling access to two wireless mediums increases the MAC protocol complexity. Most sensor networks have the nature that data messages are too small, which leads to decrease in the benefits behind the separation of the data and control channels. However, ideas such as those proposed through PAMAS may work for sensor networks with large data messages like multimedia sensor networks.

9.5.1.2 Application-Oriented MAC Protocols

The application characteristics may be used to enhance the MAC protocols efficiency in terms of energy conservation. For example, a monitoring-based sensor network will

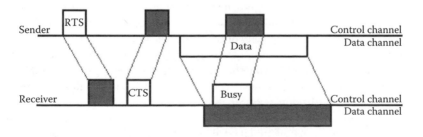

Figure 9.5 PAMAS data transfer.

have very little traffic most of the time, but may produce relatively large volumes of data when an event of interest occurs. MAC protocols that operate based on the assumption of constant traffic generation would waste energy when the sensor network has no data to manipulate. A recent and typical protocol that utilizes the application characteristics to save a considerable amount of energy is CC-MAC.

9.5.1.2.1 The CC-MAC: Collaborative MAC protocol

CC-MAC protocol attempts to conserve energy while fulfilling application requirements, by utilizing the fact that sensor nodes located near each other generate correlated measurement [41]. To achieve energy savings, CC-MAC filters measurements from highly correlated sensor nodes in an effort to reduce the traffic the sensor network must handle. Minimizing traffic leads to a reduction in the contention of wireless medium, and hence minimizes collisions.

CC-MAC consists of two components: the Event MAC (E-MAC), which filters sensor node measurements to reduce traffic and Network MAC (N-MAC), which forwards the filtered measurement to the sensor network sink. E-MAC reduces the traffic generated in an area by allowing only sensor nodes separated by at least the correlation distance to generate measurements. Other nodes periodically sleep to save energy and awake to forward messages. Correlated sensor nodes rotate the role of generating measurements to balance energy consumption throughout the network. N-MAC forwards messages from sensor nodes generating measurements to the sensor network sink, but since the E-MAC protocol has removed most of the redundancy present in multiple measurements, the forwarded traffic becomes more important.

The main disadvantage of CC-MAC is that it requires sensor nodes to possess or obtain ranging information about their neighbors in order for E-MAC to be able to filter data from correlated sensor nodes. Furthermore the complexity of the CC-MAC protocol may limit the applications of the protocol. In addition, as the number of sensing events increases, especially if the sensing conditions change with time, the overhead associated with computing the correlation radius and distributing throughout the network increases. For large networks, this overhead may become significant.

9.5.1.3 Multi-Path Data Propagation MAC Protocols

In this class of MAC protocols, a MAC protocol utilizes the back-off mechanism to decrease the chance for collisions. The MAC protocol transmits only multiple copies of data messages after a delay, while removing the overhead caused by control messages and carrier sensing. Transmitting multiple copies of data messages increases the probability of data delivery. In spite of the simplicity of the protocols of this class, many disadvantages and inefficiencies arise. As the transmissions occur without any coordination, collisions are likely to happen. This problem may be solved through increasing the back-off interval, but this results in increasing message latencies. Additionally, although sensor nodes do not exchange any handshaking messages about the success of data delivery, the protocol wastes energy through transmitting multiple copies of the same message via multiple paths without any guarantee about data delivery. However, this type of protocol may be an advantageous solution for sensor networks that generate light traffic and only require a limited number of messages to arrive at the destination. A typical example is presented in Ref. [90].

9.5.1.3.1 SRBP, ARBP, and RARBP

In Ref. [90] the authors have proposed three protocols where each one of them enhances its predecessor. *Simple Random Back-off Protocol (SRBP)* is the first protocol, which works simply by transmitting a message after an initial random back-off. Sensor nodes neither sense the channel nor exchange any control messages. *Adaptive Random Back-off Protocol (ARBP)*, enhances the performance of SRBP by adjusting the maximum back-off interval according to the sensor nodes density and current traffic conditions. Acquiring information about sensor nodes density and traffic conditions is done through utilizing two subprotocols: the density sensing protocol ($P_{density}$), and and message traffic sensing protocol ($P_{traffic}$). More details are given in Ref. [90]. The last protocol is the *Range Adaptive Random Back-off Protocol (RARBP)*, RARBP is using information regarding the distance between the sender and receiver in order adjust the random back-off interval. Nodes that are distant from the sender are given a higher chance to select small back-off values, and hence be able to send earlier. This way reduces the latency introduced by the back-off mechanism.

However, these protocols inherit many of disadvantages listed above.

9.5.1.4 Rendezvous-Based MAC Protocols

Communication between any two nodes is possible only if both of them are powered simultaneously. Hence a method to put nodes wishing to communicate on time is necessary. This method is typically called a rendezvous scheme. There are many ways to accomplish rendezvous between wireless nodes. The most popular strategy is called cycled receiver. In this scheme, nodes are powered on and off periodically, and a beaconing approach is used to express the desire or willingness to communicate. An example of this type of protocols is presented in Ref. [44].

9.5.1.4.1 TICER and RICER Protocols

The Transmitted Initiated Cycled Receiver (TICER) and Receiver Initiated Cycled Receiver (RICER) are two similar protocols [44]. The TICER protocol makes sensor nodes with data to periodically transmit RTS control packet followed by a sensing period. Receivers periodically listen to the wireless channel and if they detect an RTS message, they reply with a CTS message. The sensor nodes can then transfer the data message. RICER reverses the operation, so receivers periodically transmit beacons when they awake from their normally scheduled sleep time. A sensor node with data to transmit stays awake and monitors the channel until it hears a wake-up beacon from the intended destination. Upon reception, it starts transmitting the data packet. The session ends with an acknowledgement (ACK) signal transmitted from the destination node to the source node, after correctly receiving the data packet. Authors of the protocol mentioned that some protocol parameters such as time between control messages and the channel characteristics play an important role in the protocol overall performance. Figures 9.6 and 9.7 show the TICER and RICER schemes respectively.

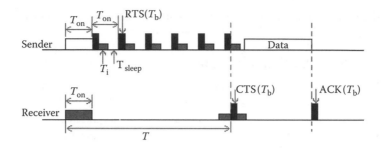

Figure 9.6 TICER scheme.

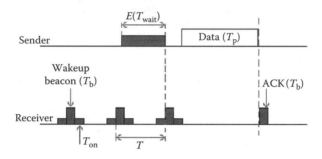

Figure 9.7 RICER scheme.

9.5.1.5 Preamble-Based MAC Protocols

In unscheduled MAC protocols, sensor nodes might not know the sleeping schedule of their neighbors, so they must somehow probe with messages until the neighbor awakes. As the communicating sensor nodes capture the messages of each other on time, they can begin the message transfer. The energy savings provided by preamble technique come from only synchronizing nearby sensor nodes when needed and only for the duration of the transmission. However, energy conservation of preamble-based protocols is greatly affected by the traffic patterns. Additionally, long preambles used in most preamble-based protocols may cause performance degradation through the increase in latency which limits the deployment of this type of protocols on real-time or latency-sensitive applications.

9.5.1.5.1 Berkeley MAC (B-MAC)

B-MAC protocol sensor nodes independently follow a sleeping schedule based on the target duty cycle for the sensor network [43]. As the sensor nodes operate on independent schedules, B-MAC uses very long beacons or preambles for message transmission. The source sensor node transmits a beacon long enough that the destination, which periodically senses the channel, has enough time to wake up and sense activity. Sensor nodes that sense activity on the channel remain awake to receive the message following

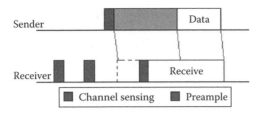

Figure 9.8   B-MAC message transfer.

the beacon or return to sleep if they do not detect activity on the channel. Figure 9.8 shows the message transfer in B-MAC.

B-MAC is flexible; that means through the protocol interface, the network designer can tweak many operating variables in the protocol, such as delay and back-off values. On the other hand, B-MAC does not provide any protection mechanism against traditional wireless problems, such as the hidden terminal problem. Furthermore long preambles in B-MAC protocol may introduce an additional latency; this problem could be considered and controlled through the B-MAC interface at the sensor network design.

9.5.2   Scheduled MAC Protocols

Scheduled MAC protocols attempt to reduce energy consumption by coordinating sensor nodes with a common schedule. Most proposed protocols use some form of TDMA since other forms of multiple access, such as frequency (FDMA) or code division (CDMA), would increase the cost and power requirement of the sensor nodes. By producing a schedule, the MAC protocol clarifies which sensor nodes should utilize the channel at any time and thus limits or eliminates collisions, idle listening, and overhearing. Nodes not participating in message communication may enter a sleep mode until they have work to perform or need to receive a message. Additionally, the MAC protocol can share traffic or status information so that individual sensor nodes can optimize energy consumption over a collection of sensor nodes instead of at just a single sensor node. However, these advantages come at the cost of increased messages to create and maintain a schedule. Node mobility, node redeployment, and node death all complicate schedule maintenance. Sensor nodes that enter the network must wait until they learn, and possibly join, the schedule to use the channel. Additionally, some delay exists between the time a sensor node dies and the time a neighboring sensor node reassigns its resources; so some resources may go unused and lead to unnecessary delays or packet loss. Then, synchronization becomes an important problem for a scheduled protocol and may occur through a periodic beacon, which increases the transceiver utilization, or by using higher precision oscillators, which increases the sensor node cost. Scheduled MAC protocols must also minimize the effect of added latency and limited throughput. Typically, added sensor node can only access the wireless channel for a fraction of the possible time. With a TDMA-based MAC protocol the time a sensor node may access the channel depends heavily on the time slot length. Typically, only one sensor node may transmit during the interval, so any unused time goes to waste. Reducing the time slot length may decrease the waste, but also decreases the maximum message length without fragmentation. Several schedule-based MAC protocols attempt to overcome the limitations on throughput and latency at the cost of sharing additional information in messages or higher duty cycle.

9.5.2.1 Slotted Contention-Based MAC Protocols

Slotted contention MAC protocols attempt to conserve energy by having nodes agree on a common sleep/listen pattern allowing them to use the radio transceiver at arbitrarily low duty cycles. Slotted protocols divide time into frames, and each frame is subdivided into a certain number of slots. Sensor nodes that have data to send wake up at the beginning of each frame and contend for the channel. This channel contention leads to a high probability of packet collisions because all communications are grouped into the listen part of the slot. To overcome this problem and enhance the performance of slotted MAC protocols, collision avoidance technique like RTS/CTS handshaking is used. As a typical example of this class of protocols, we present and discuss S-MAC protocols [32,33]. Other proposed protocols which extend S-MAC protocol such as DMAC, TMAC, DSMAC, MS-MAC, and ACMAC could be found on Refs. [45–48,64], respectively.

9.5.2.1.1 Sensor MAC Protocol (S-MAC)

The basic idea behind the Sensor-MAC (S-MAC) protocol is based on periodic sleep–listen schedules and locally managed synchronization [32,33]. Neighboring nodes form virtual clusters to set up a common sleep schedule. If two neighboring nodes reside in two different virtual clusters, they wake up at listen periods of both clusters. A drawback of S-MAC algorithm is the possibility of following two different schedules, which results in more energy consumption via idle listening and overhearing. Schedule exchanges are accomplished by periodical SYNC packet broadcasts to immediate neighbors. Collision avoidance is achieved through carrier sensing. Furthermore, RTS/CTS packet exchanges are used for unicast-type data packets. Figure 9.9 represents a sample sender (A)/receiver (B) communication.

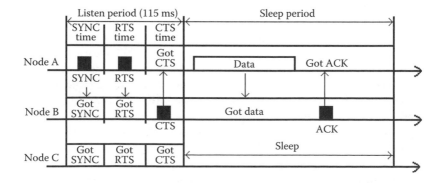

Figure 9.9   S-MAC message transfer.

An important feature of S-MAC is the concept of message-passing where long messages are divided into frames and sent in a burst. With this technique, one may achieve energy savings by minimizing communication overhead at the expense of unfairness in medium access. Periodic sleep may result in high latency especially for multi-hop routing algorithms, since all immediate nodes have their own sleep schedules. Adaptive listening technique is proposed to improve the sleep delay, and thus the overall latency. In this technique, the node which overhears its neighbor's transmissions wakes up for a short time at the end of the transmission. Hence, if the node is the next-hop node, its neighbor could pass data immediately. The end of the transmissions is known by the duration field of RTS/CTS packets.

The important advantage that is offered by sleeping schedules is the reduction on idle listening time, which consequently leads to more energy conservation in sensor nodes. But this comes at the cost of latency. In addition, the adaptive listening which is proposed to improve the sleeping delay may cause overhearing or idle listening if the packet is not destined to the listening node. Furthermore, sleep and listen periods are predefined and constant, which decreases the efficiency of the algorithm under variable traffic load.

9.5.2.2   Time Division-Based MAC Protocols

TDMA provides a tempting solution for sensor network MAC protocols because reducing collision and idle listening can save considerable amounts of energy. TDMA divides the channel into N time slots as shown in Figure 9.10. In each slot, only one node is allowed to transmit. The N slots comprise a frame, which repeats cyclically.

When designing a TDMA-based protocol, many complications arise. Time slot assignment becomes difficult because sensor nodes cannot coordinate on large scales without introducing large overhead. Synchronization functionality must exist to correct timing errors caused by clock drift within each sensor node. Strict TDMA protocols also suffer from utilization problems during periods of light traffic generation.

TDMA scheme has some disadvantages that limit its use in WSNs. TDMA normally requires nodes to form clusters, analogous to the cells in the cellular communication systems. One of the nodes within the cluster is selected as the cluster head, and acts as the base station. This hierarchical organization has several implications. More importantly, TDMA-based protocols have limited scalability and adaptability to the changes on the number of nodes. When new nodes join or old nodes leave a cluster, the base station must adjust the frame length or slot allocation. In addition, frame length and static slot

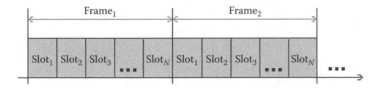

Figure 9.10   TDMA divides the time into frames of N time slots.

allocation can limit the available throughput for any given node, and the maximum number of active nodes in any cluster may be limited. Finally, TDMA-based protocols depends on distributed, fine-grained time synchronization to align slot boundaries. Many variations on this basic TDMA protocols are possible. Rather than scheduling slots for node transmissions, slots may be assigned for reception with some mechanism for contention within each slot. The base station may dynamically allocate slot assignments on a frame-by-frame basis. In ad hoc settings, regular nodes may assume the role of base station, and this role may rotate to balance energy consumption [49]. In general, TDMA-based protocols can provide good energy efficiency, but they are not flexible to changes in node density or mobility, and lack of peer-to-peer communication.

9.5.2.2.1 Lightweight MAC (LMAC)

Lightweight MAC protocol (LMAC) is based on the TDMA paradigm. Time is divided into frames, each frame is divided into time slots, which nodes can use to transfer data without having to content for the medium or having to deal with energy wasting collisions of transmissions. Every node gets periodically a time slot in which it is allowed to control the wireless medium to carry out its transmission. When a node has data to send it waits interference with other transmitting nodes. Figure 9.11 illustrates the frame format of the LMAC protocol.

Unlike traditional TDMA-based systems, the time slots in LMAC protocol are not divided among the networking nodes by a central manager. Instead a distributed algorithm is used. During its time slot, a node will always transmit a message which consists of two parts: a control message and a data unit. The control message has a fixed size and is used for several purposes. It carries the ID of the time slot controller; it indicates the distance of the node to the gateway in hops for simple routing to a gateway in the network; and it addresses the intended receiver and reports the length of the data unit. Additionally, the control message is also used to maintain synchronization between the nodes and therefore the nodes also transmit the sequence number of their time slot in the frame.

All neighboring nodes put effort in receiving the control messages of their neighboring nodes. When a node is not addressed in that message or the message is not addressed as an omnicast message, the nodes will switch off their power-consuming transceivers only to wake up at the next time slot. If a node is addressed, it will listen to the data

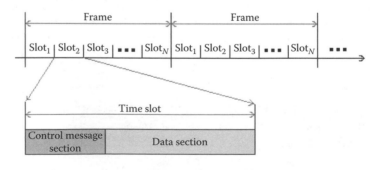

Figure 9.11   LMAC frame format.

unit which might not fill the entire remainder of the time slot. Both transmitter and receiver(s) turn off their transceivers after the message transfer has completed. A short time-out interval ensures that nodes do not waste energy for idle listening in time slots that are not controlled. In this protocol, it is only possible for a node to transmit a single message per frame.

9.5.2.3 Reservation-Based MAC Protocols

Time-based medium access has the potential of capturing most of the opportunities for energy optimization in sensor networks. Energy wastage due to overhearing, collision, idle mode, and transitions between different states can be minimized if the medium access is shared on a time basis. In addition, time-based medium arbitration can enhance delay predictability and limit packet drops due to interference and buffer overflow. However the problem of scheduling access to the medium is NP-hard (i.e., at least as hard as Non-deterministic Polynomial time Problem [NP-Problem]) making the scalability of time-based MAC scheme a major concern. Moreover, distributed time-based medium arbitration typically introduces excessive overhead. In addition, maintaining clock synchrony among nodes is essential to enforce the schedule which is a nontrivial problem for the resource-constrained sensor nodes. Most of the time-based MAC protocols proposed in the literature have focused on addressing these issues either using reservation requests over preset data routes or pursuing simplified heuristics to tackle the complexity of medium access scheduling.

9.5.2.3.1 Energy-Efficient TDMA MAC Protocol Scheduling

The use of reservation requests has been explored for tackling the scalability of time-based medium arbitration [91]. Nodes that have data to transmit make a reservation request to a base station, which responds with a traffic control message indicating medium access schedule. Nodes that are not included in the traffic control message can turn off their radio transceivers. The nodes that have been assigned slots transmit in the order scheduling by the base station. The base station trades off latency with energy efficiency. Although it is better to bundle all transmissions from a node in consecutive time slots, the transmission of other nodes will be delayed.

9.5.2.4 Priority-Based MAC Protocols

By using a random function, the access to the wireless channel is controlled by assigning priorities to sensor nodes or links to destinations. Sensor node with highest priority is given the chance to access the channel. Sensor node IDs and time slot numbers provide an input to a random function that establishes the priority within a two neighborhood. One example of this type of protocols is the series of protocols proposed by Bao and Garcia-Luna-Aceves [53].

9.5.2.4.1 NAMA, LAMA, and PAMA MAC Protocols

The Node Activation Multiple Access (NAMA) protocol uses TDMA with time divided into frames; each frame is subdivided into sections. Each section is further divided into

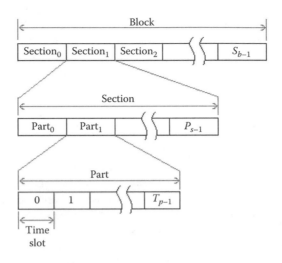

Figure 9.12   Time division structure of NAMA protocol.

parts, where each part is composed of a number of time slots. Figure 9.12 depicts the time division structure of the NAMA protocol. Each node selects a single part, chosen to balance channel utilization across the parts, and contends with the other sensor nodes that select the same part. NAMA reserves the last section of each frame for signaling messages that allow sensor nodes to join the network. Each sensor node computes its priority along with the priority of its neighbors and uses these to determine who has access to the current time slot within the sensor node's chosen part. A sensor node gets assigned a particular slot within a section based on its priority. If a sensor node has the highest priority among its two-hop neighbors for the given time slot, then the sensor node may transmit. If no sensor node's priority maps to a time slot, then the sensor node with the highest priority may use the time slot. For more details about NAMA protocol refer to Ref. [53]. *Link Activation Multiple Access* (LAMA) protocol uses receiver-oriented direct sequence spread spectrum (DSSS) in addition to time slotting (i.e. a transmitter chooses a code corresponding to the receiver node) to activate links to destination sensor nodes. Each sensor node gets a code assigned from a finite set of pseudonoise codes. During each time slot the sensor node with the highest priority in a two hop neighborhood may activate a link by using the code assigned to the receiver. *The Pairwise-link Activation Multiple Access* (PAMA) protocol activates links between sensor nodes by assigning priorities to the links and by varying the codes and priorities of links based on the current time slot. A communication link between two sensor nodes gets activated if the link has the highest priority among all other links which link the two sensor nodes. Similar to LAMA, the use of DSSS allows nodes to communicate on different codes without interruption and the protocol algorithm prevents collisions on the same code.

The main disadvantage of the NAMA, LAMA, and PAMA protocols is the computation overhead of sensor nodes priorities which rapidly consumes energy and shortens the network lifetime. Additionally, LAMA and PAMA require the sensor nodes to have radios with spread spectrum capabilities, which increases sensor node cost. Dynamic slot assignment also prevents sensor nodes from developing a regular sleep schedule since the priorities vary based on the current slot number.

9.5.3 Hybrid MAC Protocols

Each of CSMA-, TDMA-, FDMA-, and CDMA-based schemes offers some advantages and suffers some shortcomings with respect to the requirements of MAC protocols for sensor networks. Hybrid MAC protocols combines the strengths of the contention-based protocols and scheduled protocols while offsetting their weakness to better address these requirements. The greatest advantage of the hybrid MAC protocols comes from its easy and rapid adaptability to traffic conditions which can save a large amount of energy. Protocols presented in Refs. [55,57,92,93] are good examples of this class of protocols. In this section, we briefly discuss a sample of such protocols.

9.5.3.1 Preamble-Based Hybrid MAC Protocols

By combing the strengths of scheduled and unscheduled MAC protocols, preamble-based protocol may be made more energy efficient and more sensitive to changes in traffic type. The reduction in energy consumption made by preamble hybrid techniques comes from two sources: first, synchronizing nearby sensor nodes when needed and only for the duration of the transmission and second, by utilizing the fact the preamble-based protocol are very sensitive to traffic patterns. Then the protocol can alter its operation based on changes in traffic to make a significant save in energy resources. However, the most important drawback of most preamble-based protocols is their use of long preambles which causes increase in latency. High latencies inherited in these types of protocols limits the deployment of this type of protocols on real-time or latency-sensitive applications. Additionally, as these protocols use mixed technique of scheduled and unscheduled protocols, the complexity of protocol becomes very high.

9.5.3.1.1 Wireless MAC Protocol (WiseMAC)

WiseMAC protocol is a hybrid TDMA/CSMA with preamble sampling, where all sensor nodes are defined to have two communication channels [54]. Data channel is accessed with TDMA method, whereas the control channel is accessed with CSMA method. WiseMAC protocol uses nonpersistent CSMA (NP-CSMA) with preamble sampling to decrease idle listening. In the preamble sampling technique, a preamble precedes each data packet for alerting the receiving node. All nodes in a network sample the medium with a common period, but their relative schedule offsets are independent. If a node finds the medium busy after it wakes up and samples the medium, it continues to listen until it receives a data packet or the medium becomes idle again. The size of the preamble is initially set to be equal to the sampling period.

To reduce the power consumption incurred by the redetermined fixed-length preamble, WiseMAC offers a method to dynamically determine the length of the preamble. That method uses the knowledge of the sleep schedules of the transmitter node's direct neighbors. Another parameter affecting the choice of the wake-up preamble length is the potential clock drift between the source and the destination.

The main drawback of WiseMAC is that decentralized sleep–listen scheduling results in different sleep and wake-up times for each neighbor of a node. This is especially an important problem for broadcast type of communication, since broadcasted packet will be buffered for neighbors in sleep mode and delivered many times as each neighbor wakes up. However, this redundant transmission will result in higher latency and power consumption. Moreover, the hidden terminal problem comes along with WiseMAC, this is because WiseMAC is based on nonpersistent CSMA.

9.5.3.2   Reservation-Based Hybrid Protocols

By combining the contention and time division schemes, the performance of reservation-based protocols could be improved. PARMAC protocol is based on this idea to gain a significant reduction in energy.

9.5.3.2.1   The Power-Aware Reservation-Based MAC (PARMAC) Protocol

PARMAC is an energy-aware protocol primarily designed for ad hoc networks and is applicable to sensor networks as well [92]. The approach is actually a combination of contention and reservation-based medium arbitration schemes. The network is divided into grids and each node is assumed to reach all the other nodes within its grid. Time is divided into fixed frames. Grids are assigned distinct frames. Each frame is composed of Reservation Period (RP) and Contention Free Period (CFP). In each RP, nodes within a grid cell exchange three messages to reserve the slots for data transmission and reception and the exchange of acknowledgments. Data is then sent in the CFP. The clocks of all nodes are assumed to be synchronized. The protocol saves energy by minimizing the idle time of the nodes allowing the nodes to sleep during a CFP. Moreover, intragrid control packets overhead and packet retransmissions are minimal, achieving significant energy savings. However, intergrid contention is still possible and the efficiency of this approach can significantly diminish if the application requires data exchange among nodes in different grids.

9.5.3.3   Traffic-Sensitive Protocols

Traffic types and conditions in the sensor network have a direct effect on the energy consumption at the sensor node. A MAC protocol can utilize this fact to make a significant save in energy resource by adapting itself to network conditions. Sensor networks that generate large volume of traffic provide a good case for MAC protocols that adapt their operation based on traffic conditions. Additionally, the differentiation in traffic characteristics between control traffic and data traffic could be utilized to make the MAC protocol to alter its operation according to the traffic type to provide a considerable reduction in energy resource consumption. However, to realize and implement these benefits come from the nature of the traffic, MAC protocols should keep track of traffic characteristics within the sensor network and share traffic information among sensor nodes within the sensor network. Here are two examples.

9.5.3.3.1 Traffic Adaptive Medium Access (TRAMA) Protocol

TRAMA protocol attempts to balance the benefits of scheduled and unscheduled protocols by providing scheduled slots with no contention for longer data messages and random access slots for small periodic control messages [57]. Additionally, sensor nodes adapt to traffic and network conditions by sharing traffic needs with neighbors and learning the two-hop topology of their neighbors. TRAMA protocol consists of three subprotocols: the Neighbor Protocol (NP), which shares the topology information; the Scheduled Exchange Protocol (SEP), which allows nodes to share what traffic they have queued; and Adaptive Election Algorithm (AEA), which selects the slots to use for data transfer based on topology and traffic conditions.

Frames within TRAMA protocol consists of several slots, where the random access control slots occur together at the beginning of the frame and the scheduled data slots occur at the end as illustrated in Figure 9.13.

The major advantage of TRAMA protocol is that a higher percentage of sleep time and lower collision probability is achieved compared to contention-based protocols. The main drawback of TRAMA protocol is that the transmission slots are set to be seven times longer than the random access period, which means that the duty cycle is very long, and hence energy resources are used more intensely than other protocols. Moreover, TRAMA protocol has a high level of complexity compared to other protocols, which limits its use.

9.5.3.3.2 Zebra MAC (Z-MAC) Protocol

Z-MAC uses CSMA as the baseline MAC scheme, but uses a TDMA schedule as a hint to enhance contention resolution [55]. The main feature of Z-MAC is its adaptability to the level of contention in the network so that under low contention, it behaves like CSMA, and under high contention, like TDMA, which saves a large amount of energy. It is also robust to dynamic topology changes and time synchronization failures commonly occurring in sensor networks. Z-MAC assigns sensor nodes TDMA slots, but easily allows sensor nodes to utilize slots they do not own through CSMA with prioritized back-off times.

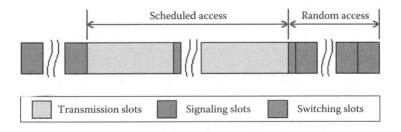

Figure 9.13 TRAMA time slot organization.

The largest disadvantage of Z-MAC is the large overhead caused by the TDMA structure. In event-based sensor networks, Z-MAC will take time to distribute control messages to switch to TDMA mode, and hence latency becomes an issue. The latency problem in Z-MAC comes from the fact that Z-MAC uses explicit congestion notification (ECN) messages to limit the effect of hidden terminals during periods of high contention. When a sensor node detects high contention it transmits an ECN messages to the neighbor. The neighbors broadcast the ECN message to its neighbors, all of whom enter a high contention level (HCL) state. Sensor nodes return to a low contention level (LCL) state in a time period if they do not receive further ECN messages.

9.5.3.4 Clustering-Based MAC Protocols

Gathering sensor nodes into clusters offers many advantages. First, clustering enables to differentiate between local traffic from global traffic to conserve energy. Second, sharing information locally provides a trade-off between global state distributions, which would consume too much energy for the dynamic nature of sensor networks, and greedy algorithms that optimize sensor node behavior independent of other sensor nodes. Third, clustering also allows protocols to scale more easily because the protocol might view a cluster as a single entity. Finally, clustering may allow sensor nodes to perform some functionality, such as synchronization, on a local scale that would consume too much energy on a global scale. However, these advantages come at the cost of coordination messages overhead. The Cluster head which manages the cluster must coordinate the sensor nodes to ensure that the cluster reduces energy on average. Protocols often rotate the role of the cluster head among sensor nodes to evenly distribute the additional energy consumption caused by managing operation. Node dynamics further complicates clustering protocols because cluster formation and cluster head assignment algorithms must adapt to redeployment or sensor node death.

9.5.3.4.1 LEACH: Low-Energy Adaptive Clustering Hierarchy

Low-Energy Adaptive Clustering Hierarchy (LEACH) is a clustering-based protocol that minimizes energy dissipation in sensor networks [109]. The purpose of LEACH is to randomly select sensor nodes as cluster heads, so that the high-energy dissipation in communicating with the base station is spread to all sensor nodes in the sensor network. The operation of LEACH is separated into two phases, the setup phase, and the steady phase. The duration of the steady phase is longer than the duration of the setup phase to minimize the overhead.

During the setup phase, a sensor node chooses a random number between 0 and 1. If this random number is less than a certain threshold, the sensor node is selected as a cluster head. After the cluster heads are selected, the cluster heads advertise to all sensor nodes in the network that they are the new cluster heads. Once the sensor nodes receive the advertisement, they determine the cluster that they want to belong based on the signal strength of the advertisement from the cluster heads to the sensor nodes. The sensor nodes inform the appropriate cluster heads that they will be a member of the cluster. Afterwards, the cluster heads assign the time on which the sensor nodes can send data to the cluster heads based on a TDMA approach.

During the steady phase, the sensor nodes can begin sensing and transmitting data to the cluster heads. The cluster heads also aggregate data from the nodes in their cluster before sending data to the base station. After a certain period of time spent on the steady phase, the network goes into the setup phase again entering into another round of selecting the cluster heads.

9.5.3.4.2 GANGS: An Energy-Efficient MAC Protocol for Sensor Networks

GANGS protocol gathers sensor nodes into clusters [52]. GANGS protocol utilizes an unspecified contention protocol for intracluster communication and TDMA-based communication protocol for transmissions between clusters. GANGS does not assume that the sensor nodes can communicate with base station, therefore the cluster heads must form a routing backbone in the sensor network using a separate routing protocol. Clusters formation in GANGS is done through two phases: cluster head election and clusters connection. As the cluster heads perform their operation they will eventually have lower energy resources than other nearby sensor nodes because of their increased functionality. When this occurs, the sensor nodes perform the cluster formation procedure again so that sensor nodes equalize energy consumption throughout the network. To assign slots, cluster heads perform a distributed algorithm that results in each cluster head having a slot to transmit in and knowing the slots used by each neighbor. After the cluster heads determine the TDMA schedule, they distribute the information within the cluster so that the other sensor nodes may use the unassigned slots at the end of the frame for sending their data.

GANGS protocol has the disadvantage that cluster formation and reconstruction consumes energy resources and takes time. Furthermore, the slot organization in GANGS also introduces wasted resources since not all slots may get used. Despite these disadvantages, GANGS protocol provides contention-free traffic flow for forwarded traffic while retaining the flexibility and simplicity of a random access protocol within the cluster.

9.5.4 Quality-of-Service-Specific MAC Protocols

The concepts of latency, throughput, and delay jitter were not primary concerns in most of the presented work on sensor networks. However, the increasing interest in real-time applications of sensor networks has posed additional challenges on protocol design. For example, handling real-time traffic of emergent event triggering in monitoring-based sensor network requires that end-to-end delay is within acceptable range and the variation of such delay is acceptable [94]. Such performance metrics are usually referred to as quality of service (QoS) of the communication network. Therefore, collecting sensed real-time data requires both energy and QoS-aware MAC protocol to ensure efficient use of the energy resources of the sensor node and effective delivery of the gathered measurements.

However, achieving QoS guarantees in sensor network is a challenging task, because of the strict resource constraints (limited battery power, and data memory) of the sensor node, and the hostile environments in which they must operate [95].

QoS provisioning in WSNs is increasingly acquiring the attention of many researchers. Recently many MAC protocols that support some type of quality of service in WSNs have emerged [96–102]. Here are some examples.

9.5.4.1   QoS Control for Sensor Networks

The authors explicitly exploit node redundancy [96]. They developed an adaptive scheme for each sensor to determine independently whether to transmit or not so that a fixed total number of transmissions occur in each slot. The protocol accomplishes its task by allowing the base station to communicate QoS information to each sensor node within the network through a broadcasting channel, and by use of the Gur Game mathematical paradigm to dynamically adjust the optimum number of active sensors. The protocol makes trade-offs between the required number of sensors that should be powered-up so that enough data is being collected to meet the required QoS and the number of sensors that should be turned-off to save a considerable amount of battery power of sensor nodes, and hence maximizing the network's lifetime. Here, the concept of QoS is defined as the total number of transmissions that should occur in each slot to gather enough data (that is, information quality).

9.5.4.2   An Energy-Efficient QoS-Aware MAC Protocol for Wireless Sensor Networks (Q-MAC) Protocol

The Q-MAC scheme attempts to minimize the energy consumption in a multi-hop WSN while providing QoS by differentiating network services based on priority levels [97]. The priority levels reflect the criticality of data packets originating from different sensor nodes. The Q-MAC accomplishes its task through two steps; intranode and internode scheduling. The intranode scheduling scheme adopts a multi-queue architecture to classify data packets according to their application and MAC layer abstraction. Internode scheduling uses a modified version of MACAW [103] protocol to coordinate and schedule data transmissions among sensor nodes.

9.5.5   Cross-Layer MAC Protocols

All MAC protocols discussed so far improve the energy efficiency to a certain extent by exploiting the collaborative nature of WSNs and its correlation characteristics. However, the main commonality of these protocols is that they follow the traditional layered protocols architecture. Whereas these protocols may achieve very high performance in terms of the metrics related to each of these individual layers, they are not jointly optimized to maximize the overall network performance while minimizing the energy expenditure. Considering the scarce energy and processing resources of WSNs, joint optimization and design of networking layers, that is, cross layer design, becomes the most promising alternative to inefficient traditional-layered protocol architecture. Recently, significant work on the cross-layer development of WSN protocols has emerged. In fact, recent research on WSNs reveals that cross-layer integration and design techniques result in significant improvement in terms of energy conservation. Generally, many reasons lay behind this improvement. First, the stringent energy, storage, and processing capabilities of wireless sensor nodes necessitate such an approach. The significant overhead of layered protocols results in high inefficiency. Furthermore, recent empirical studies necessitate that the properties of low-power radio transceivers and the wireless channel conditions be considered in protocols design. Finally, the event-centric approach of wireless sensor networks requires application-aware communication protocols. In this section, we review some recent communication protocols devised for WSNs that focus on cross-layer design approach. The protocols are classified according to the interactions between different layers of OSI network stack. The classification presented in Ref. [31] is followed, and other protocols are added.

9.5.5.1 MAC + PHY

In Ref. [65], the energy consumption analysis for physical and MAC layers is performed for three different MAC protocols. The authors provide analysis of energy consumption and conclude that single-hop communication can be more efficient if real radio models are used. Despite this interesting result, the analysis is based on a linear network, which may not be practical in realistic scenarios.

A cross-layer solution among MAC and PHY layers is proposed in Ref. [66], a new cross-layer-based carrier-sensing mechanism for alleviating exposed/hidden node problem, referred as MP scheme. This scheme uses MAC-address-based physical carrier sensing to determine if the medium is busy. In MP, the addresses of transmitter and receiver of a packet are incorporated into the PHY header. Making use of this address information for its carrier-sensing operation, a node can drastically reduce the detrimental effects of exposed/hidden node. Results show that the proposed scheme is more efficient and more effective than the previous schemes.

9.5.5.2 MAC + Network

The MAC and Routing cross-layer interaction for receiver-based routing has been investigated in many research papers [67–69]. In these papers authors discuss the energy efficiency, latency, and multi-hop performance of the algorithm. Reference [70] expends the work proposed in Refs. [68,69] for a single radio node.

In Ref. [71], the MACRO protocol is proposed, where the routing decision is performed as a result of successive competitions at the medium access level. More specifically, the next hop is selected based on a weighted progress factor and the transmit power is increased successively until the most efficient node is found. Moreover, on–off schedules are used.

The MAC-CROSS is proposed in Ref. [72], MAC-CROSS protocol minimizes the number of nodes that should be awake to complete the communication process. The protocol utilizes the routing information to awake only the nodes that are involved in the routing path. All other nodes that are not in the routing path can stay in their sleep mode until the beginning of the next duty cycle. However, the improvement achieved in energy conservation comes at the cost of protocol's latency.

A joint scheduling and routing scheme is proposed in Ref. [73] for periodic traffic in WSNs. In this scheme, the nodes form distributed on–off schedules for each flow in the network while the routes are established such that the nodes are only awake when necessary. Since the traffic is periodic, the schedules are then maintained to favor maximum efficiency. The authors also investigate the trade-off between on–off schedules and the connectivity of the network.

The usage of on–off schedules in a cross-layer routing and MAC frame work is also investigated in Ref. [74]. In this work, a TDMA-based MAC scheme is devised, where nodes distributively select their appropriate time slots based on local topology information. The routing protocol also exploits this information for route establishment.

The performance evaluations of all these solutions present the advantages of cross-layer approach at the routing and MAC layer.

Multi-hop Infrastructure Network Architecture (MINA) is another work for integrating MAC and routing protocols [60]. Ding et al. propose a layered multi-hop network architecture where the network nodes with the same hop count to the base station are grouped into the same layer. Channel access is a TDMA-based MAC protocol combined with CDMA or FDMA. The superframe is composed of a control packet, a beacon frame, and a data transmission frame. Beacon and data frames are time slotted. In the clustered network architecture, all members of a cluster submit their transmission requests in beacon slots. Accordingly, the cluster-head announces the schedule of the data frame.

The routing protocol is a simple multi-hop protocol where each node has a forwarder node at one nearer layer to the base station. The forwarding node is chosen from candidates based on the residual energies. Ding et al. then formulate the channel allocation problem as an NP-complete problem and propose a suboptimal solution. Moreover, the transmission range of sensor nodes is a decision variable, since it affects the layering of the network (hop-counts change). Simulations are run to find a good range of values for a specific scenario.

9.5.5.3   Network + PHY

A cross-layer optimization of network throughput for multi-hop wireless networks is presented in Ref. [75]. The authors split the throughput optimization problem into two subproblems, that is, multi-hop flow routing at the network layer and power allocation at the physical layer. The throughput is tied to the per-link data flow rates, which in turn depend on the link capacities and hence, the per-node radio power level. On the other hand, the power allocation problem is tied to interference as well as the link rate. Based on this solution, a CDMA/OFDM-based solution is provided such that the power control and the routing are performed in a distributed manner.

In Ref. [76], new forwarding strategies for geographic routing are proposed. The authors provide expressions for the optimal forwarding distance for networks with automatic repeat request (ARQ) and without ARQ. Moreover, two forwarding strategies for these cases are provided. The forwarding algorithms require the packet reception rate of each neighbor for determination of the next hop and construct routes accordingly. Although the new forwarding metrics illustrate the advantages of cross-layer forwarding techniques in WSNs, the analysis for the distribution of optimal hop distance is based on a linear network structure.

9.5.5.4 Transport + PHY

In Ref. [77], a cross-layer optimization solution for power control and congestion control is considered. The authors provide analytical analysis of power control and congestion control, and the trade-off between layered and cross-layer approach is presented. Based on this framework, a cross-layer communication protocol based on CDMA is proposed, where the transmission power and the transmission rate is controlled. However, the proposed solution only applies to CDMA-based wireless multihop networks, which may not apply to WSNs where CDMA technology may not be feasible.

9.5.5.5 3-Layer Solutions

In addition to the proposed protocols that focus on pairwise cross-layer interaction, more general cross-layer approaches among three protocol layers exist. In Ref. [78], the optimization of transmission power, transmission rate, and link schedule for TDMA-based WSNs is proposed. The optimization is performed to maximize the network lifetime, instead of minimizing the total average power consumption.

In Ref. [79], adaptation strategies that maximize the network lifetime through joint routing, scheduling, and link layer optimization is proposed. Authors propose a variable-length TDMA scheme where the slot length is optimally assigned according to the routing requirement while minimizing the energy consumption across the network. The optimization problem considers energy consumption that includes both transmission energy and circuit processing energy. Based on this analysis, it is shown that single-hop communication may be optimal in some cases where the circuit energy dominates the energy consumption instead of transmission energy. Although the optimization problems presented in the paper are insightful, no communication protocol for practical implementation is proposed. Moreover, the transport layer issues such as congestion and flow control are not considered.

9.6 IEEE 802.15.4/ZigBee MAC Protocol

ZIGBEE [104] is a new wireless technology guided by the IEEE 802.15.4 Personal Area Networks standard [105]. It is primarily designed for the wide ranging automation applications and to replace the existing nonstandard technologies. It currently operates in the 868 MHz band at a data rate of 20 kbps in Europe, 914 MHz band at 40 kbps in the United States, and the 2.4 GHz ISM bands worldwide at a maximum data rate of 250 kbps.

ZigBee is expected to provide low cost and low-power connectivity for equipment that needs battery life as long as several months to several years but does not require high data transfer rates. Some of its primary features are standard-based wireless technology, low data rates, low power consumption, simple and low-cost wireless networking, etc. Sometimes, people confuse IEEE 802.15.4 with ZigBee, an emerging standard from the ZigBee alliance. ZigBee uses the services offered by IEEE 802.15.4 and adds network construction (star networks, peer-to-peer/mesh networks, cluster-tree networks), security, application services, and more. The IEEE 802.15.4 MAC protocol is a hybrid

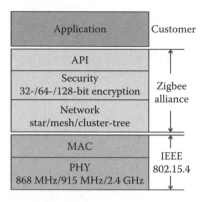

Figure 9.14   IEEE 802.15.4/ZigBee protocol stack architecture.

protocol that combines both scheduled-based as well as contention-based schemes. In this section we briefly describe the IEEE 802.15.4/Zig-Bee standards with more emphasis on the MAC layer.

9.6.1   IEEE 802.15.4/ZigBee Protocol Stack Architecture

Following the standard OSI reference model, ZigBee protocol stack is structured in layers as shown in Figure 9.14. The first two layers, physical (PHY) and media access (MAC), are defined by IEEE 802.15.4 standard [105]. The layers above them are defined by ZigBee Alliance [104].

9.6.2   ZigBee Network Architecture

Two different device types can participate in a ZigBee network, a full-function device (FFD) and a reduced-function device (RFD). The FFD can operate in three modes serving as a personal area network (PAN) coordinator; a coordinator; or a device. An FFD can talk to RFDs or other FFDs, while an RFD can talk only to an FFD. A system conforming to IEEE 802.15.4 consists of several components. The most basic mode is the device. A device can be an RFD or an FFD. Two or more devices within a Personal Operating Space (POS) communicating on the same physical channel constitute a Wireless Personal Area Network (WPAN). However, a network shall include at least one FFD, operating as the PAN coordinator. Depending on the application requirements, the Low Range Wireless Personal Area Network (LR-WPAN) may operate in either of two topologies: the star topology or the peer-to-peer topology. In the star topology all communication and resource reservation occurs through the PAN coordinator. Within the peer-to-peer topology, devices communicate with each other directly without passing through the PAN coordinator, but they must be associated with the PAN coordinator prior to participating in the network. Peer-to-peer topology allows more complex network formations to be implemented, such as mesh networking topology. Applications such as industrial control and monitoring, WSNs, asset and inventory tracking, intelligent agriculture, and security would benefit from such a network topology. A peer-to-peer network can be ad hoc, self-organizing and self-healing. It may also allow multiple hops to route messages from any device to any other device on the network. Such functions can be added to the network layer. The IEEE 802.15.4 focuses on the star-topology networks, but leaves many options and functionality of peer-to-peer networks undefined. Consequently, in the remainder of this section, we focus on the data exchange between coordinator and devices in a star network.

9.6.3 Superframe Structure

The coordinator of the star network operating in the beacon-enabled mode organizes channel access and data transmission with the help of a superframe. Superframes are typically utilized within the context of low-latency devices, whose associations must be kept even if inactive for long periods of time. The structure of the superframe is depicted in Figure 9.15. All superframes have the same length. The coordinator starts each superframe by sending a frame beacon packet. The frame beacon includes a superframe specification describing the length of the various components of the following superframe:

- The superframe is subdivided into an active period and an inactive period. During the inactive period, all nodes including the coordinator can switch off their transceivers and go into sleep state. The nodes have to wake up immediately before the inactive period ends to receive the next beacon.
- The active period is subdivided into 16 slots. The first time slot is occupied by the beacon frame and the remaining time slots are partitioned into a Contention Access Period (CAP) followed by a number (seven at maximum) of contiguous Guaranteed Time Slots, (GTSs), where the application can use one or more of them to transfer data.

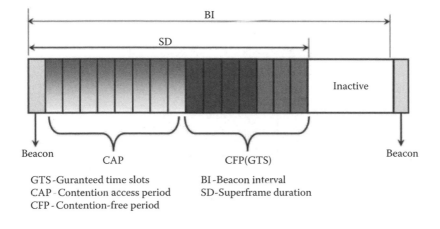

Figure 9.15   IEEE 802.15.4 superframe structure.

The lengths of the active period, inactive period, and a single time slot as well as the usage of GTS slots are configurable.

The coordinator is active during the entire active period. The associated devices are active in the GTS phase only in time slots allocated to them; in all other GTS slots they can enter sleep mode. In the CAP, a device can shut down its transceiver if it has neither any own data to transmit nor data to fetch from the coordinator.

9.6.4 Data Transmission

There can be three different types of data transmission possible. They are: transmission from a device to the coordinator, transmission from the coordinator to the device, and finally transmission between any two devices. In a star topology only the first two transmission techniques are possible. Transmission between any two devices is not supported, whereas in a peer-to-peer network all the three types of transmissions are possible. The transmissions can be carried out again in either of two modes, depending on if the beacon transmissions are allowed or not. We focus only on a beacon-enabled network. The beacon period of the superframe must be sufficient to give service to the network structure and its devices. Data transfers to the coordinator require a beacon synchronization phase, if applicable, followed by CSMA/CA transmission (by means of slots if superframes are in use); acknowledgment is optional. Data transfers from the coordinator usually follow device requests: if beacons are in use, these are used to signal requests; the coordinator acknowledges the request and then sends the data in packets which are acknowledged by the device. The same is done when superframes are not in use, only in this case there are no beacons to keep track of pending messages.

Point-to-point networks may either use unslotted CSMA/CA or synchronization mechanisms; in this case, communication between any two devices is possible, whereas in structured modes one of the devices must be the network coordinator. In general, all implemented procedures follow a typical request-confirm/indication-response classification. Figures 9.16 and 9.17 show the data transfer procedure in IEEE 802.15.4.

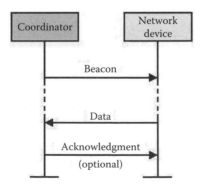

Figure 9.16   IEEE 802.15.4 data transfer: data transfer to coordinator.

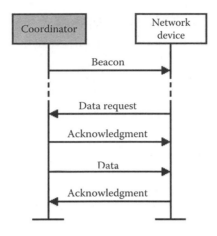

Figure 9.17 IEEE 802.15.4 data transfer: data transfer from coordinator.

Although, the IEEE 802.15.4 focuses on applications similar to sensor networks, many disadvantages exist that limit its use in the sensor networks area. First, the standard defines only the communication mechanisms for star topologies, while the operation of peer-to-peer topology is not clearly defined. As well, the standard does not allow interoperation between different coordinators.

9.6.5 Bluetooth

The Bluetooth system is designed as a WPAN with one major application, the connection of devices to a personal computer. It already has been used as a means for prototyping wireless network applications. The physical layer is based on the frequency hopping spread spectrum (FHSS) scheme having a hopping frequency of 1.6 kHz and an appropriate allocation of hopping sequences. The nodes are organized into piconets with one master and up to seven active slave nodes. The master chooses the hopping sequence, which the slaves have to follow. Furthermore, there can be several passive slave nodes in a piconet. The master polls the active slaves continuously. Two major drawbacks of Bluetooth are the need to constantly have a master node, spending much energy on polling this slaves, and rather limited number of active slaves per piconet. This is not compatible with the case of dense WSNs where a huge number of master nodes would be needed. An active slave must always be switched on since it cannot predict when it will be polled by the master. A passive slave has to apply at the master to become an active slave. This fails if there are seven active nodes already. Furthermore, it is required that each node is able to take the role of masters or slaves and thus bears considerable complexity. Also, the fast frequency hopping operations require tight synchronization between the nodes in a piconet [110].

9.7 Open Research Directions

Although many medium access schemes have been proposed for sensor networks, MAC layer protocol design is still largely open to research. In this section, we identify these open issues, and present future research directions.

Standardization: There is a lack in the MAC protocol standardization for sensor networks. Up to now there is no protocol accepted as standard. The main reason behind this is that the choice of MAC protocol, in general, is application dependant, which means that there will be many standard MAC for sensor networks. Another reason is the lack in standardization in lower layers (physical layer) and the physical sensor hardware.

TDMA has a natural advantage of collision-free medium access. However it includes clock drift problems and decrease throughput at low traffic loads due to idle slots. The difficulty with TDMA systems are the synchronization of the nodes and adaptation to topology changes where these changes are caused by insertion of new nodes, exhaustion of battery capacities, broken links because of interference, sleep schedules of relay nodes, and scheduling caused by clustering algorithms. The slot assignments, therefore, should be done regarding such possibilities. However, it is not easy to change the slot assignment within a decentralized environment for traditional TDMA, since all nodes must agree on the slot assignment.

FDMA is another scheme that offers a collision-free medium though, it brings an additional circuitry requirement to dynamically communicate with different radio channels. This increases the cost of the sensor nodes which is contrary to the objective of the sensor network systems.

CDMA also offers collision-free medium, but its high computational requirement is a major obstacle for less energy consumption objective of the sensor node. If it is shown that the high computational complexity of CDMA could be traded with its collision avoidance feature, CDMA protocols could also be considered as good solutions for sensor networks.

CSMA-based protocols have a lower delay and promising throughput potential at low traffic loads, which generally happens to the case in WSNs. However, additional collision avoidance or collision detection methods should be employed to handle the collision possibilities. Lack of comparison of TDMA, CSMA, or other medium access protocols in a common framework is a crucial deficiency of the literature.

To date, the primary design goal for sensor networks, in general, and MAC in particular has been energy efficiency. However, as new applications of sensor networks emerge other optimization criteria (or QoS parameters) such as latency and compliance with real-time constraints, or reliable data delivery may gain importance. One particular issue is that many applications need to be optimized for multiple, conflicting criteria. Hence, applications need a way to implement particular trade-offs between these conflicting goals.

Mobility management: Considering mobility in MAC protocol design has been well identified as an open research challenge in sensor networks for quite some time and yet even the most recent MAC protocols appearing in the literature do not explicitly consider mobility at the MAC layer except few ones such as Refs. [48,63,106]. Recently, there has been an increased interest in medical care and disaster response applications of sensor networks and these environments make use of mobile sensor nodes. So, there is much room for research in this area.

Scalability issues are most notable among possible future research directions given the growing ambition for very large deployments of tiny sensors and the widening scope of the use in many applications.

Adaptability to traffic and topology changes: To conserve more energy MAC protocol should adapt to changes in both network topology and traffic characteristics. However adaptability to changes usually increases the protocol complexity, which brings other disadvantages and leads to consuming sensor node resources. More research is still required to address the adaptability issue.

QoS handling: Despite the approaches discussed above, the area of QoS in WSNs is still rather an unexplored field. The problem here is how to balance between application's QoS requirement and energy constraints [29].

Hybrid approach in protocol design is a promising methodology. Although there has been significant work in this domain, extensive research is still required to address the wireless sensor MAC protocol requirement discussed above.

Cross-layer interaction: Despite the existing research on developing new communication protocol based on cross-layer interactions, there is still much to be gained by rethinking the protocol functions of network layers in a unified way so as to provide a single communication module for efficient communication in WSNs. There are several open research problems toward the development of systematic techniques for cross-layer design of WSN protocols. For more details refer to Ref. [31].

Realistic simulation models: Assumptions made in most simulation environments do not necessarily reflect the real-world conditions (e.g., a radio's transmission area is circular, all radios have equal range, ..., etc). To fully understand the complexity of designing a MAC protocol and to develop solution which works in real life, it is necessary to develop more realistic radio and energy simulation models. It is important to revisit Kotz's mistaken axioms of wireless network research (see [108]), to understand why MAC protocols that yield extremely accurate results in simulation fail in real-life deployments.

To obtain more realistic insight into MAC layer performance, sensor networks researchers should move from simulation to *prototype or real-world experiments*. In summary, existing wireless MAC protocols focus on optimizing system energy considering the MAC performances, yet still do not adequately consider all of the requirements of sensor networks. The key challenge remains to provide predictable delay or prioritization guarantees while minimizing overhead packets and energy consumption.

9.8 Conclusion

MAC protocols in WSNs have attracted a lot of attention in the recent years and introduced unique challenges compared with traditional MAC in other wireless networks. In this chapter, we have discussed the special requirements of MAC protocols for WSN; classified the existing research on MAC in WSNs and covered many proposed protocols; discussed and identified the open research issues; and presented a set of future research directions. Although many MAC schemes have been proposed for WSNs, the area is still largely open to research and there is no clear single direction in which future efforts should be directed, as discussed in the previous section. It remains an open question of great interest, "Does a general, and flexible MAC protocol exists that supports various applications and operating environments while consuming minimal power and offering acceptable traffic characteristics?" However, along with the current research of MAC protocols in WSNs, we encourage more insight into the problems and more development in solutions to the open research issues as described in this chapter.

References

1. I. F. Akyildiz et al., Wireless sensor networks: A survey, *Computer Networks*, 38, 393–422, March 2002.
2. R. Min et al., Low power wireless sensor networks, in *Proceedings of International Conference on VLSI Design*, Bangalore, India, January 2001.
3. S. Tilak, N. B. Abu-Ghazaleh, and W. Heinzelman, A taxonomy of wireless microsensor network models, in *ACM SIGMOBILE Mobile Computing and Communications Review*, 6(2), 28–36, April 2002, ACM, New York.
4. J. M. Rabaey et al., PicoRadio supports ad-hoc ultra low power wireless networking, in *IEEE Computer*, 33, 42–48, July 2000.
5. G. J. Pottie and W. J. Kaiser, Wireless integrated network sensors, *Communication of the ACM*, 43(5), 51–58, 2000.
6. R. H. Katz, J. M. Kahn, and K. S. J. Pister, Mobile networking for smart dust, in *Proceedings of the 5th Annual ACM/IEEE International Conference on Mobile Computing and Networking, MobiCom'99*, Seattle, WA, August 1999.
7. T. Bokareva, W. Hu, S. Kanhere, B. Ristic, N. Gordon, T. Bessell, M. Rutten, and S. Jha, Wireless sensor networks for battlefield surveillance, in *Proceedings of the Land Warfare Conference, LWC-2006*, Brisbane, Queensland, Australia, October 2006.
8. A. Mainwaring, J. Polastre, R. Szewczyk, D. Culler, and J. Anderson, Wireless sensor networks for habitat monitoring, in *Proceedings of the 1st ACM International Workshop on Wireless Sensor Networks and Applications*, pp. 88–97, September 2002, Atlanta, GA.
9. G. Tolle, J. Polastre, R. Szewczyk, N. Turner, K. Tu, S. Burgess, D. Gay, P. Buonadonna, W. Hong, T. Dawason, and D. Culler, A macroscope in the redwoods, in *Proceedings of the 3rd International Conference on Embedded Networked Sensor Systems (SenSys'05)*, pp. 51–63, November 2005, San Diego, CA.
10. N. Xu, S. Rangwala, K. Chintalapudi, D. Ganesan, A. Broad, R. Govindan, and D. Estrin, A wireless sensor network for structural monitoring, in *Proceedings of the 2nd International Conference on Embedded Networked Sensor Systems (SenSys'04)*, pp. 13–24, November 2004, Baltimore, MD.
11. K. Srinivasan, M. Ndoh, H. Nie, H. Xia, K. Kaluri, and D. Ingraham, Wireless technologies for condition-based maintenance (CBM) in petroleum plants, in

Proceedings of the 1st International Conference on Distributed Computing in Sensor Systems (DCOSS'05), Poster Session, June 30–July 1, 2005, Marina del Rey, CA.

12. K. Sohrabi et al., Protocols for self-organization of a wireless sensor network, *IEEE Personal Communications*, 7(5), 16–27, October 2000.

13. A.L. Buczak, V.R. Jamalabad, Self-organization of a heterogeneous sensor network by genetic algorithms, in *Intelligent Engineering Systems through Artificial Neural Networks*, eds., C. H. Dagli, M. Akay, A. L. Buczak, O. Ersoy, B. R. Fernandez, vol. 8, ASME Press, New York, 1998, pp. 259–264.

14. N. Abramson, The ALOHA system—another alternative for computer communications, in *Proceedings of AFIPS Conference*, vol. 36, pp. 295–298, 1970.

15. L. Kleinrock and F. Tobagi, Packet switching in radio channels: Part I -carrier sense multiple access modes and their throughput-delay characteristics, *IEEE Transactions on Communications*, 23(12), 1400–1416, December 1975.

16. L. Roberts, ALOHA packet system with and without slots and capture, Stanford Research Institute, Advanced Research Projects Agency, Network Information Center, Tech. Rep. ASS Note 8, 1972.

17. I. Demirkol, C. Ersoy, and F. Alagoz, MAC protocols for wireless sensor networks: A survey, *IEEE Communications Magazine*, 44(4), 115–121, April 2006.

18. W. Ye and J. Heidemann, Medium access control in wireless sensor networks, USC/ISI Technical Report ISI-TR-580, October 2003.

19. K. Kredo II and P. Mohapatra, Medium access control in wireless sensor networks, *Journal of Computer Networks*, 51, 961–994, 2007.

20. A. Bohm, State of the art on energy-efficient and latency constrained networking protocols for wireless sensor networks, HH, Technical Report, IDE0749, June 2007.

21. K. Langendoen, Meduim access control in wireless sensor networks, Book chapter, *Practice and Standards*, vol. II, eds., H. Wu and Y. Pan, 2007.

22. P. Baronti, P. Pillai, V. W. C. Chook, S. Chessa, A. Gotta, and Y. Fun Hu, Wireless sensor networks: A survey on the state of the art and the 802.15.4 and ZigBee standards, *Journal of Computer Communication*, 30, 1655–1695, 2007.

23. P. P. Czapski, A survey: MAC protocols for applications of wireless sensor networks, in *Proceedings of IEEE TENCON 2006, A Technical Conference of the IEEE Region*, vol. 10, pp. 1–4, Hong Kong, November 14–17, 2006.

24. K. Langendoen and G. Halkes, Energy-efficient medium access control, Book chapter in the *Embedded Systems Handbook*, ed., R. Zurawski, CRC Press, ISBN: 9780849328244, August 2005.

25. Jurdak, R., C. V. Lopes, and P. Baldi, A survey, classification and comparative analysis of medium access control protocols for ad hoc networks, in *IEEE Communications Surveys*, 2004.

26. P. Naik and K. M. Sivalingam, A survey of MAC protocols for sensor networks, Book chapter, *Wireless Sensor Networks*, Kluwer Academic Publishers, Norwell, MA, pp. 93–107, 2004.

27. D. Ganesan, A. Cerpa, W. Ye, Y. Yu, J. Zhao, and D. Estrin, Networking issues in wireless sensor networks, *Journal on Parallel and Distributed Computing*, 64, 2004.

28. J. N. Al-Karaki and A. E. Kamal, Routing mechniques in wireless sensor networks: A survey, *IEEE Journal on Wireless Communications*, 11(6), 6–28, December 2004.

29. D. Chen and P. K. Varshney, QoS support in wireless sensor networks: A survey, in *Proceedings of the International Conference on Wireless Networks 2004, ICWN '04*, Las Vegas, NV, June 2004.

30. M. Ali, U. Saif, A. Dunkels, T. Voigt, K. Rmer, K. Langendoen, J. Polastre, and Z. A. Uzmi, Medium access control issues in sensor networks, *ACM SIGCOMM Computer Communication Review*, 36(2), 33–36, April 2006.

31. T. Melodia, M. C. Vuran, and D. Pompili, The state of the art in cross-layer design for wireless sensor networks, *Network Architect. 2005*, LNCS 3883, pp. 78–92, Springer-Verlag, Berlin, Heidelberg, 2006.

32. W. Ye, J. Heidemann, and D. Estrin, Medium access control with coordinated adaptive sleeping for wireless sensor networks, *IEEE/ACM Transactions on Networking*, 12(3), 493–506, June 2004.

33. W. Ye, J. Heidemann, and D. Estrin, An energy-efficient MAC protocol for wireless sensor networks, in *Proceedings of IEEE INFOCOM*, New York, June 2002.

34. C. Jones, K. Sivalingam, P. Agrawal, and J. C. Chen, A survey of energy efficient network protocols, *Wireless Networks*, 7, 2001.

35. P. Lettieri and B. Srivastava, Advances in wireless terminals (I), *IEEE Personal Communications Magazine*, 6(1), 6–19, February 1999.

36. P. Havinga, G. Smit, and M. Bos, Energy efficient wireless ATM design, *Mobile Networks and Applications*, 5(2), 147–155, Kluwer Academic Publishers, Hingham, MA, June 2000.

37. M. Miller and N. Vaidya, Minimizing energy consumption in sensor networks using a wakeup radio, in *Proceedings of the IEEE International Conference on Wireless Communications and Networks, WCNC'04*, Atlanta, GA, March 2004.

38. E. Shih, S. Cho, N. Ickes, R. Min, A. Sinha, A. A. Wang, and A. Chandrakasan, Physical layer driven protocol and algorithm design for energy efficient wireless sensor networks, in *Proceedings of the 7th ACM/IEEE Conference on Mobile Computing and Networks MOBICOM'01*, Rome, Italy, July 2001.

39. L. Kleinrock and F. Tobagi, Packet switching in radio channels: Part II—The hidden terminal problem in CSMA and busy tone solutions, *IEEE Transactions on Communications*, 23(12), 1417–1433, 1975.

40. S. Singh and C. Raghavendra, PAMAS—Power aware multi-access protocol with signaling for ad hoc networks, *SIGCOMM Computer Communications Review*, 28(3), 5–26, 1998.

41. M. C. Vuran and I. F. Akyildiz, Spatial correlation-based collaborative medium access control in wireless sensor networks, *IEEE/ACM Transactions on Networking*, 14(2), 316–329, IEEE Press, Piscataway, NJ, April 2006.

42. K. Jamieson, H. Balakrishnan, and Y. C. Tay, Sift: A MAC protocol for event-driven wireless sensor networks, MIT Laboratory for Computer Science, Tech. Rep. 894, May 2003.

43. J. Polastre, J. Hill, and D. Culler, Versatile low power media access for wireless sensor networks, in *Proceedings of the International Conference on Embedded Networked Sensor Systems, SenSys'04*, pp. 95–107, 2004.

44. E.-Y. A. Lin, J. M. Rabaey, and A. Wolisz, Power-efficient Rendez–Vous schemes for dense wireless sensor networks, in *Proceedings of the IEEE International Conference on Communications, ICC'04*, vol. 7, pp. 3769–3776, 2004.

45. G. Lu, B. Krishnamachari, and C. S. Raghavendra, An adaptive energy efficient and low-latency MAC for data gathering in wireless sensor networks, in *Proceedings of the 18th International Parallel and Distributed Processing Symposium (IPDPS'04)*, pp. 224, April 26–30, 2004, Santa Fe, NM.

46. T. V. Dam and K. Langendoen, An adaptive energy-efficient MAC protocol for wireless sensor networks, *The First ACM Conference on Embedded Networked Sensor Systems, Sensys'03*, Los Angeles, CA, November, 2003.

47. P. Lin, C. Qiao, and X. Wang, Medium access control with a dynamic duty cycle for sensor networks, in *Proceedings of the IEEE Wireless Communications and Networking Conference (WCNC'04)*, vol. 3, pp. 1534–1539, March 21–25, 2004, Atlanta, GA.

48. H. Pham and S. Jha, An adaptive mobility aware MAC protocol for sensor networks, in *The 1st IEEE International Conference on Mobile Ad Hoc and Sensor Networks*, October 24–27, Tampa, FL, 2004.

49. Y. E. Sagduyu and A. Ephremides, The problem of medium access control in wireless sensor networks, *IEEE Wireless Communications*, 11(6), 44–53, 2004.

50. L. van Hoesel and P. Havinga, A lightweight medium access protocol (LMAC) for wireless sensor networks, in *1st Int. Workshop on Networked Sensing Systems INSS'04*, Tokyo, Japan, June 2004.

51. L. van Hoesel and P. Havinga, Poster abstract: A TDMA-based MAC protocol for WSNs, in *Proceedings of the 2nd International Conference on Embedded Networked Sensor Systems (SenSys'04)*, pp. 303–304, 2004, Baltimore, MD.

52. S. Biaz and Y. D. Barowski, GANGS: An energy efficient MAC protocol for sensor networks, in *Proceedings of the 42nd Annual Southeast Regional Conference*, pp. 82–87, 2004, Huntsville, AL.

53. L. Bao and J. Garcia-Luna-Aceves, A new approach to channel access scheduling for ad hoc networks, in *Proceedings of the 7th Annual International Conference on Mobile Computing and Networking, MobiCom'01*, pp. 210–221, July 2001, Rome, Italy.

54. A. El-Hoiydi, Spatial TDMA and CSMA with preamble sampling for low power ad hoc wireless sensor networks, in *Proceedings of ISCC 2002, The Seventh International Symposium on Computers and Communications*, pp. 685–692, July 1–4, 2002.

55. I. Rhee, A. Warrier, M. Aia, and J. Min, Z-MAC: A hybrid MAC for wireless sensor networks, Technical report, Department of Computer Science, North Carolina State University, Raleigh, NC, April 2005.

56. I. Rhee, A. Warrier, and L. Xu, Randomized dining philosophers to TDMA scheduling in wireless sensor networks, Technical report, Computer Science Department, North Carolina State University, Raleigh, NC, 2004.

57. V. Rajendran, K. Obraczka, and J. Garcia-Luna-Aceves, Energy-efficient collision free medium access control for wireless sensor networks, in *Proceedings of the 1st International Conference on Embedded Networked Sensor Systems, SenSys'03*, pp. 181–192, November 2003, Los Angeles, CA.

58. V. Ekanayake, C. Kelly, and R. Manohar, An ultra low-power processor for sensor networks, *ACM SIGPLAN Notices*, 29(11), 27–36, November 2004.

59. S. Cui, R. Madan, A. J. Goldsmith, and S. Lall, Joint routing, MAC, and link layer optimization in sensor networks with energy constraints, in *Proceedings of the IEEE International Conference on Communications, ICC'05*, vol. 2, pp. 725–729, Korea, May 16–20, 2005.

60. J. Ding, K. Sivalingam, R. Kashyapa, and L. J. Chuan, A multi-layered architecture and protocols for large-scale wireless sensor networks, in *IEEE 58th Vehicular Technology Conference, VTC'03*, vol. 3, pp. 1443–1447, October 6–9, 2003, Orlando, FL.

61. M. Zorzi, A new contention-based MAC protocol for geographic forwarding in ad hoc and sensor networks, in *IEEE International Conference on Communications (ICC'04)*, vol. 6, pp. 3481–3485, June 20–24, 2004, Paris, France.

62. R. Rugin and G. Mazzini, A simple and efficient MAC-routing integrated algorithm for sensor network, in *IEEE International Conference on Communications (ICC'04)*, vol. 6, pp. 3499–3503, June 20–24, 2004, Paris, France.

63. M. Ali, T. Suleman, and Z. A. Uzmi, MMAC: A mobility-adaptive, collision-free MAC protocol for wireless sensor networks, in *Proceedings of the 24th IEEE IPCCC'05*, Phoenix, AZ, April 2005.

64. J. Ai, J. Kong, and D. Turgut, An adaptive coordinated medium access control for wireless sensor networks, in *Proceedings of the 9th International Symposium on Computers and Communications (ISCC'04)*, vol. 1, pp. 214–219, June 28–July 1, 2004, Alexandria, Egypt.

65. J. Haapola, Z. Shelby, C. Pomalaza-Raez, and P. Mahonen, Cross-layer energy analysis of multi-hop wireless sensor networks, in *Proceedings of the 2nd European Workshop in Wireless Sensor Networks (EWSN'05)*, pp. 33–44, January 31–February 2, 2005, Istanbul, Turkey.

66. A. Chan and S. Chang Liew, Merit of PHY-MAC cross-layer carrier sensing: A MAC-address-based physical carrier sensing scheme for solving hidden-node and exposed-node problems in large-scale Wi-Fi networks, in *Proceedings of the 31st IEEE Conference on Local Computer Networks, LCN'06*, pp. 871–878, November 14–16, 2006, Tampa, FL.

67. P. Skraba, H. Aghajan, and A. Bahai, Cross-layer optimization for high density sensor networks: Distributed passive routing decisions, in *Proceedings of Ad-Hoc Now04*, Vancouver, British Columbia, Canada, July 2004.

68. M. Zorzi and R. Rao, Geographic random forwarding (GeRaF) for ad hoc and sensor networks: Multihop performance, *IEEE Trans. Mobile Computing*, 2(4), 337–348, October–December 2003.

69. M. Zorzi and R. Rao, Geographic random forwarding (GeRaF) for ad hoc and sensor networks: Energy and latency performance, *IEEE Trans. Mobile Computing*, 2(4), 349–365, October–December 2003.

70. M. Zorzi, A new contention-based MAC protocol for geographic forwarding in ad hoc and sensor networks, in *Proceedings of the IEEE International Conference on Communications (ICC'04)*, vol. 6, pp. 3481–3485, June 20–24, 2004, Paris, France.

71. D. Ferrara et al., MACRO: An integrated MAC/routing protocol for geographical forwarding in wireless sensor networks, in *Proceedings of IEEE INFOCOM'05*, vol. 3, pp. 1770–1781, March 13–17, 2005, Miami, FL.

72. C. Suh, Y. Ko, and D. Son, An energy efficient cross-layer MAC protocol for wireless sensor networks, *LNCS Book Series: Advanced Web and Network Technologies, and Applications Book*, ISBN:978-3-540-31158-4, Springer, Berlin/Heidelberg.

73. M. L. Sichitiu, Cross-layer scheduling for power efficiency in wireless sensor networks, in *Proceedings of the 23rd Annual Joint Conference of the IEEE Computer and Communications Societies (INFOCOM'04)*, vol. 3, pp. 1740–1750, March 7–11, 2004, Hong Kong.

74. L. van Hoesel, T. Nieberg, J. Wu, and P. J. M. Havinga, Prolonging the lifetime of wireless sensor networks by cross-layer interaction, *IEEE Wireless Communications*, 11(6), 78–86, December 2004, ISSN 1536-1284. Available online at: http://doc.utwente.nl/55651/1/01368900.pdf.

75. J. Yuan, Z. Li, W. Yu, and B. Li, A cross-layer optimization framework for multicast in multi-hop wireless networks wireless Internet, in *Proceedings of the 1st International Conference on Wireless Internet, WICON05*, pp. 47–54, July 10–14, 2005, Budapest, Hungary.

76. K. Seada, M. Zuniga, A. Helmy, and B. Krishnamachari, Energy-efficient forwarding strategies for geographic routing in lossy wireless sensor networks, in *Proceedings of the 2nd International Conference on Embedded Networked Sensor Systems, Sensys04*, pp. 108–121, November 2004, Baltimore, MD.

77. M. Chiang, Balancing transport and physical layers in wireless multihop networks: Jointly optimal congestion control and power control, *IEEE Journal on Selected Areas in Communications (IEEE JSAC)*, 23(1), 104–116, January 2005.

78. R. Madan, S. Cui, S. Lall, and A. Goldsmith, Cross-layer design for lifetime maximization in interference-limited wireless sensor networks, in *Proceedings of IEEE INFOCOM'05*, vol. 3, pp. 1964–1975, March 13–17, 2005, Miami, FL.

79. S. Cui, R. Madan, A. Goldsmith, and S. Lall, Joint routing, MAC, and link layer optimization in sensor networks with energy constraints, in *Proceedings of the IEEE International Conference on Communications, ICC'05*, vol. 2, pp. 725–729, May 16–20, 2005, Seoul, Korea.

80. A. Bohn, State of the art on energy efficient and latency constrained networking protocols for wireless sensor networks, Technical report, IDE0749, Halmstad University, Halmstad, Sweden, June 2007.

81. M. Younis and T. Nadeem, Energy efficient MAC protocols of ad hoc networks, Book chapter in *Wireless Ad-Hoc and Sensor Networks*, Chapter 9, ed., A. Safwat, Pub: Kluwer Academic Publishers, Hingham, MA. Available online at: http://tmrnadeem.ifastnet.com//Web_Page/html_css/papers/energy_chapter.pdf.

82. H. Karl and A. Wiling, Protocols and architectures for wireless sensor networks, Book Published by: John Wiley & Sons, Ltd, April 2005, ISBN: 13-978-0-470-09510-2.

83. A. Chandrakasan, S. Sheng, and R. Brodersen, Low power CMOS digital design, *IEEE Journal of Solid State Circuits*, 27(4), 473–484, 1992.

84. X. Xia and Q. Liang, Latency and energy efficiency evaluation in wireless sensor networks, in *IEEE 62nd Conference on Vehicular Technology, VTC-2005-Fall*, pp. 1594–1598, September 25–28, 2005, Dallas, TX.

85. A. Wang et al., Energy-efficient modulation and MAC for asymmetric microsensor systems, in *Proceedings of ISLPED 2001*, Huntington Beach, CA, August 2001.

86. V. Raghunathan, C. Schurgers, S. Park, and M. B. Srivastava, Energy-aware wireless microsensor networks, *IEEE Signal Processing Magazine*, 19(2), 40–50, 2002.

87. S. S. Kulkarni, TDMA services for sensor networks, in *Proceedings of 24th International Conference on Distributed Computing Systems Workshops*, (ICDCS 2004 Workshops), pp. 604–609, March 23–24, 2004, Hachioji, Tokyo, Japan.

88. A. Chandra, V. Gummalla, and J. O. Limb, Wireless medium access control protocols, *IEEE Communication Surveys and Tutorials*, 3(2), 2–15, 2000.

89. S. Singh and C. S. Raghavendra, PAMAS: Power aware multi-access protocol with signaling for ad hoc networks, *ACM Computer Communications Review*, 28(3), 526, 1998.

90. I. Chatzigiannakis, A. Kinalis, and S. Nikoletseas, Wireless sensor networks protocols for efficient collision avoidance in multi-path data propagation, in *Proceedings of the 1st ACM International Workshop on Performance Evaluation of Wireless Ad Hoc, Sensor, and Ubiquitous Networks*, pp. 8–16, October 4–4, 2004, Venezia, Italy.

91. P. Havinga and G. Smit, Energy-efficient TDMA medium access control protocol scheduling, in *Proceedings of the Asian International Mobile Computing Conference, AMOC*, 31 October–3 November 2000, Penang, Malaysia.

92. M. Adamou, I. Lee, and I. Shin, An energy efficient real-time medium access control protocol for wireless ad-hoc networks, in *WIP Session of IEEE Real-Time Systems Symposium, RTSS01*, London, U.K., December 2001.

93. M. Salajegheh, H. Soroush, and A. Kalis, HYMAC: Hybrid TDMA/FDMA medium access control protocol for wirless sensor networks, in *The 18th Annual IEEE International Symposium on Personal, Indoor and Mobile Radio Communications, PIMRC07*, pp. 1–8, September 3–7, 2007, Athens, Greece.

94. M. Younis, K. Akkaya, M. Eltoweissy, A. Wadaa, On handling QoS traffic in wireless sensor networks, in *Proceedings of the 37th Annual Hawaii International Conference on Computer Sciences, (HICSS'04)*, January 5–8, 2004, Big Island, HI.

95. D. Chen and P. K. Varshney, QoS support in wireless sensor networks: A survey, in *Proceedings of the 2004 International Conference on Wireless Networks (ICWN 2004)*, pp. 227–233, June 21–24, 2004, Las Vegas, NV.

96. R. Iyer and L. Kleinrock, QoS control for sensor networks, in *Proceedings of the 38th Annual IEEE International Conference on Communications, ICC'03*, vol. 1, pp. 517–521, May 11–15, 2003, Anchorage, AK.

97. Y. Liu, I. Elhanany, and Q. Hairong, An energy-efficient QoS-aware media access control protocol for wireless sensor networks, in *Proceeding of the 2nd IEEE International Conference on Mobile Ad-hoc and Sensor Systems, MASS'05*, Poster Session, November 7–10, 2005, Washington, DC.

98. W. L. Lee, A. Datta, and R. Cardell-Oliver, QMAC: A quality of service oriented medium access control protocol for data gathering in wireless sensor networks, Tech. Rep. UWA-CSSE-05-005, The University of Western Australia, 2005.

99. Q. Zhao and L. Tong, QoS specific medium access control for wireless sensor networks with fading, ACSP technical report TR-06-03-01, School of Electrical and Computer Engineering, Cornell University, June 2003.

100. Y. Wu, S. Fahmy, and N. B. Shroff, Optimal sleep/wake scheduling for time synchronized sensor networks with QoS guarantees, in *Proceedings of 14th IEEE International Workshop on Quality of Service (IEEE-IWQoS'06)*, pp. 102–111, June 19–21, 2006, New Haven, CT.

101. J. Frolik, QoS control for random access wireless sensor networks, in *Proceedings of the 5th IEEE Wireless Communications and Networking Conference, WCNC'04*, vol. 3, pp. 1522–1527, March 21–24, 2004, Atlanta, GA.

102. B. Yahya and J. Ben-Othman, An energy efficient hybrid medium access control scheme for wireless sensor networks with quality of service guarantees, Accepted and will appear in the *GLOBECOM'08 Proceedings, Ad-hoc and Sensor Networks Symposium*, New-Orleans, LA, 29 November–4 December 2008.

103. V. Bharghavan et al., MACAW: A media access protocol for wireless LANS, in *Proceedings of the ACM SIGCOMM '94 Conference on Communications Architectures, Protocols and Applications*, 24(4), 1994, August 31–September 2, 1994, London, U.K.

104. See http://www.zigbee.org/; A brife slide set on ZigBee entitled ZigBee overview, can be found under http://www.zigbee.org/en/resources.

105. LAN/MAN Standards Committee of the IEEE Computer Sciety. IEEE Standard for Information technology—Telecommunications and information exchange between systems—Local and metropolitan area networks—Specfic requirements—Part 15.4: Wireless medium access (MAC) and physical layer (PHY) specifications for low rate wireless personal area networks (LT-WPANs), October 2003.

106. A. Jhumka and S. Kulkarni, On the design of mobility-tolerant TDMA-based media access control (MAC) protocol for mobile sensor networks, in *LNCS Book Series Distributed Computing and Internet Technology Book*, ISBN:978-3-540-77112-8, Springer, Berlin/Heidelberg, November 2007.

107. J. F. Shi, X. X. Zhong, and S. Chen, Study on communication mode of wireless sensor networks based on effective results, in *International Symposium on Instrumentation Science and Technology*, 8–12 August 2006, Harbin, China, *Journal of Physics*, Conference series 48, 1317–1321, IOP Publishing Ltd, 2006. Available online at: http://www.iop.org/EJ/toc/1742-6596/48/1.

108. D. Kotz, C. Newport, R. S. Gray, J. Liu, Y. Yuan, and C. Elliott, Experimental evaluation of wireless simulation assumptions, in *Proceedings of the 7th ACM International Symposium on Modeling, Analysis and Simulation of Wireless and Mobile Systems (MSWiM04)*, pp. 78–82, October 4–6, 2004, Venice, Italy.

109. M. J. Handy, M. Haase, and D. Timmermann, Low energy adaptive clustering hierarchy with deterministic cluster-head selection, in *Proceedings of the Fourth IEEE Conference on Mobile and Wireless Communications Networks*, pp. 368–372, Stockholm, September 2002.

110. M. Leopold, M. B. Dydensborg, and P. Bonnet, Bluetooth and sensor networks: A reality check, in *Proceedings of 1st ACM Conference on Embedded Networked Sensor Systems (SenSys 2003)*, pp. 103–113, New York, November 2003.

Chapter 10

Localization in Wireless Sensor Networks

Yu Wang and Lin Li

CONTENTS

Localization is a fundamental problem in designing wireless sensor networks. Location information can be used in many wireless sensor network applications, such as event detecting, target tracking, environmental monitoring, and network deployment. On the other hand, location information can also be used in different networking protocols to enhance the performance of sensor networks, such as routing packets using position-based routing, controlling the network topology and coverage using geometric methods, or achieving better load balancing in routing. Manual configuration of locations or providing each sensor with a global positioning system (GPS) is expensive and not feasible for large-scale sensor networks. Therefore, it is important to develop localization methods in which sensor nodes can compute their positions by exchanging information with some nodes* that have known locations. This chapter gives an overview of localization methods designed for wireless sensor networks. Both theoretical basics and research challenges for localization will be introduced. Recent solutions from range-based methods to range-free methods will be discussed and compared. With specific hardware, range-based schemes typically achieve higher accuracy based on either node-to-node distance measurements or angle measurements. On the other hand, range-free mechanisms support the coarse-positioning accuracy with less expense. At the end of the chapter, the reader will understand the principal design and new research trends of localization techniques in wireless sensor networks.

10.1 Introduction

The main task of localization in wireless sensor networks is to obtain the absolute or relative, or precise or approximate location of each sensor. The following obtained localization information is useful or even necessary in many cases for wireless sensor networks:

- Providing location information of a certain event when it is detected by sensors
- Providing a sequence of locations for tracking a moving object using sensor networks
- Providing location stamps for each measurement obtained by each individual sensor
- Determining the quality of coverage of all active sensors using their positions
- Controlling the network topology based on geometric techniques to maintain connectivity and save energy
- Making routing decisions based on the positions of the current node, its neighbors, and the destination using position-based routing to reduce the routing overhead
- Achieving load balancing in routing protocols by spreading the traffic based on location information
- Providing user location information for various location-aware services, such as finding nearby servers or printers
- Supporting geocast [22] and mobicast [16] services by which a user can send a message to a specific area and at a specific time

---

*In this chapter, we use reference nodes, anchor nodes, or beacon nodes to refer to these nodes whose positions are known.

However, localization in wireless sensor networks is a very challenging task. Configuring locations manually and providing each sensor with a GPS system are two intuitionistic methods. In a small-scale, stationary wireless sensor network, the location of every sensor node is fixed and can be manually deployed. However, when we consider large-scale sensor networks, sensor networks with random or ad hoc deployment, or mobile sensor networks, manual configuration of locations will not be available. Satellite-based GPS system is the most well-known positioning system for outdoor environments. However, it is too expensive for every tiny sensor device to be armed with a GPS system. In addition, usually each sensor node has limited battery power, which cannot afford the GPS system. Finally, the GPS system is not feasible in indoor environments because buildings can block satellite signals. Therefore, it is necessary and important to develop new localization methods that are feasible and affordable for wireless sensor networks.

Recently, the localization problem in sensor networks has attracted many researchers' attention and a great number of localization methods have been proposed. These methods can be mainly classified into two categories: range-based localization methods and range-free localization methods. Range-based methods utilize received signal strength (RSS), time of arrival (ToA), time difference of arrival (TDoA), or angle of arrival (AoA) to measure distances or angles between nodes, and then use these distance or angle estimations to compute positions of nodes. Range-free methods do not employ the above measurement techniques, instead they use alternative methods, such as hop-count-based methods or area, to locate nodes. In this chapter, we will briefly review some of the newly developed localization methods belonging to both categories. Notice that to fulfill the requirements of localization for large-scale wireless sensor networks, localization methods should be accurate (i.e., having small differences between estimated positions and true positions), distributed (i.e., not dependent on global infrastructure), robust (i.e., tolerant to node failures and measure errors), and efficient (i.e., having little computation and communication overheads). However, it is very hard to achieve all these requirements at the same time. Current methods usually can satisfy one or some of these requirements.

Section 10.2 introduces some theoretical basics of the localization problem. Sections 10.3 and 10.4 discuss existing range-based methods and range-free methods, respectively. Section 10.5 provides a summary by comparing all these localization methods and pointing out some open issues in localization.

10.2 Theoretical Fundamentals

Before discussing the localization methods in detail, we first introduce several theoretical basics, such as basic distance-measurement methods, trilateration, triangulation, and the rigidity theory on localizability.

10.2.1 Range Measurement

To determine the locations of a sensor node, the first basic step is measuring the ranges or the angles between the node and its neighboring nodes (or reference nodes). Various measurement methods have been proposed in the literature for wireless sensor networks.

Time of arrival (ToA): The ToA method estimates the distance between any two neighboring nodes by measuring the time the signal takes to travel between the two nodes. If the propagation velocity of the signal is known, the distance can be calculated by multiplying the velocity and the time. The ToA method usually employs the signal whose propagation speed is low, such as sound wave and ultrasound signals. Wireless radio waves can be used in ToA but the resolution is not very accurate. Also the propagation speed of the signal depends on external factors, such as temperature, pressure, and humidity. This may affect the accuracy of the ToA measurement.

Usually, ToA methods need time synchronization. To avoid time synchronization, ToA methods need to measure a ping-pong style round trip delay. For example, we want to estimate the distance between two nodes, a and b, and the propagation speed of the signal is v. As shown in Figure 10.1a, at time t_a, a sends a signal to b; b receives the signal at time t_b and replies the signal at time t_b'; and a receives the replied signal at time t_a'. Here, t_a and t_a' are times at node a while t_b and t_b' are times at node b. Then, a can calculate the distance $|ab|$ between a and b according to the following equation:

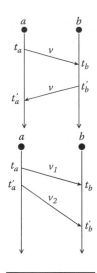

Figure 10.1 Range measure-ments: (a) ToA and (b) TDoA.

$$|ab| = \frac{((t_a' - t_a) - (t_b' - t_b)) \times v}{2}.$$

Time difference of arrival (TDoA): The TDoA method is similar to the ToA method, because both of them utilize the time that the signal takes to travel between two nodes. Different from ToA, however, TDoA employs two kinds of signals traveling at different speeds, such as ultrasound and RF signals, to overcome the need for explicit synchronization or ping-pong style two-way handshake. Let v_1 and v_2 be the propagation speeds for these two kinds of signals, respectively. Once again, suppose that we want to estimate the distance between two nodes, a and b. As shown in Figure 10.1b, a sends two kinds of signals at time t_a and t_a', respectively; and b receives them at time t_b and t_b', respectively. We know that the distance between a and b equals to $v_1(t_b - t_a)$ and $v_2(t_b' - t_a')$. In other words, $\frac{|ab|}{v_1} = t_b - t_a$ and $\frac{|ab|}{v_2} = t_b' - t_a'$. To avoid synchronization, we take the difference between these two equations, $|ab|(\frac{1}{v_1} - \frac{1}{v_2}) = (t_b' - t_b) - (t_a' - t_a)$. Therefore, the distance can be decided by the following formula:

$$|ab| = ((t_b' - t_b) - (t_a' - t_a)) \times \left(\frac{v_1 \times v_2}{v_1 - v_2}\right).$$

Comparing with ToA, TDoA usually can achieve better accuracy [27,29] but needs at least two types of senders and receivers on each node.

Notice that the definition of TDoA here is different from the TDoA used in emitter location techniques [1]. In the emitter location problem, TDoA is coupled with Frequency difference of arrival (FDoA) to passively estimate the location of a fixed emitter. It involves the measurement of the difference between the times of arrival at two receivers that are set apart. However, TDoA here is the value measured as the difference of the times of arrival of two different kinds of signals at the same receiver.

Received signal strength (RSS): When signals travel in media, their strength reduces with the increase of the distance that they travel. According to this fact, the received signal strength at the receiver can be used for computing the distance between the receiver and the transmitter. In the ideal case, the path loss model for wireless radio propagation can be given by the following equation:

$$P_r = c \frac{P_t}{d^\alpha}$$

where

P_r is the received signal strength at the receiver

P_t is the transmission signal strength at the transmitter

α is the path loss coefficient, which is usually a constant between 2 and 5 dependend on the environment

c is a constant

d is the distance between the transmitter and the receiver

Thus, given the measurement of the received signal strength, c, P_t, and α, we can calculate the distance according to the following formula:

$$d = \sqrt[\alpha]{\frac{cP_t}{P_r}}.$$

The advantage of the RSS-based method is that sensor nodes do not need additional hardware and any additional communication overhead. But the accuracy of the RSS-based method is limited by the accuracy of the path loss model. Usually, using the above simple path loss model, the RSS-based method may incur significant errors, because the received signal strength is much more complex, and even for a fixed transmitter and receiver it can be varied and can oscillate. This is caused by various uncontrollable effects such as multipath fading, shadows, and terrain. Therefore, more complex techniques need to be used if a certain accuracy is required for RSS-based methods.

Angle of arrival (AoA): The above three methods focus on directly measuring the distance, but another alternative method is measuring angles to the reference nodes (whose positions are known) or orientations of received signals, instead of distances to determine the position of the node. AoA-based methods usually use directional antennas or antenna arrays to measure the angle of a connecting line between the device and the reference node. In a two-dimensional space, angulation needs two angle measurements (angles to two reference nodes) and one length measurement (such as the distance between two reference nodes) to locate a node.

Frequency difference of arrival (FDoA): FDoA is one of the techniques for achieving the precision emitter location [1]. It involves the measurement of the difference between the received frequency at two receivers from a single transmitter. Because the difference in the received frequency is caused by differences in the Doppler shift, FDoA is also called the differential Doppler. Recently, there is a trend to apply FDoA for sensor localization [3,19,20,23]. In Ref. [19], sensors estimate their location by measuring the acoustic Doppler shift in a tone that is emitted from a mobile anchor. It assumes that the sensor node knows the location and heading of the anchor as well as the frequency of the acoustic tone. In Ref. [23], two anchors, a master and an assistant, transmit a pure sine wave at slightly different frequencies, which interfere, resulting in a low envelope beat frequency whose phase can be measured by two receivers at a designated time. The difference in phase is a linear combination of the distances between transmitters and receivers. In Refs. [3,20], a similar technique is used for tracking mobile sensors. But instead of measuring the phase, it measures the Doppler shift in frequency, which occurs when the source of a transmitted signal is moving relative to an observer. Because the Doppler shift is determined by the relative velocities of the transmitter and receiver, absolute translational velocity of a mobile node can be determined using a prior knowledge of the transmitted frequency and the frequencies observed by stationary sensors at known positions.

10.2.2 Trilateration

Trilateration is the most basic technique for positioning a system and has been used for thousands of years. Trilateration uses the known locations of two or more reference nodes, and the measured distances between the nodes need to be located at each reference node. To accurately and uniquely determine the relative location of a node in a 2D plane using trilateration alone, generally, at least 3 reference nodes are needed.

The basic idea of trilateration is as follows. To locate the position (x_d, y_d) of node d, we need three reference nodes, a, b, and c, whose positions, (x_a, y_a), (x_b, y_b), and (x_c, y_c), respectively, are known. As illustrated in Figure 10.2a, the position of d should be located at the intersection of the three circles centered at a, b, and c, respectively. Thus, the following equations hold:

$$|ad|^2 = (x_d - x_a)^2 + (y_d + y_a)^2,$$
$$|bd|^2 = (x_d - x_b)^2 + (y_d + y_b)^2,$$
$$|cd|^2 = (x_d - x_c)^2 + (y_d + y_c)^2.$$

By solving the above equations, d's location can be obtained. Notice that here the reference nodes, a, b, and c, cannot be on a straight line.

Trilateration can also work in three-dimensional networks. The only difference is that to locate a node in 3D networks needs four references nodes instead of three.

10.2.3 Triangulation

In trigonometry and geometry, triangulation is the process of finding coordinates of a point by using measurements of angles and sides of a triangle formed by that point and two or three other known reference points, using the law of sines or the law of cosines. Unlike trilateration, which uses only distance measurements, triangulation also uses angle measurements, such as those obtained by AoA-based methods. For example, in 2D networks, we can easily calculate the location of any node if the three angles

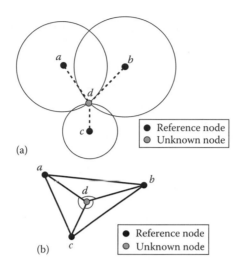

Figure 10.2   Basics of localization algorithms: (a) trilateration and (b) triangulation.

between the node and other three known reference nodes are known. Figure 10.2b shows an example. Again we suppose there are three reference nodes, *a, b,* and *c.* The unknown node, *d,* can measure the angles ∠*adb,* ∠*bdc,* and ∠*adc.* Because the locations of reference nodes are known, the distances between each pair of them are also known. Then, we can obtain the following equations:

$$|ad|^2 = (x_d - x_a)^2 + (y_d + y_a)^2,$$
$$|bd|^2 = (x_d - x_b)^2 + (y_d + y_b)^2,$$
$$|cd|^2 = (x_d - x_c)^2 + (y_d + y_c)^2,$$
$$|ab|^2 = |ad|^2 + |bd|^2 - 2|ad||bd| \cos \angle adb,$$
$$|bc|^2 = |bd|^2 + |cd|^2 - 2|bd||cd| \cos \angle bdc,$$
$$|ac|^2 = |ad|^2 + |cd|^2 - 2|ad||cd| \cos \angle adc.$$

By solving the above six equations, *d*'s location, (x_d, y_d), can be obtained.

10.2.4   Theory of Network Localization: Localizability and Rigidity Theory

Both trilateration and triangulation can be used to compute the position of a sensor node using range or angle measurements via reference nodes when enough measurements are available. However, in practice, it is possible that many nodes' positions cannot be uniquely determined in wireless sensor networks with limited measurements. The important question is what are the precise conditions under which the network localization problem is solvable (i.e., each node has a unique position solution). There is a strong connection between the problem of unique network localization and a mathematical topic known as rigidity theory [15]. Recently, there are several new results [4,10,14] published in the sensor network area.

The network localization problem with the distance information is to determine the locations, p_i, of all nodes, $v_i \in V$, in real d-dimensional space, R^d, given the graph of the network $G = (V, E)$, the position of the reference nodes p_j in R^d, and the distance between each neighbor pair $(i, j) \in E$. The localization problem is said to be solvable or the network is said to be localizable if there is exactly one set of positions in R^d for all unknown nodes that is consistent with all available information about distances and positions. The localization problem can also be formed as a point formation, $F_p = (\{p_1, p_2, \cdots, p_n\}, L)$, where p_i is node i's position and L is a set of links whose internode distance is given (including both the distances measured from the unknown node to the reference node and the distances among reference nodes). Then, in Refs. [4,10], the following theorem gives the conditions for a network to be localizable.

Theorem 10.1

[4,10] *For a network in R^d, $d = 1, 2$, and 3, if there are at least $d + 1$ reference nodes in the general position; the network is uniquely localizable if and only if the point formation for the graph G is globally rigid.*

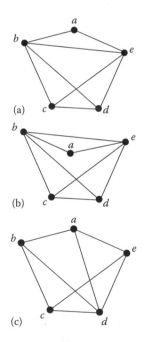

We then give a rough definition of global rigidity. Consider a point formation (and its corresponding graph) with edges connecting some of them to represent distance constraints. If there is no other point formation which consists of different points and preserves all the distance constraints, then we call this point formation or its corresponding graph globally rigid. Figure 10.3 shows examples of nonglobally rigid and globally rigid graphs in a 2D plane.

Results from rigidity theory give efficient way (polynomial algorithms) to check whether a graph is globally rigid. The following theorem gives the sufficient and necessary condition for the global rigidity test in a 2D space.

Theorem 10.2

[17] *A graph with $n \geq 4$ vertices is generically globally rigid in R^2 if and only if it is redundantly rigid and three-connected in two dimensions, R^2.*

Figure 10.3 Examples of two nonglobally rigid graphs ((a) and (b)) and a globally rigid graph (c).

Here a graph is redundantly rigid if the removal of any single edge results in a graph that is also generically rigid. To test the generic rigidity, the following theorem can be used.

Theorem 10.3

[21] *In R^2, a graph with n vertices and $2n - 3$ edges is generically rigid if and only if no subgraph has more than $2n' - 3$ edges, where n' is the number of vertices of the subgraph.*

Notice that both of the above theorems are true only for a 2D space and cannot be extended to higher dimensions.

Even though the global rigidity can be determined efficiently, the problem of realizing globally rigid weighted graphs (the network localization problem) is NP hard. Aspnes et al. [4] proved this by giving a polynomial-time reduction of the set-partition problem to the globally rigid weighted graph realizing problem. For more details about the complexity of the localization problem, please refer to Ref. [4].

10.3 Range-Based Localization Methods

From this section, we begin to review several existing localization methods for wireless sensor networks. We first focus on range-based localization methods where precise distance or angle measurement techniques such as ToA, TDoA, AoA, and RSS, are used to estimate the distance among nodes and the distance between sensor nodes and reference nodes. Hereafter, we also call these reference nodes, whose positions are known beforehand and have the capacity to send beacon messages to other nodes, as anchor nodes or beacon nodes. We categorize the range-based methods into four classes: single hop to anchors methods, multiple hops to anchors methods, mobile anchors employed methods and methods without anchor nodes.

10.3.1 Single Hop to Anchors Methods

If the anchor nodes are powerful enough such that every sensor node can directly communicate with them, then the localization method is straightforward. Both trilateration and triangulation can be used to locate a sensor node, if it can hear signals from at least three anchors in a 2D plane or four anchors in a 3D space. Here, we just review some example implementations of such methods.

GPS: The GPS system is the most well-known outdoor positioning system that uses single hop to anchors methods. Utilizing a constellation of at least 24 Medium Earth Orbit satellites that transmit precise microwave signals, the system enables a GPS receiver to determine its location, speed, direction, and time. GPS receivers can use four GPS satellite signals to get their positions in a 3D space via simple trilateration. Currently, the accuracy of GPS is around 3–20 m. However, GPS has some oblivious drawbacks for wireless sensor networks. Arming each sensor node with a GPS receiver is too expensive and the GPS system is not feasible in many cases, such as indoor environments and underwater environments.

RADAR system: The RADAR system [5] implements a location service utilizing the RF signal strength information at multiple base stations positioned to provide an overlapping coverage in the area of interest. It uses the RF signal strength (RSS) as an indicator of the distance between a transmitter and a receiver. This distance information is then used to locate a user by trilateration. In addition, the RADAR system improves the accuracy by searching in recorded scenes from an off-line phase. During the off-line phase, the system builds a database of the RF signal strength at a set of fixed receivers, for known transmitter positions. During the normal operation, the RF signal strength of a transmitter, as measured by the set of fixed receivers, is sent to a central computer, which examines the signal-strength database to obtain the best fit for the current transmitter position.

Cricket system: The cricket system [27] is a location-support system for in-building, mobile, and location-dependent applications. It uses multiple anchor nodes spread throughout the building, and each anchor node generates beacons using both radio and ultrasonic signals. When each node hears the beacons from multiple anchors, it uses the TDoA method to estimate the distance from each anchor and then compute the location information via trilateration. In Ref. [27], the authors also consider interference and collision among the beacons from different anchors and describe a randomized algorithm to overcome this effect.

AoA method: Nasipuri and Li [25] present a technique by which sensor nodes can determine their positions in a sensor network by obtaining angular bearings relative to a set of fixed beacon nodes. They assume that the beacon nodes are equipped with special transmission capabilities for sending wireless beacon signals throughout the network, and each beacon signal consists of a continuous RF carrier signal on a narrow-directional beam that rotates with a constant angular speed. Then a sensor node can record the time when it receives the different beacon signals and evaluate its angular bearings and location with respect to the beacon nodes by triangulation. At least three beacon nodes are required for the localization technique to work in an ideal case, with additional beacon nodes needed for resolving errors from multipath reflections. The proposed localization scheme exhibits excellent accuracy and requires very little additional complexity at the sensor nodes. The main source of error is due to the beam width of the directional beacon signals. However, the location error has been found to be small for beam widths within 15 degrees. In addition, the performance of this method is not affected by the density or number of sensor nodes in the network.

Methods for underwater sensor networks: The localization problem in underwater environments [2] is more difficult than that in terrestrial environments. Here, the RF signal cannot be used because it will be absorbed by water. Therefore, an acoustic signal is usually the choice in underwater environments. Acoustic channels have the following characteristics: low bandwidth, high delay, and high bit error rate. Because the velocity of the acoustic signal can change with salinity, pressure, and temperature, it is difficult to get quite precise ranges between nodes underwater. Several localization methods [8,9,34] have been proposed for underwater sensor networks. Most of them use the same principles as used in localization methods for terrestrial sensor networks, except for (1) the use of acoustic signals and (2) the need for one more anchor node (because the underwater sensor network is a 3D network).

10.3.2 Multiple Hops to Anchors Methods

The previous section discusses how to locate a sensor node if it can directly hear from anchors via a single hop connection. However, in most sensor networks (especially large-scale sensor networks), the network is formed by multihop links, and many sensor nodes cannot directly communicate with anchor nodes. Therefore, multiple hops to anchors methods should be employed.

10.3.2.1 Iterative and Collaborative Multilateration

In a 2D network, trilateration can locate a sensor if it can hear from at least three anchors. However, many sensors cannot directly communicate with enough anchor nodes to compute their positions. Multilateration methods have been proposed to solve this problem. The basic idea of multilateration is that nodes measure distances to their neighbors and share their position information with their neighbors to collaboratively compute their position. In Refs. [29,30], Savvides et al. propose both iterative multilateration and collaborative multilateration methods.

During the first phase of multilateration, every sensor node measures distances to its neighbors by employing ToA, TDoA, or RSS. If there are enough neighbors, the positions of which are known, trilateration can be used to compute the node's position. Thus, at the end of this phase, only those nodes that are able to directly communicate with enough anchors can obtain their positions.

Then, in the iterative multilateration method [29], the sensor nodes, whose positions have already been uniquely determined in the first phase, send their positions to their neighbors. These nodes can be treated as new anchor nodes. If a node that does not know its position, now has sufficient neighbors whose positions are known, it can obtain its position and send its position to all of its neighbors. This iterative process continues until there are no nodes that can be further localized (i.e., every node that does not know its position yet, has less than three neighbors whose positions are known in a 2D network). The drawback of the iterative multilateration is that the use of localized nodes as new anchor nodes can introduce substantial cumulative errors in the localization.

To further solve this problem, so that at the end of iterative multilateration some nodes may not have sufficient anchor nodes to be localized, collaborative multilateration [29] has been introduced. Basically, it attempts to estimate positions by using location information over multiple hops. It determines collaborative subgraphs within the network that contain anchors and nodes, whose positions are unknown, such that their positions and internode distances can be written as a set of quadratic equations, which can be solved by certain optimization algorithms. In Ref. [30], the authors also introduce a method using leastsquares estimation (such as the Kalman Filter) to refine the node positions.

10.3.2.2 Sweeping Method

Recently, Goldenberg et al. [13] introduced a localization method for sparse networks using the sweep technique. Recall that in actual sensor networks only a part of sensor nodes can be uniquely localizable, and there are still some nodes whose positions cannot be uniquely decided. However, most such nodes can be localized up to a set of possible locations. In many cases, it is useful for nodes to know the set of all possible positions. This kind of localization can be called finite localization instead of unique localization. The idea of the sweeping method is similar to the iterative or collaborative multilateration.

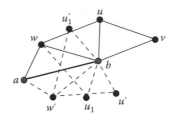

Figure 10.4   Illustration of eliminating possible positions in the sweeping method (From Goldenberg, D.K. et al., Localization in sparse networks using sweeps, in *Proceedings of the 12th Annual International Conference on Mobile Computing and Networking*, pp. 110–121, New York, 2006. ACM. With permission.)

At each stage of the algorithm, not only the localizable nodes are localized, but also a set of possible positions are generated for each node that can be finitely localized. This information will be exchanged among neighbors and will be used to eliminate some of the possible positions. One drawback of the sweeping method is that the possible positions could increase exponentially with the large number of nodes.

Figure 10.4 shows an example of reducing possible positions of sensors under sweep. Suppose nodes a and b are anchors. Node w has already measured the distances to a and b. Thus, node w has two possible positions: w and w'. After another node u obtains the distances to w and b, u has four possible positions: u and u', which are derived from w and b, and u'_1 and u_1, which are derived from w' and b. Then if node v knows its position and broadcasts it to its neighbors, u will receive v's position and can remove impossible positions (u'_1 and u_1).

10.3.2.3 Multidimensional Scaling (MDS)

Multidimensional scaling is a technique used for the analysis of dissimilarity of data on a set of objects. Recently, it has been used for recovering the positions of the adjacent sensor nodes in a geometric space, based on their pairwise distances [18,31]. Basically, MDS compares the estimated positions of anchor nodes with their actual positions and refines the estimated position of all sensor nodes by iterative adjustment. The MDS method can also be used in a distributed manner. This method described in Ref. [31] consists of the following phases.

In the first phase, for each node a local map of its adjacent sensor nodes is formed. Here, the local map could be a map including all sensors within constant hops from the node. Based on the local map, all shortest path distances between every pair of nodes in the local map are computed. These distances then form the distance matrix for multidimensional scaling. The time complexity of this step is $O(k^3)$, where k is the average number of sensor nodes in a local map. Multidimensional scaling uses the distance matrix to obtain the largest eigenvalues and eigenvectors, which can be employed to construct a local map. And the time complexity of this step is also $O(k^3)$.

In the second phase, multiple local maps are merged into a global map. There are many ways to merge local maps. For example, a local map can be randomly chosen as the core map. And then the core map is stitched with local maps of neighboring nodes to a new core map. This process will iteratively continue until the core map covers the entire network. The time complexity of this step is $O(k^3n)$, where n is the number of sensor nodes. After the end of this phase, we obtain global positions but with possible errors.

In the third phase, estimated positions can be refined by using the least squares minimization. In the last phase, if there are sufficient anchor nodes, the global positions of nodes can be transformed into absolute positions.

10.3.3 Mobile Anchors Employed Methods

In both single hop and multihop to anchors localization methods, some sensor nodes may not be uniquely determined because there are no sufficient anchor nodes. Thus, some localization methods deploy a great number of anchors to avoid such a situation. However, this incurs a lot of cost. One efficient way to address this problem is to introduce mobile anchor nodes (also called mobile beacons). These mobile anchors can move around in the network and know their positions at a certain/any time (either by a GPS or by predefined positions). The beacon signals sent by the mobile anchor include its position. When a sensor hears the beacon signals from mobile anchors, more information can be obtained for localization, thus improving the accuracy of localization. Different methods [11,12,32,38] have been proposed using mobile anchors. Here we briefly review a method used in underwater sensor networks. In Ref. [12], an autonomous underwater vehicle (AUV) is used to aid the localization of underwater sensors. When AUV is on the surface of water, it can receive signals from GPS and can get its position. Then AUV dives into water and follows a known trajectory. While AUV is moving among sensor nodes, it broadcasts some information containing the present position of AUV. This kind of information can be referred as beacons. When a sensor node receives a signal from AUV, it can measure the distance to AUV and get the position of AUV. If the node obtains enough distances to AUV and positions of AUV, it can compute its position by using trilateration. AUV usually follows the lattice trajectory or the spiral trajectory or other trajectories. However, it is hard to determine which trajectory of AUV is optimal, because the positions of the unknown nodes are not known. A similar method (called Dive "N" Rise, where the mobile beacons sink and rise in water and broadcast their positions) is proposed in [11].

In addition, the FDoA-based methods [3,19,20] also often use mobile anchors.

10.3.4 Methods without Anchor Nodes

All localization methods, we discussed so far, either use GPS to locate positions of several initial nodes or assume that anchor nodes or mobile beacons know their positions. There is another set of methods [6,24] that do not use any anchor nodes and only aim to obtain relative positions instead of absolute positions. Localization methods without anchor nodes (such as [24]) usually consist of three main phases. In the first phase, local coordinate systems should be built. Each sensor node is the center of a cluster and it measures the distances to its neighboring nodes. Neighboring nodes also measure distances to other nodes and send the information containing these distances to their neighboring nodes. The second phase refines the estimated positions of sensor nodes inside each cluster. Different optimization techniques, such as spring relaxation or Newton–Raphson, with the full set of measured distance constraints can be used. In the third phase, the transformations between the local coordinate systems of neighboring clusters are computed by finding the set of nodes in common between two clusters and solving for the rotation, translation, and possible reflection that best aligns the cluster. The advantage of the methods without anchor nodes is that every node has a local coordinate system with itself as the origin. In addition, sensor nodes only measure the distances to their neighboring nodes and broadcast this information, so these methods are totally distributed.

10.4 Range-Free Localization Methods

The range-based localization methods discussed above can provide sensor nodes with precise positions. However, in some applications, the cost and limitations of the hardware on sensing nodes prevent the use of range-based localization schemes that depend on absolute point-to-point distance estimates. Because coarse accuracy is sufficient for most sensor network applications, solutions in range-free localization are being pursued as a cost-effective alternative to more expensive range-based approaches. Range-free methods can be classified into two categories: hop-count-based methods and area-based methods.

10.4.1 Hop-Count-Based Methods

In hop-count-based methods, anchor nodes are usually placed along the boundaries or only at the corners of an area.

10.4.1.1 Distance-Vector-Based Localization

Similar with the distance-vector routing, in distance-vector (DV)-based localization methods, each sensor can only communicate with its immediate neighbors, and each sensor estimates the distances to anchor nodes and sends these distances to its neighbors.

The most basic DV-based localization method is the DV-hop propagation method [26], consisting of three phases. First, the classical distance-vector exchange is employed and hence each sensor node can estimate the distance in hops to anchor nodes. Second, each anchor node can compute the average distance for one hop. Remember that after the first step, anchor nodes also get hop distances to any other anchor node. Because the positions of anchor nodes are known beforehand, each anchor node can compute the average length for one hop. Then anchor nodes broadcast this information through the distance-vector exchange. Finally, after receiving the size of one hop, sensor nodes can estimate distances in meters to anchor nodes. Then sensors can get their positions by performing trilateration. When an anchor node, j, computes the average size of one hop, it follows the formula:

$$c_i = \frac{\Sigma_j \sqrt{(x_i - x_j)^2 + (x_i - x_j)^2}}{\Sigma_j\, h_{ij}}, \quad \text{for all other anchor nodes, } j,$$

where

c_i is the average size of one hop to anchor i
$(x_i,\ y_i)$ and $(x_j,\ y_j)$ are the positions of anchors i and j
h_{ij} is the hop count between i and j

Figure 10.5 shows an example where there are three anchor nodes, *a, b*, and *c*. These anchor nodes know the distances in meters to each other because the positions of anchor nodes are known beforehand. After the distance-vector exchange, *a* can know that the numbers of hops to *b* and *c* are 2 and 6, respectively. So, according to the formula above, *a* can compute its average size of one hop, that is, $(100 + 40)/(6 + 2) = 17.5$. Similarly, anchor nodes *b* and *c* can compute their average sizes for one hop, and they are 16.42 and 15.90 m, respectively. Anchor nodes broadcast the sizes through the distance-vector exchange. Sensor nodes can use these sizes to estimate the distances to each anchor node. The advantage of the DV-hop propagation method is that it is very simple but it can only work well in isotropic networks where the properties in all directions are the same.

There are three other methods of hop-to-hop distance propagation in DV-based methods, proposed in Ref. [26], besides DV hop. The second propagation method is the DV-distance propagation method, which is very similar to the DV-hop method. The only difference between both the methods is that the DV-distance uses the distance measurement instead of hop counts during the distance-vector exchange. Here, the distance between nodes can be measured by ToA, TDoA, RSS, or AoA. Clearly, DV-distance is less coarse than DV-hop. But this method is not really range-free anymore. In the third method, the Euclidean propagation method, instead of distance measurements, true Euclidean distances to anchor nodes are propagated. These true distances could be obtained from the GPS system. The last propagation method is the DV-coordinate propagation method where the coordinates of sensor nodes are propagated. This method is similar to collaborative multilateration. Before propagating coordinates, each node should establish its local coordinate system. When receiving coordinates from neighboring nodes, each node should transform coordinates into its own local coordinate system.

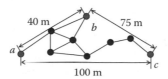

Figure 10.5 DV-hop: estimation of the length of one hop. (From Niculescu, D. and Nath, B., *J. Telecommun. Syst.,* **22, 267, 2003. With permission.)**

10.4.1.2 Other Improvements

Positions obtained through hop-count-based methods are comparatively coarse. Based on basic DV-based localization methods, many further improvements [28,33,36,37] have been proposed. Here, we briefly review some of them.

In Ref. [28], a two-phase positioning algorithm is proposed. The first phase basically is the DV-hop method. After each node gets its estimated position, a refinement algorithm is run iteratively. Each node broadcasts its estimated position, and it receives the neighboring nodes' positions and corresponding range estimates. Each node can obtain its new position by computing a least squares triangulation solution. The constraints imposed by the distances to the neighboring location will force the new position toward the true position. After several iterations, the position update becomes small and the algorithm terminates.

For many hop-count-based localization methods, errors of estimated positions come from the distances from sensor nodes to anchor nodes. Usually, anchor nodes are deployed along the boundary of the networks. Many researchers find that the localization accuracy is usually the highest in the center of the network if the network is uniformly distributed. Based on this finding, selective iterative multilateration (SIM) [33] is proposed. SIM chooses sensor nodes with high localization accuracy as anchor nodes. Basically, SIM first locates the center of the network based on the average of locations of all anchor nodes. Then nodes around the center are good candidates for acting as new anchor nodes. Simulations showed this method can improve the accuracy of DV-hop localization.

Remember that DV-hop can work well in isotropic networks. However, when sensor nodes are unevenly distributed, the actual size for one hop changes greatly among neighboring nodes. This leads to errors in distance estimation which tend to accumulate with the increase of path length. In Ref. [37], Wong et al. propose the density-aware hop-count localization (DHL) method. DHL considers two potential issues: density and path length, during the distance estimation. In DV-hop, the distance from a node to an anchor is computed as hop counts times the average length per hop. However, in DHL, the length of each hop is estimated based on the local density. Because of the cumulative error in distance estimation, an estimated distance computed from a fewer number of hops is more accurate, and hence, can be given a higher confidence rating. Thus, DHL uses the distances to the nearest anchor nodes that have high confidence levels to estimate the location. Similar to this idea, in Ref. [36], Wang and Xiao propose a method to detect the incorrect distance measurements caused by concave networks when multiple distance measurements are available.

10.4.2 Area-Based Methods

In many actual applications, there is no need to get coordinates of sensor nodes and we only need to determine the area in which a sensor node lies. Thus, area-based localization methods focus on providing the areas information of sensor nodes.

Area localization scheme (ALS) [7] is a centralized range-free scheme providing the areas in which sensor nodes lie. In ALS, there are three kinds of nodes: anchor nodes, sensor nodes, and sinks. ALS usually divides the network into grids and deploys anchor nodes along the boundaries of grids. Each anchor node periodically broadcasts beacon signals with different power levels. Because signals with different power levels can reach different distances, that is, the signal with a higher power can reach a greater distance, anchor nodes can compute the power levels required to reach different distances. In ALS, each anchor node devises a set of power levels and signals with the highest power level that can cover the whole grid. When each anchor node broadcasts a signal, the signal contains the ID of the anchor node and the power level at which the signal is sent. Each sensor node just listens and records the power levels of signals received from each anchor node. When receiving a signal, each sensor node can obtain the ID of the anchor node who sent the signal and the power level of the signal. And each sensor node records the lowest power level received from each anchor node and the ID of this anchor node, and then send this information to sinks. Sinks will decide the areas in which sensor nodes lie.

Figure 10.6 shows an example where four anchor nodes are at the corners of a square region. Every anchor node can broadcast signals with three different power levels. The lowest power level is denoted as integer 1 and the highest power level is denoted as integer 3, which can reach the whole network. $<3, 1, 3, 3>$, $<1, 3, 3, 3>$ and so on are signal coordinates. A signal coordinate is denoted as $<S_1, S_2, \cdots, S_n>$, where S_i is the lowest power level received from the anchor node i. In Figure 10.6, the contour lines are the farthest distances that signals at each power level can reach. Clearly, the contour lines of each sensor node can divide the region into three areas and all contour lines divide the grid into nine areas. And each area is associated with a different signal coordinate. Obviously, the lowest power levels received by the sensor nodes in Area A from anchor nodes 1, 2, and 3 are all 3 and the lowest power level received from anchor node 4 is 2. Thus, the signal coordinate of Area A is $<3, 3, 3, 2>$.

Sinks can receive signal coordinates of sensor nodes. According to this information, sinks can estimate the area in which a sensor node lies and send the results back to sensor

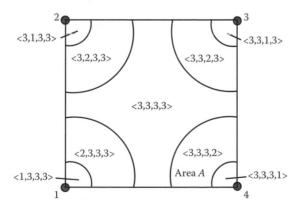

Figure 10.6 Illustration of the ALS. (From Chardrasekhar, V. and Seoh, W., An area localization scheme for underwater sensor networks. In OCEANS 2006—Asia Pacific, 2007. With permission.)

nodes. ALS assumes that sinks know the positions of anchor nodes and their power levels beforehand. Sinks have higher capabilities in computing.

In Ref. [35], a range-free ad hoc localization algorithm called urban pedestrians localization (UPL) is proposed for providing the area of presence of each mobile node in the urban district. The UPL considers two things: the information about obstacles such as walls and mobility of nodes. The following is the basic idea of the UPL. The UPL assumes that mobile nodes cannot receive signals from anchor nodes frequently and mobile nodes are fully connected in the urban district. Each mobile node records the area in which it lies and obtains other areas when it receives information from other mobile nodes. According to the transmission range, we can compute the area of presence of the node by intersecting two or more areas. Because of the mobility of each node, as time passes, we can compute the area of presence according to the speed of the node and the obstacle information.

10.5   Conclusion

Localization, which determines the positions of all sensors, is an essential and fundamental problem in wireless sensor networks. In this chapter, we introduce some fundamental techniques for localization, such as range measurement, trilateration, and triangulation, and then review several representative localization methods from range-based methods to range-free methods. These methods are quite different in precision or coarseness, distribution or centralization, placement of anchors, and percentage of anchors. Table 10.1 summarizes the comparison between them.

Table 10.1   Comparison of Localization Methods

Methods	Accuracy	Algorithm	Placement of Anchors	Percent of Anchors
GPS	Precise	Distributed	On satellites	Low (24 anchors)
Single hop to anchors	Precise	Distributed	On boundary	Low or median
Multiple hops to anchors	Precise	Distributed	Random or on boundary	Median
Mobile anchors employed	Precise	Distributed	Mobile	Very low (one or a few anchors)
Hop-count-based	Coarse	Distributed	On boundary	Low
Area-based	Coarse	Centralized	On boundary or on gird	High

Range-based methods can usually provide precise position information while range-free methods can only give coarse position information. Due to the distributed nature of wireless sensor networks, most localization methods are distributed except area-based methods. Single-hop to anchors methods either have a few powerful anchor nodes, which can cover the most or even the entire area, or have enough anchors so that each sensor can directly communicate with several of them. To reduce the number of anchors or avoid powerful anchors, multiple-hops to anchors methods can be applied. In such methods, the percentages of anchors are usually much lower and anchors are often randomly deployed. If mobile anchors are employed, the number of anchors can be further reduced to a few or even to one. For hop-count-based methods, anchors are usually deployed along the boundary of the network and their percentage is low. For area-based methods (such as the ALS), a large number of anchors are needed to guarantee the coverage and a certain accuracy.

Another important metric for localization algorithms is the message overhead, that is, the number of messages interchanged among sensor nodes or anchors. It is clear that for GPS or single hop to anchors solutions the number of messages is minimized, because each sensor only needs to listen to the signal from anchors without relaying any information. The only messages are the broadcasted beacon messages from anchors. However, in the solutions for multiple hops to anchors or hop-count-based methods, certain information (such as location, estimated distance, or hopcount) needs to be exchanged and propagated among multi-hop neighbors. This must incur a certain amount of message overhead. For methods with mobile anchors or area-based methods, the message overhead depends on how the system is designed and implemented (such as the whether sensor relays messages for other sensors or the whether sensor needs to send information to a sink node for processing).

Even though many localization methods have been proposed and studied, to have a robust and accurate localization algorithm under rough environments where wireless sensor networks are usually deployed is still a very challenging task. There are many open research problems in localization that need to be addressed. For example: How to preserve the accuracy of localization under erroneous measurements? How to play the trade-off among accuracy, number of anchors, and time, message, or energy overheads? Is there a sufficient and necessary condition for rigidity testing in 3D or higher dimensions? Finally, the design of the localization algorithm heavily depends on the sensor network application. For different applications with various constraints, suitable localization algorithms need to be carefully designed.

References

1. David Adamy. *EW 102: A Second Course in Electronic Warfare*. Artech House, Boston, 2004.
2. I.F. Akyildiz, D. Pompili, and T. Melodia. Underwater acoustic sensor networks: Research challenges. *Ad Hoc Networks*, 3(3): 257–279, 2005.
3. I. Amundson, X. Koutsoukos, and J. Sallai. Mobile sensor localization and navigation using RF doppler shifts. In *Proceedings of 1st ACM International Workshop on*

Mobile Entity Localization and Tracking in GPS-less Environments (MELT'08), San Francisco, CA, 2008.

4. J. Aspnes, T. Eren, D.K. Goldenberg, A.S. Morse, W Whiteley, Y.R. Yang, B.D.O. Anderson, and P.N. Belhumeur. A theory of network localization. *IEEE Transaction on Mobile Computing*, 5(12): 1–15, 2006.

5. P. Bahl and V. N. Padmanabhan. RADAR: An in-building RF-based user location and tracking system. In *Proc. of 19th Annual Joint Conference of the IEEE Computer and Communications Societies (INFOCOM 2000)*, Tel Aviv, Israel, 2000.

6. S. Capkun, M. Hamdi, and J.-P. Hubaux. GPS-free positioning in mobile ad-hoc networks. In *Proc. of the 34th Annual Hawaii International Conference on System Sciences (HICCS)*, Maui, HI, 2001.

7. V. Chandrasekhar and W. Seah. An area localization scheme for underwater sensor networks. In *Proc. of the IEEE OCEANS Asia Pacific Conference*, Singapore, 2006.

8. V. Chandrasekhar, W.K.G. Seah, Y.S. Choo, and H. Voon Ee. Localization in underwater sensor networks: Survey and challenges. In *WUWNet '06: Proceedings of the 1st ACM International Workshop on Underwater Networks*, pp. 33–40, New York, 2006. ACM.

9. X. Cheng, H. Shu, and Q. Liang. A range-difference based self-positioning scheme for underwater acoustic sensor networks. In *Proc. of International Conference on Wireless Algorithms, Systems and Applications (WASA 2007)*, Chicago, IL, 2007.

10. T. Eren, D.K. Goldenberg, W Whiteley, Y.R. Yang, A.S. Morse, B.D.O. Anderson, and P.N. Belhumeur. Rigidity, computation, and randomization in network localization. In *Proc. of 23rd Annual Joint Conference of the IEEE Computer and Communications Societies (INFOCOM 2004)*, Hong Kong, China, 2004.

11. M. Erol, L.F.M. Vieira, and M. Gerla. Localization with dive'n'rise (DNR) beacons for underwater acoustic sensor networks. In *WUWNet '07: Proceedings of the Second Workshop on Underwater Networks*, pp. 97–100, New York, 2007. ACM.

12. M. Erol, L.F.M. Vieira, and M. Gerla. AUV-aided localization for underwater sensor networks. In *Proc. of International Conference on Wireless Algorithms, Systems and Applications (WASA 2007)*, Chicago, IL, 2007.

13. D.K. Goldenberg, P. Bihler, M. Cao, J. Fang, B.D.O. Anderson, A. Stephen Morse, and Y. Richard Yang. Localization in sparse networks using sweeps. In *MobiCom '06: Proceedings of the 12th Annual International Conference on Mobile Computing and Networking*, pp. 110–121, New York, 2006. ACM.

14. D.K. Goldenberg, A. Krishnamurthy, W.C. Maness, Y.R. Yang, A.S. Morse, and A. Savvides. Network localization in partially localizable networks. In *Proc. of 24th Annual Joint Conference of the IEEE Computer and Communications Societies (INFOCOM 2005)*, Miami, FL, 2005.

15. J. Graver, B. Servatius, and H. Servatius. *Combinatorial Rigidity*. Graduate Studies in Math., AMS, 1993.

16. Q. Huang, C. Lu, and G.-C. Roman. Design and analysis of spatiotemporal multicast protocols for wireless sensor networks. *Telecommunication Systems*, 26(2–4): 129–160, 2004.

17. B. Jackson and T. Jordan. Connected rigidity martoids and unique realizations of graphs. *Journal of Combinatorial Theory, Series B*, 94(1): 1–29, 2005.

18. X. Ji and H. Zha. Sensor positioning in wireless ad-hoc sensor networks using multidimensional scaling. In *Proc. of 23rd Annual Joint Conference of the IEEE Computer and Communications Societies (INFOCOM 2004)*, Hong Kong, China, 2004.

19. R.J. Kozick and B.M. Sadler. Sensor localization using acoustic doppler shift with a mobile access point. In *Proceedings of IEEE/SP 13th Workshop on Statistical Signal Processing*, Bordeaux, France, 2005.

20. B. Kusy, A. Ledeczi, and X. Koutsoukos. Tracking mobile nodes using RF Doppler shifts. In *Proceedings of the ACM 5th International Conference on Embedded Networked Sensor Systems*, Sydney, Australia, 2007.

21. G. Laman. On graphs and rigidity of plane skeletal structures. *Journal of Engineering Mathematics*, 4(4): 331–340, 1970.

22. C. Maihofer. A survey of geocast routing protocols. *IEEE Communications Surveys and Tutorials*, 6(2): 32–42, 2004.

23. M. Maroti, P. Volgyesi, S. Dora, B. Kusy, A. Nadas, A. Ledeczi, G. Balogh, and K. Molnar. Radio interferometric geolocation. In *Proceedings of the ACM 3rd International Conference on Embedded Networked Sensor Systems*, Boulder, CO, 2005.

24. D. Moore, J. Leonard, D. Rus, and S. Teller. Robust distributed network localization with noisy range measurements. In *SenSys '04: Proceedings of the 2nd International Conference on Embedded Networked Sensor Systems*, pp. 50–61, New York, 2004. ACM.

25. A. Nasipuri and K. Li. A directionality based location discovery scheme for wireless sensor networks. In *WSNA '02: Proceedings of the 1st ACM International Workshop on Wireless Sensor Networks and Applications*, pp. 105–111, New York, 2002. ACM.

26. D. Niculescu and B. Nath. DV based positioning in ad hoc networks. *Journal of Telecommunication Systems*, 22(1–4): 267–280, 2003.

27. N.B. Priyantha, A. Chakraborty, and H. Balakrishnan. The cricket location-support system. In *MobiCom '00: Proceedings of the 6th Annual International Conference on Mobile Computing and Networking*, pp. 32–43, New York, 2000. ACM.

28. C. Savarese, J.M. Rabaey, and K. Langendoen. Robust positioning algorithms for distributed ad-hoc wireless sensor networks. In *ATEC '02: Proceedings of the General Track of the Annual Conference on USENIX Annual Technical Conference*, pp. 317–327, Berkeley, CA, 2002. USENIX Association.

29. A. Savvides, C.-C. Han, and M.B. Strivastava. Dynamic fine-grained localization in ad-hoc networks of sensors. In *MobiCom '01: Proceedings of the 7th Annual International Conference on Mobile Computing and Networking*, pp. 166–179, New York, 2001. ACM.

30. A. Savvides, H. Park, and M.B. Srivastava. The bits and flops of the n-hop multilateration primitive for node localization problems. In *WSNA '02: Proceedings of the 1st ACM International Workshop on Wireless Sensor Networks and Applications*, pp. 112–121, New York, 2002. ACM.

31. Y. Shang and W. Ruml. Improved mds-based localization. In *Proc. of 23rd Annual Joint Conference of the IEEE Computer and Communications Societies (INFOCOM 2004)*, Hong Kong, China, 2004.

32. K.-F. Ssu, C.-H. Ou, and H. Christine Jiau. Localization with mobile anchor points in wireless sensor networks. *IEEE Transctions on Vehicular Technology*, 54(3): 1187–1198, 2005.

33. J.H.S. Tay, V.R. Chandrasekhar, and W.K.G. Seah. Selective iterative multilateration for hop count-based localization in wireless sensor networks. In *Proc. of 7th International Conference on Mobile Data Management (MDM 2006)*, Nara, Japan, 2006.

34. C. Tian, W. Liu, J. Jin, Y. Wang, and Y. Mo. Localization and synchronization for 3D underwater acoustic sensor networks. In *Proc. of 4th International Conference on Ubiquitous Intelligence and Computing (UIC 2007)*, Hong Kong, China, 2007.

35. A. Uchiyama, S. Fujii, K. Maeda, T. Umedu, H. Yamaguchi, and T. Higashino. Ad-hoc localization in urban district. In *Proc. of 26th IEEE International Conference on Computer Communications (INFOCOM 2007)*, Anchorage, AK, 2007.

36. C. Wang and L. Xiao. Locating sensors in concave areas. In *Proc. of 25th IEEE International Conference on Computer Communications (INFOCOM 2006)*, Barcelona, Spain, 2006.

37. S.Y. Wong, J.G. Lim, S.V. Rao, and W.K.G. Seah. Multihop localization with density and path length awareness in non-uniform wireless sensor networks. In *Proc. of IEEE 61st Vehicular Technology Conference (VTC 2005-Spring)*, Stockholm, Sweden, 2005.

38. C.-H. Wu, W. Sheng, and Y. Zhang. Mobile sensor networks self localization based on multi-dimensional scaling. In *Proc. of 2007 IEEE International Conference on Robotics and Automation*, Roma, Italy, 2007.

Chapter 11

Data Aggregation in Wireless Sensor Networks

Suat Ozdemir

CONTENTS

Wireless sensor networks often consist of a large number of low-cost sensor nodes that have strictly limited sensing, computation, and communication capabilities. Due to resource-restricted sensor nodes, it is important to minimize the amount of in-network data transmission so that the average sensor lifetime and the overall bandwidth utilization

297

are improved. In wireless sensor networks, data aggregation is the process of summarizing and combining sensor data in order reduce the amount of data transmission while increasing the reliability of the data. This chapter surveys the current state-of-the-art data aggregation techniques in wireless sensor networks. In the first part of the chapter, data aggregation protocols are categorized into two parts regarding their network topology. Then, the interaction between data aggregation and security protocols is investigated as they both are essential for wireless sensor networks. In the last part of the chapter, the open research areas and future research directions in data aggregation problem are presented.

11.1   Introduction

Wireless sensor networks that are composed of hundreds or thousands of inexpensive, low-powered sensing devices with limited computational and communication resources are quickly gaining popularity [1]. These networks offer potentially low-cost solutions to an array of problems in both military and civilian applications, including battle-field surveillance, target tracking, environmental and health care monitoring, wildfire detection, and traffic regulation. The most straightforward application of wireless sensor network technology is to monitor remote environments by collecting data at low frequencies. For example, a border region could be easily monitored using thousands of sensor nodes. In this example, sensor nodes automatically form a wireless interconnection network and immediately report the detection of any illegal border crossing to a central base station. Unlike traditional wired networks, deployment cost of wireless sensor networks should be very low. To achieve this low-cost deployment requirement, sensor nodes have simple hardware and severe resource constraints in terms of power, processing capability, memory, and storage. Due to such constraints, it is a challenging task to provide efficient solutions to data gathering problem. Among these constraints, power is the most limiting factor in designing wireless sensor network protocols. Hence, to reduce the average power consumption of sensor nodes, several mechanisms are proposed such as radio scheduling, control packet elimination, topology control algorithms, and data aggregation [2]. In this chapter, we focus on the data aggregation techniques which aim to combine and summarize data packets from several sensor nodes by looking at their contents so that energy efficiency is achieved. An example data aggregation scheme is shown in Figure 11.1 where a group of sensor nodes report temperature measurements from the target region. When the base station queries the network to obtain the average temperature of the target region, instead of sending each sensor node's data to base station, one of the sensor nodes, called data aggregator, collects the temperature readings from the sensor nodes, aggregates them (i.e., computes the average), and sends the computed average temperature value to the base station. As illustrated by the example, data aggregation reduces the number of data transmissions thereby improving the bandwidth and energy utilization in the network.

In wireless sensor networks, the benefit of data aggregation increases if the intermediate sensor nodes perform data aggregation incrementally when data is being forwarded to the base station. However, while this continuous data aggregation operation improves

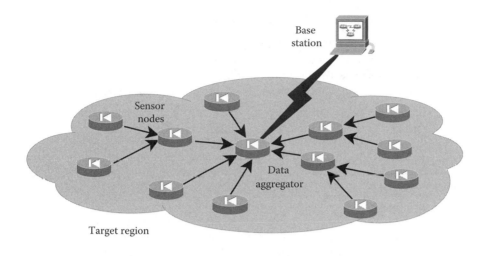

Figure 11.1   Data aggregation in a wireless sensor network.

the bandwidth and energy utilization, it may negatively affect other performance metrics such as delay, accuracy, fault-tolerance, and security [2]. Among these metrics, security is the most important one because the majority of wireless sensor network applications require a certain level of network security. In addition, there is a strong conflict between security and data aggregation protocols. Security protocols require sensor nodes to encrypt and authenticate any sensed data prior to its transmission and prefer data to be decrypted by the base station [28,30]. On the other hand, data aggregation protocols tend to implement data aggregation at every intermediate node so that energy efficiency is maximized. Due to these two conflicting goals, data aggregation and security protocols must be designed together during the system design to allow data aggregation without sacrificing security. The necessity of implementing data aggregation and security have led many researchers to work on *secure data aggregation* problem and hence it deserves a special attention.

In this chapter, we give a broad overview of data aggregation paradigm in wireless sensor networks. The organization of the chapter as follows. We start with a brief summary on wireless sensor networks to give the motivation behind data aggregation paradigm. Then, the current state of the art in data aggregation research is presented in two subsections based on the network topology. Then, we investigate the interaction between data aggregation and security protocols as they both are essential for wireless sensor networks. Secure data aggregation protocols are also presented in two parts. First, we present the solutions in which sensor data is decrypted before data aggregation and special mechanisms are used to secure the aggregation of that plain data. Then, in the second part, secure data aggregation protocols that can perform data aggregation over encrypted data is explained. We conclude the chapter by providing the open research areas and future research directions in data aggregation problem.

11.2 Wireless Sensor Network Overview

A wireless sensor network consists of a large number of sensor nodes spreading across a geographical area. Each sensor node has limited wireless communication capability and significantly low-level computation capability for signal processing and networking of the data. Sensor nodes are powered by small batteries which are assumed to be nonchangeable. Therefore, sensor nodes are also constrained in terms of energy source. The aforementioned resource constrains of sensor nodes and the large scale of wireless sensor nodes introduce a new set of research issues and challenges that previous research did not need to address. One such research issue is data aggregation which aims to reduce the amount of data transmission. Before giving the details of data aggregation paradigm, in what follows, we summarize the unique characteristics of wireless sensor networks and present how they are in relation with data aggregation protocols.

Large scale: Typical application areas of wireless sensor networks (e.g., battlefields, habitat monitoring) require a large geographic coverage. At the same time, a high node density is required to work against the high failure rate of sensor nodes, the low confidence in individual sensor readings, and the limited communication range of sensor nodes. Certain applications require *k-coverage* schemes in which events are detected by at least *k* sensor nodes. Due to such reasons, wireless sensor networks are expected to scale up to thousands of nodes. Data produced by such large number of sensor nodes must be summarized before being transmitted to base station so that energy consumption due to data transmission is reduced. Data aggregation techniques are shown to be a very effective way to summarize the collected information without requiring all the pieces of data.

Constrained resource: Because of the low-cost deployment requirement of wireless sensor networks, sensor nodes have a simple hardware which severely limits the processing and communication ability of sensor networks. For example, one common sensor type (TelosB) has a 16 bit, 8 MHz RISC CPU with only 10K RAM, 48K program memory, and 1024K flash storage [3]. In addition, once the network is deployed, the batteries of sensor nodes cannot be easily replaced or recharged. Hence, the lifetime of the entire sensor network depends on battery charge of sensor nodes. Data aggregation techniques significantly improve the resource utilization of sensor nodes by reducing the amount of data transmission in the network.

Redundancy: The highly unpredictable nature of wireless sensor networks and short communication range of sensor nodes necessitate a high node redundancy. Sensor nodes are normally deployed with a high degree of connectivity to cope with sensor node failures. With such redundancy, the failure of a single node has a negligible impact on overall capacity of the sensor networks. In addition, redundancy is required to support quality of service and reliability because data of a single sensor node may be misleading. Hence, target region must be covered by sensor nodes that have overlapping sensing ranges. Then, the data of a sensor node can be calibrated by the data of other sensor nodes that sense the same event. However, such node redundancy increases the amount of data to be transmitted from sensor nodes to base station which greatly reduces the network lifetime. Therefore, high data redundancy in wireless sensor networks must be eliminated by data aggregation protocols.

Security sensitive: Many wireless sensor network applications, such as surveillance, military tracking, or biomedicine, are highly security sensitive. Due to constrained resources, it is not possible to deal with all possible security issues, yet wireless sensor networks are vulnerable to node capture attack. The node capturing attack does not exist in traditional networks; hence security solutions of traditional ad hoc networks cannot be employed in wireless sensor networks. From the data aggregation point of view, security is a big obstacle for design and development of efficient data aggregation protocols. Security protocols require sensor nodes to encrypt the sensor data prior to its transmission and prefer data to be decrypted by the base station. However, data aggregation protocols prefer data to be available at every intermediate node to implement data aggregation. Therefore, security protocol designers for wireless sensor networks must consider the limitations of sensor nodes, node capturing attacks, and data aggregation.

Data centric processing: Data centric processing is an intrinsic characteristic of wireless sensor networks. The IDs of the sensor nodes are of no interest to the applications; therefore naming schemes in sensor networks are usually data oriented. For example, an environmental monitoring system requests the temperature readings through queries such as "collect temperature readings in the region bound by the rectangle (x_1, y_1, x_2, y_2)," instead of queries such as "collect temperature readings from a set of nodes with the sensor node IDs x, y, and z." Such data centric processing provides an excellent environment for data aggregation protocols. Data centric processing property of wireless sensor networks are in favor of data aggregation process. Because IDs of the sensor nodes that transmit the data is not important for the base station, data aggregation protocols can combine and compress the collected data easily.

Real-time constraints: Because wireless sensor nodes deal with the real world processes, it is often necessary for communication to meet real-time constraints. In border surveillance systems, for example, communication delays within sensing and actuating loops directly affect the quality of target tracking. On the other hand, sensor node failures are common due to the large number of sensor nodes, low-cost sensor hardware, climate conditions, and hostile environment. Also, the wireless medium shared by densely deployed sensor nodes is subject to heavy congestion and jamming. Moreover, high bit error ratio, low bandwidth and asymmetric channel make the communication highly unpredictable. Due to the nature of the wireless communication and unpredictable traffic pattern, it is infeasible to guarantee hard real-time constrains. Data aggregation is another factor that affects the real-time constraints negatively because data processing adds more delay to transmission time. Hence, the data aggregation protocol designer must trade-off between energy efficiency and real-time constraints. For example, in applications where human life is at stake, data aggregation may not be employed at all to meet the real-time constraint of the application.

11.3 Data Aggregation

In a typical wireless sensor network, a large number of sensor nodes collect application-specific information from the environment and this information is transferred to a central base station where it is processed, analyzed, and used by the application. In these resource-constrained networks, general approach is to jointly process the data generated by different sensor nodes while being forwarded toward the base station. Such in-network distributed processing of data is generally referred to as *data aggregation* and involves combining the data that belong to the same phenomenon or processing (i.e., computing average) of the sensor data. The main objective of data aggregation is to increase the network lifetime by reducing the resource consumption of sensor nodes (such as battery energy and bandwidth). While increasing network lifetime, data aggregation protocols may degrade two important quality of service metrics in wireless sensor networks: *data accuracy* and *latency*. Therefore, the design of an efficient data aggregation protocol is an inherently challenging task because the protocol designer must trade off between energy efficiency, data accuracy, and latency. To achieve this trade-off, data aggregation techniques are tightly coupled with how packets are routed through the network, hence the architecture of the sensor network plays a vital role in the performance of different data aggregation protocols. There are several protocols that allow routing and aggregation of data packets simultaneously. These protocols can be categorized into two parts: tree-based data aggregation protocols and cluster-based data aggregation protocols. Earlier work on data aggregation focused on improving the existing routing algorithms in such a way as to make data aggregation possible. As a result, many data aggregation protocols based on the shortest path tree structure have been proposed. To reduce the latency due to tree-based data aggregation, recent work on data aggregation tends to group sensor nodes into clusters so that data is aggregated in each group for improved efficiency. Other than these two groups, there are also multipath-based data aggregation protocols [4]. In addition, there are hybrid schemes that are a mixture of tree-based and multipath solutions. Although the efficiency of multipath-based and hybrid protocols is shown to be less than tree and cluster-based data aggregation protocols, we will also give a brief overview of these schemes at the end of this section.

11.3.1 Tree-Based Data Aggregation Protocols

The simplest way to achieve distributed data aggregation is to determine some data aggregator nodes in the network and ensure that the data paths of sensor nodes include these data aggregator nodes. Such tree-based data aggregation techniques have been extensively studied in the literature [5–18]. The main issue of tree-based data aggregation protocols is the construction of an energy-efficient data aggregation tree. Figure 11.2 illustrates an example of tree-based data aggregation. Greedy Incremental Tree (GIT) [5] is a data centric routing protocol that allows data aggregation based on Directed Diffusion [6]. In Ref. [7], GIT is compared with two other data-centric routing schemes, namely Center at Nearest Source (CNS) and Shortest Path Tree (SPT). The simulation results show that GIT performs the best in terms of average number of transmissions. Another SPT-based data aggregation protocol that promotes the parent energy-awareness is proposed in Ref. [8]. In this protocol, parent selection is based on sensor nodes' distance to the base station and their residual energy level. There are also data aggregation protocols that consider information theory as routing metric. For example, in Ref. [9], a centralized approach that routes the packet based on their joint entropies is proposed. However, this

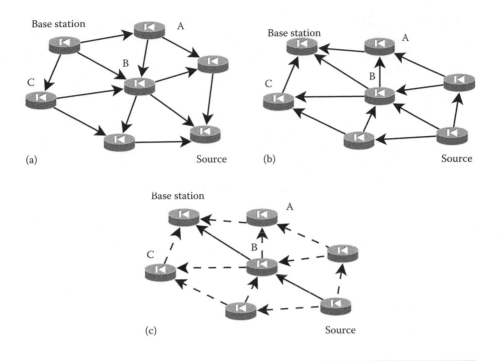

Figure 11.2   Illustrative example of Directed Diffusion. (a) Interest dissemination, (b) gradient set up, and (c) data delivery over the reinforced path.

protocol is not feasible for wireless sensor networks as it depends on the global knowledge of theinformation entropy of each sensor node as well as the joint entropy of each node pair. In the rest of this subsection, we present some of the important work in tree-based data aggregation in detail.

In Rcf. [10] Madden et al. proposed a data-centric data aggregation framework, called Tiny AGgregation (TAG) Service, which is based on SPT routing. TAG is specifically designed for monitoring applications and allows an adjustable sleep schedule for sensor nodes. To achieve this, parent nodes let their children know about the waiting time for transmission. Also, parent nodes cache their children's data to prevent from data loss. TAG performs data aggregation in two phases. In the first phase, called distribution phase, base station-oriented queries are disseminated to the sensor nodes, then in the second phase, called collection phase, aggregated sensor readings are routed up the aggregation tree. During the distribution phase, a message is broadcasted by the base station requiring sensor nodes to organize a routing tree so that the base station can send its queries. Each message has a field that specifies the level or distance from the root of the sending node (the level of the root is equal to zero). When a node that does not belong to any level receives this message, it sets its own level by incrementing the current level in the message by one and assigns the sender as its parent. This process continues until all sensor nodes in the network joins the tree and have a parent. This messaging is periodically repeated to keep the tree structure updated. The routing messages are periodically broadcast by the sink in order keep the tree structure updated. Once the tree is formed, the base station queries the network via the aggregation tree. Sensor nodes use their parents when replying base station queries. TAG employs an SQL-like language to query the network. Each query specifies the quantity that needs to be collected, aggregation function, and the sensor nodes that needs to perform the data collection.

Directed Diffusion [11] is a reactive data-centric protocol which takes places in three phases (i) interest dissemination (ii) gradient setup, and (iii) path reinforcement and forwarding. In the first phase, the base station propagates an interest message (interest dissemination) that describes the type of data that needs to be collected and the operational mode for the collection. Upon reception of this message, each sensor node broadcasts it to its neighbors. Sensor nodes also prepare interest gradients which are basically the vectors containing the next hop that has to be used to propagate the result of the query back to the base station (gradient setup). For each type of data a different gradient may be set up. At the end of gradient setup phase, for a certain type of data, only a single path is used to route packets toward the sink (path reinforcement and forwarding). An illustrative example of Directed Diffusion protocol is presented in Figure 11.3. Data aggregation is performed during data forwarding phase. The base station periodically refreshes the data gathering tree which is formed by the reinforced paths. However, this is an expensive operation and it may overcome the gain by the

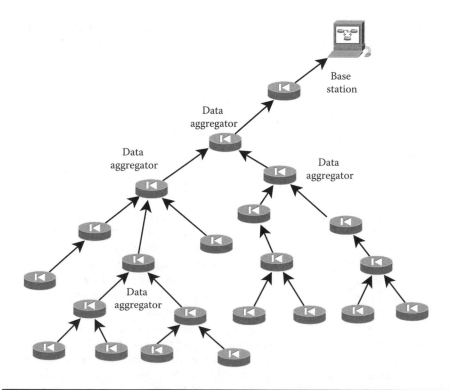

Figure 11.3   Tree-based data aggregation.

data aggregation if the network has a dynamic topology. A modified version of directed diffusion, called Enhanced Directed Diffusion (EDD), is proposed in Ref. [12] which integrates directed diffusion with a cluster-based architecture so that the efficiency of the local interactions during gradient setup phase increases. Another similar protocol is proposed in Ref. [13].

Power-Efficient GAthering in Sensor Information Systems (PEGASIS) that organizes sensor nodes in a chain for the purpose of data aggregation is proposed in Ref. [14]. Each data aggregation chain has a leader that is responsible to transmit aggregated data to the base station. To evenly distribute the energy expenditure in the network, nodes take turns to act as the chain leader. The chain forming can be achieved either in centralized manner by the base station or in a decentralized manner by using a greedy algorithm at each node. Both approaches require the global knowledge of the network. The chain building process starts from the node furthest from the base station and continues toward the base station. Positions of the nodes in a chain are shifted by one at each data transmission so that errors are minimized. In a sensor node chain, each sensor node receives data from a neighbor and aggregates it with its own reading by generating a single packet that has the same length with the received data. This process is repeated along the chain and the leader adds its own data into the packet and sends it to the base station directly. Two major drawbacks of PEGASIS have been observed. First, PEGASIS require each sensor node to have a complete view of the network topology so that chains can be formed properly. In addition, all nodes must be able to transmit directly to the base station. Second, if the distances between sensor nodes in a chain are too big, then the energy expenditure of sensor nodes can be significantly high.

A data aggregation tree construction protocol that only relies on local knowledge of the network topology is proposed in Ref. [15]. The proposed protocol, called EADAT, is based on an energy-aware distributed heuristic. The base station is the root of the aggregation tree hence it initiates the tree forming by broadcasting a control message which has the following five fields: ID, parent, power, status, and hopcnt. This message is forwarded among sensor nodes until each node broadcasts the message once and the result is an aggregation tree rooted at the base station. By considering energy level of sensor nodes, the algorithm gives higher chance to sensor nodes with higher residual power to become a nonleaf tree node so that data forwarding task is performed by the sensor nodes that have high energy level. Simulation results show that EADAT prolongs network lifetime and saves more energy in comparison with routing methods without aggregation. It is also observed that the average energy level of sensor nodes decreases much more slowly compared to the scenario without data aggregation. Therefore, EADAT can be used to construct energy-efficient data aggregation trees when it is essential to maximize the network lifetime.

There are many additional solutions that solve the problem of efficiently constructing data aggregation trees in wireless sensor networks. A different approach, called Delay Bounded Medium Access Control (DBMAC), that integrates routing and MAC protocols to perform data aggregation is proposed in Ref. [16]. The main objective of the proposed DBMAC scheme is both to minimize the latency for delay bounded applications and to increase energy efficiency by taking advantage of data aggregation mechanisms. DBMAC employs a carrier sense multiple access with collision avoidance (CSMA/CA) medium access scheme based on a request to send/clear to send/data/acknowledgment

(RTS/CTS/DATA/ACK) handshake. By taking advantage of CTS messages of other nodes, sensor nodes can select the relay node among those nodes that already have some packets to transmit in their queue. This process increases the data aggregation efficiency in the network as all the information stored along the path is aggregated into a singe data packet. DBMAC is an excellent example of how routing and data aggregation may influence each other by showing that energy-efficient data aggregation solutions are obtained by a cross-layer design. In another work [17], Dynamic Convoy Tree-Based Collaboration (DCTC) is proposed. DCTC aims to reduce the energy consumption by balancing the aggregation tree in the target region. However, DCTC incurs a heavy message exchange and assumes that sensor nodes have the knowledge of distance to the center of the event, which may not be feasible to compute with the sensed information in all applications. In Ref. [18] a power-efficient data gathering and aggregation protocol (PEDAP) is proposed to maximize the lifetime of the network in terms of number of data transmission rounds. In the proposed protocol each round corresponds to aggregation of data transmitted from different sensor nodes to the base station. PEDAP is a minimum spanning tree-based protocol and outperforms protocols like LEACH and PEGASIS when the base station is located inside the target region.

11.3.2 Cluster-Based Data Aggregation Protocols

In cluster-based data aggregation protocols, sensor nodes are subdivided into clusters. In each cluster, a cluster head is elected in order aggregate data locally and transmit the aggregation result to the base station. Cluster heads can communicate with the sink directly via long range radio transmission; however this is quite inefficient for energy-constrained sensor nodes. Therefore, cluster heads usually form a tree structure to transmit aggregated data by multihopping through other cluster heads which results in significant energy savings. Figure 11.4 presents an example of cluster-based data aggregation. Recently, several cluster-based data aggregation protocols have been proposed [19–25].

In Ref. [19], a self-organizing and adaptive clustering protocol, called Low-Energy Adaptive Clustering Hierarchy (LEACH) is proposed. LEACH takes advantage of randomization to evenly distribute the energy expenditure among the sensor nodes. LEACH is a clustered approach where cluster heads act as data aggregation points. The protocol consist of two phases. In the first phase, cluster structures are formed and then, in the second phase, cluster heads aggregate and transmit the data to the base station. LEACH's cluster head election process is based on a distributed probabilistic approach. In each data aggregator selection round, sensor nodes calculate the following threshold:

$$T(n) = \begin{cases} \frac{P}{1 - P(R_{mod}(1/P))} & \text{if } n \in G, \\ 0 & \text{otherwise} \end{cases}$$

Here

P is the desired percentage of cluster heads

R is the round number

G is the set of nodes that have not been cluster heads during the last $1/P$ rounds

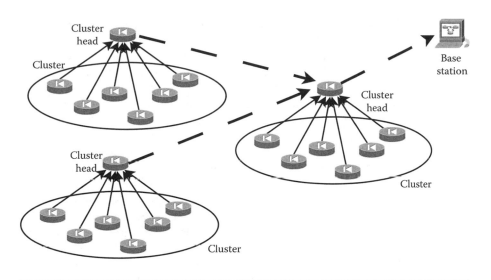

Figure 11.4   Cluster-based data aggregation.

To be a cluster head, a sensor node n picks a random random number between [0,1] and becomes a cluster head if this number is lower than $T(n)$. Cluster head advertisements are broadcasted to sensor nodes and sensor nodes join the clusters based on the signal strength of the advertisement messages. Based on the number of cluster members, each cluster head schedules its cluster based on TDMA to optimally manage the local transmissions. In the second phase, sensor nodes send their data to cluster heads according to the established schedule. Optionally, sensor nodes may turn off their radios until their scheduled TDMA transmission slot. LEACH requires cluster heads to send their aggregated data to the base station over a single link. However, this is a disadvantage of LEACH because single link transmission may be quite expensive when the base station is far away from the cluster head. LEACH is completely distributed as it does not require any global knowledge regarding network structure. It is also an adaptive protocol in terms of cluster head selection. On the other hand, there may be high control message overhead if the network topology is dynamic due to mobile nodes.

Another cluster-based data aggregation protocol, called HEED, is proposed in Ref. [20]. For cluster head selection, HEED benefits from the availability of multiple power levels at sensor nodes. In fact, a combined metric is composed of the node's residual energy and the node's proximity to its neighbors. HEED defines the average of the minimum power level required by all sensor nodes within the cluster to reach the cluster head. This is called Average Minimum Reachability Power (AMRP). AMPR is used to estimate the communication cost in each cluster. To select cluster heads, each sensor node computes its probability of becoming the cluster head as follows:

$$P_{(\text{CH})} = C \times \frac{E_{\text{residual}}}{E_{\text{max}}}$$

where the initial percentage of cluster heads is denoted by C and $E_{residual}$ and E_{max} represent the current residual and initial energy of the sensor node, respectively. Each sensor node broadcasts a cluster head message; sensor nodes select their cluster head as the node with the lowest AMRP in the set of received cluster head messages. This process recursively continues until every node is assigned to a cluster head. Similar to LEACH, in HEED, cluster heads directly communicate with the base station. Simulation results show that HEED extends the network lifetime and results in geographically balanced set of cluster heads.

In Ref. [21], a clustering scheme that performs periodic perhop data aggregation is proposed. The proposed scheme, called Cougar, is suitable for applications where sensor nodes continuously generate correlated data. Once cluster heads aggregate their cluster data, they send the local aggregated data to a gateway node. Similar to LEACH, Cougar is negatively affected from dynamic network topologies, however Cougar has a unique cluster head election procedure. Cougar selects the cluster heads based on more than one metric and allows sensor nodes to be more than one hop away from their cluster heads. This calls for routing algorithms to exchange packets within clusters. Cougar employs the Ad Hoc On Demand Distance Vector (AODV) protocol for inter cluster relaying. In Cougar, a synchronization mechanism is used to correctly aggregate the data. The cluster head is synchronized with all sensor nodes in the cluster and it does not report its aggregated data to the gateway node until all sensor nodes send their data.

Clustered Diffusion with Dynamic Data Aggregation (CLUDDA) [22] is a hybrid approach that combines clustering with diffusion mechanisms. CLUDDA includes query definitions inside interest messages which are initiated by the base station. Each interest message contains the definition of the query that describes the operations that need to be performed on the data components to generate a proper response. Interest transformation reduces the processing overhead by utilizing the existing knowledge of queries. CLUDDA combines Directed Diffusion [11] and clustering during the initial phase of interest propagation. Using clustering mechanism, it is ensured that only cluster heads that perform inter cluster communication are involved in the transmission of interest messages. As the regular sensor nodes do not transmit any data unless they are capable of servicing a request, CLUDDA conserves energy. In CLUDDA, any cluster head that has the knowledge of query definition can perform data aggregation, hence the aggregation points are dynamic. Also, each cluster head maintains a query cache to present the different data components that were aggregated to obtain the final data. Cluster heads also keep a list of the addresses of neighboring nodes from which the data messages originated. These addresses are used to propagate interest messages directly to specific nodes instead of broadcasting.

Some other cluster-based protocols for data aggregation have been proposed in the literature. Some of them are improvements of existing protocols. In Ref. [24], a cross-layer approach is adopted by integrating MAC design into data aggregation concept. A location-based clustering scheme where the sensors self-organize to form static clusters is developed in Ref. [25]. Sensor nodes send their data to cluster head along shortest paths, and in-network aggregation is performed at the intermediate nodes. Cluster heads perform aggregation and send aggregated data to the base station over multihop paths.

However, during aggregated data transmission from cluster heads to the base station no further aggregation is performed.

11.3.3 Multipath-Based Data Aggregation Protocols

Other than tree and cluster-based data aggregation protocols, there are also multipath-based data aggregation protocols [26] in which sensor nodes divide their aggregated data into several parts and send those data parts to a single parent over multiple paths. The main idea of these schemes is to send duplicate small data parts to the sinks over multiple paths to improve the robustness of the network. However, due to sending duplicate data, they incur high communication overhead compared to tree-based data aggregation protocols. Multipath-based data aggregation protocols usually employ ring topology in which sensor nodes are divided into several levels based on the distance from the base station in terms of number of hops. For example, in Synopsis Diffusion [26], data aggregation is performed over multiple paths as packets move level by level toward the base station. Both tree-based and multipath approaches have their own drawbacks such as communication overhead or link failures. Therefore, the scheme proposed in Ref. [27] aims to overcome the problems of both tree and multipath approaches by combining the best features of both schemes. The scheme is a hybrid protocol called Tributaries and Deltas. In Tributaries and Deltas protocol data aggregation structures may simultaneously run in different parts of the network. The motivation behind Tributaries and Deltas protocol is to take advantage of data aggregation trees when the packet loss rate is low and employ multipath approach in case of high packet loss rates. To achieve this goal, Tributaries and Deltas divides sensor nodes into two categories: nodes using a tree-based approach to forward packets and nodes using a multipath scheme. Data of different regions are combined using some correction rules. To summarize this section, we give a comparison of data aggregation protocols in Figure 11.5.

11.4 Secure Data Aggregation

The resource-constrained nature of sensor networks poses great challenges for security. In addition to the military applications, security is critical in premise surveillance, building monitoring, burglar alarms, and critical systems such as airports and hospitals. However,

	TAG [10]	D. Diffusion [11]	PEGASIS [14]	DBMAC [16]	LEACH [19]	COUGAR [21]	S. Diffusion [26]	T. and Deltas [27]
Aggregation method	Tree	Tree	Chain	Distributed	Cluster	Cluster	Multipath	Multipath/tree
Maintenance overhead	High	High	High	Low	Medium	Medium	Medium	Medium
Mobility support	Low	Medium	Low	High	Low	Low	High	Medium
Link failure support	Medium	Medium	Low	Medium	Low	Medium	High	High

Figure 11.5   Comparison of data aggregation protocols.

there is a strong conflict between security and data aggregation protocols. Secure communication requires sensor nodes to encrypt any sensed data prior to its transmission. In addition, it is desirable to have end-to-end security with the data decrypted only at base station to avoid security problems as much as possible. However, data aggregation protocols demand intermediate nodes to process packets to identify the redundant ones which requires data packets to be decrypted. These two conflicting goals require data aggregation algorithms to be designed together with secure communication algorithms.

Security requirements of wireless sensor networks can be satisfied using symmetric key or asymmetric key cryptography. Due to resource constraints of sensor nodes, symmetric key cryptography is preferable over asymmetric key cryptography in wireless sensor networks. Using symmetric key cryptography algorithms, confidentiality and data aggregation can be achieved together in a hop-by-hop fashion. However, in this case, data aggregators must decrypt every message they receive, aggregate the messages according to the corresponding aggregation function, and encrypt the aggregation result before forwarding it. Hence, using hop-by-hop encryption, it is not possible to achieve end-to-end confidentiality. Moreover, this scheme requires aggregators and forwarding nodes to establish secret keys with their immediate neighboring nodes. The necessity of implementing data aggregation and security using symmetric key cryptography algorithms have led many researchers to work on secure data aggregation problem [28–36]. All of these schemes but Ref. [30] and Ref. [34] require data aggregators to decrypt sensor data for data aggregation. Recently, a set of data aggregation protocol is proposed to implement data aggregation without requiring decryption of the sensor data for aggregation [37,40,43,44]. Some of these protocols employ asymmetric key cryptography primitives [38,39] that are suitable for resource-constrained sensor nodes. Downside of the data aggregation protocols that do not require decryption of data is that they are applicable to only a set of aggregation functions. In what follows, we present secure data aggregation protocols in two subsections: secure data aggregation protocols that perform data aggregation over plain data and secure data aggregation protocols that use encrypted data for aggregation.

11.4.1   Secure Data Aggregation over Plain Data

Early work on secure data aggregation is focused on aggregation of plain data. In Ref. [28], the security mechanism detects node misbehaviors such as dropping or forging messages and transmitting false data. In Ref. [29], random sampling mechanisms and interactive proofs are used to check the correctness of the aggregated data at base station. In Ref. [31], the witness nodes of data aggregators also aggregate data and compute MACs to help verify the correctness of the aggregators' data at base station. Because the data validation is performed at base station, the transmission of false data and MACs up to base station affects adversely the utilization of sensor network resources. In Ref. [32], sensor nodes use the cryptographic algorithms only when a cheating activity is detected. Topological constraints are introduced to build a secure aggregation tree (SAT) that facilitates the monitoring of data aggregators. In SAT, any child node is able to listen to the incoming data of its parent node. When the aggregated data of a data aggregator are questionable, a weighted voting scheme is employed to decide whether the data aggregator is properly behaving or cheating. In Ref. [35], a Secure Hop-by-hop Data Aggregation Protocol (SDAP) is proposed. Authors of SDAP are motivated by the fact that, compared to low-level sensor nodes, more trust is placed on the high-level nodes (i.e., nodes closer to the root) during a normal hop-by-hop aggregation process in a tree topology. This is because aggregated data calculated by a high-level node represents the data of a large number of low-level sensor nodes. If a compromised node is closer to the base station, the false aggregated data produced by this compromised node will have a larger impact on the final result computed by the base station. Because all sensor nodes have simple hardware that is prone to compromise none of those low-cost sensor nodes should be more trustable than others. Hence, SDAP aims to reduce the approach of reducing the trust on high-level nodes by following the divide-and-conquer principle. SDAP dynamically partitions the topology tree into multiple logical groups (subtrees) of similar sizes using a probabilistic approach. In this way, fewer nodes are located under a high-level sensor node in a logical subtree resulting in reduced potential security threat by a compromised high-level node. An example of a grouped tree is shown in Figure 11.6.

In Ref. [36], authors argue that cryptographic primitives alone cannot provide a sufficient enough solution to secure data aggregation problem as compromised nodes have access to cryptographic keys that are used to secure the aggregation process. Based on this observation, authors propose a *S*ecure and r*EL*iable *D*ata *A*ggregation protocol, called SELDA. Protocol SELDA makes use of a web of trust to overcome the shortcomings of

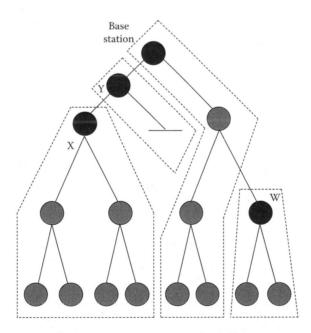

Figure 11.6   An example of the aggregation tree in SDAP. The nodes X, Y, and W with the color black are leader nodes, and the base station as the root is a default leader.

cryptography-based secure data aggregation solutions and it is based on trustworthiness of sensor nodes and data aggregators. As Beta family of distribution functions [41] can be used to predict the posteriori probabilities of sensor node actions [42], in protocol SELDA trustworthiness of sensor nodes is evaluated via Beta distribution function of sensor node misbehaviors. The basic idea behind protocol SELDA is that sensor nodes observe actions of their neighboring nodes to develop trust levels (trustworthiness) for the environment and the neighboring nodes. Sensor nodes employ monitoring mechanisms to detect node availability, sensing and routing, and misbehaviors of their neighbors. These misbehaviors are quantified as trust levels using Beta distribution function. Figure 11.7 illustrates sensor node misbehavior detection in protocol SELDA. Sensor nodes exchange their trust levels with neighboring nodes to form a web of trust that allows them to determine secure and reliable paths to data aggregators. Based on these trust levels, sensor nodes transmit their data to data aggregator(s) over one or more secure paths. During data aggregation, data aggregators weight the data used based on the trust levels of the sender nodes. The simulation results show that protocol SELDA increases the reliability of the aggregated data at the expense of a tolerable communication overhead.

All of the above protocols use actual sensor data for data aggregation and hence require decryption of sensor data at aggregators. In the rest of this section we present two more data aggregation protocols that do not need actual data so that security and data aggregation can be achieved together.

In Ref. [30] authors present an Energy-efficient and Secure Pattern-based Data Aggregation (ESPDA) protocol which considers both data aggregation and security concepts together in cluster-based wireless sensor networks. ESPDA is the first study to consider data aggregation techniques without compromising security and it uses pattern codes to perform data aggregation. The pattern codes are basically representative data

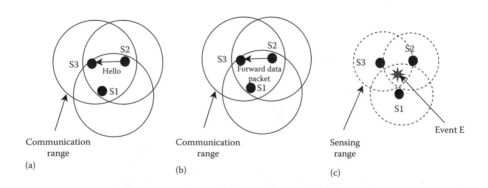

Figure 11.7 **(a) S1 detects node availability misbehavior of S3, if S3 does not responds S2's Hello messages over a period of time. (b) S1 detects routing misbehavior of S3, if S3 does not forward S2's data packets properly. (c) Event E is detected by S1, S2, and S3, if event E is reported falsely by any one of these nodes, the sensing misbehavior of that is detected by the other two nodes.**

items that are extracted from the actual data in such a way that every pattern code has certain characteristics of the corresponding actual data. The extraction process may vary depending on the type of the actual data. For example, when the actual data image of human beings sensed by the surveillance sensors, the key parameter values for the face and body recognition are considered as the representative data depending on the application requirements. When a sensor node consists of multiple sensing units, the pattern codes of the sensor node are obtained by combining the pattern codes of the individual sensing units. Instead of transmitting the whole sensed data, sensor nodes first generate and then send the pattern codes to cluster heads. Cluster heads determine the distinct pattern codes and then request only one sensor node to send the actual data for each distinct pattern code. This approach makes ESPDA both energy and bandwidth efficient. ESPDA is also secure because cluster heads do not need to decrypt the data for data aggregation and no encryption/decryption key is broadcast. Additionally, the proposed nonblocking OVSF block hopping technique further improves the security of ESPDA by randomly changing the mapping of data blocks to NOVSF time slots. ESPDA assumes that sensor nodes are usually deployed in high density in order cope with node failures due to harsh environments, yielding highly overlapping sensing regions of the nodes. Random deployment of the network also results in many areas to be covered by more than one sensor node. Therefore, it is highly desirable to ensure that an area is covered by only one sensor node at any time, so that no more than one sensor node senses the same data. This leads to an improvement for the efficiency of data aggregation because the redundant data is not even sensed.

In ESPDA, sensor nodes are deployed randomly over a target area to be monitored and organize themselves into clusters after the initial deployment. A cluster head is chosen from each cluster to handle the communication between the cluster nodes and the base station. Cluster heads are changed dynamically based on residual energy in order have uniform power consumption among all sensor nodes. Because data transmission is a major cause of energy consumption, ESPDA first reduces transmission of redundant data from sensor nodes to cluster heads with the help of sleep-active mode coordination protocol. Then, data aggregation is used to eliminate redundancy and to minimize the number of transmissions for saving energy. In conventional data aggregation methods, cluster heads receive all the data from sensor nodes and then eliminate the redundancy by checking the contents of the sensor data. ESPDA uses pattern codes instead of sensed data to perform data aggregation; therefore, the contents of the transmitted data do not have to be disclosed at the cluster-heads. This enables ESPDA to work in conjunction with the security protocol. In security protocol, sensor data, which is identified as nonredundant by the cluster heads, is transmitted to the base station in encrypted form. The pattern codes are generated using a secret pattern seed broadcast by the cluster head periodically. Pattern seed is a random number used for improving the confidentiality of the pattern codes by not allowing the same pattern codes to be produced all the time. As the pattern seed is changed, pattern generation algorithm produces a different pattern code for the same sensor data. Thus, redundancy is eliminated even before the sensor data is transmitted from the sensor nodes.

In another work [34] that incorporates both data aggregation and security concepts together in cluster-based wireless sensor networks Secure Reference-Based Data Aggregation (SRDA) protocol is proposed. Similar to ESPDA, SRDA also realizes the fact that data aggregation protocols should work in conjunction with the data communication security protocols, as any conflict between these protocols might create loopholes in network security. In SRDA, raw data sensed by sensor nodes are compared with reference data values and then only the difference data is transmitted. Reference data is taken as the average value of a number of previous sensor readings. The motivation behind SRDA is that it is critical to reduce the number of bits in transmission due to the fact that radio is a major energy consumer in a sensor node. Although data aggregation reduces the number of packets, decreasing the *size* of the transmitted packets will further improve the energy savings. In conventional data aggregation algorithms, sensor nodes transmit their raw sensed data to the cluster heads. This causes waste of energy and bandwidth because a certain range of the data may remain the same in each packet. However, SRDA transmits the differential data rather than the raw sensed data. That is, the raw data sensed by sensor nodes are compared with reference data and then only the difference data is transmitted. As an example, let $102\,°F$ denote the temperature measurement of a sensor node. If $100\,°F$ is considered as reference temperature by the cluster head, the sensor node can send only the difference (i.e., $2\,°F$) of the current measurement from the reference value in the transmission. Consequently, differential aggregation has great potential to reduce the amount of data to be transmitted from sensor nodes to cluster heads. The basic motivation behind differential data aggregation is that significant changes in sensor measurements occur only when an important event (e.g., a fire event for a temperature network) happens in the environment. In general, these so-called important events occur much less frequently than ordinary events in sensor networks.

SRDA is implemented in every session of data transmission, where session refers to the time interval from the moment the communication is established between a sensor node and the cluster head until the communication terminates. Each session is expected to have a large number of packets. SRDA is independent of the clustering scheme and can be applied on top of any clustering algorithm. Both receiving and sending nodes benefit from this technique because reception also has significant energy consumption compared to transmission for low range wireless sensor nodes. Reference-based aggregation can be applied at all levels of the clustering hierarchy. The efficiency of this technique is greater when the reference value is larger compared to the differential value due to the fact that this raw data is not transmitted in packets. Another important factor for the performance of this technique is the variance of the value of successive packet contents, as the variance gets smaller the gains achieved by differential data aggregation increases because smaller number of bits are needed to represent the differential data values.

11.4.2 Secure Data Aggregation over Encrypted Data

Using traditional symmetric key cryptography algorithms, end-to-end confidentiality and data aggregation cannot be achieved together. If the application of symmetric key-based cryptography algorithms is combined with the requirement of efficient data aggregation, then the messages must be encrypted hop-by-hop. But, this means that, in order to perform data aggregation, intermediate nodes have to decrypt each received message, then aggregate the messages according to the corresponding aggregation function, and finally encrypt the aggregation result before forwarding it. In addition, this process requires neighboring data aggregators to share secret keys for decryption and encryption. To achieve end-to-end data confidentiality and data aggregation together without requiring secret key sharing among data aggregators privacy homomorphic cryptography has been used in the literature [37,40,43,44].

A privacy homomorphism is an encryption transformation that allows direct computation on encrypted data. Let E denotes encryption and D denotes decryption. Also let $+$ denotes addition and $\times$ denotes multiplication operation over a data set Q. Assume that K_{pr} and K_{pu} are the private and public keys of the base station, respectively. An encryption transformation is accepted to be additively homomorphic, if

$$a + b = D_{K_{pr}}(E_{K_{pu}}(a) + E_{K_{pu}}(b)) \quad \text{where} \ a, b \in Q.$$

and it is accepted to be multiplicatively homomorphic, if

$$a + b = D_{K_{pr}}(E_{K_{pu}}(a) \times E_{K_{pu}}(b)) \quad \text{where} \ a, b \in Q.$$

Because, additively and multiplicatively homomorphic cryptographic functions support additive and multiplicative operations on encrypted data, respectively, data aggregators can perform addition and multiplication-based data aggregation over the encrypted data.

In CDA [37], sensor nodes share a common symmetric key with the base station that is kept hidden from middle way data aggregators. The major contribution of this work is the provision of end-to-end encryption for reverse multicast traffic between the sensor nodes and the base station. In the proposed approach, data aggregators carry out aggregation functions that are applied to ciphertexts (encrypted data). This provides the advantage that intermediate aggregators do not have to carry out costly decryption and encryption operations. Therefore, data aggregators do not have to store sensitive cryptographic keys which ensures an unrestricted data aggregator node election process for each epoch during the wireless sensor network's lifetime. Unrestricted data aggregator selection is impossible for hop-by-hop encryption because only the nodes which have stored the key can act as a data aggregator. As the privacy homomorphic encryption function, the proposed protocol employs the function proposed by Domingo-Ferrer [38]. Domingo-Ferrer's encryption function is probabilistic in the sense that the encryption transformation involves some randomness that chooses the ciphertext corresponding to a given plaintext from a set of possible ciphertexts.

The public parameters of Domingo-Ferrer's encryption function are a positive integer $d \geq 2$ and a large integer g that must have many small divisors. In addition, there should be many integers less than g that can be inverted modulo g. The secret key is computed as $k = (r, g')$. The value $r \in \mathbb{Z}_g$ is chosen such that $r^{-1} \bmod g$ exists where $\log_{g'} g$ indicates the security level provided by the function. The set of plaintext is $\mathbb{Z}_{g'}$ and the set of ciphertext is $(\mathbb{Z}_g)^d$. Encryption and decryption processes are defined as follows:

Encryption: Randomly split $a \in \mathbb{Z}_{g'}$ into a secret $a_1 \ldots a_d$ such that $\sum_{j=1}^{d}(a_j \bmod g')$ and $a \in \mathbb{Z}_{g'}$. Compute

$$E_k(a) = (a_1 r^1 \bmod g, a_2 r^2 \bmod g, \ldots, a_d r^d \bmod g)$$

Decryption: Compute the jth coordinate by r^{-j} mod g to retrieve a_j mod g. To obtain a compute

$$D_k(E_k(a)) = \sum_{j=1}^{d}(a_j \bmod g')$$

The ciphertext operation $\times$ is performed by cross multiplying all terms in Z_g, with the d_1-degree term by a d_2-degree term yielding a t-degree term. Then, the terms having the same degree are added up. The ciphertext operation $+$ is relatively easy compared to $\times$ operation and is performed componentwise.

As it is seen from the above definitions, Domingo-Ferrer's asymmetric key-based privacy homomorphism is computationally expensive for resource-constrained sensor nodes. Authors of Ref. [37] compared the clock cycles required by asymmetric key-based privacy homomorphism and symmetric key-based encryption solutions. The results show that encryption, decryption, and addition that are necessary to perform Domingo-Ferrer's function are by far more expensive compared to those necessary to perform symmetric key-based RC5. However, authors argue that this disadvantage is acceptable as CDA advantageously balance the energy consumption. Using symmetric key-based encryption solutions to perform hop-by-hop data aggregation results in shorter lifetime for data aggregator nodes. Therefore, as data aggregators are the performance bottlenecks when maintaining a connected wireless sensor network backbone, it is preferable to employ CDA's asymmetric key-based privacy homomorphism to balance the energy consumption of data aggregators.

In Ref. [43], a secure data aggregation protocol, called CDAP, takes advantage of asymmetric key-based privacy homomorphic cryptography to achieve end-to-end data confidentiality and data aggregation together. Authors point out that asymmetric cryptography based privacy homomorphism incurs high computational overhead; encryption and aggregation cost of network cannot be afforded by regular sensor nodes with scarce resources. Therefore, CDAP protocol employs a set of resource-rich sensor nodes, called aggregator nodes (AGGNODEs), for privacy homomorphic encryption and aggregation of the encrypted data. In CDAP, after the network deployment each AGGNODE establishes pairwise keys with its neighboring nodes so that neighboring nodes can send their sensor readings to the AGGNODE securely. In data collection phase of protocol CDAP, each AGGNODE queries its neighboring nodes for sensor readings, such as temperature or humidity. Each neighboring node encrypts its data using a symmetric key encryption algorithm (RC5) and sends it to the AGGNODE. The AGGNODE decrypts all data received from its neighbors, aggregates them, and encrypts the aggregated data using the privacy homomorphic encryption algorithm. Once the data is encrypted with the privacy homomorphic encryption algorithm, only the base station can decrypt them using its private key. However, due to homomorphic property, intermediate AGGNODEs can aggregate those encrypted data even though they do not have the private key of the base station. Therefore, the data collected by the sensor nodes are aggregated by AGGNODEs as they travel toward the base station. The base station decrypts the final aggregated data using its private key. An illustrative example of CDAP is given in Figure 11.8. Due to the computational overhead of privacy homomorphic encryption algorithms, in CDAP, only AGGNODEs are allowed to encrypt and aggregate the collected data using privacy

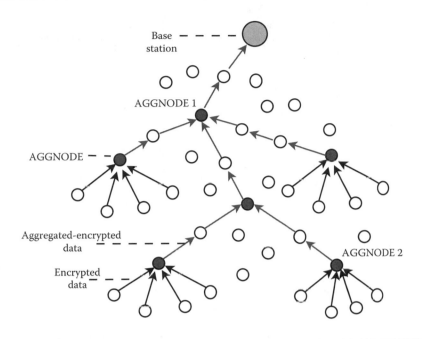

Figure 11.8 **The aggregation scenario of protocol CDAP. AGGNODEs collect information from their neighborhood and encrypted data is aggregated at AGGN-ODEs while data travels toward the base station.**

homomorphic algorithms. Therefore, in the first data collection phase of the protocol CDAP, sensor nodes use symmetric key algorithms for encryption. Due to the symmetric encryption, a compromised AGGNODE may disclose the secrecy of its neighboring nodes' data or inject false data into the data. However, authors argue that the effect of this attack is local and, hence, it can be tolerated.

In Ref. [44] a simple and provably secure additively homomorphic stream cipher that allows efficient aggregation of encrypted data is proposed. The proposed technique is based on an extension of the one-time pad encryption technique using additive operations over modulo n. The main idea of the proposed scheme is to replace the Exclusive OR operation of stream ciphers with modular addition $(+)$. The encryption and decryption processes can be summarized as follows. Represent message m as integer $m \in [0, M-1]$ where M is a large integer. Also, let k be a randomly generated key stream, where $k \in [0, M-1]$. Then, ciphertext c is computed as $c = \text{enc}(m, k, M) = m + k(\text{mod} M)$. To decrypt ciphertext c, perform $\text{Dec}(c, k, M) = c - k(\text{mod } M)$. Based on these functions, ciphertexts are added as follows: Let $c_1 = \text{Enc}(m_1, k_1 M)$ and $c_2 = \text{Enc}(m_2, k_2, M)$, then for $k = k_1 + k_2$, $\text{Dec}(c_1 + c_2, k, M) = m_1 + m_2$. It is assumed that the message m is $0 \leq m < M$ and because addition possesses the commutative property, the proposed scheme is additively homomorphic. The proposed scheme significantly reduces the energy consumption of sensor nodes due to encryption process. However, in the proposed scheme, each aggregate message is coupled with the list of nodes that failed to contribute

	Data Confidentiality	Data Integrity	Source Authentication	Availability
Hu et al. [28]		●	●	
SIA [29]	●	●	●	
ESPDA [30]	●	●	●	
Du et al. [31]		●	●	
Wu et al. [32]		●	●	
SRDA [34]	●	●	●	
SDAP [35]	●	●	●	
SELDA [36]		●	●	●
CDA [37]	●			
Ozdemir [40]	●			
CDAP [43]	●			
Castellucia et al. [44]	●			

Figure 11.9   Comparison of secure data aggregation protocols.

to the aggregation because of node or communication failures. In Ref. [40], this problem has been solved by adapting a hierarchical data aggregation model. To summarize this section, we give a comparison of secure data aggregation protocols in Figure 11.9.

11.5   Open Research Issues and Future Research Directions

In this chapter, a comprehensive overview of data aggregation concept in wireless sensor networks was given. We presented state-of-the-art data aggregation protocols and categorized them based on network topology and security. Although the presented research addresses the many problems of data aggregation, there are still many research areas that needs to be associated with the data aggregation process.

The relation between routing mechanisms and data aggregation protocols have been well studied as they are correlated topics. In addition to diffusion and tree-based data aggregation protocols, many cluster-based data aggregation protocols that route aggregated data over cluster heads have been proposed. Although, these protocols are shown to be very efficient in static networks in which the cluster structures do not change for a sufficiently long time, in dynamic networks they perform quite poorly. Hence, data aggregation in dynamic environments is a possible future research direction. In addition, the application of source coding theory for data aggregation has drawn little attention so far. Considering that sensor data is highly correlated, data aggregation can be achieved by employing source coding techniques. Existing research in this area focuses on only theoretical results and there are no practical algorithms applicable to wireless sensor networks yet. Therefore, there is significant scope for future work in source coding-based data aggregation. Security is another important issue in data aggregation protocols. Although many protocols have been proposed in this context, there are still unanswered problems such as compromised data aggregators that inject false data during data aggregation. Because data aggregation usually results in alterations in data, false data injected by compromised data aggregators are hard to detect. Detecting false data injected by a compromised data aggregator is an interesting problem for future research. The impact of sensor node heterogeneity over the data aggregation protocols is another unexplored research area. The protocols that use powerful sensor nodes as data aggregators presented promising results. However, determining locations of these powerful nodes for the best data aggregation results needs further research.

11.6   Conclusion

In this chapter, we have provided a detailed review of data aggregation concept in wireless sensor networks. To give the motivation behind data aggregation, we first explained how the unique properties of wireless sensor networks, such as resource constraints and data centric processing, are in relation with data aggregation concept. Then, we have given the state of the art in data aggregation protocols. The trade-off between security and data aggregation is examined in detail and secure data aggregation protocols are explained rigorously. We have concluded the chapter by providing the open research issues and future research directions in data aggregation concept.

References

1. I.F. Akyildiz, W. Su, Y. Sankarasubramaniam, and E. Cayirci, A survey on sensor networks, *IEEE Communications Magazine*, 40(8), 102–114, Aug. 2002.
2. K. Akkaya, M. Demirbas, and R. S. Aygun, The impact of data aggregation on the performance of wireless sensor networks, *Wiley Wireless Communications and Mobile Computing (WCMC) Journal*, 8, 171–193, 2008.
3. Crossbow Technologies, Inc., http://www.xbow.com
4. E. Fasolo, M. Rossi, J. Widmer, and M. Zorzi, In-network aggregation techniques for wireless sensor networks: a survey, *IEEE Wireless Communications*, 14(2), 70–87 Apr. 2007.
5. C. Intanagonwiwat, D. Estrin, R. Govindan, and J. Heidemann, Impact of network density on data aggregation in wireless sensor networks, in *Proc. of 22nd International Conference on Distributed Computing Systems*, pp. 457–458, Vienna, Austria, Jul. 2002.
6. C. Intanagonwiwat, R. Govindan, D. Estrin, J. Heidemann, and F. Silva, Directed diffusion for wireless sensor networking, *IEEE/ACM Transactions on Networking*, 11(1), 2–16, Feb. 2003.
7. B. Krishnamachari, D. Estrin, and S. Wicker, The impact of data aggregation in wireless sensor networks, in *Proc. of 22nd International Conference on Distributed Computing Systems Workshops*, pp. 575–578, Vienna, Austria, Jul. 2002.

8. M. Ding, X. Cheng, and G. Xue, Aggregation tree construction in sensor networks, in *Proc. of the 58th IEEE Vehicular Technology Conference*, vol. 4, pp. 2168–2172, Oct. 2003.

9. R. Cristescu, B. Beferull-Lozano, and M. Vetterli, On network correlated data gathering, in *Proc. of the 23rd Annual Joint Conference of the IEEE Computer and Communications Societies*, vol. 4, pp. 2571–2582, Mar. 2004.

10. S. Madden et al., TAG: A tiny aggregation service for ad hoc sensor networks, in *OSDI 2002*, Boston, MA, Dec. 2002.

11. C. Intanagonwiwat et al., Directed diffusion for wireless sensor networking, *IEEE/ACM Trans. Netw.*, 11(1), 2–16, Feb. 2002.

12. B. Zhou et al., A hierarchical scheme for data aggregation in sensor network, in *IEEE ICON 04*, Singapore, Nov. 2004.

13. M. Lee and V. W. S. Wong, An energy-aware spanning tree algorithm for data aggregation in wireless sensor networks, in *IEEE PacRrim 2005*, Victoria, British Columbia, Canada, Aug. 2005.

14. S. Lindsey, C. Raghavendra, and K. M. Sivalingam, Data gathering algorithms in sensor networks using energy metrics, *IEEE Trans. Parallel Distrib. Sys.*, 13(9), 924–935, Sept. 2002.

15. M. Ding, X. Cheng, and G. Xue, Aggregation tree construction in sensor networks, in *IEEE VTC 03*, Orlando, FL, vol. 4, pp. 2168–2172, Oct. 2003.

16. G. Di Bacco, T. Melodia, and F. Cuomo, A MAC protocol for delay-bounded applications in wireless sensor networks, in *Med-Hoc-Net 2004*, Bodrum, Turkey, June 2004.

17. W. Zhang and G. Cao, DCTC: Dynamic convoy tree-based collaboration for target tracking in sensor networks, *IEEE Transactions on Wireless Communications*, 3(5), 1689–1701, Sep. 2004.

18. H. O. Tan and I. Korpeoglu, Power efficient data gathering and aggregation in wireless sensor networks, *SIGMOD Record*, 32(4), 66–71, Dec. 2003.

19. W. B. Heinzelman, A. P. Chandrakasan, and H. Balakrishnan, An application-specific protocol architecture for wireless microsensor networks, *IEEE Trans. Wireless Commun.*, 1(4), 660–670, Oct. 2002.

20. O. Younis and S. Fahmy, HEED: A hybrid, energy-efficient, distributed clustering approach for ad hoc sensor networks, *IEEE Transactions on Mobile Computing*, 3(4), 366–379, Dec. 2004.

21. Y. Yao and J. Gehrke, The Cougar approach to in-network query processing in sensor networks, *ACM SIGMOD Record*, 31(3), 9–18, Sept. 2002.

22. S. Chatterjea and P. Havinga, A dynamic data aggregation scheme for wireless sensor networks, in *Proc. Program for Research on Integrated Systems and Circuits*, Veldhoven, the Netherlands, Nov. 2003.

23. V. Mhatre and C. Rosenberg, Design guidelines for wireless sensor networks: Communication, clustering and aggregation, *Elsevier Ad Hoc Networks Journal*, 2(1), 45–63, Jan. 2004.

24. P. Popovski et al., MAC-layer approach for cluster-based aggregation in sensor networks, in *IEEE IWWAN 04*, Oulu, Finland, May 2004.

25. S. Pattem, B. Krishnamachari, and R. Govindan, The impact of spatial correlation on routing with compression in wireless sensor networks, in *ACM/IEEE IPSN 04*, Berkeley, CA, Apr. 2004.

26. S. Nath, P. B. Gibbons, S. Seshan, and Z. R. Anderson, Synopsis diffusion for robust aggregation in sensor networks, in *ACM SenSys 2004*, Baltimore, MD, Nov. 2004.

27. A. Manjhi, S. Nath, and P. B. Gibbons, Tributaries and deltas: Efficient and robust aggregation in sensor network stream, in *ACM SIGMOD 2005*, Baltimore, MD, June 2005.

28. L. Hu and D. Evans, Secure aggregation for wireless networks, in *Proc. of Workshop on Security and Assurance in Ad hoc Networks*, Orlando, FL, Jan. 28 2003.

29. B. Przydatek, D. Song, and A. Perrig, SIA : Secure information aggregation in sensor networks, in *Proc. of SenSys'03*, pp. 255–265, New York, 2003.

30. H. Çam, S. Ozdemir, P. Nair, and D. Muthuavinashiappan, and H.O. Sanli, Energy-efficient and secure pattern based data aggregation for wireless sensor networks, in *Special Issue of Computer Communications on Sensor Networks*, pp. 446–455, Feb. 2006.

31. W. Du, J. Deng, Y. S. Han, and P. K. Varshney, A witness-based approach for data fusion assurance in wireless sensor networks, in *Proc. of IEEE Global Telecommunications Conference (GLOBECOM '03)*, pp. 1435–1439, San Francisco, 2003.

32. K. Wu, D. Dreef, B. Sun, and Y. Xiao, Secure data aggregation without persistent cryptographic operations in wireless sensor networks, *Ad Hoc Networks*, 5(1), 100–111, 2007.

33. R. Rajagopalan and P.K. Varshney, Data aggregation techniques in sensor networks: A survey, *IEEE Communications Surveys and Tutorials*, 8(4), 4th Quarter 2006.

34. H. O. Sanli, S. Ozdemir, and H. Çam, SRDA: Secure reference-based data aggregation protocol for wireless sensor networks, in *Proc. of IEEE VTC Fall Conference*, Los Angeles, CA, 7, pp. 4650–4654, Sep. 2004.

35. Y. Yang, X. Wang, S. Zhu, and G. Cao, SDAP: A secure hop-by-hop data aggregation protocol for sensor networks, in *Proc. of ACM MOBIHOC'06*, Florence, Italy, May 2006.

36. S. Ozdemir, Secure and reliable data aggregation for wireless sensor networks, *LNCS 4836*, H. Ichikawa et al. (Eds.), pp. 102–109, 2007.

37. D. Westhoff, J. Girao, and M. Acharya, Concealed data aggregation for reverse multicast traffic in sensor networks: Encryption, key distribution and routing adaptation, *IEEE Transactions on Mobile Computing*, 5(10), 1417–1431, October 2006.

38. J. Domingo-Ferrer, A provably secure additive and multiplicative privacy homomorphism, in *Proc. Information Security Conf.*, pp. 471–483, Sao Paulo, Brazil, Oct. 2002.

39. T. Okamoto and S. Uchiyama, A new public-key cryptosystem as secure as factoring, in *Advances in Cryptology—EUROCRYPT'98*, pp. 208–318, Espoo, Finland, 1998.

40. S. Ozdemir, Secure data aggregation in wireless sensor networks via homomorphic encryption, *Journal of The Faculty of Engineering and Architecture of Gazi University*, 23(2), 365–373, 2008.

41. A. Josang and R. Ismail, The beta reputation system, in *Proc. 15th Bled Conf. Electronic Commerce*, Bled, Slovenia, 2002.
42. S. Ganeriwal and M. B. Srivastava, Reputation-based framework for high integrity sensor networks, in *Proc. of the 2nd ACM Workshop on Security of Ad Hoc and Sensor Networks*, Washington, DC, pp. 66–77, 2004.
43. S. Ozdemir, Concealed data aggregation in heterogeneous sensor networks using privacy homomorphism, in *Proc. of ICPS'07 : IEEE International Conference on Pervasive Services*, Istanbul, Turkey, pp. 165–168, 2007.
44. C. Castelluccia, E. Mykletun, and G. Tsudik, Efficient aggregation of encrypted data in wireless sensor networks, in *Proc. of Conference on Mobile and Ubiquitous Systems: Networking and Services*, pp. 109–117, Boston, MA, 2005.

Chapter 12

Clustering in Wireless Sensor Networks

Basilis Mamalis, Damianos Gavalas, Charalampos Konstantopoulos, and Grammati Pantziou

CONTENTS

The use of wireless sensor networks (WSNs) has grown enormously in the last decade, pointing out the crucial need for scalable and energy-efficient routing and data gathering and aggregation protocols in corresponding large-scale environments. Hierarchical clustering protocols (as opposed to direct single-tier communication schemes) have extensively been used toward the above directions. Moreover, they can greatly contribute to overall system scalability, lifetime, and energy efficiency. In this chapter the state of the art in corresponding hierarchical clustering approaches for large-scale WSN environments is presented. The need for clustering in WSNs is first motivated and a brief description of the implied hierarchical network pattern is given. The basic advantages, objectives, and design challenges are also briefly explored. A set of appropriate taxonomy parameters as well as a global classification scheme is then introduced. In the main body of the chapter, the most significant of the existing WSN clustering algorithms are concisely presented and commented according to the previously stated parameters and classification scheme. The chapter is concluded by stating some general remarks as well as some open research issues in the field.

12.1   Introduction

In most wireless sensor network (WSN) applications nowadays the entire network must have the ability to operate unattended in harsh environments in which pure human access and monitoring cannot be easily scheduled or efficiently managed or it is even not feasible at all [1]. Based on this critical expectation, in many significant WSN applications the sensor nodes are often deployed randomly in the area of interest by relatively uncontrolled means (i.e., dropped by a helicopter) and they form a network in an ad hoc manner [2,3]. Moreover, considering the entire area that has to be covered, the short duration of the battery energy of the sensors and the possibility of having damaged nodes during deployment, large populations of sensors are expected; it is a natural possibility that hundreds or even thousands of sensor nodes will be involved. In addition, sensors in such environments are energy constrained and their batteries usually cannot be recharged. Therefore, it is obvious that specialized energy-aware routing and data gathering protocols offering high scalability should be applied in order that network lifetime is preserved acceptably high in such environments.

Naturally, grouping sensor nodes into *clusters* has been widely adopted by the research community to satisfy the above scalability objective and generally achieve high energy efficiency and prolong network lifetime in large-scale WSN environments. The corresponding hierarchical routing and data gathering protocols imply cluster-based organization of the sensor nodes in order that data fusion and aggregation are possible, thus leading to significant energy savings. In the hierarchical network structure each cluster has a leader, which is also called the cluster head (CH) and usually performs the special tasks referred above (fusion and aggregation), and several common sensor nodes (SN) as members.

The cluster formation process eventually leads to a two-level hierarchy where the CH nodes form the higher level and the cluster-member nodes form the lower level. The sensor nodes periodically transmit their data to the corresponding CH nodes. The

CH nodes aggregate the data (thus decreasing the total number of relayed packets) and transmit them to the base station (BS) either directly or through the intermediate communication with other CH nodes. However, because the CH nodes send all the time data to higher distances than the common (member) nodes, they naturally spend energy at higher rates. A common solution in order balance the energy consumption among all the network nodes, is to periodically re-elect new CHs (thus rotating the CH role among all the nodes over time) in each cluster. A typical example of the implied hierarchical data communication within a clustered network (assuming single-hop intracluster communication and multi-hop intercluster communication) is further illustrated in Figure 12.1.

The BS is the data processing point for the data received from the sensor nodes, and where the data is accessed by the end user. It is generally considered fixed and at a far distance from the sensor nodes. The CH nodes actually act as gateways between the sensor nodes and the BS. The function of each CH, as already mentioned, is to perform common functions for all the nodes in the cluster, like aggregating the data before sending it to the BS. In some way, the CH is the sink for the cluster nodes, and the BS is the sink for the CHs. Moreover, this structure formed between the sensor nodes, the sink (CH), and the BS can be replicated as many times as it is needed, creating (if desired) multiple layers of the hierarchical WSN (multi-level cluster hierarchy).

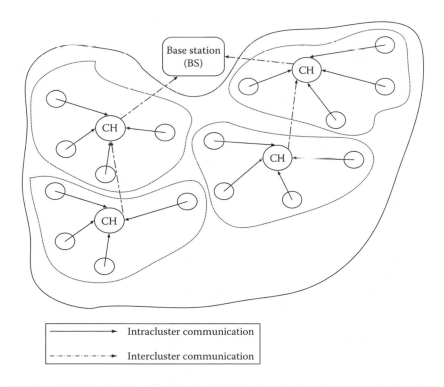

Figure 12.1   Data communication in a clustered network.

12.1.1 Main Objectives and Design Challenges of Clustering in WSNs

As was mentioned at the beginning, hierarchical clustering in WSNs can greatly contribute to overall system scalability, lifetime, and energy efficiency. Hierarchical routing is an efficient way to lower energy consumption within a cluster, performing data aggregation and fusion in order decrease the number of transmitted messages to the BS. On the contrary, a single-tier network can cause the gateway to overload with the increase in sensors density. Such overload might cause latency in communication and inadequate tracking of events. In addition, the single-tier architecture is not scalable for a larger set of sensors covering a wider area of interest because the sensors are typically not capable of long-haul communication. Hierarchical clustering is particularly useful for applications that require scalability to hundreds or thousands of nodes. Scalability in this context implies the need for load balancing and efficient resource utilization. Applications requiring efficient data aggregation (e.g., computing the maximum detected radiation around a large area) are also natural candidates for clustering. Routing protocols can also employ clustering [9,27]. In Ref. [50], clustering was also proposed as a useful tool for efficiently pinpointing object locations.

In addition to supporting network scalability and decreasing energy consumption through data aggregation, clustering has numerous other secondary advantages and corresponding objectives [1]. It can localize the route setup within the cluster and thus reduce the size of the routing table stored at the individual node. It can also conserve communication bandwidth because it limits the scope of intercluster interactions to CHs and avoids redundant exchange of messages among sensor nodes. Moreover, clustering can stabilize the network topology at the level of sensors and thus cuts on topology maintenance overhead. Sensors would care only for connecting with their CHs and would not be affected by changes at the level of inter-CH tier. The CH can also implement optimized management strategies to further enhance the network operation and prolong the battery life of the individual sensors and the network lifetime. A CH can schedule activities in the cluster so that nodes can switch to the low-power sleep mode and reduce the rate of energy consumption. Furthermore, sensors can be engaged in a round-robin order and the time for their transmission and reception can be determined so that the sensors reties are avoided, redundancy in coverage can be limited, and medium access collision is prevented.

WSNs also present several particular *challenges* in terms of design and implementation. Similar challenges and design goals have also been faced earlier in the field of mobile ad hoc networks (MANETs), and naturally a lot of related ideas (considering clustering protocols etc.) have been borrowed from that field. In WSNs, however (in which the support of mobility even if it is applicable, it is not critical), the limited capabilities (battery power, transmission range, processing hardware and memory used, etc.) of the sensor nodes combined with the special location-based conditions met (not easily accessed in order recharge the batteries or replace the entire sensors) make the energy efficiency and the scalability factors even more crucial. Moreover, the challenge of prolonging network lifetime under the above restrictions is difficult to be met by using only traditional techniques. Consequently, it becomes unavoidable to follow alternative techniques (i.e., see

Section 12.3) leading to more efficient protocols with a lot of differences compared to the ones designed for MANETs.

Beyond the typical (however vital) challenges mentioned above (limited energy, limited capabilities, network lifetime) some additional important considerations in the design process of clustering algorithms for WSNs should be the following: *Cluster formation*: The CH selection and cluster formation procedures should generate the best possible clusters (well balanced, etc.). However they should also preserve the number of exchanged messages low and the total time complexity should (if possible) remain constant and independent to the growth of the network. This yields a very challenging trade-off. *Application dependency*: When designing clustering and routing protocols for WSNs, application robustness must be of high priority and the designed protocols should be able to adapt to a variety of application requirements. *Secure communication*: As in traditional networks, the security of data is naturally of equal importance in WSNs too. The ability of a WSN clustering scheme to preserve secure communication is ever more important when considering these networks for military applications. *Synchronization*: Slotted transmission schemes such as TDMA allow nodes to regularly schedule sleep intervals to minimize energy used. Such schemes require corresponding synchronization mechanisms and the effectiveness of this mechanisms must be considered. *Data aggregation*: Because this process makes energy optimization possible it remains a fundamental design challenge in many sensor network schemes nowadays. However its effective implementation in many applications is not a straightforward procedure and has to be further optimized according to specific application requirements.

12.2   Classification of Clustering Algorithms

12.2.1   Clustering Parameters

Before documenting on the possible classification options of WSNs clustering algorithms as well as on the algorithms themselves in more details, it is worth reporting on some important parameters with regard to the whole clustering procedure in WSNs. These parameters also serve as the basic means for further comparison and categorization of the presented clustering protocols throughout this chapter.

- Number of clusters (cluster count). In most recent probabilistic and randomized clustering algorithms the CH election and formation process lead naturally to variable number of clusters. In some published approaches, however, the set of CHs are predetermined and thus the number of clusters are preset. The number of clusters is usually a critical parameter with regard to the efficiency of the total routing protocol.
- Intracluster communication. In some initial clustering approaches the communication between a sensor and its designated CH is assumed to be direct (one-hop communication). However, multi-hop intracluster communication is often (nowadays) required, that is, when the communication range of the sensor nodes is limited or the number of sensor nodes is very large and the number of CHs is bounded.

■ Nodes and CH mobility: If we assume stationary sensor nodes and stationary CHs we are normally led to stable clusters with facilitated intracluster and intercluster network management. On the contrary, if the CHs or the nodes themselves are assumed to be mobile, the cluster membership for each node should dynamically change, forcing clusters to evolve over time and probably need to be continuously maintained.

■ Nodes types and roles: In some proposed network models (i.e., heterogeneous environments) the CHs are assumed to be equipped with significantly more computation and communication resources than others. In most usual network models (i.e., homogeneous environments) all nodes have the same capabilities and just a subset of the deployed sensors are designated as CHs.

■ Cluster formation methodology: In most recent approaches, when CHs are just regular sensors nodes and time efficiency is a primary design criterion, clustering is being performed in a distributed manner without coordination. In few earlier approaches a centralized (or hybrid) approach is followed; one or more coordinator nodes are used to partition the whole network off-line and control the cluster membership.

■ Cluster-head selection: The leader nodes of the clusters (CHs) in some proposed algorithms (mainly for heterogeneous environments) can be preassigned. In most cases however (i.e., in homogeneous environments), the CHs are picked from the deployed set of nodes either in a probabilistic or completely random way or based on other more specific criteria (residual energy, connectivity, etc.).

■ Algorithm complexity: In most recent algorithms the fast termination of the executed protocol is one of the primary design goals. Thus, the time complexity or convergence rate of most cluster formation procedures proposed nowadays is constant (or just dependent on the number of CHs or the number of hops). In some earlier protocols, however, the complexity time has been allowed to depend on the total number of sensors in the network, focusing in other criteria first.

■ Multiple levels: In several published approaches the concept of a multi-level cluster hierarchy is introduced to achieve even better energy distribution and total energy consumption (instead of using only one cluster level). The improvements offered by multi-level clustering are to be further studied, especially when we have very large networks and inter-CH communication efficiency is of high importance.

■ Overlapping: Several protocols give also high importance on the concept of node overlapping within different clusters (either for better routing efficiency or for faster cluster formation protocol execution or for other reasons). Most of the known protocols, however, still try to have minimum overlap only or do not support overlapping at all.

According to the above parameters, we then try to introduce and further compare most of the algorithms presented in this chapter. This brief initial presentation is given in Table 12.1. The reader should refer to this table in combination with the global classification scheme given in the next section to gain a more clear view of the presented algorithms.

Table 12.1 Comparison of the Presented Clustering Algorithms

Clustering Approaches	Time Complexity	Node Mobility	Cluster Overlap	In-Cluster Topology	Cluster Count	Clustering Process	CHs Selection	CHs Rotation	Multi Level
LBC [5]	N/A	No	No	1-hop	Fixed	Centralized	Preset	No	No
MSNDP [6]	N/A	No	No	1-hop	Variable	Centralized	Preset	No	No
LCA [7]	Variable	Possible	No	1-hop	Variable	Distributed	ID-based	No	No
AC [9]	Variable	Yes	No	1-hop	Variable	Distributed	ID-based	No	No
DCATT [10]	N/A	No	No	1-hop	Fixed	Manual	Preset	No	No
LEACH [11]	Constant	Limited	No	1-hop	Variable	Distributed	Prob/random	Yes	No
EEHC [13]	Variable	No	No	k-hop	Variable	Distributed	Prob/random	Yes	Yes
HEED [14]	Constant	Limited	No	1-hop	Variable	Distributed	Prob/energy	Yes	No
LEACHC [12]	N/A	Limited	No	1-hop	Variable	Centralized	Prob/random	Yes	No
TLEACH [15]	Constant	Limited	No	1-hop	Variable	Distributed	Prob/random	Yes	Yes
MOCA [16]	Constant	Limited	Yes	k-hop	Variable	Distributed	Prob/random	Yes	No
TCCA [17]	Variable	No	No	k-hop	Variable	Distributed	Prob/energy	Yes	No
EECS [18]	Constant	No	No	1-hop	Constant	Distributed	Prob/energy	Yes	No
EEMC [19]	Variable	No	No	k-hop	Variable	Distributed	Prob/energy	Yes	Yes
RCC [21]	Variable	Yes	No	k-hop	Variable	Hybrid	Random	No	No

(continued)

Table 12.1 (continued) Comparison of the Presented Clustering Algorithms

Clustering Approaches	Time Complexity	Node Mobility	Cluster Overlap	In-Cluster Topology	Cluster Count	Clustering Process	CHs Selection	CHs Rotation	Multi Level
CLUBS [22]	Variable	Possible	Yes	2-hop	Variable	Distributed	Random	No	No
FLOC [23]	Constant	Possible	No	2-hop	Variable	Distributed	Random	No	No
RECA [24]	Constant	No	No	1-hop	Variable	Distributed	Random	Yes	No
HCC [27]	Variable	Possible	Yes	k-hop	Variable	Distributed	Connectivity	No	Yes
HC [28]	Variable	Possible	No	1-hop	Variable	Distributed	Connectivity	No	No
MMDC [29]	Variable	Yes	No	k-hop	Variable	Distributed	Connectivity	No	No
EEDC [30]	Variable	No	No	1-hop	Variable	Centralized	Connectivity	No	No
CAWT [31]	Constant	No	No	2-hop	Variable	Distributed	Connectivity	No	No
EACLE [32]	Variable	No	No	2-hop	Variable	Distributed	Proximity	Yes	No
ACE [33]	Constant	Possible	Yes	k-hop	Variable	Distributed	Connectivity	No	No
WCA [38]	Variable	Yes	No	1-hop	Variable	Distributed	Weight-based	No	No
DWEHC [39]	Constant	No	No	k-hop	Variable	Distributed	Weight-based	No	No
TASC [40]	Variable	No	No	2-hop	Variable	Distributed	Weight-based	No	No
GS3 [25]	Variable	Possible	Yes	k-hop	Constant	Distributed	Preset	No	No
GROUP [26]	Variable	No	No	k-hop	Controlled	Hybrid	Proximity	No	No

12.2.2 Taxonomy of Clustering Protocols

There have been several different ways (based directly on the above-mentioned parameters or not) to initially distinguish and further classify the algorithms used for WSNs clustering, [4]. Two of the most early and common classifications in the bibliography are (i) clustering algorithms for homogeneous or heterogeneous networks and (ii) centralized or distributed clustering algorithms.

The first of the above classifications is based on the characteristics and functionality of the sensors in the cluster, whereas the other one is based on the method used to form the cluster. In *heterogeneous* sensor networks (i.e., [6,10]), there are generally two types of sensors, sensors with higher processing capabilities and complex hardware (used generally to create some sort of backbone inside the WSN—being preset as the CH nodes—and also serve as data collectors and processing centers for data gathered by other sensor nodes), and common sensors, with lower capabilities, used to actually sense the desired attributes in the field. In *homogeneous* networks, all nodes have the same characteristics, hardware and processing capabilities (i.e., this is the typical case when the sensors are deployed in battle fields). In this case (which is the most usual in nowadays applications) every sensor can become a CH. Moreover, the CH role can be periodically rotated among the nodes in order achieve better load balancing and more uniform energy consumption.

Also, when all the nodes have the same capabilities (homogeneous environments), a *distributed* CH election and formation process is the most appropriate technique to gain increased flexibility and fast execution-convergence times independent of the number of nodes of the WSN. There are also a few approaches using *centralized* or *hybrid* techniques (i.e., [5,6,12]—where one or more coordinator nodes or the BS is responsible to partition the whole network off-line and control the cluster membership), however they are naturally not suitable for practical general-purpose large-scale WSNs applications (they may be suitable only for special purpose limited-scale applications where high-quality connectivity and network partitioning is required). Here we mainly focus on distributed (which are the most efficient, especially for large networks) clustering protocols for homogeneous environments (which are the most general purpose and widely used nowadays).

Another common classification is between *static* and *dynamic* clustering. A cluster formation procedure is regarded as dynamic (otherwise as static) when it includes regular (periodic or event driven) CH reelection or cluster reorganization procedures, either to effectively react to network topology changes and adjust appropriately the cluster topology, or simply aiming at the appropriate rotation of the CH role among the nodes to gain in energy efficiency. Dynamic cluster architectures make a better use of the sensors in a WSN and naturally lead to improved energy consumption management and network lifetime.

Most of the known clustering algorithms for WSNs can be further distinguished into two main categories (as presented in details in Sections 12.3 and 12.4), depending on cluster formation criteria and parameters used for CH election:

- Probabilistic (random or hybrid) clustering algorithms
- Nonprobabilistic clustering algorithms

In the category of probabilistic selection clustering algorithms [11–24], a priori probability assigned to each sensor node is used to determine the initial CHs (or some other type random election procedure is scheduled). The probabilities initially assigned to each node often serve as the primary (random) criterion in order for the nodes to decide individually on their election as CHs (in a flexible, uniform, fast and completely distributed way); however other secondary criteria may also be considered either during CH election process (i.e., the residual energy) or during the cluster formation process (i.e., the proximity or the communication cost) in order achieve better energy consumption and network lifetime. Beyond the high energy efficiency (which is facilitated also from the periodic CH reelection scheme usually adopted), the clustering algorithms of this category usually achieve faster execution/convergence times and reduced volume of exchanged messages.

In the category of nonprobabilistic clustering algorithms [25–43], more specific (deterministic) criteria for CH election and cluster formation are primarily considered, which are mainly based [25–36] on the nodes' proximity (connectivity, degree, etc.) and on the information received from other closely located nodes. The cluster formation procedure here is mainly based on the communication of nodes with their neighbors (one or multi-hop neighbors) and generally requires more intensive exchange of messages and probably graph traversing in some extent, thus leading sometimes to worse time complexity than probablistic/random clustering algorithms. On the contrary, these algorithms are usually more reliable toward the direction of extracting robust and well-balanced clusters. In addition to node proximity, some algorithms [37–40] also use a combination of metrics such as the remaining energy, transmission power, and mobility (forming corresponding combined weights) to achieve more generalized goals than single-criterion protocols. In the same category we also address a relatively new and quite challenging class of clustering algorithms for WSNs, namely, the biologically inspired protocols [41–43] (based on swarm intelligence) which are probably the most promising alternative approaches for clustering in WSNs nowadays.

Furthermore, in Section 12.5, we refer separately to a special-purpose class of clustering protocols, those that are suitable for Reactive Networks [44–49]. These protocols have clearly different objectives compared to the most common category of proactive clustering algorithms to which all the other above-mentioned protocols belong. They are specifically oriented to applications with timing restrictions and usually take advantage of user queries for the sensed data or of specific triggering events that occur in the WSN. It should also be noted that throughout this chapter the concept of sensor nodes or CHs mobility has not been considered in any particular way, as the number of applications that require mobile nodes are still considerably rare; also there is not much specialized work in the literature till now. The reader may find some relevant information and specific related work in Refs. [51,52].

Finally, before the detailed presentation of the main clustering categories introduced above, we will refer briefly to the former protocols used for clustering in WSNs (i.e., even before the last decade). The first clustering algorithms for WSNs were naturally inspired from (or entirely based on) corresponding algorithms already studied and used in the field of wired sensor networks or, later, in the field of mobile ad hoc networks.

Uniformly assigned unique identifiers were usually the key parameter for selecting CHs in those algorithms.

One of the first such clustering algorithms (initially developed for wired sensor networks) was the Linked Cluster Algorithm (LCA—[7]). LCA was a distributed ID-based, one-hop, static clustering algorithm, trying to maximize network connectivity. The main disadvantage of LCA was that usually led to excessive number of clusters. An improved LCA-based approach (generating smaller number of clusters) was given in Ref. [8] (LCA2). Both algorithms [7,8] had limited scope as clustering algorithms for WSNs because they did not consider the problem of limited energy of WSNs. Additionally, both protocols construct one-hop clusters and their time complexity is $O(n)$ which is rather unacceptable for large size WSNs. Similarly, an early example of clustering protocols initially developed for mobile ad hoc networks and then applied also to WSNs, was the adaptive clustering algorithm presented in Ref. [9]. Other classical paradigms of clustering algorithms designed initially for MANETs, were the MAX-MIN [29], HC [28], and WCA [38] algorithms; we will briefly refer to them later (due to their specific characteristics) in Section 12.4. Finally, some of the initial clustering schemes proposed for WSNs were based on some sort of manual formation of the clusters (mostly applicable to heterogeneous environments). Such a representative case can be found in Ref. [10] (DCATT). These manual-based clustering formation schemes are not applicable to general-purpose WSNs of our days, unless specific conditions are met.

12.3   Probabilistic Clustering Approaches

As the need for efficient use of WSNs on large regions increased in the last decade dramatically, more specific clustering protocols were developed to meet the additional requirements (increased network lifetime, reduced and evenly distributed energy consumption, scalability, etc.). The most significant and widely used representatives of these focused on WSN clustering protocols (LEACH, EEHC, and HEED) and their most valuable extensions are presented in the main part of this section. They are all probabilistic in nature and their main objective was to reduce the energy consumption and prolong the network lifetime. Some of them (such as LEACH, EEHC, and their extensions) follow a random approach for CH election (the initially assigned probabilities serve as the basis for the random election of the CHs), whereas others (like HEED and similar approaches) follow a hybrid probabilistic methodology (secondary criteria are also considered during CH election—i.e., the residual energy). Some additional energy-efficient random selection approaches with good performance (like RECA) are also examined at the end of the section.

12.3.1   Popular Probabilistic Clustering Protocols

12.3.1.1   Low Energy Adaptive Clustering Hierarchy (LEACH)

One of the first and most popular clustering protocols proposed for WSNs was LEACH (Low Energy Adaptive Clustering Hierarchy) [11,12]. It is probably the first dynamic clustering protocol which addressed specifically the WSNs needs, using homogeneous stationary sensor nodes randomly deployed, and it still serves as the basis for other improved clustering protocols for WSNs. It is an hierarchical, probabilistic, distributed, one-hop protocol, with main objectives (a) to improve the lifetime of WSNs by trying to evenly distribute the energy consumption among all the nodes of the network and (b) to reduce the energy consumption in the network nodes (by performing data aggregation and thus reducing the number of communication messages). It forms clusters based on the received signal strength and also uses the CH nodes as routers to the BS. All the data processing such as data fusion and aggregation are local to the cluster.

LEACH forms clusters by using a distributed algorithm, where nodes make autonomous decisions without any centralized control. All nodes have a chance to become CHs to balance the energy spent per round by each sensor node. Initially a node decides to be a CH with a probability "p" and broadcasts its decision. Specifically, after its election, each CH broadcasts an advertisement message to the other nodes and each one of the other (non-CH) nodes determines a cluster to belong to, by choosing the CH that can be reached using the least communication energy (based on the signal strength of each CH message). In Figure 12.2 the cluster formation scheme is given in a more clear view.

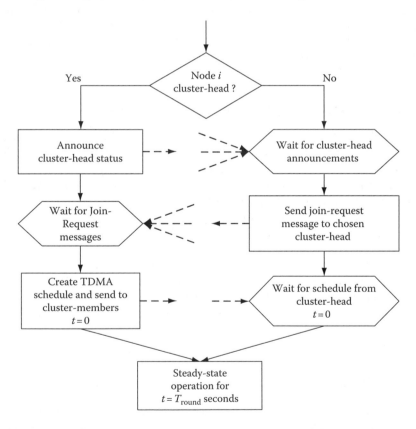

Figure 12.2 Flowchart of the cluster formation process of LEACH. (Redrawn from Heinzelman, W.B. et al., *IEEE Trans. Wireless Commun.*, 1, 660, 2002.)

The role of being a CH is rotated periodically among the nodes of the cluster to balance the load. The rotation is performed by getting each node to choose a random number "T" between 0 and 1. A node becomes a CH for the current rotation round if the number is less than the following threshold:

$$T(i) = \begin{cases} \dfrac{p}{1 - p * \left(r \bmod \frac{1}{p}\right)} & \text{if } i \in G \\ 0 & \text{otherwise} \end{cases}$$

where
  p is the desired percentage of CH nodes in the sensor population
  r is the current round number
  G is the set of nodes that have not been CHs in the last $1/p$ rounds

The clusters are formed dynamically in each round (through the process of Figure 12.2) and the time to perform the rounds are also selected randomly.

Generally, LEACH can provide a quite uniform load distribution in one-hop sensor networks. Moreover, it provides a good balancing of energy consumption by random rotation of CHs. Furthermore, the localized coordination scheme used in LEACH provides better scalability for cluster formation, whereas the better load balancing enhances the network lifetime. However, despite the generally good performance, LEACH has also some clear drawbacks. Because the decision on CH election and rotation is probabilistic, there is still a good chance that a node with very low energy gets selected as a CH. Due to the same reason, it is possible that the elected CHs will be concentrated in one part of the network (good CHs distribution cannot be guaranteed) and some nodes will not have any CH in their range. Also, the CHs are assumed to have a long communication range so that the data can reach the BS directly. This is not always a realistic assumption because the CHs are usually regular sensors and the BS is often not directly reachable to all nodes. Moreover, LEACH forms in general one-hop intracluster and intercluster topology where each node should transmit directly to the CHs and thereafter to the BS, thus normally it cannot be used effectively on networks deployed in large regions.

12.3.1.2 Energy-Efficient Hierarchical Clustering (EEHC)

Another significant probabilistic clustering algorithm was earlier proposed in Ref. [13] (Energy Efficient Hierarchical Clustering—EEHC). The main objective of this algorithm was to address the shortcomings of one-hop random selection algorithms such as LEACH by extending the cluster architecture to multiple hops. It is a distributed, k-hop hierarchical clustering algorithm aiming at the maximization of the network lifetime. Initially, each sensor node is elected as a CH with probability "p" and announces its election to the neighboring nodes within its communication range. The above CHs are now called the "volunteer" CHs. Next, all the nodes that are within "k"-hops distance from a "volunteer" CH, are supposed to receive the election message either directly or through intermediate forwarding. Consequently, any node that receives such CH election message and is not itself a CH, becomes a member of the closest cluster. Additionally, a number of 'forced' CHs are elected from nodes that are neither CHs nor belong to a cluster. Specifically, if the election messages do not reach a node within a preset time interval t, the node becomes a "forced" CH assuming that it is not within k hops of all volunteer CHs.

However, the most challenging feature of the EEHC algorithm is the direct extension to a corresponding multi-level clustering structure. The initial clustering process is recursively repeated at the level of CHs making it possible to build multiple levels of cluster hierarchy. Assuming that an "h"-level cluster hierarchy has been constructed in that way (with corresponding preset CH election probabilities $p_1, p_2, \ldots, p_h$ for each level), the algorithm ensures the efficient "h"-level communication between common sensor nodes and the BS, as follows (assuming that level h is the highest): Common sensor nodes transmit their collected data to the corresponding first-level (level # 1) CHs, the CHs of the first-level clusters transmit the aggregated data to the second-level CHs and so on, till the top (#h) level of the clustering hierarchy is reached; the CHs of those h-level clusters transmit their final aggregated data reports to the BS. This multi-level protocol has a time complexity of $O(k_1 + k_2 + \cdots + k_h)$, where k_i is the corresponding parameter (for each level) to the above-mentioned "k" parameter (number of hops in the basic-initial procedure). That was a significant improvement over the $O(n)$ time complexity that many of the existing algorithms till then (like LCA) had, and made this algorithm quite suitable for large networks.

Considering the overall performance of EEHC, the energy consumption for network operations (data gathering, aggregation, transmission to the BS, etc.) clearly depends on the parameters p and k of the algorithm. The authors derive mathematical expression for the values of p and k that achieve minimal energy consumption and they show via simulation results that by using the optimal parameter values energy consumption in the network can be reduced significantly. Also the simulation results validate the worth of use multiple levels (instead of single-level) of cluster hierarchy, as it is presented in Figure 12.3 for different values of communication radii r and spatial density λ.

12.3.1.3 Hybrid Energy-Efficient Distributed Clustering (HEED)

Another improved and very popular energy-efficient protocol is HEED (Hybrid Energy-Efficient Distributed Clustering [14]). HEED is a hierarchical, distributed, clustering scheme in which a single-hop communication pattern is retained within each cluster, whereas multi-hop communication is allowed among CHs and the BS. The CH nodes are chosen based on two basic parameters, residual energy and intracluster communication cost. Residual energy of each node is used to probabilistically choose the initial set of CHs. On the other hand, intracluster communication cost reflects the node degree or node's proximity to the neighbor and is used by the nodes in deciding to join a cluster or not. Thus, unlike LEACH, in HEED the CH nodes are not selected randomly. Only sensors that have a high residual energy are expected to become CH nodes. Also, the probability of two nodes within the transmission range of each other becoming CHs is small. Unlike LEACH, this means that CH nodes are well distributed in the network. Moreover, when choosing a cluster, a node will communicate with the CH that yields the lowest intracluster communication cost. In HEED, each node is mapped to exactly

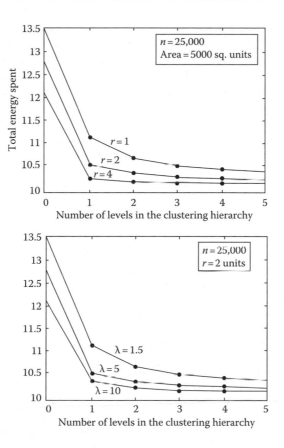

Figure 12.3 Energy consumption of multi-level EEHC. (Redrawn from Bandyopadhya, S. and Coyle, E., An energy efficient hierarchial clustering algorithm for wireless sensor network, in *22nd Annual Joint Conference of the IEEE Computer and Communication Societies (INFOCOM 2003)*, San Francisco, CA, April 2003.)

one cluster and can directly communicate with its CH. Also, energy consumption is not assumed to be uniform for all the nodes. The algorithm is divided into three stages.

At the beginning, the algorithm sets an initial percentage of CHs among all sensors. This percentage value, C_{prob}, is used to limit the initial CHs announcements to the other sensors. Each sensor sets its probability of becoming a CH, CH_{prob}, as follows: $CH_{prob} = C_{prob} * E_{residual}/E_{max}$, where $E_{residual}$ is the current energy in the sensor, and E_{max} is the maximum energy, which corresponds to a fully charged battery. CH_{prob} is not allowed to fall below a certain threshold p_{min}, which is selected to be inversely proportional to E_{max}.

The main body of the algorithm consists of a (constant) number of iterations. Every sensor goes through these iterations until it finds the CH that it can transmit to with the least transmission power (cost). If it hears from no CH, the sensor elects itself to be a CH and then sends an announcement message to its neighbors informing them about the change of status. Finally, each sensor doubles its CH_{prob} value and goes to the next iteration of this phase. It stops executing this phase when its CH_{prob} reaches 1. Therefore, there are two types of CH status that a sensor could announce to its neighbors: (a) The sensor becomes a "tentative" CH if its CH_{prob} is less than 1 (it can change its status to a regular node at a later iteration if it finds a lower cost CH). (b) The sensor "permanently" becomes a CH if its CH_{prob} has reached 1.

At the end, each sensor makes a final decision on its status. It either picks the least cost CH or announces itself as CH. Note also that for a given sensor's transmission range, the probability of CH selection can be adjusted to ensure inter-CH connectivity.

Generally, HEED's mechanism to select the CHs and form the clusters produces a uniform distribution of cluster heads across the network through localized communications with little overhead. It also clearly outperforms LEACH with regard to the network lifetime and the desired distribution of energy consumption. However, synchronization is required and the energy consumed during data transmission for far away cluster heads is significant, especially in large-scale networks. Also, a knowledge of the entire network is normally needed to determine reliably the intracluster communication cost and configuration of those parameters might be difficult in practical world.

12.3.2 Extensions and Other Similar Approaches

On the basis of the probabilistic nature of LEACH, several other protocols were developed aiming at better energy consumption and overall performance. First, the LEACH-C and the LEACH-F protocols were proposed in Ref. [12], introducing slight modifications to the initial LEACH cluster formation procedure. LEACH-C is a centralized version of LEACH, in the sense that the responsibility of the cluster creation is transferred to the BS. Each node is initially obligated to perform a direct communication with the BS in order that a global view of the network is formed. As a result an improved cluster formation procedure is performed and a slightly better overall performance of the network is achieved. LEACH-F is also a centralized protocol and is based initially on the same global clustering scheme as in LEACH-C. The main difference lies on the fact that all clusters are fixed once when they are formed, thus reducing the overhead of cluster formation in the network. However, the above design directive prevents the use of the protocol in networks with any kind of mobility.

A valuable extension to LEACH has been proposed in Ref. [15] (two-level LEACH), where the key idea of probabilistic CH election is extended (similar to the EEHC protocol but keeping one-hop intracluster topology) to construct a two-level clustering scheme. The outer level consists of the "primary" CHs where as the inner level consists of the "secondary" CHs. The "primary" CHs in each outer-level cluster communicate directly with the corresponding "secondary" CHs and the "secondary" CHs in each inner-level cluster communicate directly with the corresponding nodes in that subcluster. Data fusion as well as communication within a cluster is performed like in LEACH, in TDMA schedules. The selection of the "primary" and the "secondary" CHs is performed also in the same way as in LEACH, by setting corresponding a priori probabilities for each node. The "primary" CHs are selected first and the "secondary" CHs are selected next from the remaining nodes. The probability to become a "primary" CH is normally less than the probability to become a "secondary" CH. Generally, the two-level clustering scheme of this algorithm achieves a significant reduction on the percentage of nodes that have to transmit data to the BS in each round. Thus, it is normally expected to reduce the total energy spent.

Also, most of the published probabilistic clustering algorithms construct "disjoint" clusters. On the contrary, in Ref. [16] the authors argue that allowing some degree of overlap among clusters can be quite effective for many tasks like intercluster routing, topology discovery and node localization, recovery from CH failure, etc. Specifically, they introduce a probabilistic (randomized), distributed Multi-hop Overlapping Clustering Algorithm (MOCA) for organizing the sensors into overlapping clusters. The goal of the clustering process is to ensure that each node is either a CH or within "k" hops from at least one CH, where k is a preset cluster radius. The algorithm initially assumes that each sensor in the network becomes a CH with probability "p." Each CH then advertises itself to the sensors within its radio range. This advertisement is forwarded to all sensors that are no more than k hops away from the CH. A node sends a request to all CHs that it heard from to join their clusters. In the join request, the node includes the ID of all CHs it heard from, which implicitly implies that it is a boundary node. The CH election probability (p) is used to control the number of clusters in the network and the degree of overlap among them. The authors also provide extensive simulation work to validate appropriate values of "p" to achieve particular cluster count and overlapping degree.

Beyond the pure use of a priori probabilities to elect the initial CHs, another significant parameter additionally used (like in HEED) is the residual energy of each node. Two such recent algorithms (similar also to LEACH with regard to the overall clustering process) were proposed in Refs. [17,18]. In Ref. [17] (Time Controlled Clustering Algorithm—TCCA), the whole operation is divided (similar to LEACH) into rounds trying to achieve better load distribution among sensor nodes. In each round initially the CH selection procedure takes place and overall cluster formation process follows. Each node decides to elect itself as a CH or not based on the suitable combination of two basic criteria, its residual energy and a preset probability "p." Actually in this step TCCA applies a direct combination of LEACH and HEED algorithms by having the (HEED inspired) energy fraction $E_{residual}/E_{max}$ participating directly in the computation of the (LEACH inspired) CH-election threshold T_i in each round. When a CH is selected, it announces its selection to the neighboring nodes by sending a message which includes its node id, initial time-to-live, its residual energy, and a time stamp. The time-to-live parameter is selected according to the residual energy and it is used to restrict the size of the clusters that are formed. On the other hand, in Ref. [18] (Energy Efficient Clustering Scheme—EECS), a constant number of CHs are elected (i) based on their residual energy (as the main criterion) and (ii) using localized competition process without iteration to complete the cluster formation process. Specifically, the candidate CHs compete for their chance to be elected at any given round by broadcasting their residual energy to neighboring candidates. If a given node does not find a node with more residual energy, it becomes the CH. Additionally, clusters are then formed by retaining variable sizes dynamically, mainly depending on the distance of each cluster from the BS. As a result, the corresponding algorithm can effectively lead to better energy consumption and uniform load distribution (having a clearly better behavior compared to LEACH in simulated experiments), based on the fact that clusters at a greater distance from the BS require more energy for transmission than those that are closer.

Also, considering the HEED algorithm a slight (however effective) modification was also proposed in Ref. [20]. Specifically, the difference here is the treatment of nodes that eventually did not hear from any CH (orphaned nodes); during the finalization phase of the initial protocol all these nodes become CHs themselves. On the contrary, in Ref. [20] the authors claim that re-executing the algorithm for just those orphaned nodes could lead to significant improvements. Furthermore, this slight modification was shown to significantly decrease the CHs' count which then leads to reduced size (shorter paths) of the routing tree needed during inter-CH communication which finally results in faster data gathering procedures.

Similarly, considering the multi-level EEHC algorithm, a valuable extension (that includes additional CH election criteria) is proposed in Ref. [19] (EEMC), where the expected number of CHs at each level is previously determined by analytical formulas. The authors generalize the analysis given in Ref. [13] and present results about the optimal number of CHs at a certain level. Considering the formation process, they follow a top-down approach starting from the formation of level-1 clusters. The CHs at each level are randomly selected according to a certain probability. The probability of a node becoming a CH is proportional to the residual energy of the node as well as the distance of this node to the sink node (or to the CH it belongs to at lower levels). The distance is taken into account as each CH should transmit the aggregated data on behalf of its member nodes to its next level CH and a large distance between these two nodes contributes to fast energy consumption in the transmitting CH. The probabilities are also normalized so that the expected number of CHs at each level is according to the optimal values determined in their analysis. Extensive simulation work is also provided, in which the EEMC protocol is shown to achieve longer network lifetime and less latency compared to LEACH and EEHC protocols.

Finally, some random selection protocols have also been developed that follow an even more clear random CH election procedure (i.e., by randomly waiting or by generating a random competition, etc.). Such an early proposed algorithm was RCC [21], which was initially designed for MANETs and applies the 'First Declaration Wins' rule. In Ref. [22], another completely randomized clustering algorithm (CLUBS) was proposed, where each node participates in the election procedure by choosing a random number from a fixed integer range and then it counts down from that number silently. Two more recent and quite efficient (converging in constant time) completely randomized protocols were proposed in Refs. [23,24]. In Ref. [23] (Fast Local Clustering service—FLOC), a distributed protocol that produces approximately equal sized clusters with minimum overlap is presented.

On the other hand, in Ref. [24] (Ring-structured Energy-efficient Clustering Algorithm—RECA) sensors are grouped into one-hop bridgeable clusters during the initial cluster formation phase. Firstly, the expected number of nodes in one cluster is estimated a priori as $\gamma = N \times \pi \times R^2/A$, where N is the total number of nodes in the network, A is the area that the network covers, and R is the minimum transmission range. A node may elect itself to be a CH or a cluster member. In each slot, each node, which has neither elected itself a CH nor associated itself with a cluster, generates a random number in the $[0,1]$ interval and compares the generated number to a threshold $h = \min(2^r \times 1/\gamma, 1)$, where r is the current slot number. If the generated number is less than h, the node becomes a CH and announces this information to nodes which reside within its area, otherwise the node waits and listens to other CH announcements. Upon receiving CH announcements, the node associates itself with the best signal-to-noise ratio (SNR) cluster. After approximately $\log_2 \gamma$ time slots, each node in the network is either a CH or a cluster member. RECA uses also a deterministic algorithm to rotate the role of CH within a cluster. In each round, cluster nodes take turns to be CHs, based on their position in the logical ring. The simulation results provided (comparing to LEACH and HEED protocols) show that RECA achieves even distribution of energy consumption among the nodes of each cluster and outperforms LEACH and HEED in terms of expected network lifetime.

In conclusion, the probabilistic (clearly random or hybrid) protocols can be regarded as the leading class of clustering algorithms for WSNs due to their simplicity and their high energy efficiency. Simple protocols like LEACH, EEHC, and HEED introduced such alternative techniques with low complexity time and improved energy efficiency, whereas later many other extensions and similar probabilistic approaches were developed (mostly by extending or combining the advantages of the above basic protocols) that presented very satisfactory total performance. The EEMC, EECS, MOCA, TCCA, and RECA protocols can be regarded as the most valuable of such recent probabilistic WSN clustering approaches, in terms of limited and balanced energy consumption and increased network lifetime. The basic disadvantage (however not critical in large-scale environments) of these protocols is that due to their probabilistic nature, the CHs are not always distributed well and the CH role is not always rotated uniformly, which sometimes influences the distribution of energy consumption. The hybrid probabilistic protocols (like HEED and its extensions) behave better with regard to this aspect, however they usually lead (due to the extra processing needed) to increased total time compared to clearly random protocols.

12.4   Nonprobabilistic Clustering Approaches

Alternatively to the probabilistic (randomized or not) algorithms presented in the previous section, another basic class of clustering algorithms for WSNs primarily adopt more specific (deterministic) criteria for CHs election and cluster formation, which are mainly based on the nodes' proximity (connectivity, degree, etc.) and on the information received from other closely located nodes. The cluster formation procedure here is mainly based on the communication of nodes with their neighbors (one or multiple hops neighbors) and generally requires more intensive exchange of messages and probably graph traversing in some extent. The use of additional metrics (including the remaining energy, transmission power, mobility, etc.) in the form of combined weighted values is also a quite promising technique followed to achieve more generalized goals than other single-metric protocols. Furthermore, an even more challenging and promising nonprobabilistic clustering approach is based on the use of swarm intelligence and has led to the construction of corresponding biologically inspired clustering protocols that already have been shown to extend network lifetime in WSNs. Finally, in a few other approaches (not described here in details), beyond the application of typical proximity–connectivity criteria, the proposed protocols are primarily guided by specific (however virtual) structure-based sensors organizations and then by progressive cluster formation steps which normally lead to clusters with more controlled/predictable characteristics. Such examples can be found in Refs. [25,26].

12.4.1 Node Proximity and Graph-Based Clustering Protocols

Such a proximity-traversing-based algorithm was earlier proposed in Ref. [27] (Hierarchical Control Clustering—HCC). It is a distributed multi-hop hierarchical clustering algorithm which also efficiently extends to form a multi-level cluster hierarchy. Any node in the WSN can initiate the cluster formation process. The algorithm proceeds in two phases, "Tree Discovery" and "Cluster Formation." The tree discovery phase is basically a distributed formation of a Breadth-First-Search (BFS) tree rooted at the initiator node. Each node, u, broadcasts a signal once every p units of time, carrying the information about its shortest hop distance to the root, r. A node v that is neighbor of u will choose u to be its parent and will update its hop distance to the root, if the route through u is shorter. The broadcast signal carries the parent ID, the root ID, and the sub tree size. Every node updates its sub tree size when its children sub tree size change. The cluster formation phase starts when a sub tree on a node crosses the size parameter, k. The node initiates cluster formation on its sub tree. It will form a single cluster for the entire sub tree if the sub tree size is less than $2k$, or else, it will form multiple clusters. The cluster size and the degree of overlap are also considered. In Figure 12.4 the proposed multi-level

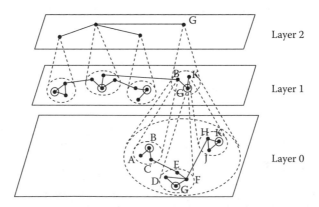

Figure 12.4 HCC Three-layer cluster hierarchy. (Redrawn from Banerjee, S. and Khuller, S., A clustering scheme for hierarchial control in multi-hop wireless networks, in *Proceedings of the 20th Joint Conference of the IEEE Computer and Communication Societies (INFOCOM 01)*, Anchorage, AK, April 2001.)

hierarchy is further illustrated. This approach has a time complexity of $O(n)$, however it has been shown to achieve quite balanced clustering as well as to handle dynamic environments very well.

Two other early proposed algorithms of this category (designed primarily for MANETs, however applicable in WSNs too) can be found in Ref. [28] (Highest Connectivity–HC) and [29] (Max-Min D-Cluster algorithm). In Ref. [28] a connectivity-based heuristic is proposed, in which the sensor node with maximum number of one-hop neighbors is elected as a CH in its neighborhood. The formation of one-hop clusters and the clock synchronization requirement limit the practical usage of this algorithm nowadays. On the other hand, in Ref. [29], a distributed algorithm is proposed, in which the clusters consist of nodes that are no more than d-hops away from the CH. The algorithm has complexity $O(d)$, it does not require clock synchronization and it provides a better load balancing compared to LCA and HC algorithms.

Other more recent examples of proximity-connectivity and neighbors' information-based algorithms have been proposed in Refs. [30–32]. In Ref. [30] a typical centralized, graph-based clustering approach (EEDC) is presented. To minimize the number of clusters and therefore maximize the energy saving, EEDC models the cluster creation process as a clique-covering problem and uses the minimum number of cliques to cover all vertices in the graph. The sink also dynamically adjusts the clusters based on spatial correlation and the received data from the sensors. The algorithm produces robust and well-balanced clusters, however it is centralized and thus not suitable for large-scale WSNs.

In Ref. [31] (Clustering Algorithm via Waiting Timer—CAWT), a distributed proximity-connectivity-based algorithm for constructing cluster hierarchy has been proposed for homogeneous sensors with the same transmission range. Once sensors are deployed, each sensor broadcasts a "hello" message to show its presence to the neighbors while listening to the others. The sensors that hear a significant number of "hello" messages (meaning that are nodes with high connectivity) organize into clusters while others are waiting to form clusters. The performance of the algorithm was evaluated using simplified simulations leading to quite good results with regard to network lifetime. However, as it is clearly observed, the generalization of the algorithm is subject to detailed evaluation with respect to load balancing, CH reelection, and energy usage across the network.

Similarly, in Ref. [32] (EACLE) a distributed clustering procedure, which beyond the proximity takes also in account the residual energy of each node, is followed. It is mainly based on the information of 2-hop neighbors with a practical transmission power control scheme, and then builds a broadcast tree only by cluster heads. Initially, each sensor is in a 'waiting' state and waits for time T_1 which is a monotonous decreasing function on the residual energy of the node. When the timer expires, the waiting node becomes a CH and broadcasts two packets with different (power-high and power-low) transmission power each, which contain the list of the neighbor-IDs received before broadcasting. When a waiting node receives a power-low packet it becomes a member node, whereas when it receives a power-high packet, it compares its own neighbor list with the list of IDs in the receiving packet, to decide if it should continue waiting or become a CH. Also, each node executes the clustering process periodically. Once a node becomes a CH in a specific round, its timer is then set to a longer value to avoid becoming a CH again in the next round.

Also, a quite valuable alternative approach was given in Ref. [33] (Algorithm for Cluster Establishment, ACE). Unlike other distributed clustering schemes, ACE employs an emergent algorithm. Emergent algorithms much like artificial neural networks evolve to optimal solution through a mix of local optimization steps. Initially, a node decides to become a "candidate" CH, and then it broadcasts an invitation message. Upon getting the invitation, a neighboring sensor joins the new cluster and becomes a follower of the new CH. At any moment, a node can be a follower of more than one cluster. Next, the migration phase takes place in order the best candidate for being CH to be selected. Each CH periodically checks the ability of its neighbors for being a CH and decides to step down if one of these neighbors has more followers than it does. A node that has the largest number of followers and the least overlap with other clusters will be considered as the best final candidate for CH. The algorithm converges in time $O(d)$ where d is the node density per unit disk. Experimental validation of ACE indicated that it achieves low variance and high average of cluster sizes when compared to node-ID-based schemes like Refs. [7,8].

Finally, toward the direction of efficient data gathering and aggregation, some other alternative solutions, without direct clustering, have also been proposed [34,35]. These approaches are mainly based on graph traversing heuristics and they have shown worth telling improvements compared to typical cluster-based implementations like LEACH. Similarly, in Ref. [36] the authors propose a corresponding "hybrid" clustering protocol (PEACH) which builds an adaptive clustering hierarchy without incurring the clustering formation overhead commonly met in other direct clustering algorithms. Instead, by overhearing the packets transmitted and received by neighboring nodes, each node can determine its role (CH or not) and possibly join the cluster of other node, thereby creating a clustering hierarchy. In PEACH protocol, a node becomes a CH when it hears a packet destined for the node itself. Otherwise, when the packet is destined for a different node, the node that overheard the packet joins the cluster of the destination node. By means of simulation, the PEACH protocol was compared to other competitive approaches, such as LEACH, HEED, and PEGASIS, and showed lower energy consumption and higher network lifetime mainly because the clustering hierarchy is created on the fly based on the information overheard by nodes.

In conclusion, the node-proximity and graph traversing clustering protocols achieve quite balanced and stable clusters (quite uniform distribution of CHs in the entire area, low intracluster communication cost, etc.), however they present several disadvantages. They usually lead to increased complexity time to satisfy more qualitative criteria, whereas important parameters like the number of clusters and the size of each cluster cannot easily be controlled without qualitative cost. Additionally, most of these protocols do not apply effective CHs rotation procedures (or they do not apply such procedures at all) leading to reduced energy efficiency and worse network lifetime than probabilistic protocols. The HCC, ACE, CAWT, and EACLE protocols can be regarded as the most valuable approaches in this category, however they do not present (in terms of energy consumption and network lifetime) good results as the corresponding probabilistic clustering protocols. On the contrary, special attention has to be paid to some recent hybrid (without direct clustering) clustering protocols, like the PEACH protocol, that seem to achieve lower energy consumption by effectively avoiding part of the cluster formation overhead.

12.4.2 Weight-Based Clustering Protocols

In addition to node proximity, some other known algorithms use a combination of metrics such as the remaining energy, transmission power (thus forming corresponding combined weights) to achieve more generalized goals than single-criterion protocols. Several algorithms following this directive were initially borrowed from the field of mobile ad hoc networks, that is, [37,38]. As a typical example, in Ref. [38] (WCA), a corresponding weight-based protocol was proposed where the CH election process is based on the computation of a "combined weight" W_v for each node, which takes into account several system parameters such as the node degree, the transmission power, mobility, and the remaining energy of the node: $W_v = w1 T_v + w2D_v + w3M_v + w4P_v$. The combined weight is calculated and broadcasted by each node. The node with the smallest weight in its neighborhood is chosen as a CH. This is a nonperiodic procedure for CH election; it is invoked on demand every time a reconfiguration of the network's topology is unavoidable. This algorithm attempted to provide better load balancing through reduced number of sensors in a cluster but the requirement of clock synchronization limits its applications.

Two more recent weight-based protocols were proposed in Refs. [39,40]. In Ref. [39] [Distributed Weight-Based Energy-Efficient Hierarchical Clustering (DWEHC)] a corresponding distributed algorithm is given, which aims at high energy efficiency by generating balanced cluster sizes and optimizing the intracluster topology. Each sensor calculates its weight after locating the neighboring nodes in its area. The weight is a function of the sensors residual energy and the proximity to the neighbors. In a neighborhood, the node with largest weight would be elected as a CH and the remaining nodes become members. At this stage the nodes are considered as first-level members because they have a direct link to the CH. A node progressively adjusts such membership to reach a CH using the least amount of energy. Basically, a node checks with its non-CH neighbors to find out their minimal cost for reaching a CH. Given the node's knowledge of the distance to its neighbors, it can assess whether it is better to stay a first-level member or become a second-level one reaching the CH over a two-hop path. Figure 12.5, illustrates the structure of the intracluster topology. Compared to HEED, the DWEHC algorithm has been shown to generate more well-balanced clusters as well as to achieve significantly lower energy consumption in intracluster and intercluster communication.

Similarly, in Ref. [40] (Topology Adaptive Spatial Clustering—TASC), the authors propose another distributed algorithm that partitions the network into a set of locally isotropic, nonoverlapping clusters without prior knowledge of the number of clusters, cluster size, and node coordinates. This is achieved by deriving a set of weights that include distance, connectivity, and density information within the locality of each node. The derived weights form the terrain for holding a coordinated leader election procedure in which each node selects the node closer to the center of mass of its neighborhood to become its leader.

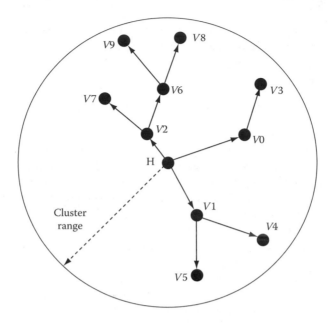

Figure 12.5 DWEHC multi-hop intracluster topology. (Redrawn from Ding, P. et al., Distributed energy efficient hierarchical clustering for wireless sensor networks, in *Proceedings of the IEEE International Conference on Distributed Computing in Sensor Systems(DCOSS05)***, Marina Del Rey, CA, June 2005.)**

Generally, the weight-based clustering protocols have been shown to produce well-balanced and stable clusters like the node-proximity and graph-based protocols, in a more systematic and deterministic way. Additionally they can achieve better distribution of energy consumption because most of them consider the residual energy as part of the computed weights during the CH election process. However, they normally suffer from the same disadvantages (increased communication time, no CHs rotation, etc.) as the node-proximity and graph-based protocols that were examined in the previous paragraph.

12.4.3 *Biologically Inspired Clustering Approaches*

Finally, the last few years some new algorithms have also been proposed based on swarm intelligence techniques which model the collective behavior of social insects such as ants. They have shown very promising results in simulated experiments (compared to protocols like LEACH and HEED) with regard to network lifetime. In Ref. [41] the authors propose such a swarm intelligence-based clustering algorithm based on the ANTCLUST method. ANTCLUST is a model of an ant colonial closure to solve clustering problems. In colonial closure model, when two objects meet together they recognize whether they belong to the same group by exchanging and comparing information about them. In the case of a WSN, initially the sensor nodes with more residual energy become CHs independently. Then, randomly chosen nodes meet each other, exchange information, and clusters are created, merged, and discarded through these local meetings and comparison of their information. Each node with less residual energy chooses a cluster based on specific criteria, like the residual energy of the CH, its distance to the CH, and an estimation of the cluster size. Eventually, energy efficient clusters are formed that result in an extension of the lifetime of the WSN.

Another related approach that ensures the good distribution of CHs and high energy efficiency, can be found in Ref. [42]. Also, in Ref. [43], a protocol that has the objective of minimizing the intracluster distance and optimizing the energy consumption of the network using Particle Swarm Optimization (PSO) is presented and evaluated via simulations. Generally, biologically inspired clustering algorithms show that they can dynamically control the CH selection while achieving quite uniform distribution of CHs and energy consumption. However, they have to be studied further as it is pointed out in the literature.

12.5 Clustering Algorithms for Reactive Networks

All the algorithms presented in the previous sections (having been considered as "proactive" clustering algorithms) are based on the assumption that the sensors always have data to send, hence, they should all be considered during the cluster formation. In contrast, "reactive" algorithms take advantage of user queries for the sensed data or of specific triggering events that occur in the WSN. Namely, nodes may react instantly to sudden and drastic changes in the value of a sensed attribute. This approach is useful for time-critical applications, but not particularly suited for applications where data retrieval is required on a regular basis.

The Threshold sensitive Energy Efficient sensor Network protocol (TEEN) [44] forms a hierarchical clustered structure, grouping nearby nodes within the same cluster. The protocol focuses on information aggregation rather than on cluster formation, which is very similar to LEACH. The protocol defines two thresholds: the hard threshold is a threshold (absolute) value for the sensed attribute, while the soft threshold is a threshold (small change) value of the sensed attribute. The concept of threshold is highly significant in a variety of WSN applications, such as fire alarm and temperature monitoring. The nodes transmit sensor readings only when they fall above the hard threshold and change by given amount (soft threshold). In TEEN, sensor nodes sense the medium continuously, but data transmission is done less frequently which favors the energy saving. However, if the thresholds are not crossed, the nodes will never communicate, namely TEEN does not support periodic reports.

The Adaptive Periodic-TEEN (APTEEN) [45] is a variation of TEEN which addresses the latter's main shortcoming. It is a hybrid routing protocol wherein the nodes still react to time-critical situations, but also give an overall picture of the network at periodic intervals in an energy efficient manner. The CH selection in APTEEN is based on the mechanism used in LEACH-C. The clusters are valid for an interval called the cluster period. At the end of this period, the BS performs reclustering. Assuming that adjacent nodes register similar data, APTEEN forms pairs of nodes where only one of them responds to queries. These two nodes can alternate in the role of handling queries, thereby saving resources. Simulations have indicated that APTEEN's performance lies in between TEEN and LEACH in terms of average energy dissipation events and network lifetime. That is a reasonable conclusion that derives from APTEEN's hybrid proactive-reactive nature. The main drawbacks of TEEN and APTEEN lie on the control overhead associated with the formation of multiple-level clusters, the method of implementing threshold functions, and none of them exploits spatial and temporal data correlation to improve efficiency.

Decentralized Reactive Clustering (DRC) has been proposed in Ref. [48]. Similarly to other reactive algorithms, the clustering procedure is initiated only in the case of events detection. Four different operation phases are defined: the postdeployment phase, followed by a cluster-forming phase which is when clusters are constructed, then an intra-cluster data processing phase and finally a CH-to-processing center phase. DRC uses power control technique to minimize energy usage in cluster formation. Unfortunately, simulations only compare DRC against LEACH and therefore do not highlight its performance gains or disadvantages against other reactive clustering protocols.

More recently, the Clustered AGgregation (CAG) [46], mechanism was proposed, which utilizes the spatial correlation of sensory data to further reduce the number of transmissions by providing approximate results to aggregate queries. CAG guarantees the result to be within a user-specified error-tolerance threshold. Cluster formation is performed while queries are disseminated to the network (query phase), where clusters group nodes sensing similar values. Subsequently, CAG enters the response phase wherein only one aggregated value per cluster is transmitted up the aggregation tree. In effect, CAG is a lossy clustering algorithm (most sensory readings are never reported) which trades a lower result precision for a significant energy, storage, computation, and communication saving.

The Updated CAG algorithm [47] extends CAG defining two operation modes, depending on the dynamics of the environment. In the interactive mode, users issue a one-shot query and the network generates a single response. This is appropriate for scenarios where the environment changes dynamically, or when users desire to change the approximation granularity or query attributes interactively. On the other hand, in the streaming mode, the CHs transmit a stream of response for a query that is issued just once. This mode of operation is well suited for static environments where sensor readings do not change frequently and the query remains valid for a certain period of time. Note that the interactive mode only exploits the spatial correlation of the sensor data to form clusters, whereas the streaming mode leverages both temporal and spatial correlations. The latter adjusts clusters locally as the data and topology change over time. Overall, CAG and Updated CAG approaches provide efficient data aggregation and energy saving solutions at the expense of a precision error bounded by a user-provided threshold.

In an even more recent approach, Guo and Li proposed Dynamic-Clustering Reactive Routing (DCRR) algorithm [49]. DCRR borrows ideas from biological neuron networks, following the observation that the latter also employ a many-to-one (neurons-to-brain) communication paradigm, similarly to the nodes of a WSN. In DCRR, once an incident emerges, the CH is dynamic selected in the incident region according to the nodes' residual energy. DCRR defines a TEEN-inspired "action threshold" for firing data to the sink. That threshold is dynamically adjusted to trace the changing speed of the incident. The action threshold is also intentionally fluctuated outside the incident region to enable all network nodes to send data in periodic basis (so that they are not misconceived as failed ones).

In conclusion, reactive clustering algorithms relax the network from the control overhead and the time required to perform the clustering process in the WSN. The use of attributes thresholds and the exploitation of the spatial and temporal correlation of sensory data, inherent in many networking environments, compress the magnitude of event-driven data transmissions. These algorithms are also highly responsive to radical changes in the monitored environment and are particularly useful for time-sensitive applications. However, they do not suit applications which require periodic retrieval of sensory readings, wherein the construction of stable and energy-efficient cluster structures is of critical importance.

12.6 Conclusion

Generally, clustering in WSNs has been of high interest in the last decade and there is already a large number of related published works. Throughout this chapter we tried to present the main characteristics of the most significant protocols that were proposed till now in the literature. As it was pointed out, grouping nodes into clusters, thus leading to hierarchical routing and data gathering protocols, has been regarded as the most efficient approach to support scalability in WSNs. The hierarchical cluster structures facilitate the efficient data gathering and aggregation independent to the growth of the WSN, and generally reduce the total amount of communications as well as the energy spent.

The main objective of most of the existing protocols lies on how to prolong the lifetime of the network and how to make a more efficient use of the critical resources, such as battery power. Furthermore, the combined need for fast convergence time and minimum energy consumption (with regard to the cluster formation procedure) led to appropriate fast distributed probabilistic (clearly random or hybrid) clustering algorithms which quickly became the most popular and widely used in the field. In these algorithms the nodes are assumed to make fast decisions (i.e., to become CHs or not) based on some probability or other local information only (i.e., on their residual energy) and usually the desired quality of the final cluster output is considered as a secondary parameter only. Another critical feature of most of these algorithms (leading to more uniform distribution of the energy consumption) is the periodic reelection of CHs (rotation of the CH role) among all the nodes of the network. Clustering algorithms that adopt as primary election criteria other classical parameters like connectivity, nodes' proximity, distance, etc., have also been developed and relevant protocols are still being used, leading probably to more qualitative output (well balanced clusters, etc). However the time complexity of these algorithms is difficult to be kept low as in leading probabilistic/random clustering algorithms.

Moreover, because the size of the WSNs (number of sensors, area covered, etc.) used in real applications become larger and larger, the extension in multi-hop communication patterns (with regard to intracluster and/or intercluster communication) is unavoidable, whereas the multi-level cluster hierarchies have also been regarded as a promising option preserving energy efficiency independent of the growth of the network. Significant progress has also been noted in specialized protocols for timely critical applications, where nodes should react instantly to sudden and drastic changes in the value of a sensed attribute.

Finally, several additional issues should be further studied in future research. Some of the most challenging of these issues include the development of a generic method for finding the optimal number of clusters in order maximize the energy efficiency, the estimation of the optimal frequency of CH rotation/reelection to gain better energy distribution, however, keeping the total overhead low, the efficient support of nodes and CHs mobility as well as the support of mobile sinks, the incorporation of several security aspects (i.e., enhanced protection needed in hostile environments when cluster-based protocols are used), the further development of efficient recovery protocols in case of CHs failure, etc.

References

1. A.A. Abbasi and M. Younis, A survey on clustering algorithms for wireless sensor networks, *Computer Communications*, 30, 2826–2841, 2007.
2. K. Sohrabi et al., Protocols for self-organization of a wireless sensor network, *IEEE Personal Communications*, 7(5), 16–27, 2000.
3. R. Min et al., Low power wireless sensor networks, in *Proceedings of International Conference on VLSI Design*, Bangalore, India, January 2001.
4. M. Liliana, C. Arboleda, and N. Nidal, Comparison of clustering algorithms and protocols for wireless sensor networks, in *Proceedings of IEEE CCECE/CCGEI Conference*, Ottawa, Ontario, Canada, pp. 1787–1792, May 2006.
5. G. Gupta and M. Younis, Load-balanced clustering in wireless sensor networks, in *Proceedings of the International Conference on Communication (ICC 2003)*, Anchorage, AK, May 2003.
6. E.I. Oyman and C. Ersoy, Multiple sink network design problem in large scale wireless sensor networks, in *Proceedings of the IEEE International Conference on Communications (ICC 2004)*, Paris, June 2004.
7. D.J. Baker and A. Ephremides, The architectural organization of a mobile radio network via a distributed algorithm, *IEEE Transactions on Communications*, 29(11), 1694–1701, 1981.
8. A. Ephremides, J.E. Wieselthier, and D.J. Baker, A design concept for reliable mobile radio networks with frequency hopping signalling, *Proceedings of IEEE*, 75(1), 56–73, 1987.
9. C.R. Lin and M. Gerla, Adaptive clustering for mobile wireless networks, *IEEE Journal on Selected Areas Communications*, 15(7), 1265–1275, 1997.
10. W.P. Chen, J.C. Hou, and L. Sha, Dynamic clustering for acoustic target tracking in wireless sensor networks, in *Proceedings of the 11th IEEE International Conference on Network Protocols (ICNP'03)*, pp. 284–294, November 4–7, Atlanta, GA, 2003.

11. W.R. Heinzelman, A.P. Chandrakasan, and H. Balakrishnan, Energy efficient communication protocol for wireless microsensor networks, in *Proceedings of the 33rd Hawaiian International Conference on System Sciences (HICSS-33)*, pp. 3005–3014, January 4–7, Maui, HI, 2000.

12. W.B. Heinzelman, A.P. Chandrakasan, and H. Balakrishnan, An application specific protocol architecture for wireless microsensor networks, *IEEE Transactions on Wireless Communications*, 1(4), 660–670, 2002.

13. S. Bandyopadhyay and E. Coyle, An energy efficient hierarchical clustering algorithm for wireless sensor networks, in *22nd Annual Joint Conf. of the IEEE Computer and Communications Societies (INFOCOM 2003)*, San Francisco, CA, April 2003.

14. O. Younis and S. Fahmy, HEED: A hybrid, energy-efficient, distributed clustering approach for Ad Hoc sensor networks, *IEEE Transactions on Mobile Computing*, 3(4), 366–379, 2004.

15. V. Loscri, G. Morabito, and S. Marano, A two-level hierarchy for low-energy adaptive clustering hierarchy, in *Proceedings of IEEE VTC Conference 2005*, Vol. 3, pp. 1809–1813, 2005.

16. A. Youssef, M. Younis, M. Youssef, and A. Agrawala, Distributed formation of overlapping multi-hop clusters in wireless sensor networks, in *Proceedings of the 49th Annual IEEE Global Communication Conference (Globecom06)*, San Francisco, CA, November 2006.

17. S. Selvakennedy and S. Sinnappan, An adaptive data dissemination strategy for wireless sensor networks, *International Journal of Distributed Sensor Networks*, 3(1), 23–40, 2007.

18. M. Ye, C. Li, G. Chen, and J. Wu, EECS: An energy efficient clustering scheme in wireless sensor networks, in *Proceedings of IEEE International Performance Computing and Communications Conference (IPCCC'05)*, pp. 535–540, April 7–9, Phoenix, AZ, 2005.

19. Y. Jin, L. Wang, Y. Kim, and X. Yang, EEMC: An energy-efficient multi-level clustering algorithm for large-scale wireless sensor networks, *Computer Networks Journal*, 52, 542–562, 2008.

20. H. Huang and J. Wu, A probabilistic clustering algorithm in wireless sensor networks, in *Proceedings of IEEE 62nd VTC Conference*, Dallas, TX, September 2005.

21. K. Xu and M. Gerla, A heterogeneous routing protocol based on a new stable clustering scheme, in *Proceeding of IEEE Military Communications Conference (MILCOM 2002)*, Anaheim, CA, October 2002.

22. R. Nagpal and D. Coore, An algorithm for group formation in an amorphous computer, in *Proceedings of the 10th International Conference on Parallel and Distributed Systems (PDCS98)*, Las Vegas, NV, October 1998.

23. M. Demirbas, A. Arora, and V. Mittal, FLOC: A fast local clustering service for wireless sensor networks, in *Proc. of Workshop on Dependability Issues in Wireless Ad Hoc Networks and Sensor Networks (DIWANS04)*, Florence, Italy, June 2004.

24. G. Li and T. Znati, RECA: A ring-structured energy-efficient clustering architecture for robust communication in wireless sensor networks, *International Journal Sensor Networks*, 2(1/2), 34–43, 2007.

25. H. Zhang and A. Arora, GS3: Scalable self-configuration and self-healing in wireless networks, in *Proceedings of the 21st ACM Symposium on Principles of Distributed Computing (PODC 2002)*, Monterey, CA, July 2002.

26. L. Yu, N. Wang, W. Zhang, and C. Zheng, GROUP: A grid-clustering routing protocol for wireless sensor networks, in *Proceedings of International Conference on Wireless Communications, Networking and Mobile Computing (WiCOM'06)*, pp. 1–5, September 22–24, Wuhan City, China, 2006.

27. S. Banerjee and S. Khuller, A clustering scheme for hierarchical control in multi-hop wireless networks, in *Proceedings of the 20th Joint Conference of the IEEE Computer and Communications Societies (INFOCOM'01)*, pp. 1028–1037, April 22–26, Anchorage, AK, 2001.

28. A.K. Parekh, Selecting routers in ad-hoc wireless networks, in *Proceedings of SBT/IEEE International Telecommunications Symposium (ITS'94)*, pp. 420–424, August 22–25, Rio de Janeiro, Brazil, 1994.

29. A. Amis, R. Prakash, T. Vuong, and D. Huynh, Max-min D-cluster formation in wireless Ad Hoc networks, in *Proceedings of the 19th Joint Conference of the IEEE Computer and Communications Societies (INFOCOM'00)*, pp. 32–41, March 26–30, Tel-Aviv, Israel, 2000.

30. C. Liu, K. Wu, and J. Pei, A dynamic clustering and scheduling approach to energy saving in data collection from wireless sensor networks, in *Proceedings of the 2nd Annual IEEE Conference on Sensor and Ad Hoc Communications and Networks (SECON'05)*, pp. 374–385, September 26–29, Santa Clara, CA, 2005.

31. C. Wen and W. Sethares, Automatic decentralized clustering for WSNs, *EURASIP Journal on Wireless Communications and Networking*, 5(5), 686–697, 2005.

32. K. Yanagihara, J. Taketsugu, K. Fukui, S. Fukunaga, S. Hara, and K.I. Kitayama, EACLE: Energy-aware clustering scheme with transmission power control for sensor networks, *Wireless Personal Communications*, 40(3), 401–415, 2007.

33. H. Chan and A. Perrig, ACE: An emergent algorithm for highly uniform cluster formation, in *Proceedings of the 1st European Workshop on Sensor Networks (EWSN)*, Berlin, Germany, January 2004.

34. S. Lindsey and C.S. Raghavendra, PEGASIS: Power-efficient gathering in sensor information networks, Computer Systems Research Department, the Aerospace Corporation, Vol. 3, pp. 3-1125–3-1130, Los Angeles, CA, 2002.

35. H.O. Tan and I. Korpeoglou, Power efficient data gathering and aggregation in wireless sensor networks, *Issue on Sensor Networks Technology, ACM SIGMOD Record*, 32(4), 66–71, 2003.

36. S. Yi, J. Heo, Y. Cho, and J. Hong, PEACH: Power-efficient and adaptive clustering hierarchy protocol for WSNs, *Computer Networks*, 30, 2842–2852, 2007.

37. S. Basagni, Distributed clustering for ad hoc networks, in *Proceedings of International Symposium on Parallel Architectures, Algorithms & Networks (ISPAN'99)*, pp. 310–315, June 23–25, Fremantle, Australia, 1999.

38. M. Chatterjee, S.K. Das, and D. Turgut, WCA: A weighted clustering algorithm for mobile ad hoc networks, *Clustering Computing*, 5, 193–204, 2002.

39. P. Ding, J. Holliday, and A. Celik, Distributed energy efficient hierarchical clustering for wireless sensor networks, in *Proceedings of the IEEE International Conference on*

Distributed Computing in Sensor Systems(DCOSS05), Marina Del Rey, CA, June 2005.

40. R. Virrankoski and A. Savvides, TASC: Topology adaptive clustering for sensor networks, in *Proceedings of the Second IEEE International Conference on Mobile Ad-Hoc and Sensor Systems, (MASS 2005)*, Washington, DC, November 2005.

41. J. Kamimura, N. Wakamiya, and M. Murata, A distributed clustering method for energy-efficient data gathering in sensor networks, *International Journal on Wireless and Mobile Computing*, 1(2), 113–120, 2006.

42. S. Selvakennedy, S. Sinnappan, and Y. Shang, A biologically inspired clustering protocol for wireless sensor networks, *Computer Communications* 30, 2786–2801, 2007.

43. N.M. Abdul Latiff, C.C. Tsimenidis, and B.S. Sharif, Energy-aware clustering for wireless sensor networks using particle swarm optimization, in *IEEE Intl. Symposium PIMRC'07*, pp. 1–5, Athens, Greece, September 2007.

44. A. Manjeshwar and D.P. Agrawal, TEEN: A routing protocol for enhanced efficiency in wireless sensor networks, in *Proceedings of the 15th International Parallel and Distributed Processing Symposium*, San Francisco, CA, 2001.

45. A. Manjeshwar and D.P. Agrawal, APTEEN: A hybrid protocol for efficient routing and comprehensive information retrieval in wireless sensor networks, in *Proceedings of International Parallel and Distributed Processing Symposium (IPDPS'02)*, pp. 195–202, April 15–19, Fort Lauderdale, FL, 2002.

46. S. Yoon and C. Shahabi, Exploiting spatial correlation towards an energy efficient Clustered AGgregation technique (CAG), in *Proceedings of IEEE International Conference on Communications (ICC'05)*, pp. 82–98, May 16–20, Seoul, Korea, 2005.

47. S. Yoon and C. Shahabi, The Clustered AGgregation (CAG) technique leveraging spatial and temporal correlations in wireless sensor networks, *ACM Transactions on Sensor Networks*, 3(1), Article #3, March 2007.

48. Y. Xu and H. Qi, Decentralized reactive clustering for collaborative processing in sensor networks, in *Proceedings of the 10th International Conference on Parallel and Distributed Systems (ICPADS'04)*, pp. 54–61, July 7–9, Newport Beach, CA, 2004.

49. B. Guo and Z. Li, A dynamic-clustering reactive routing algorithm for wireless sensor networks, *Wireless Networks Journal*, Springer, 15(4), 423–430, 2009.

50. D. Estrin, R. Govindan, J. Heidemann, and S. Kumar, Next century challenges: scalable coordination in sensor networks, in *Proceedings of the ACM/IEEE MOBICOM Intl. Conference*, Boston, MA, pp. 6–11, August 2000.

51. C.M. Liu and C.H. Lee, Power efficient communication protocols for data gathering on mobile sensor networks, in *IEEE 60th VTC*, Los Angeles, CA, pp. 4635–4639, 2004.

52. X. Zhang, H. Wang, and A. Khokhar, An energy-efficient data collection protocol for mobile sensor networks, in *IEEE 64th VTC*, Montreal, Quebec, Canada, pp. 1–5, 2006.

Chapter 13

Energy-Efficient Sensing in Wireless Sensor Networks

Lijun Qian, John Attia, Xiangfang Li, and Deepak Kataria

CONTENTS

In a wireless sensor network (WSN), energy efficiency is one of the primary concerns because the sensor nodes typically rely on batteries to operate. A sensor node is usually composed of two parts: the sensor module for taking measurements and the radio module for wireless communications. Although many energy saving schemes have been proposed for the wireless communications module, very few have addressed the issue of energy savings of the sensing module. Recent measurement results show that the sensor nodes spend a lot of energy when taking many measurements, and the energy expenditure of sensing is comparable to that of data communications. As a result, it is necessary to study energy saving schemes for sensing.

WSNs are usually deployed to monitor static or dynamic events. Static events (such as temperature and humidity) are easy to capture. On the contrary, dynamic events are typically noncooperative, and thus, are not easy to measure. The movement of an enemy vehicle in a battlefield and migration of whales in the ocean are examples of noncooperative events. They are not easily monitored as they come and go. As a result, they can be observed only if the sensors are constantly monitoring the environment. However, energy efficiency is a serious concern if the sensors are always ON. In this chapter, various sensing schemes are reviewed and the trade-off between energy efficiency and events coverage in a WSN is studied. We begin with fixed sensing and sleep-time schemes for a single wireless sensor node and quantify the performance results in terms of the event miss-rate and the normalized average power consumption. If the statistics of the event is unknown, an adaptive scheme based on an additive increase/multiplicative decrease (AIMD) rule is proposed to adapt the sleep schedule with the sensing data. Both fixed and adaptive schemes are extended to multiple sensor nodes and the coverage of the entire area of a network is taken into consideration in addition to energy efficiency. The interactions between the sensing module and the radio module are also discussed. Discrete-event simulations are carried out to demonstrate the effectiveness of the proposed schemes.

13.1 Introduction

The wireless sensor network (WSN) is one of the recent rising technologies in wireless communications with a wide range of applications, such as environment surveillance, health care, intelligent buildings, and battlefield control [1]. A typical WSN consists of resource-constraint sensors responsible for monitoring physical phenomena and reporting to access points or fusion centers. One of the primary characteristics of a WSN is its longtime functionality. Because of the large number of sensor nodes that may be deployed and the large and diverse geographical area that may be covered, replacing the battery is usually not an option. WSNs must utilize the minimal possible energy while operating over a wide range of operating scenarios.* Therefore, the network lifetime becomes a critical issue when designing WSNs.

Although many previous works have proposed energy-efficient wireless communications schemes (medium access control and routing schemes) for WSNs, energy-efficient sensing has not received much attention. However, the sensing module consumes energy comparable to that of the radio communication module in a typical sensor node. Furthermore, the frequency of sensing is usually higher than that of wireless communications. Hence, energy-efficient sensing techniques are reviewed and studied in this chapter. We start by reviewing various energy saving schemes in Section 13.2, especially energy-efficient sensing. Then a systematic analysis of energy-efficient sensing is provided in later sections. In Section 13.3, alternate sensing schemes of a single sensor node are studied, where both fixed and adaptive schemes are considered. Simulation results are provided for both schemes in Section 13.4. A networkwide coverage is investigated in Section 13.5. Open problems and issues are discussed in Section 13.6. Section 13.7 concludes the chapter and provides suggestions for future research work in this area.

13.2 Review of Energy Saving Schemes

To achieve a prolonged network lifetime, many energy saving schemes are proposed for WSNs. Three categories of the proposed approaches in the literature are reviewed in this section, namely, hardware power management, energy-efficient sensing techniques, and energy-efficient wireless communications techniques, with the emphasis on energy-efficient sensing. Examples of these techniques are highlighted in Figure 13.1.

13.2.1 Hardware Power Management

Advances in microelectromechanical systems (MEMS) technology, electronics, and wireless communications have led to the development of low-cost sensor nodes. A sensor node usually consists of a sensing unit (an A/D converter and an embedded sensor), a communication unit (RF circuits), a power unit (battery), and a processing unit (processor with memory). For example, sensor nodes using Mica2/Micaz motes from XBOW compose of a sensor board, a communication board, two AA batteries, and 8 bit microcontrollers [8]. To define a system architecture for a wireless sensor, designers typically start by selecting a microcontroller or a sensor computer [2].

13.2.1.1 Dynamic Voltage Scaling

For a CMOS-based processor, energy consumption is classified into switching and leakage. One of the techniques for improving energy efficiency of microprocessors is using dynamic voltage scaling (DVS). DVS is the active adjustment of the supply voltage and the clock frequency in response to fluctuations in the utilization of a processor [3].

---

*In some cases, energy scavenging is possible [9]. Nonetheless, energy-efficient operations of WSNs are highly desirable in all cases [10].

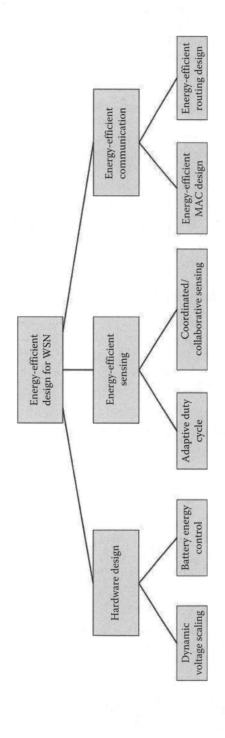

Figure 13.1   Energy saving schemes.

DVS is being implemented in commercially available microprocessors, and it takes advantage of the changes in the workload of the processor and constraints on latency by realizing this trade-off in energy at the circuit level. DVS reduces both leakage and switching of energy. Sensor nodes that implement DVS consume up to 60 percent less energy than sensor nodes with fixed voltage processors [3,4]. An example is the μAMPS sensor nodes [5,6], where DVS is implemented.

13.2.1.2 Power Source Management

With most wireless sensors, the battery is the main limiting factor that inhibits the sensor node from operating without maintenance for several years. The emerging trend is using super-capacitors as a central element in power subsystems of wireless embedded sensing systems [2]. Placing super-capacitors in parallel with the battery may be part of the solution for this problem. This is done so that transient power is delivered by the capacitor as opposed to the battery. Another technique targets the energy source of the sensors in battery energy control, where researchers have proposed battery-friendly discharge patterns [7].

13.2.2 Energy-Efficient Wireless Communications

A number of energy-efficient protocols have been proposed for medium access control (MAC) and routing in WSNs, for example, S-MAC [13], T-MAC [14], STEM [12], and Span [17] for the MAC design, and directed diffusion [18], rumor routing [19], and SPIN [20] for routing. Because alternate sensing has a similar sleep/listen mechanism that has been adopted by the sensor MAC design as well, we will briefly review some relevant energy-efficient MAC designs.

13.2.2.1 Contention-Based MAC

S-MAC [13] is a contention-based MAC protocol for energy saving in WSNs. It uses a simple scheduling scheme that allows neighbors to sleep for long periods and synchronizes wakeup. During the wakeup-period state, nodes use CSMA/CA for communication. It reduces the energy wasted in idle listening significantly, but it increases latency because the data arrived during sleep is queued until the next active cycle. T-MAC [14] tries to improve S-MAC by introducing the adaptive duty cycle by dynamically ending the active part of it. This also reduces the amount of energy wasted in idle listening. The STEM [12] protocol provides a way to establish communications when the nodes are sleeping. The protocol uses two radio architectures, data radio and wakeup radio. The sender sends data to the target nodes by using the data radio. The wakeup radio uses an ultralow power radio to wake up the target nodes. If a node wants to establish communication, it starts sending beacons polling a specific user. Within a bounded time, the polled node will wake up and start communicating. An alternative approach is adopted in Ref. [15] that allows a sensor to wake up a neighbor with a busy tone instead of using beacons. However, it introduces a problem: Each busy tone must wake up a node's entire neighborhood because the intended receiver's ID is not encoded on the wake-up channel. The main contribution of Ref. [15] is selectively waking up the data radio at nodes that have previously engaged in communication via rate estimation. Span [17] reduces the energy consumption of a multihop ad hoc wireless network by selecting a connected set of nodes as coordinators and turning the rest of the nodes off. Span coordinators stay awake and perform a multihop packet routing.

13.2.2.2 TDMA-Based MAC

TDMA-Wakeup (TDMA-W) is a scheduled TDMA-based protocol [16]. In a typical TDMA scheme, each node is assigned a time slot and can only receive and transmit during its allocated time slot. In TDMA-W, each node has two slots: transmit/send slot (S-slot) and wakeup slot (W-slot). A node listens to a channel on its S-slot. If addressed by another node, it starts listening when the S-slot associates with the sender. When nodes are more than two-hop distances apart, a W-slot can be shared by one or more nodes [16]. One disadvantage is collisions occur when more than one wakeup messages are received in a single TDMA-W frame. This leads to the node starting to listen to all its neighbors to determine the source of the data.

13.2.3   Energy-Efficient Sensing

Although many energy saving schemes (for example, [12–14,17]) have been proposed for the wireless communications module, very few have been considered specifically for the sensing module. Recent measurement results show that the sensor nodes spend a lot of energy when taking measurements [11] and the energy expenditure of sensing is comparable to that of data communications (transmissions and receptions). Moreover, the frequency of sensing is usually higher than that of wireless communications. For instance, a sensor that monitors fire in a forest only sends an alarm signal when the measured temperature level exceeds a threshold. As a result, it is necessary to study energy saving schemes for sensing, in other words, put the sensor board to sleep when no active sensing is in process. In this section, we will review energy-efficient sensing techniques. Then we will focus on energy-efficient sensing for detecting dynamic noncooperative events, in the rest of the chapter.

In general, there are two types of approaches considered for energy-efficient sensing:

1. Adaptive sensing duty cycles: In the adaptive sensing approach, adaptive adjust-ment of sleep/sensing duty cycles is proposed to best estimate/fit the pattern of event occurrence.
2. Coordinated/Collaborative sensing: In coordinated/collaborative sensing, collab-oration schemes are proposed to explore spatial–temporal coordination of sensors that monitor a common area or events.

13.2.3.1   Adaptive Sensing Duty Cycles

WSNs are usually deployed to monitor static or dynamic events. Static events (such as temperature and humidity), are easy to measure. On the contrary, dynamic events are typically noncooperative and thus are not easy to catch. The movement of an enemy's vehicle in a battlefield, migration of whales in the ocean, are examples of noncooperative events. They are not easily monitored as they come and go. As a result, they can be observed only if the sensors are constantly monitoring the environment. However, energy efficiency is a serious concern if the sensor boards are always ON. Hence, it is critical to study energy saving schemes for sensing of dynamic noncooperative events.

There are several works on monitoring static events, such as ELECTION [21]. In ELECTION [21], sensor nodes sense the environment periodically and turn sensors off during sleep. It also adjusts sleep cycles according to the measurements. To measure time-critical data, nodes switch to the radio mode only when the measurement exceeds a threshold value and report to the base station.

Algorithms for optimal data collection and data fusion in a large environmental sensor network are reported in Ref. [27]. An inductive model using the exponential back-off policy is used to collect optimal amount of data. However, the double/half algorithm for controlling the sensing cycle is restrictive. The response to environmental changes may be slow. Indeed, the authors also pointed out that "the technique of doubling and halving the sleep times in the data collection algorithm can be improved by tuning to varying degrees of increase or decrease depending on the data dynamics." Another issue of Ref. [27] is that all sensor nodes sense the environment at the same time. This may be improved by coordinating with neighboring sensor nodes. The problem of data fusion is addressed by the introduction of a novel concept of a super-sensor, based on self-organization and collaboration among sensors [27].

The proposed method in Ref. [27] is improved in DANCE [28], where both the measured data and neighbor sensing information are considered to determine the sensing schedule. In DANCE, each sensor node changes its sensing time if neighbor nodes have a similar sensing schedule. Otherwise, if there are no neighbors which try to sense the phenomena, the sensor node performs the sensing task. This reduces the inefficiency of energy consumption while increasing the reliability of the sensing phenomena because DANCE scatters the point of sensing time and minimizes the delay between change of field and detection [28].

A distributed data-stream architecture is considered in Ref. [25] for large numbers of sensors that deliver continuous data to a central server. The rate at which the data is sampled at each sensor affects the communication resource and the computational load at the central server. A Kalman filter (KF) is employed such that the sensor can use the KF estimation error to adaptively adjust its sampling rate within a given range. The sensor and the server negotiate a new sampling rate when the current rate exceeds the range.

An adaptive control mechanism is proposed in Ref. [26] to change the frequency of measurements by each sensor node. The accuracy of the sensing period increases with more available data and the variation in the sensing period will adapt to the sensed environment. However, the difficulty resides in modeling the realistic environment.

13.2.3.2 Coordinated/Collaborative Sensing

A major goal in sensor networks is to preserve the coverage of the sensing network for the maximum possible time. Network coverage problem determines how well a sensor network is monitored by the sensors. In the past years, a number of researches have been done on the area of coverage for a sensor network.

Meguerdichian et al. [29] define a sensor-coverage metric called surveillance that can be used as a measurement of quality of service provided by a particular sensor network. An optimal polynomial time algorithm that uses graph-theoretical and computational geometry constructs was proposed for solving for best- and worst-case coverage. However, their algorithms rely heavily on specific geometric structures such as the Delaunay triangulation and the Voronoi diagram, which cannot be constructed locally or even efficiently in a distributed manner. More efficient distributed algorithms are provided in Ref. [32] by improving the method proposed in Ref. [29]. An energy-efficient surveillance system (ESS) is proposed by He et al. [52], in which the trade-off between energy consumption and surveillance performance is explored by adaptively adjusting the sensitivity of the system.

The sleep scheduling problem is formulated as a maximization problem with constraints on battery lifetime and sensing coverage in Ref. [50]. A centralized and a distributed algorithm are also presented to maximize network lifetime while achieving k-coverage. The proposed scheme can guarantee a specific degree of sensing coverage assuming that the sensor density is high.

In Ref. [53], a distributed scheduling mechanism called lightweight deployment-aware scheduling (LDAS) is proposed. It is assumed that sensor nodes are not equipped with GPS or other devices to obtain location information. Because it is difficult to determine whether a nodes sensing area is absolutely covered by other nodes without location information, LDAS can only achieve a specific level of sensing coverage in a statistical sense.

PEAS proposed in Ref. [22] consists of two simple algorithms: Probing Environment and Adaptive Sleeping that determine (1) which sensors should work and how a wakeup sensor makes the decision of whether to go back to sleep, and (2) how the average sleep times of sensors are dynamically adjusted to keep a relatively constant wakeup rate, respectively. PECAS is proposed in Ref. [51] and it is an extension of the PEAS protocol. It advertises the remaining working time in its reply messages to its neighbors' probe messages so that blind spots can be prevented. In summary, PECAS achieves better coverage than PEAS at the expense of higher energy consumption due to the higher message-exchange overhead.

In Ref. [49], a simple self-scheduling mechanism called randomized independent scheduling (RIS) is proposed to extend the network lifetime while achieving an asymptotic k-coverage. RIS does not require location or distance information and it has no communication overhead.

In Refs. [24,30], a node-scheduling scheme was proposed to reduce the system's overall energy consumption by turning off some redundant nodes in sensor networks. They use probabilistic probing schemes to determine when a node can be turned off and when it should be rescheduled to become active again. To avoid neighboring nodes turning off simultaneously, a back-off-based approach was designed.

Note that schemes described in Refs. [22,24,30] cannot guarantee full coverage of an area. In these schemes, a sensor node does not perform sensing if there exist neighboring nodes that are active. However, the sensing scope of the sleeping node may not be fully covered by neighboring nodes.

A balanced-energy scheduling (BS) scheme is proposed for dense cluster-based sensor networks [23]. The BS scheme aims to evenly distribute the energy load of the sensing and communication tasks among all the nodes in the cluster, thereby extending the time until the cluster can no longer provide adequate sensing coverage.

The coverage problem that alternates between active and sleep states to conserve energy is investigated in Ref. [31]. The authors consider two types of mechanisms in the context of coverage: the random-sleep type, where each sensor keeps an active-sleep schedule independent of another, and the coordinated-sleep type, where sensors coordinate with each other in reaching an active-sleep schedule. It is shown with either type of sleep schedule the benefit of added redundancy saturates at some point in that the reduction in duty cycles starts to diminish beyond a certain threshold in deployment redundancy. It is also shown that at the expense of an extra control overhead, a coordinated sleep schedule is more robust and can achieve a higher duty cycle reduction with the same amount of redundancy compared to a random sleep schedule.

The notion of a connected sensor cover is introduced in Ref. [38] and a centralized approximation algorithm that constructs a topology involving a near-optimal-connected sensor cover is developed. Given a spatial query requesting data of interest in a geographical region, the proposed greedy algorithm recurrently selects a path of sensors that is connected to an already selected sensor until the given query region is completely covered.

An efficient method is proposed in Ref. [37] to extend the sensor-network lifetime by organizing the sensors into a maximal number of set covers that are activated successively. Only the sensors from the current active set are responsible for monitoring all targets and for transmitting the collected data, while all other nodes are in a low-energy sleep mode.

The sensing coverage problem also extends to mobile sensor networks. The goal is to maximize the coverage of a given target area with constraints on deployment time, the distance the sensors have to travel to maximize coverage and the complexity of the protocol. The gains attained by mobile sensors over static sensors, and optimal motion strategies for mobile sensors, are discussed in Ref. [46].

A potential-fields-based approach for self deployment of mobile sensor networks is proposed in Ref. [44]. It is assumed that each node is equipped with a sensor that allows it to determine the range and bearing of both nearby nodes and obstacles. The fields are constructed such that each node is repelled by both obstacles and by other nodes, thereby forcing the network to spread itself throughout the environment. The approach is both distributed and scalable in the sense that it does not require models of the environment, localization, or communication between nodes.

Heo et al. [45] proposed two schemes called Distributed Self-Spreading algorithm (DSS) and Intelligent Deployment and Clustering Algorithm (IDCA). In DSS, sensors are assumed to be randomly deployed initially, and they start moving based on partial forces exerted by the neighbors. In IDCA, local density is compared with the density expected when all nodes are uniformly distributed in the area. Then a node's movement is determined by the relative remaining energy. The idea is to reduce the variance in the remaining energy once all the nodes are uniformly distributed in the area.

In Ref. [42], Voronoi diagrams are applied to discover the coverage holes once all the sensors have been initially deployed. Ganeriwal et al. [43] proposed a protocol that uses mobility-capable sensors to repair the loss of coverage due to energy depletion in a deployed sensor network. It does not work when nodes gets physically destroyed.

The network connectivity problem is also jointly considered with sensing coverage in many proposed methods, such as those in [33–36,39–41]. When the sensors' transmission range is at least twice their sensing range, ensuring k-coverage can lead to k-connectivity [33,34]. In general, a higher degree of connectivity usually leads to a higher degree of robustness against failures. However, if the connectivity degree is too high, the data collisions among sensors may adversely affect the data delivery ratio. More recent results in Ref. [39] do not rely on the assumption that the sensors' communication ranges are no less than twice their sensing ranges. Shakkottai et al. [36] consider a grid-based sensor network consisting of sensors that may fail probabilistically, and study the coverage, connectivity, and diameter of the network. An optimal deployment pattern is proposed in Ref. [40] to achieve both full coverage and 2-connectivity for all communication and sensing ranges. For more details regarding the comparison among different proposed schemes for sensing coverage and network connectivity, readers may refer to Wang and Xiao [47].

Because sensor nodes in a WSN operate in a fully distributed manner, it is hard to organize the sensor nodes switching between the sleeping mode and the active mode without a centralized control and global knowledge of the network.

To minimize the event miss-rate and at the same time minimize and balance the energy consumption of the sensor nodes, a distributed scheme is needed to put the sensor nodes to sleep and wake them up when needed. The performance analysis of various distributed alternate sensing schemes including both fixed and adaptive schemes and their implications on the network coverage of dynamic noncooperative events will be the focus of the rest of this chapter. Specifically, the trade-off between the energy efficiency and dynamic noncooperative events coverage in a WSN is formulated as a joint optimization problem. We begin with fixed listen- and sleep-time schemes for a single sensor node and quantify the performance results in terms of the event miss-rate and the normalized average power consumption. If the statistics of the event is unknown, an adaptive scheme for a single sensor node is proposed to adapt the sleep interval with the sensing/measurement data. Both fixed and adaptive schemes are extended to multiple nodes and the coverage of the entire area of the network is taken into account in addition to the energy efficiency.

13.3   Alternate Sensing Schemes

Although the previous works make an effort to minimize energy consumption at the data collection level, none of them has emphasized on the sensing mechanism for noncooperative events. Dynamic noncooperative events can be continuous or discrete in time. Instead of recording a continuous data flow or data aggregation, the detection of dynamic noncooperative events requires a decision feedback (typically binary) to be taken into account for determining the sensing behaviors of the sensor nodes, such as duty cycles.

Figure 13.2 An alternate sensing scheme.

In this study, alternate sensing schemes are adopted in WSNs to save energy. One example of such schemes is shown in Figure 13.2. During a listen interval, the sensor nodes turn their sensor boards on and take measurements. During a sleep interval, sensors are switched off to save energy. In spite of these schemes' popularity, how to design the listen and sleep intervals appropriately to monitor dynamic noncooperative events is still an open problem.

The following notations are used in this study:

- L: Listen interval in seconds
- S: Sleep interval in seconds
- λ: Poisson rate of noncooperative events
- E_S: Unit energy consumption for sensing one event
- P_S: Average power consumption during sleep period
- P_L: Average power consumption during listen period

The design of the alternate sensing schemes may be formulated as a joint optimization problem:

$$\arg \text{Min}_{L,S} \ J \tag{13.1}$$

where

$$J = \{E[R_{\text{miss}}] + E[\bar{P}]\} , \tag{13.2}$$

the expected event miss-rate, $E[R_{\text{miss}}]$, is given by

$$E[R_{\text{miss}}] = E\left\{ \frac{\lambda S}{\lambda S + \lambda L} \right\} = \frac{E[S]}{E[S] + E[L]} , \tag{13.3}$$

and the normalized average power consumption, $E[\bar{P}]$, is given by

$$E[\bar{P}] = \frac{E[L] + \frac{P_S}{P_L} E[S]}{E[L] + E[S]} . \tag{13.4}$$

It is assumed that the detection of events from measurement data is perfect. In other words, whenever an event happens within the sensor node's sensing range, and the sensor is ON, then the event will be detected without error. The performance index, J, addresses the fundamental trade-off between the event miss-rate (due to sensor sleep) and energy consumption. A sequence of listen and sleep intervals need to be determined to minimize the sum of the expected event miss-rate and the normalized average power consumption.

If the final state of the network is known, a dynamic programming approach may be applied to solve this optimization problem. However, this is not the case in practice and other suboptimal design methods (heuristics) are needed.

Two types of alternate sensing schemes are considered here: fixed timer (FT) schemes and adaptive timer (AT) schemes. In an FT scheme, both the sleep interval and the listen interval are predetermined and remain fixed once they have been set. In an AT scheme, each node tries to adjust its sleep schedule dynamically according to the (estimated) frequency of the events and some design parameters specified by users.

FT schemes may be useful when the statistics of the events is known or can be estimated in advance. AT schemes are more flexible than FT schemes and have the capability of estimating the statistics of the events and adjusting the trade-off between the event miss-rate and energy consumption on the fly, which are desirable in practice.

In this chapter, we propose the following AT scheme where the duration of the sleep interval changes according to the AIMD rule [55]

$$S(k) = \begin{cases} S(k-1) + \delta S(k) & \text{if } I(L(k-1)) = 0 \\ \beta(k)S(k-1) & \text{if } I(L(k-1)) = 1 \end{cases}$$

where $\delta S(k) > 0$ is the incremental step size and $0 < \beta(k) < 1$ is the decreasing factor at a time step k. Both of them are design parameters. $I(.)$ is an indicator function and defined by

$$I(t) = \begin{cases} 1 & \text{if event is detected during } t \\ 0 & \text{otherwise} \end{cases}$$

The essence of the proposed AT scheme is to probe the statistics of the events dynamically and tune the sleep interval accordingly. The two design parameters provide the capability of a trade-off between the event miss-rate and energy consumption. For example, large β saves more energy but increases the miss-rate.

Readers may notice that the proposed AT scheme uses the same principle as that used in the TCP congestion control [56]. Indeed, the essence of the TCP congestion control is to measure the path throughput (congestion level) and tune the transmissions accordingly. Similarly, the essence of the proposed AT scheme is to estimate the trend of event occurrence and adapt the sleep schedule accordingly.

Note that the proposed AIMD scheme will drive the network to the optimal operating point, where the best trade-off between efficiency and fairness is obtained [56]. The efficiency is represented by the inverse of the average event miss-rate, while fairness is reflected by the average power consumption of the sensor nodes. It is worth pointing out that unlike AIMD, many previous approaches employing an exponential back off, multiplicative increase/multiplicative decrease (MIMD), may not converge to the optimal operating point, as proved in Ref. [56].

Although the proposed AIMD scheme is very simple, it works well in practical sensor networks. Current sensors have a very limited memory space for storing programs. For instance, the MICA2 mote has only 8 kB of memory [8]. Moreover, sensors usually have very limited computational power and they are difficult to debug. Therefore, simple mechanisms are more likely to be deployed in WSNs.

13.4 Performance Analysis

In this section, discrete-event simulations are performed to compare the FT scheme and the proposed AT scheme. OPNET [59] is selected as the simulation tool. The performance of the pure FT scheme can be found in Ref. [55] and is omitted here. In summary, the event miss-rate and energy consumption depend on the sleep and listen intervals of a node using the FT scheme. Hence, to minimize the energy consumption and miss-rate, the network designer can choose an appropriate operating point for the sensor network on the miss-rate and the energy trade-off curve that best suits the needs of the application.

In this simulation study, the noncooperative events are generated according to a Poisson process with the Poisson rate, λ, and λ varies along time. $\lambda = 1$ during the first 20 seconds, then increases to 10 from 20 to 40 seconds, then it drops to 0.1 from 40 to 100 seconds, and then returns to 1 thereafter. The purpose of the changing event rate is to create a realistic (dynamic) environment and compare the FT scheme with the proposed AT scheme.

The event miss-rate and normalized average energy consumption using two FT schemes (with $S = 1$ and $L = 1$, and $S = 0.1$, and $L = 1$, respectively) and the proposed AT scheme are shown in Figure 13.3. It is observed that the FT scheme with an inappropriate setting (with $S = 1$ and $L = 1$) gives an unacceptable performance. An FT

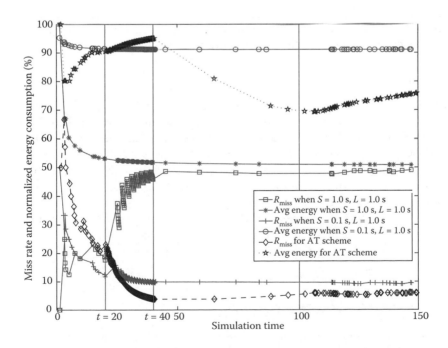

Figure 13.3   Event miss-rate and normalized average energy consumption using two FT schemes and the proposed AT scheme (with $\delta S = 0.1$ and $\beta = 0.5$).

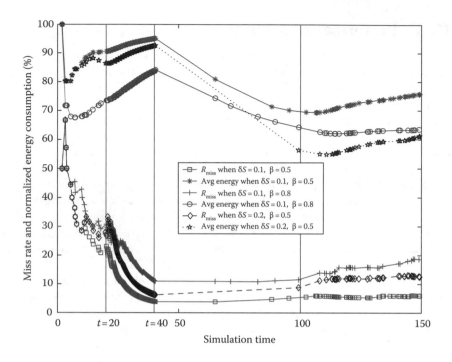

Figure 13.4 **Event miss-rate and normalized average energy consumption using the proposed AT scheme with different parameter settings.**

scheme with a better setting (with $S = 0.1$ and $L = 1$) can achieve a good performance. However, setting parameters appropriately in the FT scheme require the knowledge of the event process a priori. On the contrary, no knowledge of the event process is needed for the proposed AT scheme. More importantly, the proposed AT scheme outperforms the FT scheme in both the event miss-rate and normalized average energy consumption.

The event miss-rate and normalized average energy consumption using the proposed AT scheme with different parameter settings are shown in Figure 13.4. It is observed that the proposed AT scheme performs well under a wide range of parameter settings. It also demonstrates the flexibility of the proposed AT scheme.

13.5 Networkwide Coverage

In the previous sections, both the FT scheme and the AT scheme have been studied for a single node to monitor dynamic noncooperative events. However, in practical applications, it is often necessary to monitor the entire coverage area of a WSN. For example, a WSN may be deployed to monitor a battlefield for possible enemy movements.

Hence, it is interesting to study how the FT scheme and the AT scheme will perform networkwide together with various network coverage schemes.

In general, it is very difficult to coordinate sensor nodes without a centralized control. A big overhead penalty will be introduced for a distributed control. Furthermore, for applications whereby coordination of sleeping among sensors is not possible or is inconvenient, random sleeping is the only option. Thus, sensor nodes will adopt an asynchronous scheme in this work, where sensors do not need to synchronize with each other and each sensor will independently determine its sleep schedule according to the proposed AT scheme discussed in Section 13.3. Contrary to most previous works on network coverage, we will combine the design of adaptive duty cycles with geographical coverage and derive the results in an integrated fashion with minimum assumptions and signaling requirements. The event miss-rate will be determined by both the event occurrence and the adaptive duty cycle of each individual sensor, not simply the geographical locations of the sensors.

13.5.1 Theoretical Results

The following notations are used in our analysis:

- (X, Y): The two-dimensional coordinates (location) of a sensor node
- N: Number of sensor nodes in the network
- r: Sensing range of a sensor node
- A: The area of the network

It is assumed that random noncooperative events happen according to a Poisson process with the Poisson rate, λ. The location of the event is randomly chosen and is assumed to be uniformly distributed across the network.

Theorem 13.1
Assuming that there are N sensor nodes uniformly distributed in a WSN of area A, and each sensor node has sensing range r, the miss-rate (the probability that a random event will be missed) is given by

$$R_{\text{miss}} = e^{-\frac{E[L]}{E[S]+E[L]}\frac{N}{A}\pi r^2} \tag{13.5}$$

Proof   The probability that a random event has N_0 sensor nodes in the range is

$$P_0 = \frac{(\pi r^2 N/A)^{N_0}}{N_0} e^{-\pi r^2 N/A} \tag{13.6}$$

The expected number of sensor nodes that are in the range of a random event is

$$E[N_0] = \frac{N}{A}\pi r^2 \tag{13.7}$$

The expected number of sensor nodes that are in the range of a random event and in the listen mode is

$$E[N_0^L] = \frac{E[L]}{E[S] + E[L]}\frac{N}{A}\pi r^2 \tag{13.8}$$

Then the probability that a random event will be sensed by some senor node is $1 - e^{-\frac{E[L]}{E[S]+E[L]}\frac{N}{A}\pi r^2}$. The theorem follows.

This theorem states that the event miss-rate is a function of the average number of nodes per unit area and the sleep schedule. Discrete-event simulations are performed to evaluate the performance using FT and AT schemes under different average number of nodes per unit area.

It is worth pointing out that only the 1-coverage result is derived here [55]. In a recent independent work [48], the stationary *k*-coverage probability and the expected coverage periods for asynchronous random sensing are derived. For surveillance applications, the detection probability and detection delay distribution are also given in Ref. [48].

13.5.2 Simulation Results

There are *N* sensor nodes in the WSN and they are uniformly distributed in the network coverage area. Each node has a binary event detector with sensing range *r*. It is assumed that the network covers a 100 m × 100 m area. A sample sensor location map (*N* = 100) is given in Figure 13.5. Note that the nodes' locations are not optimized for coverage

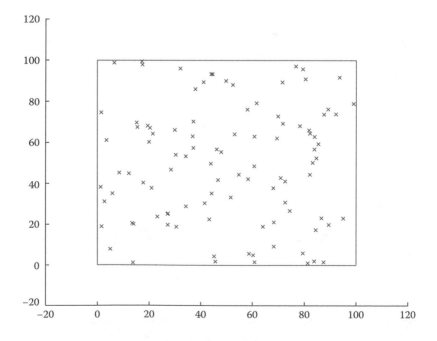

Figure 13.5   A sample WSN: node locations.

here because of practical constraints. For example, sensors are deployed by unmanned aircraft.

It is assumed that random noncooperative events happen according to a Poisson process described in Section 13.4. The location of the event is randomly chosen and is assumed to be uniformly distributed across the network. In other words, both coordinates of a random event, X and Y, are uniformly distributed in [0, 100].

To evaluate the network coverage of the FT scheme, all nodes are assumed to have the same energy budget and the event miss-rate is used as the performance criterion. A missed event means an event happens in the place where no sensor node is in the range or all sensors in the range are in the sleep mode.

The parameters for the FT scheme are $S = 0.1$ and $L = 1$, and the parameters for the proposed AT scheme are $\delta S = 0.1$ and $\beta = 0.5$, respectively. The FT scheme is compared with the proposed AT scheme in networks with different number of nodes and different sensing ranges. The results are summarized in Figure 13.6 (for different N) and Figure 13.7 (for different r). It is observed that the proposed AT scheme has a significantly less event miss rate than that of the FT scheme in all tests. Figure 13.6 also shows that when N increases the difference between the FT scheme and the proposed AT scheme starts to decrease due to the increase in the node density. When the sensing range r is very small, the geometric coverage dominates the event miss-rate. Thus, there is a small difference between the FT scheme and the proposed AT scheme. As the sensing range r increases, the sensing schemes start to play an important role in event coverage and the difference between the FT scheme and the proposed AT scheme becomes larger.

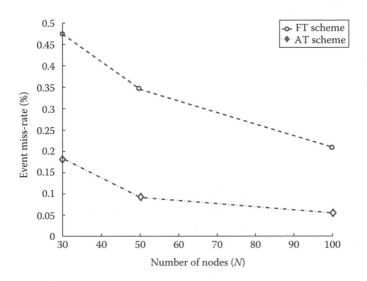

Figure 13.6   Event miss-rate vs. number of sensor nodes in the network (r is fixed and $r = 15$ m).

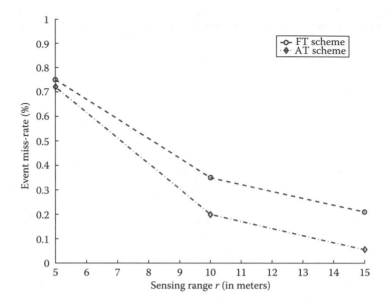

Figure 13.7   Event miss-rate vs. sensing range (*N* is fixed and *N* = 100).

13.6   Open Problems and Issues

In most of the previous approaches for an energy-efficient WSN design, sensing and communication energy costs are not clearly identified or separated. Usually an entire sensor node is assumed either sleeping or awake. Even when sensing and communication functions and energy costs are considered separately, their sleep schedules are not treated in an integrated manner. Furthermore, in many cases, the impact of different sensing applications is not taken into account when designing sleep schedules.

Previous works on networkwide coverage mainly focus on geometric analysis. Although there have been many attempts to consider the sensing range and the communication range together, the dynamics of sensing schedules and traffic loads are usually not included in the analysis.

The ultimate design goals of WSNs have two somewhat conflicting aspects: on one hand, the event miss-rate needs to be minimized across the network coverage area; on the other hand, sensor nodes have limited resources, such as energy, and they have to achieve a prolonged lifetime. To fulfill these requirements, sensing, radio communication, and data processing algorithms (such as the recently intensively pursued compressive sensing [57,58]) have to be jointly designed such that the best trade-off can be obtained. Furthermore, the energy source model has to be taken into account to accurately represent the energy flow within a sensor node. Finally, it is important to consider the problem from the networkwide coverage point of view rather than that of each individual sensor node, while integrating the dynamics of sensing schedules and communication needs in the design. The above suggestions are highlighted in Figure 13.8.

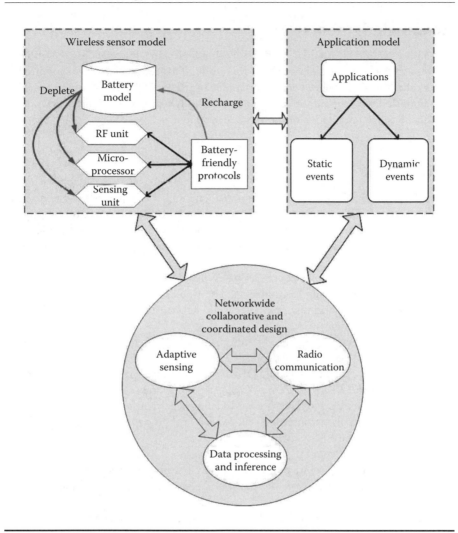

Figure 13.8   Integrated WSN design.

13.7   Conclusions and Future Work

In this chapter, the trade-off between energy efficiency and fault-tolerant event monitoring is considered through flexible and dynamic scheduling of sensing. The impact of the dynamically changing parameters in FT and AT schemes on the performance of energy-efficient sensing is investigated. Detailed simulation studies provide guidance for practical deployment. Network coverage results are also presented.

Note that the proposed schemes may be generalized such that both the incremental step size and the decreasing factor can be determined systematically in real time. Moreover, it would be interesting to compare our proposal with other approaches, such as in [54], for tracking specific targets, and it will be one of our future efforts.

Acknowledgments

The authors would like to thank OPNET Technologies, Inc. for providing the OPNET software to perform the simulations in this study. This research work is supported in part by the National Science Foundation under award 0531507. The contents are solely the responsibility of the authors and do not necessarily represent the official views of the National Science Foundation or the U.S. Government.

References

1. I. F. Akyildiz, W. Su, Y. Sankarasubramanian, and E. Cayirci, A survey on sensor networks, *IEEE Communications Magazine*, 40(8), 102–114, Aug. 2002.
2. P. H. Chou and C. Park, Energy efficient platform designs for real-world wireless sensing applications, *IEEE International Conference on Computer Aided Design*, pp. 912–919, San Jose, CA, 2005.
3. A. Hac, *Wireless Sensor Network Designs*, Wiley, New York, 2003.
4. T. Pering, T. Burd, and R. Broderson, The simulation and evaluation of dynamic voltage scaling algorithms, *Proc. Int. Symp. on Low Power Electronics and Design*, pp. 76–81, Monterey, CA, 1998.
5. R. Min, M. Bhardwaj, S. Cho, E. Shih, A. Sinha, A. Wang, and A. Chandrakasan, Low power wireless sensor networks, *VLSI Design*, Bangalore, India, Jan. 2001.
6. A. Sinha and A. Chandrakasan, Dynamic power management in wireless sensor networks, *IEEE Design and Test of Computers*, 18(2), 62–74, Apr. 2001.
7. M. Doyle, T. Fuller, and J. Newman, Modeling of galvanostatic charge and discharge of the lithium/polymer/insertion cell, *Journal of the Electrochemical Society*, 140(6), 1526–1533, June 1993.
8. http://www.xbow.com
9. S. Roundy, P. K. Wright, and J. M. Rabaey, *Energy Scavenging for Wireless Sensor Networks: With Special Focus on Vibrations*, Kluwer Academic Publishers, Boston, MA, 2003.
10. A. Kansal, J. Hsu, S. Zahedi, and M. B. Srivastava, Power management in energy harvesting sensor networks, *ACM Transactions on Embedded Computing Systems*, 6(4), article 32, pp. 1–38, Sep. 2007.
11. G. Anastasi, A. Falchi, A. Passarella, M. Conti, and E. Gregori, Performance measurements of motes sensor networks, *Proc. of the 7th ACM International Symposium on Modeling, Analysis and Simulation of Wireless and Mobile Systems*, Venice, Italy, Oct. 2004.
12. C. Schurgers, V. Tsiatsis, S. Ganeriwal, and M. Srivastava, Optimizing sensor networks in the energy-space-density design space, *IEEE Transactions on Mobile Computing*, 1(1), 70–80, Jan.–Mar. 2002.
13. W. Ye, J. Heidemann, and D. Estrin, An energy-efficient MAC protocol for wireless sensor networks, *Proc. IEEE INFOCOM*, New York, pp. 1567–1576, June 2002.
14. T. V. Dam and K. Langendeon, An adaptive energy-efficient MAC protocol for wireless sensor networks, *ACM SenSys'03*, pp. 171–180, Los Angeles, CA, Nov. 2003.

15. M. Miller and N. Vaidya, A MAC protocol to reduce sensor network energy consumption using a wakeup radio, *IEEE Transactions on Mobile Computing*, 4(3), 228–242, May–Jun. 2005.
16. M. Zohaib and T. M. Jadoon, Comparison of S-MAC and TDMA-W protocol for energy efficient wireless sensor networks, *International Conference on Emerging Technologies*, pp. 486–492, Peshawar, Pakistan, Nov. 2006.
17. B. Chen, K. Jamieson, H. Balakrishnan, and R. Morris, Span: An energy-efficient coordination algorithm for topology maintenance in ad hoc wireless networks, *ACM Wireless Networks Journal*, 8, 481–494, Sep. 2002.
18. C. Intanagonwiwat, R. Govindan, and D. Estrin, Directed diffusion: A scalable and robust communication paradigm for sensor networks, *Proceedings of the Sixth Annual International Conference on Mobile Computing and Networks (MobiCom'00)*, Boston, MA, Aug. 2000.
19. D. Braginsky and D. Estrin, Rumor routing algorithm for sensor networks, *WSNA'02*, Atlanta, GA, Sep. 2002.
20. J. Kulik, W. Heinzelman, and H. Balakrishnan, Negotiation-based protocols for disseminating information in wireless sensor networks, *Wireless Networks*, 8(2/3), 169–185, 2002.
21. S. Begum, S. Wang, B. Krishnamachari, and A. Helmy, ELECTION: Energy-efficient and low-latency scheduling technique for wireless sensor networks, *The 29th Annual IEEE Conference on Local Computer Networks (LCN)*, Tampa, FL, Nov. 2004.
22. F. Ye, G. Zhong, S. Lu, and L. Zhang, PEAS: A robust energy conserving protocol for long-lived sensor networks, *Proc. IEEE Int. Conf. Network Protocols (ICNP)*, Paris, France, 2002.
23. J. Deng, Y. Han, W. Heinzelman, and P. Varshney, Balanced-energy sleep scheduling scheme for high-density cluster-based sensor networks, *Computer Communications*, 28(14), 1631–1642, Sep. 2005.
24. D. Tian and N. D. Georganas, A node scheduling scheme for energy conservation in large wireless sensor networks, *Wireless Communications and Mobile Computing Journal*, 3(2), 271–290, Mar. 2003.
25. A. Jain and E. Y. Chang, Adaptive sampling for sensor networks, *Proc. International Workshop on Data Management for Sensor Networks*, 72, 10–16, Toronto, Canada, 2004.
26. A. D. Marbini and L. E. Sacks, Adaptive sampling mechanisms in sensor networks, *London Communications Symposium*, London, U.K., 2003.
27. R. Dantu, K. Abbas, M. ONeill II, and A. Mikler, Data centric modeling of environmental sensor networks, *Proc. IEEE Globecom*, pp. 447–452, Dallas, TX, 2004.
28. J. Lee, D. Lee, J. Kim, W. Cho, and J. Pajak, A dynamic sensing cycle decision scheme for energy efficiency and data reliability in wireless sensor networks, *Lecture Notes in Computer Science*, no. 4681, pp. 218–229, 2007.
29. S. Meguerdichian, F. Koushanfar, M. Potkonjak, and M. B. Srivastava, Coverage problems in wireless ad-hoc sensor, *IEEE INFOCOM*, pp. 1380–1387, Anchorage, AK, 2001.

30. D. Tian and N. D. Geoganas, A coverage-preserving node scheduling scheme for large wireless sensor networks, *First ACM International Workshop on Wireless Sensor Networks and Applications (WSNA)*, Atlanta, GA, 2002.
31. C. Hsin and M. Liu, Network coverage using low duty-cycled sensors: Random & coordinated sleep algorithms, *3rd International Symposium on Information Processing in Sensor Networks (IPSN)*, Berkeley, CA, 2004.
32. X. Li, P. Wan, and O. Frieder, Coverage in wireless ad hoc sensor network, *IEEE Transactions on Computers*, 52(6), 753–763, June 2003.
33. X. Wang, G. Xing, Y. Zhang, C. Lu, R. Pless, and C. Gill, Integrated coverage and connectivity configuration in wireless sensor networks, *ACM SenSys'03*, Los Angeles, CA, Nov. 2003.
34. H. Zhang and J. C. Hou, Maintaining sensing coverage and connectivity in large sensor networks, *NSF International Workshop on Theoretical and Algorithmic Aspects of Sensor, Ad Hoc Wireless and Peer-to-Peer Networks*, Fort Lauderdale, FL, Feb. 2004.
35. J. Lu, L. Bao, and T. Suda, Probabilistic self-scheduling for coverage configuration in wireless ad-hoc sensor networks, *International Conference on Sensing Technology (ICST)*, Palmerston North, New Zealand, 2005.
36. S. Shakkottai, R. Srikant, and N. Shroff, Unreliable sensor grids: Coverage, connectivity and diameter, *Proc. of IEEE INFOCOM*, San Francisco, CA, 2003.
37. M. Cardei, M. Thai, Y. Li, and W. Wu, Energy-efficient target coverage in wireless sensor networks, *Proc. of IEEE INFOCOM*, Miami, FL, 2005.
38. H. Gupta, S. Das, and Q. Gu, Connected sensor cover: Self-organization of sensor networks for efficient query execution, *MobiHoc'03*, Annapolis, MD, 2003.
39. C. Huang, Y. Tseng, and H. Wu, Distributed protocols for ensuring both coverage and connectivity of a wireless sensor network, *ACM Transactions on Sensor Networks*, 3(1), article 5, pp. 1–22, Mar. 2007.
40. X. Bai, S. Kumar, D. Xuan, Z. Yun, and T. H. Lai, Deploying wireless sensors to achieve both coverage and connectivity, *MobiHoc'06*, Florence, Italy, 2006.
41. Z. Jiang, R. Kline, J. Wu, and F. Dai, A practical method to form energy efficient connected *K*-coverage in wireless sensor networks, *ICDCSW'06*, Lisboa, Portugal, 2006.
42. G. Wang, G. Cao, and T. La Porta, Movement-assisted sensor deployment, *IEEE INFOCOM*, Hong Kong, China, 2004.
43. S. Ganeriwal, A. Kansal, and M. B. Srivastava, Self aware actuation for fault repair in sensor networks, *IEEE International Conference on Robotics and Automation*, New Orleans, LA, 2004.
44. A. Howard, M. J Mataric, and G. Sukhatme, Mobile sensor network deployment using potential fields: A distributed, scalable solution to the area coverage problem, *6th International Symposium on Distributed Autonomous Robotics Systems*, Fukuoka, Japan, 2002.
45. N. Heo and P. K. Varshney, An intelligent deployment and clustering algorithm for a distributed mobile sensor network, *IEEE International Conference on Systems, Man and Cybernetics*, Washington, DC, 2003.

46. N. Bisnik, A. A. Abouzeid, and V. Isler, Stochastic event capture in mobile sensor networks subject to a quality metric, *MobiCom'06*, pp. 89–109, Los Angeles, CA, Sep. 2006.

47. L. Wang and Y. Xiao, A survey of energy-efficient scheduling mechanisms in sensor networks, *Mobile Networks and Applications*, 11, 723–740, 2006.

48. C. Hua and T. P. Yum, Asynchronous random sleeping for sensor networks, *ACM Transactions on Sensor Networks*, 3(3), article 15, pp. 1–25, Aug. 2007.

49. S. Kumar, T. Lai, and J. Balogh, On K-coverage in a mostly sleeping sensor network, *Mobicom 04)*, Philadelphia, PA, 2004.

50. P. Berman et al., Efficient energy management in sensor networks, in *Ad Hoc and Sensor Networks*, Nova Science Publishers, New York, 2005.

51. C. Gui and P. Mohapatra, Power conservation and quality of surveillance in target tracking sensor networks, *Mobicom 04*, Philadelphia, PA, 2004.

52. T. He et al., Energy-efficient surveillance system using wireless sensor networks, *MobiSys 04*, Boston, MA, 2004.

53. K. Wu et al., Lightweight deployment-aware scheduling for wireless sensor networks, *ACM/Kluwer Mobile Networks and Applications (MONET)*, 10(6), 837–852, 2005.

54. S. Pattem, S. Poduri, and B. Krishnamachari, Energy-quality tradeoffs for target tracking in wireless sensor networks, *Lecture Notes in Computer Science*, no. 2634, pp. 32–46, Springer-Verlag, 2003.

55. L. Qian, A. Quamruzzman, and J. Attia, Energy efficient sensing of non-cooperative events in wireless sensor networks, *The 40th Annual Conference on Information Sciences and Systems (CISS)*, pp. 93–98, Princeton, NJ, Mar. 2006.

56. D. Chiu and R. Jain, Analysis of the increase and decrease algorithms for congestion avoidance in computer networks, *Computer Networks and ISDN Systems*, 17, 1–14, 1989.

57. E. Candès, J. Romberg, and T. Tao, Robust uncertainty principles: Exact signal reconstruction from highly incomplete frequency information, *IEEE Trans. on Information Theory*, 52(2), 489–509, Feb. 2006.

58. R. Baraniuk, Compressive sensing, *IEEE Signal Processing Magazine*, July 2007.

59. http://www.opnet.com

Chapter 14

Mobility in Wireless Sensor Networks

Rolland Vida and Attila Vidács

CONTENTS

Both sensor and sink mobility in wireless sensor networks (WSNs) can have several advantages when compared to static setups. Mobile sensors can be used to eliminate coverage holes, can follow the monitored events, or can concentrate on interesting areas of the monitored region, to supply more detailed measurements where needed. On the other hand, mobile sink nodes can move close to areas where some event is reported (reducing thus the communication distances), can move close to isolated sensors and collect data if multi-hop transmission is not possible, or can balance the load on the relaying sensors and thus the energy consumption in the network. In this chapter, we will discuss in detail the different solutions for sensor and sink mobility, their advantages and drawbacks, and outline the impact they might have on issues such as Medium Access Control (MAC) or routing.

14.1 Introduction

When wireless sensor networks (WSNs) came into the spotlight a few years ago, most of the applications and the majority of the research papers focused on completely static scenarios. Nevertheless, people soon realized that there are several possible advantages of sensor or sink mobility that are worth exploring. It is usually assumed that sensors have very limited energy supplies which should not be wasted for mobility. However, there are many application scenarios where sensors do not consume their own energy for moving around, but are carried by different "mobile agents;" they are mounted on wild animals for habitat monitoring, on cars to increase driving safety, or they float in the water to monitor oil slicks. In these cases of uncontrolled sensor mobility, with the topology of the sensor network changing very dynamically, it is a challenge to ensure the availability of sensor readings, their localization, and efficient gathering. On the other hand, controlled sensor mobility, i.e., enabling certain sensor nodes to move on their own according to a certain algorithm, can have as well many advantages in scenarios where a static network structure might show its limitations. Mobile sensors can thus be used to eliminate coverage holes, can follow the monitored events and concentrate on the most interesting areas, or can act as relay nodes in sparsely connected environments.

As opposed to sensor nodes, the sink might have considerable energy supplies, it might carry rechargeable or replaceable batteries using their energy to support mobility, and is thus usually more acceptable. Employing mobile sink nodes can serve several purposes. On the one hand, moving the sink nodes from time to time results in changes in the data gathering paths, which leads to a beneficial load balancing in the network. In case of single-hop networks, sensors that are far from the sink get rapidly depleted, while in multi-hop networks, this happens to the sensors neighboring the sink. In both situations, relocating the sink node helps in smoothing the energy consumption in the network. This relocation might be done randomly, but there are also algorithms that monitor the energy levels of the sensors and make the sink to avoid areas that are almost depleted.

On the other hand, mobile sinks can enhance energy efficiency by moving closer to the sensors and reducing thus the communication distances. In applications that are not time critical, sensors might buffer their readings for a period of time and deliver them to the sink only when it passes by. This can occur at a random moment, when the sink moves around randomly, or at regular intervals, when the sink follows a predefined gathering path. Using mobile sinks has also the benefit that the network does not have to be entirely connected anymore; the mobile sink is able to gather sensor data even from sparsely populated areas, where multi-hop routing could not be ensured. There are, however, applications where time-critical data delivery is a must; when an unusual event is detected somewhere in the monitored region, the sink has to be alerted immediately. Mobile sinks are useful in such cases as well, as they can adaptively move near to the current events, and reduce thus the communication paths.

When employing mobile sensors or sink nodes, there are several important elements that should be taken into account. How do the mobile elements announce their new positions to the other nodes in an energy-efficient way? How do we reconfigure the communication paths according to the new positions? How does node mobility affect other energy-optimization techniques, such as clustering or data aggregation? All the above presented aspects of sensor and sink mobility will be explored in detail in this chapter, with specific application scenarios and use cases.

The rest of the chapter is organized as follows. First we present controlled and uncontrolled sensor mobility solutions. Then, we examine why it is useful to employ mobile sinks, and detail solutions of random, predictable, controlled, and adaptive sink mobility. Next, we examine cases of virtual mobility, when no networking elements physically move, but rather the different functionalities are exchanged among the nodes, resulting in scenarios that have to be handled in a very similar manner to the cases of traditional mobility. At the end of the chapter, we analyze the different consequences of sensor or sink mobility, regarding medium access control (MAC) or routing.

14.2 Sensor Mobility

Sensor mobility is commonly considered as an extra dimension of complexity in the deployment and management of WSNs. At the same time, sensor mobility also introduces another degree of control that makes a sensor network more flexible and reconfigurable, and enables interesting optimization trade-offs.

There are two main categories of sensor mobility (see Table 14.1), according to how and why the sensor nodes are mobile. If sensors move just because of external circumstances (e.g., floating sensors in water or in air), the question is not how to move them but rather how to handle this intrinsic mobility, so as to still be able to manage and operate the network. In this case, mobility is a challenge to cope with and handle carefully. On the other hand, if there is the possibility and ability of moving the sensors, or at least of having some (direct or indirect) control on the sensors' movement, sensor mobility arises as a new tool to exploit and use, so as to manage and improve the network in some ways. Mobile sensors can move around either by themselves (e.g., via wheels, robotic legs, and mini rockets), or by attaching themselves to some kinds of transporters (e.g., robots and vehicles). Sensors with mobility capabilities are able to move around in a self-adjustable manner, or can be controlled by the network administrator or the application itself. There is a transitional area in between controlled and uncontrolled mobility, that is, for sensors that are attached to some "third-party" transporter (e.g., animals

Table 14.1   Sensor Mobility Types and Tasks

Sensor Mobility	Description	Task
Uncontrolled	Sensors move passively according to some external environmental forces (e.g., floating on water).	To handle intrinsic mobility [20].
Controlled (distributed or centralized)	Sensors can move around by themselves actively (e.g., via wheels, robotic legs). They move in a self-adjustable manner (distributed) or can be controlled by the network administrator or the application itself (centralized).	To improve/maintain network coverage; To assure connectivity; To react to environmental changes (e.g., target detection and tracking) [14]; Topology control; To improve the quality of data readings or take actions in critical regions [8,26,27].
"Third-party"	Sensors travel around being attached to some kind of transporter device (such as vehicles or individuals) carrying them.	To take advantage of transporters [32].

and uncontrolled vehicles), their moving patterns are dependent on the transporters. In this case, the network administrator or the application may have little influence on the movement of sensor nodes.

When we speak about controlled mobility as a way to improve network operation, there can be different targets to aim at. Moving sensors can come in handy to improve or maintain network coverage or to assure connectivity throughout the entire sensor field. On the other hand, relocating sensor nodes can also be used to adaptively react on environmental changes or focus on areas of interests, possibly changing in time and space as well. The task of target detection and tracking can be an example for the latter case, when tracking means physical tracking of the detected event or target.

The connectivity of a sensor field can be efficiently increased if some of the mobile sensors can facilitate data delivery, acting as intermediate relay nodes. The main idea behind is to guide these mobile relay nodes to fill up communication "holes" within the network, i.e., move in between two sensor nodes that are farther away from each other than their maximum communication radius. Such mobile facilitators can introduce extra hops to reduce transmission ranges of the other nodes.

When we speak about topology control, we mean that the mobile sensors should position themselves and adjust their transmission power according to the selected sensor deployment plans, taking into consideration the current topology of stationary sensor nodes and the coverage requirements [35]. When the coverage is in focus, most of the works concentrate on algorithms to reposition sensors to achieve a static configuration with an enlarged covered area. In Ref. [15], the dynamic aspects of the coverage that depend on the sensor movement are examined in more detail.

Additionally, the task and role of mobile sensors can be different if we consider nonhomogeneous sensor networks, i.e., sensor fields with different types of sensors, with devices having different capabilities. It is simple to argue that those sensors that are capable to move around are more expensive, larger in size, and require more energy and maintenance. Thus, a viable scenario is where only a (small) subset of all the nodes are more advanced mobile sensors, used to execute special tasks. As an example, consider a WSN application where a large number of static nodes are deployed and used to monitor a large area, collecting basic information from the entire field. If the sensor readings indicate that there is some interesting event taking place in a given region at a given time, the sophisticated mobile nodes can be instructed to visit that region and take further actions (e.g., conduct more advanced or exact measurements, and take pictures).

14.2.1   Uncontrolled Mobility

When we speak about uncontrolled sensor mobility we usually assume that the sensor nodes are not static, but move passively, according to some external environmental forces. For example, sensors deployed in the air or the ocean move according to the air or the ocean currents, drifting away from their initial position of deployment. In Ref. [20], we can see the example of a drifter, which is a small computing platform of the size of a basketball floating on the ocean surface. This floating device is usually equipped with a long underwater pedal, which moves the drifter as the ocean current moves. Sensor boards can be attached to it to detect phenomena such as oil spills or marine microorganisms. As a special case of underwater communication, these small-scale sensor nodes can communicate with each other using lower-energy acoustic signals.

In Ref. [32], the authors present a different sensing model that uses uncoordinated mobile nodes. In this model, no efforts are made to coordinate the motion of the nodes; the mobile nodes travel around being attached to some kind of transporter device, such as vehicles or individuals carrying the sensors. The sensors operate in the background, and take measurements from time to time, while the carrying device or person moves along its arbitrary path. While uncontrolled mobile nodes only measure their vicinity along the path, nodes coming across each other can aggregate their estimations to cover a larger area. Such information exchange is opportunistic, but the model can be extended with a data-exchange infrastructure to facilitate efficient information sharing. Consider a number of publicly recognized data-fusion centers, which may be placed at locations where large flows of people from different areas come across (e.g., airports, train stations, and popular meeting places in downtown areas). In contrast to the effort of deploying a large number of static sensors, this uncontrolled mobility model can be more cost-efficient for long-term monitoring, being a practical candidate for monitoring and detection over a geographic area too large to be entirely covered at once. The apparent trade-off is the latency needed to complete the measurements sequentially.

14.2.2   Controlled Mobility

Adding spatiotemporal context awareness in managing sensor networks is a challenging direction. In controlled mobility, the network manager (or the application itself!) can choose a suitable strategy to guide the mobile nodes so as to fulfill the application's requirements. As mentioned earlier, mobility can serve several different purposes in a WSN scenario. The two sides of the main question are why we want to move the nodes and how do we carry this out. The following examples try to highlight some ideas regarding the why-part.

In Ref. [26], the authors try to optimally guide a subset of the sensors toward the interior of the so-called critical geographical region to ensure the quality of data reading. Each sensor is considered capable of moving where the motion may be remotely controlled. For example, sensors can be deployed throughout a forest to monitor fires when they occur. The reactive mobility of the sensors assures robustness, because other sensors will be readily available in the critical region in case of failures. On the other hand, it is also important not to leave the rest of the environment completely unattended. One solution could be that while mobile sensors move toward the area of interest they also check for the coverage of their local neighborhood. Then, they continue to move only if there are enough remaining sensors not to leave the area unattended.

The authors of Ref. [8] deal with the problem of efficient guidance of swarms, formed by nodes with higher processing capabilities than the regular sensor nodes, toward the locations of "hot" static sensors. They propose to enhance the sensor network by deploying a limited number of mobile swarms. A swarm is a group of nodes that are physically close to each other and share the same mobility pattern. These special nodes are mobile with a relatively high speed. For example, a mobile swarm can be a group of tanks in the battlefield or unmanned aerial vehicles (UAV) moving together. The swarms can be directed to the hot spots to provide detailed information on the intended area. Different swarms can talk to each other and to the command center (see details in Ref. [8]).

In Ref. [27], the authors propose an integrated mobile surveillance and wireless sensor (iMouse) system. The iMouse system consists of a large number of inexpensive static wireless sensors and a small number of more expensive mobile sensors. The static sensors are to monitor the environment, while the mobile ones can move to certain locations (e.g., potential emergency sites) and take more advanced actions (such as taking pictures of emergency scenes and conducting in-depth analysis).

In Ref. [14], a distributed mobile sensor network is used for multiple-target detection and tracking. After a target is detected, monitoring sensors either remain stationary or begin to follow their targets. The decision whether to remain stationary or to track a target is based on a priority scheme associated with the target and the coordination mechanism between the sensors.

14.2.3   Mobility-Control Solutions

In a way, using mobility in WSNs is a kind of tool to maintain the ability and opportunity to react on the changing physical environment in which the network operates. Because the changes in the environment can be rather unpredictable and, in most cases, stochastic by nature, it would be very hard—if possible at all—to plan the mobility control strategies well in advance. However, different mobility-control strategies can still be defined in advance, the appropriate one being chosen from among them, in function of the needs of the given application and the environment.

In theory, the control on mobility can be either completely centralized, totally distributed, or something in between. On the first end of the spectrum, there is a central intelligence that has full control on the sensor nodes, including the guidance of the mobile ones as well, while the other extreme is that each single mobile node has to decide on its own when and where it wants to move. If we look for a viable solution for a deployed WSN, most probably the right choice would be a distributed solution that builds on communication among the sensing nodes as well as on decisions made locally using information available from nearby nodes to ensure robustness. The following mobility-control solutions focus on coverage improvement on one hand, and on convergence on events occurring in the environment on the other.

Maybe the simplest event-based mobility-control method is the purely reactive one (see Ref. [3] as an example). It uses only the current position of the sensor and the position of an event to determine the motion of the sensor. Each sensor takes a small step toward each event. The free parameter that is also a critical factor in this strategy is the step size. Moving by a small constant distance at a time leads to clustering around the event, or around the mean of all events, depending on the time scale at which the events occur and persist. A more sophisticated method would be to define a function $f(d)$ of the distance d between the mobile sensor and the event, where the sensor moves toward the event by the amount given by f. When f is constructed properly, it can be ensured that the sensors will be separated in clusters if the events are themselves in multiple clusters (e.g., $f(\infty) = 0$), and sensors do not have to pass or cross each other when reacting to events (e.g., $d - f(d)$ is monotonic).

When compared to the purely reactive solution, the history-based control method uses not only the location(s) of the actual event(s) but also the history and the distribution of previous events. Each sensor maintains a coarse histogram as an approximation of the distribution of the event positions (see Ref. [3] for the detailed algorithm). Since the number of events should be proportional to the number of sensors present for each region of the environment, by scaling down the event distribution appropriately, the portion of sensors that should be present at each region can be calculated. Assuming an initially uniform distribution, each sensor can determine its role based on its initial position, without the need to communicate with the other sensors.

Note that an undesirable sideeffect can emerge when sensors start to move toward the occurring event(s) in an uncoordinated way, as the network may lose overall connectivity. Thus, an important requirement in practice is to maintain, first of all, connectivity, and follow the events as closely as possible only as far as connectivity is not in danger. The importance of preserving connectivity is twofold. First, network connectivity is necessary to reliably relay information back to the sink node(s). Second, maintaining connectivity can ensure coverage as well, meaning that no portion of the area is left unattended, and thus, future events appearing at any place in the monitored area will be detected. We should note here that the radio coverage is not necessarily the same as the sensing coverage. The relation between the two strongly depends on the ratio of the radio communication range and the sensing range of a given node. As a consequence, a well-connected network in the communication sense does not necessarily guarantee full network coverage, and a fully covered area does not assure full radio connectivity either.

Assuming that each sensor has a limited sensing range, r_s, and every point in the environment should be sensed by at least one sensor, a simple method, called Complete Voronoi protocol, can be applied to assure coverage. The existence of potential coverage holes can be checked by using the Voronoi diagram of the sensors' positions (see Figure 14.1 as an illustrative example). Each sensor uses the geometry of its own Voronoi region to decide whether it is required for coverage or not. If any part of the sensor's Voronoi region is farther away than r_s (note that only vertices need to be checked), it knows that no other sensor is closer to that point. As long as each sensor maintains the coverage of its Voronoi region, the network coverage is maintained. The problem with the Complete Voronoi protocol is that it can easily be computationally infeasible, since the positions of all neighboring sensors need to be known by that particular sensor node that calculates its own Voronoi region. The location estimation is even more complicated if all sensors are mobile. One solution could be to communicate the new locations each time a sensor moves, but certainly, this can be costly in terms of energy, not to mention the difficulties of localization.

A so-called space-filling coverage method can be achieved by using potential field-based algorithms in which each mobile sensor would move away from its neighbors to produce a regular pattern in the space. In potential field-based algorithms nodes are

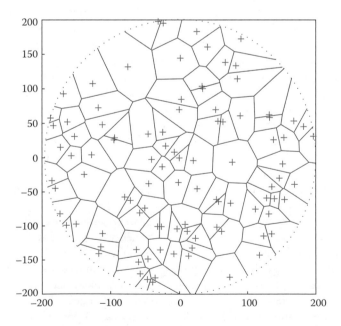

Figure 14.1   Voronoi diagram of a sensor field.

treated as virtual particles subjected to virtual forces. Typically, the virtual forces repel the nodes from each other (and from obstacles), and ensure that nodes spread out to extend coverage.

14.3 Sink Mobility

In traditional WSNs, the applications were generally based on the delivery of sensor measurements to one or more static sink nodes, in a single-hop or multi-hop manner, depending mainly on the size of the network and the radio ranges of the nodes. However, it was soon realized that mobile sinks would fit much better the needs of several application types, being also very helpful in sparing energy and extending thus the network lifetime. Let's see now some concrete arguments in favor of sink mobility.

14.3.1 Why to Move the Sink?

14.3.1.1 Data Gathering in Sparsely Populated Networks

There are many situations when covering and monitoring a very large area with a connected multi-hop WSN is not feasible or unnecessary. In case of an environmental monitoring application as an example, the task might be to monitor a large area of several square miles, but in most of the cases only some specific parts of that area are interesting. Therefore, it is reasonable to deploy most of the sensors around these sensible areas, and leave the other parts partially or even entirely uncovered. There are also cases when the harsh environmental conditions, legal restrictions, or other external factors do not allow us to deploy sensors over the entire area. As a result, it might not be possible to ensure multi-hop data gathering to a statically deployed sink node, as some parts of the network might be completely disconnected from others, i.e., there might be no wireless path between them. However, the deployed sensors can still perform their measurement tasks, and store temporarily the results, until a mobile sink passes by and fetches the data. Figure 14.2 presents such a scenario. Nodes in the two dense areas can perform multi-hop routing and deliver their data even if the sink is not next to them. Nodes in the sparse middle area are out of each otherŠs radio range; they have to store their data until the sink visits them.

14.3.1.2 Load Balancing

In a network with a static sink node, the load on different sensors might be highly unevenly distributed. In the case of a single-hop network, sensors that are far away from the sink will consume a lot of energy to deliver their data and will get depleted fast. In Figure 14.3 (left), these nodes are represented by white circles; the more energy a sensor has, the darker its circle is. As opposed to this, in a multi-hop network (Figure 14.3, middle) it is the turn of the sensors neighboring the sink to rapidly deplete their batteries, as they will act as last-hop relay nodes for the packets of all the other sensors. If that happens, the sink node might get isolated from the rest of the network, and the entire

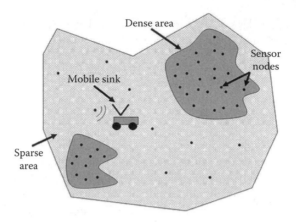

Figure 14.2   Mobile sink in an unevenly populated network.

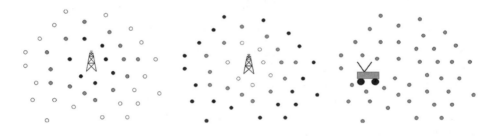

Figure 14.3   Uneven battery depletion for single-hop (left) and multi-hop (middle) networks, in case of static sink. If the sink is mobile, energy consumption is more balanced (right).

network might become useless. Finally, if a shortest path routing solution is employed in the network, sensors along those shortest paths leading to a static sink will again get rapidly depleted.

Using a mobile sink could solve many of these problems. If the sink node moves away from time to time, the load on the different sensors will change; new shortest paths will be set up and new last-hop relay nodes will be selected, and thus, the load distribution among the sensors can be much more equilibrated (Figure 14.3, right).

14.3.1.3   Shortening the Communication Paths

Sensors consume energy for sensing the field and for digitizing and processing the data, but the most penalizing task is by far the transmission of the information. In the most commonly accepted power attenuation model [21], signal power falls as $1/d^{\alpha}$, where d is the distance from the transmitter antenna and α is the attenuation exponent, a constant dependent on the wireless transmission environment, with values typically between 2 and 5. Therefore, assuming that all receivers have the same power threshold for signal

Table 14.2 Sink Mobility Strategies

Mobility Strategy	Description
Random	Sink moves randomly.
Predictable	Sink moves on a predefined path, with predefined speed(s).
Controlled	Sink moves on a predefined path, but with changing speeds that depend on the unpredictable behavior of the sensors; sensors that have much data to send control the sink by slowing it down.
Adaptive	Sink moves freely inside the network, so as to adapt to current events.

detection, typically normalized to one, the energy required to support communication between two nodes located at a distance d one from the other is d^{α}. In such conditions, it is straightforward to assert that by minimizing the distance between a sensor and a sink node, e.g., by moving the sink near the reporting sensors, we can efficiently reduce power consumption. This is true not only for single-hop networks but for multi-hop setups as well; reducing the length of the multi-hop path results in fewer or shorter hops, i.e., less energy is needed to relay data to the sink.

Having one or several mobile sink nodes moving inside the sensor network has thus many benefits: it allows data gathering from sparsely populated or disconnected network segments, and it reduces and equilibrates energy consumption among the sensors, prolonging thus the lifetime of the network. But the question is how to move those sink nodes? What kind of strategy to follow, and which parameters should influence those strategies?

The approaches based on sink mobility can be classified in four categories: random, predictable, controlled, and adaptive mobility solutions. Table 14.2 presents the main differences among these solutions. In the following, each of these categories will be detailed, presenting some specific use cases as well.

14.3.2 Random Mobility

The case of random mobility is straightforward: as its name suggests, the sink node moves around randomly inside the network, e.g., following the random waypoint mobility model (Figure 14.4). There are many solutions that are built on such a scenario.

Shah et al. use randomly moving mobile agents, called Data MULEs (Mobile Ubiquitous LAN Extension) to ensure data gathering in sparsely populated sensor networks [22]. These mules pick up data from the sensors when in close range, buffer it, and drop off the data to wired access points. By employing these mobile agents it is possible to gather measurement results even from isolated sensors or areas, from which there would be no wireless path otherwise leading to a static sink. Moreover, no sensor depends on any single mule, and hence failure of any particular mule does not disconnect the sensor from the sparse network, it only degrades the performance.

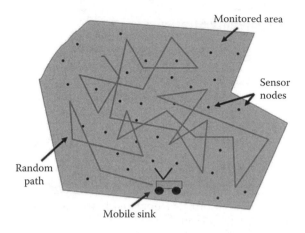

Figure 14.4   Sink following the random waypoint mobility model.

The primary disadvantage of this approach, however, is the increased latency, because sensors have to wait for a mule to approach before the transfer can occur. Thus, for some time-critical applications, where immediate alert messages have to be sent, this is not an acceptable solution. For example, if the task of the sensor network is to monitor the radiation levels in a nuclear plant, we cannot afford to wait for a mobile sink to pass by; the network has to ensure the reliable and timely delivery of an alert message, no matter which sensor detects the radiation. Nevertheless, other applications are much less sensible to delays. The underwater sensors deployed on the Great Coral Reef, near the coast of Queensland, Australia, measure the salinity, temperature, and nutrient level of the ocean's water to detect the causes of coral bleaching. In this case, delivering the measurement results is not so time critical, and a latency of a few hours might still be acceptable.

A similar approach, but for dense networks, is used by SENMA (SEnsor Networks with Mobile Agents) [25]; a mobile agent is randomly flying above the sensor field and gathers data from sensors that are triggered based on the estimated fading state of their communication with the agent. Each sensor estimates its fading state γ during a period when the mobile agent transmits a beacon message. The node then flips a biased coin with the probability, $s_n(\gamma)$, of being "head," where n is the size of the network. If the outcome is "head," the node transmits its packet. Otherwise, the node is silent for the current time slot and restarts in the next slot. Note that although all sensor nodes use the same probability mass function, because the fading conditions are different, the probability of transmission for one node is different from that of others. In practice, a sensor will start transmitting its data only when it has a favorable fading state, i.e., when the agent is flying nearby, and it will automatically stop the transmission when the agent moves away.

There are also solutions that use people's mobile phones as sink nodes to collect measurement results from sparsely deployed sensor nodes, unable to deliver the data by their own [9]. Today's mobile phones are more intelligent devices; they can use many wireless technologies (e.g., Bluetooth, ZigBee, and 802.11) to communicate with other devices in their neighborhood, and have large enough memory to store data. Thus, they can perfectly fulfil the role of a mobile sink. These solutions can also be classified as random mobility, randomness here being interpreted as "uncontrolled." People, carrying their mobile phones in their pocket, do not usually follow a random mobility pattern; they rather walk on sidewalks, drive on highways, and follow more or less predefined paths. However, their mobility is motivated by their own personal interests and tasks, and has nothing to do with the sensor network application. The network cannot rely on the fact that someone will regularly pass by, it cannot be predicted in any way when or where will a mobile phone show up. Thus, from the wireless sensors' perspective the sink nodes move "randomly."

Finally, randomly moving sink nodes are also assumed in the SEAD (Scalable Energy-efficient Asynchronous Dissemination) protocol [13], but in a scenario where there is a single sensor to transmit data to all these mobile sinks; the goal of the protocol is to build and maintain an energy-efficient multicast dissemination tree that covers all the interested nodes.

14.3.3 Predictable Mobility

In case of a predictable mobility scenario, the sink node moves along a predefined path, usually with a constant speed, or at least a predictable one. The sensors know in advance this path and the regular time intervals when the sink gets closest to them. They will try to deliver their sensor readings at these specific moments, reducing thus the energy consumption. Predictable mobility is a good model for public transportation vehicles (buses, shuttles, and trains), which can act as mobile sinks in wide-area sensor networks. Figure 14.5 shows such a scenario.

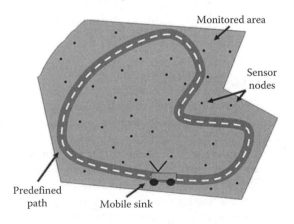

Figure 14.5   Sink moving around on a predefined path.

There are several recent research papers that build on the predictable movement of the sink node. In Ref. [5], the observer (i.e., the sink) moves along a predefined path, and pulls data from the sensors when it arrives close to them. The authors have built a prototype of the proposed model at the Rice University, with university shuttle buses carrying the mobile sinks, and the sensors being deployed on the campus buildings.

In Ref. [17], it is assumed that sensors are deployed in a circular region, and the predefined path that the sink follows is the periphery of this region. As opposed to the previous approach, where the sensors have waited for the best moment to send their report directly to the sink, here all the deployed sensors send data periodically to the sink, even if the sink is at that moment in the exactly opposite side of the region; if the data cannot be sent directly, a multi-hop path is established. The goal here was not to reduce the communication distances, but to ensure load balancing inside the network. The efficiency of the solution is compared to the case of a static sink, deployed in the middle of the circular region, and to the case of a randomly moving sink. The results show that a randomly moving sink brings already an important reduction in the energy consumption of the most heavily loaded sensors, but moving the sink in a predictable way, on the periphery of the area, enhances even more load balancing, leading to an increase of more than 500 percent in network lifetime. The drawback is again related to delivery delay: if the sink is positioned on the periphery, the data sent by a particular sensor might have to travel twice as long as if the sink would be statically deployed in the middle of the area. However, an upper bound on the delay can be easily given, as the sensor will try all the time to deliver its data, instead of waiting for an unpredictable time interval for a randomly moving sink node to pass by.

As opposed to the above cases, in Ref. [2], the mobile sink does not follow a predefined path. However, the proposed solution can still be classified as predictable mobility, as it assumes an inherent sink movement pattern that nodes in the sink's vicinity can learn over time and statistically characterize as a probability distribution. The pattern might change in time, but it is assumed that every new pattern is sustained for a significant period of time. Then, by adding a temporal dimension to the routing decision, data can be pushed toward the moving sink in an energy-efficient and robust way. The routing solution proposed in this paper is called Hybrid Learning-Enforced Time Domain Routing (HLETDR).

14.3.4 Controlled Mobility

There are also solutions based on controlled mobility. In these cases, although the sink node still moves along a predefined path, its speed is changing in time, being influenced by various external factors that cannot be predicted in advance. The mobility of the sink nodes is controlled so as to adapt to these external factors. In AIMMS (autonomous intelligent mobile micro server) [11,12], a mobile microserver moves across the network along a specific trail to route data from the deeply embedded nodes; its mobility is controlled in the sense that it spends extra time (e.g., it stops or slows down) in areas where there is a large amount of data to send, or where the current channel conditions require so. However, if the sink's movements cannot be predicted, the sensors will not know in advance when to send their data; thus, they should be alerted directly by the sink when it passes by. This fits well the query-driven data gathering model, where the sensors send their data only when queried by the sink.

In Ref. [33], a linear programming solution is given to determine the movement of the sink and its sojourn time at different points of the network, so as to maximize network lifetime, which is considered here the time until the first sensor dies. The sensors are placed on a small bidimensional grid, and generate data at a constant rate. The sink node can move only along the grid points; if a sensor is neither colocated with the sink nor on a neighboring grid point, then multi-hop routing is employed along the shortest path(s) linking the sensor to the sink.

If the solution of the LP problem indicates that the optimal sojourn time in a given grid point is zero, then the sink will not visit that point. The sink visiting order is not important, because the traveling time of the sink between nodes is considered negligible and data generation is independent of time. The results presented in the paper are quite interesting, as they show that the sink should spend most of the time in one of the grid corners or in the middle area. By placing the sink in the center of the network, we reduce the average route length from sensors to the sink; however, sensors in the middle area will have a much higher forwarding load than those near the corners. Thus, the sink should move from time to time to the grid corners and stay there for large sojourn times to exploit the large residual energy of the neighboring sensors. By following these movement patterns, all the sensors in the grid, excepting those in the corners, will deplete their batteries at nearly the same time. This leads to an increase of more than 500 percent in network lifetime, as opposed to the case of a static sink node. Note that the residual energy of the sensors in the corner points cannot be exploited at maximum: when the sink moves to a corner, all the traffic will be forwarded by the neighboring nodes, and not by the sensor colocated with the sink. On the other hand, when the sink is far away, the corner sensors will just send their own data to one of their own neighbors; no forwarding task is required from them.

This solution is however very restrictive. Linear programming does not scale for large network sizes, real networks rarely have a grid topology, and restricting the sink to move only on grid points is of course quite unrealistic as well.

14.3.5 Adaptive Mobility

The case of adaptive sink mobility is different from all the above solutions in the sense that the sink is assumed to continuously move toward an optimal position as far as energy consumption is concerned, taking into account the current events in the network. The previous solutions addressed either time-driven or query-driven scenarios. In a time-driven scenario, moving the sink node has mainly a load-balancing role: in case of multi-hop communication it would spare nearby nodes that relay packets from all the others, while in case of single-hop communication it would spare faraway nodes that have to spend much energy to reach the sink. In a query-driven scenario, the sink alerts the sensors it passes by and gathers their stored data in an off-line manner, which makes the solution suitable only for non-time-critical applications.

As opposed to these, in an event-driven scenario the adaptive mobility of the sink could result in energy-saving benefits that reach far beyond simple load balancing. If not all the sensors are active at a given moment but only those that sensed a specific event, it might be worth moving the sink toward those nodes. By doing so, the communication paths are reduced, and thus, less energy is needed for data gathering. However, if there are several simultaneous events in the monitored region, it is not obvious how to choose the optimal position of the sink.

In Refs. [30,31], authors present some possible strategies to deal with such a scenario in multi-hop networks. One solution would be to minimize the overall energy consumption in the network, given the current distribution of events. This means to find the location that gives the minimal average distance from the events, as the energy requirement of reporting an event is proportional to the distance between the event and the sink (or more precisely the reporting sensor(s) and the sink). Figure 14.6 presents this scenario. In this example, there are four zones inside the monitored area where an event was detected and the neighboring sensors start reporting to the sink. Sink S1 moves adaptively so as to minimize the minimal average distance to these events; thus, it moves close to the area where there are three nearby events.

The problem with this approach could be that, although the overall energy consumption is minimized, the energy contributions of the sensors might be rather uneven. To avoid this, one would think of minimizing the transmission energy for the most heavily loaded sensor in the network. The authors show that the maximal traffic load depends on the biggest event distance from the sink node. Thus, this strategy is equivalent with that of minimizing the maximum event distance from the sink, which is equivalent to the minimal enclosing circle problem, i.e., finding the minimum radius circle that encloses all the events. The optimal sink position is then the center point of this circle. In Figure 14.6, the mobile sink, S2, is placed at this location.

However, it is usually not realistic to assume that the sink can move directly to the optimal position; it can rather take only a step toward it in a certain period of time. Thus,

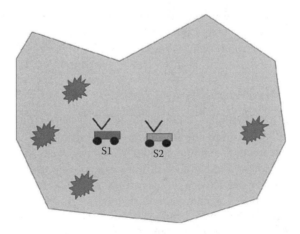

Figure 14.6 Optimal positioning of the sink node in function of the current events: S1 and S2 minimize the average and the maximum energy consumption, respectively.

the task is to continuously optimize the sink placement, and thus, energy consumption, in case of dynamically evolving events.

In Figure 14.7, we can see some simulation results showing sink position histograms in the case of a circular area of radius R of 1000 m. Events appeared in the area with a given probability, lasted for a random time period, and then disappeared. In the bottom figure, the sink moved randomly, independent of the current events. As opposed to this, the two upper figures show sink positions when adaptively moving the sink in function of the current events, and according to the different strategies. Moving the sink randomly results in a more homogenous distribution of sink positions, and a significant increase in network lifetime when compared to a stationary sink, as shown in Figure 14.8. However, a random mobility is not the best solution. The upper histograms in Figure 14.7 show that the adaptive strategies position the sink more often in the central part of the area, especially if the goal is to minimize the maximum event distance. However, this does not burden excessively the neighboring sensors, and results in a further increased network lifetime, as shown in Figure 14.8. Thus, randomly moving the sink ensures only load balancing, but might result in an equilibrated but high forwarding load. Moving the sink adaptively, in function of the current events, is more efficient, as it generates an uneven but a reduced forwarding load, and thus a longer network lifetime.

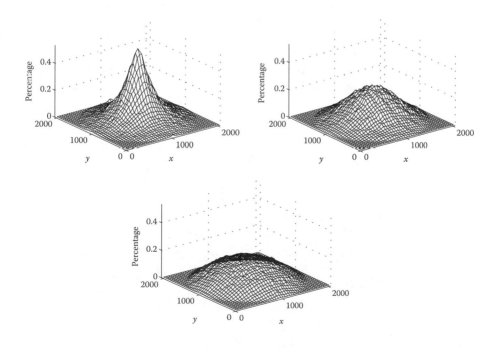

Figure 14.7 **Histogram of sink positions: when we minimize the maximum event distance (upper left), the average event distance (upper right), or just move the sink randomly (bottom).**

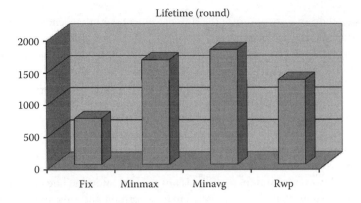

Figure 14.8 Network lifetime in case of different sink mobility strategies, when compared to a stationary sink.

In Ref. [29], the same reasoning is presented for event-driven single-hop networks. Besides the above two strategies, in a single-hop network, a third solution, based on the current energy levels of the sensors, could also be applied. The idea is to spare the energy of those sensors that have reported already many events and are near to depletion. If a new event occurs near to one such sensor, the sink will tend to spare that sensor and move closer to that event, even if other events are simultaneously present in the network. Figure 14.9 presents such a situation. In this example, there are only two parallel events; thus, the sink should be positioned somewhere at a mid-distance between these events. However, the sensors detecting the right-hand side event have less energy (shown in

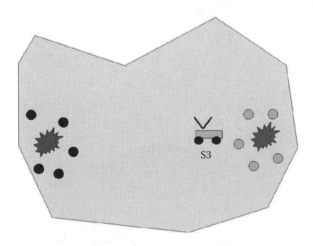

Figure 14.9 Optimal positioning of the sink if the energy levels of the reporting sensors are also considered.

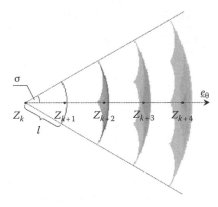

Figure 14.10   Multiple-step forecast for a mobile event.

grey) than those detecting the left-hand side event (shown in black); thus, the sink S3 will spare those sensors and move closer to them.

The situation is complicated if the events are not stationary, but move following a specific mobility pattern. Take the example of an intruder detection application, where the event is an intruder that enters the restricted area and moves with a given speed following a specific direction. In this case, the sink should not calculate its optimal position based on the current event location, but it should predict, if possible, the future positions of the event and move accordingly. Figure 14.10 shows the case of an event Z that follows a correlated random walk model; in each round it takes a step of length l, and deviates from the original direction vector e_θ, with an angle of at most σ. The grey areas show the possible locations of the event at consecutive moments.

A detailed analysis on how a mobile sink should follow a mobile event is presented in Ref. [28]. The authors analyze the case of an intruder detection and tracking application, where the goal is to find the optimal trajectory the sink should follow, so as to minimize the energy consumption of the sensors reporting about the mobile event. Optimizing the sink trajectory in the case of several parallel mobile events is of course a significantly harder task.

Table 14.3 presents a summary of the sink-mobility solutions presented in this section, together with their advantages and drawbacks.

14.4   Virtual Mobility

In the previous sections, different cases of sensor and sink mobility were discussed from different aspects. This section adds to these a new type of mobility that we term as "virtual;" this refers to the case where nothing is moving physically, but still a somehow similar movement can be observed within the network, having the same effects on the network usage and protocols.

Table 14.3 Summary of Sink-Mobility Solutions

Solution	Mobility	Description	Pros and Cons
Shah et al. [22]	Random	Data MULE picks and carries data from distant sensors.	+ Support for sparse networks, where no multi-hop path exists. – Increased latency. – Only probabilistic upper bound on delay.
Tong et al. [25]	Random	Mobile agent randomly flies above the field; sensors send data when an agent passes nearby.	+ Data sent when distance is minimal, energy is saved. – Increased latency. – Only probabilistic upper bound on delay.
Jayaraman et al. [9]	Random	People's mobile phones as sinks.	+ Support for sparse networks, where no multi-hop path exists. – Increased latency. – Only probabilistic upper bound on delay.
Chakrabarti et al. [5]	Predictable	Shuttle buses carry the sink on predefined path.	+ Data sent when distance is minimal, energy is saved. + Predictable latency. – Increased latency.
Liu et al. [17]	Predictable	Circular region, sink moves on the periphery; data is continuously sent.	+ Lower latency, data is sent periodically. + Strict upper bound on delay can be given. + Load balancing in multi-hop routing. – Long paths, larger delay, and larger energy consumption.
Baruah et al. [2]	Predictable	Path is not predefined, but nodes can learn it.	+ Efficient routing to node closest to the sink. + Robust to node failures. – Learning process and convergence time might be lengthy. – Paths longer than optimal shortest paths.

Table 14.3 (continued) Summary of Sink-Mobility Solutions

Kansal et al. [11,12]	Controlled	Sink moves along a predefined trail, spends more time where sensors have much data to send, or where bad channel quality exists.	+ Support for sparse networks, where no multi-hop path exists. + Shorter paths and lower energy consumption. + Delay bounds can be given. – Increased latency.
Wang et al. [33]	Controlled	Sensors placed on a grid; ILP solution to determine the optimal sojourn time of the sink in grid points.	+ Multi-hop routing on the shortest path, low latency. + Load balancing among the sensors and increased network lifetime. – ILP does not scale for large networks. – Grid topology is very restrictive.
Vincze et al. [30,31]	Adaptive	Sink moves to adapt to current events in the multi-hop network.	+ Lower communication distances and lower energy consumption. + Good solution for event-driven networks, which are rarely addressed. – Impossible to move to the global optimum, the sink just approaches it step by step.
Vincze et al. [29]	Adaptive	Sink moves closer to sensors that have low energy supplies in event-driven single-hop networks.	+ Spares the energy of sensors near to depletion. – Impossible to move to the global optimum, the sink just approaches it step by step.

One such example is when the location of the sink changes. As discussed earlier, changing the location of the sink nodes can be advantageous for various reasons, or even compulsory to avoid severe loss of sensor nodes due to battery depletion. An elegant solution is not to move the sink physically but to change its position by readdressing the sink roles to other sensor nodes that have favorable characteristics at the moment. From the network point of view, changing the sink location virtually results in the same changes in protocol parameters (e.g., routing tables, next-hop neighbors, and gradients) as if the sink would have moved physically to the new position. Changing cluster-head roles also results in reclustering and rebuilding intercluster communication schemes.

Sensor mobility can also be virtualized if static sensor nodes join or leave the network following a certain logic that can be beneficial from the networks point of view. Turning sensors on and off using some strategy falls within the scope of topology control. The main idea behind this is to send the unnecessary nodes to the sleep mode when the network operation does not require their presence. However, those spared nodes could come in handy later on, either when the environment changes, when the active nodes start to deplete their batteries, or simply when node failures happen. The authors of Ref. [4] propose an adaptive self-configuring WSN topology in which sensors can choose whether or not to join the network, based on the network condition. Here the sensors do not move, but the overall structure of the network has to adapt to the situation in a similar manner to the case when sensors would move inside the network, changing its topology.

Another challenging idea is to use the mobile code and mobile agents within the network, which consist of a software program transmitted from an entity to another through the network to be executed at the destination (see Ref. [19] for more details). This solution builds on one of the fundamental properties of many WSN application areas, namely, that the physical location of a sensor node might be of paramount importance regarding network operation, the whole network being organized in a location-aware way. In this case, typically the sensing and computing tasks are to be performed locally at the place where the monitored event or phenomenon takes place. In other words, the notion of a mobile agent may be easily seen as an efficient programming strategy for sensor networks, because sensing tasks may be specified as mobile code scripts that may spread across the sensor network carrying collected data items with them. In this scenario, it is the mobile code that traverses the network.

The location-aware nature of network operation is essential when the WSN application targets some kinds of events or critical regions with a well-defined physical location within the monitored area. These events or critical regions attract the attention of the entire network. In an event-driven scenario, sensors close to the events will become activated and start to send measurement data toward the sink. In a query-driven scenario, the sink directs queries toward the critical region (using mobile agents, for example). The monitored events can be anything, from real objects (e.g., vehicles, animals, or people) to some physical phenomena (e.g., spread of fire), but if they move, they move physically within the monitored region. Tracking these events introduces thus some kind of mobility in the network as well: routes from event locations toward the sink(s) have to be set up or have to be reshaped at least locally to keep up with the changing position of the focus of attention. For example, target detection and tracking is a frequently cited WSN application either in military or home security applications.

14.5 Consequences of Sensor or Sink Mobility

Using mobile sensor nodes in a WSN can, of course, have a severe impact on network operation and maintenance. First of all, mobility involves topology variations that may affect the algorithms and oblige them to tune some protocol parameters at the MAC or the networking layer, to properly perform routing and addressing tasks. The concept of the cross-layer design in sensor network planning is widely accepted and used to design performance- and cost-optimal hardware and networking-protocol solutions. Taking into account the possible presence of mobility at all layers is a must. Not doing so could result in severe network-performance degradation and application failures, as protocols designed for static networking cannot be usually applied to a continuously changing network topology. The following sections highlight the most important effects of mobility on networking protocols at different layers, pointing as well to the possible solutions to cope with mobility.

14.5.1 Medium Access Control (MAC) Solutions with Mobile Nodes

Mobility might represent an important issue for MAC protocols (see for example Ref. [19]). The main task for the MAC sublayer is to resolve the contention for the shared wireless medium used for radio communication between the nodes. MAC algorithms based on medium-reservation mechanisms may fail in the case of mobility, because the reservation procedures usually assume static nodes. A strict channel-reservation procedure that is based on allocating time slices in a TDMA (time division multiple access) system (e.g., polling or round-robin-like solutions) simply does not leave space for the newly appearing mobile nodes. One solution could be to reallocate the time slices periodically; thus, the newcomer nodes could get access to the medium in the next round. At the same time, unused slots are released by those nodes that moved away since the last allocation.

Mobility arises as an extra complexity as well when some type of clustering is introduced in the network. Partitioning the network nodes in clusters can be beneficial from many aspects and is widely used in sensor networks. The main idea behind is to distinguish inter- and intracluster communication, where intercluster communication is easier to manage and provides an efficient way for data aggregation and thus reduction of network traffic. Cluster-based protocols typically operate in two alternating phases, namely, cluster formation as the setup phase and communication period as the normal operating phase. When only static nodes are considered and the environment does not change significantly, in theory, it would be enough to set up the clusters only once at the beginning of network operation, and then use the built-up infrastructure during the entire operation phase. However, a static clustering hierarchy does not support mobility in a sense that it is problematic for a newly arriving node to join the cluster. This is especially true if the intracluster communication is solved with rigid channel partitioning as mentioned earlier. However, luckily from the mobile nodes' point of view, cluster formations are rarely (or not at all) static for the entire lifetime of network operation. For several reasons (e.g., to avoid early depletion of cluster heads and unexpected node failures) clusters are re-formed from time to time, thus giving an opportunity to the nodes to change clusters as changing their physical location. One should keep in mind that in case of high mobility the frequent reclustering can cause severe overhead for the network in terms of time and energy. In some cases, it is required that only the cluster heads must be reliable static nodes. By doing so it is easier to maintain the cluster structure even if the number of nodes in each cluster is changing dynamically.

From the mobility point of view, it would be most desirable not to have a clustered architecture nor a rigid channel-access solution. A distributed probabilistic MAC protocol, using locally available information, would be instead more beneficial, handling mobility seamlessly. (Note that this is far from being always possible, if we take into account the requirements set by the WSN application itself!) One solution could be based on the basic CSMA (carrier sense multiple access) method when all nodes listen to the shared radio channel to sense whether or not it is free. If one finds the channel empty it starts its radio transmission. However, this plain CSMA solution cannot fully avoid collisions (e.g., the so-called hidden terminal problem could arise). An efficient method to further reduce the probability of collisions is to use the RTS/CTS (Request-to-Send/Clear-to-Send) handshake before any data transmission takes place. The problem with this solution in case of mobility is that it fails because nodes can move outside the mutual coverage range after the handshake, or external nodes can get into the contention area and start transmitting unaware of any medium reservation. The communication can be further improved if it is possible to separate mobility and data transmission in a way that a given node keeps moving silently but stops completely for the time period of data communication.

To give an example from the many proposed solutions, the goal of the mobile MAC protocol called EAR (Eavesdrop-And-Register), presented in Ref. [23], is to provide the required connectivity to mobile sensors as they interact with the static network. In this scenario, some randomly distributed mobile nodes are considered as extensions to the stationary WSN. The EAR protocol attempts to provide continuous service for the mobile nodes. A positive feature of the solution is that the EAR protocol is transparent to the existing stationary protocol, by placing the mobile MAC protocol in the background. A TDMA-like frame structure is assumed for the stationary nodes, reserving the first slot for mobile sensor connections. At some semiregular intervals, an invitation message is sent out to the surrounding neighborhood with the intent of inviting appearing mobile nodes to the network.

14.5.2 Routing and Mobility

In general, it can be said that the majority of routing protocols for WSNs are designed for networks that have fixed homogeneous sensor nodes. There are literally hundreds of proposals for routing protocols in WSNs (see for example Ref. [1] for an excellent survey). Some of them support mobility in a satisfactory way, others do not. However, because the mobile nodes interact with, or are seamlessly integrated into, the network, it is highly possible that they become involved in the routing paths calculated at the network layer [23]. If the degree of mobility is relatively low, new routes can be calculated each time a particular mobile node changes its location. To avoid unnecessary recomputations, it is possible to simply recompute the routes in the neighborhood of the mobile node. Although this tends to become inefficient when the mobile node moves far away from its original location, a new routing tree has to be calculated only if necessary. There is thus an obvious trade-off between the energy spent on route setup and maintenance, and network efficiency.

Mobility may affect cluster-based algorithms because of the cost for maintaining the cluster architecture in a setup involving mobile nodes, as mentioned when discussing MAC issues. Besides intra cluster communication, at the higher hierarchy level of cluster heads, networkwide routing issues emerge. When end-to-end multi-hop paths have to be established, care must be taken to maintain the validity of the entire path by efficiently handling route breakage due to relay nodes that move away suddenly. This is especially true if one end node involved in all the network paths, i.e., the sink node, keeps moving. The simplest but least efficient solution in this case is to have all the routes torn down and rebuilt each time the sink changes its location. However, there are more sophisticated solutions to handle mobile sinks—solutions which already attracted some attention in the literature.

The first routing protocol that supports mobile sinks is TTDD (Two-Tier Data Dissemination) [16], where each data source builds up its own grid structure. To get data from a source, sinks have to access one of the corresponding grid points through an immediate agent (IA) followed by a primary agent (PA) located in the actual grid cell. If the sink moves away, a new IA and PA have to be chosen depending on how far the sink moves. Kim et al. [13] present a similar concept to TTDD in terms of having an agent, called an access node, for data forwarding. This access node represents the mobile sink during operation. After having subscribed at a source, data updates are disseminated along a tree to the access node, which forwards it to the mobile sink. The tree is updated only if one of the access nodes changes, i.e., if one of the mobile sinks changes its access node. The switch between two access nodes may be triggered by the significant increase of hop distance between a sink and its current access node.

Some other routing protocols do not demand subscription for data in advance, but restrict the mobile sinks to move on a fixed path. When a sensor is triggered by an event, the node automatically sends the corresponding information to one of the mobile sinks in the network. In Refs. [10,24], a mechanism is proposed to send sensor data to the nearest node along the path of the sink. In the initialization phase, all nodes discover the optimal route to the nearest point of the path. Then, in the second phase, nodes forward messages to the corresponding node located along the sink's path. These nodes pass the data to the mobile sink when it approaches them. The proposed solution is similar to the Delay Tolerant Network (DTN) [6] routing approach, since both mechanisms are based on the store-and-forward principle.

The MobiRoute solution [18] extends the Berkeley MintRoute [34] approach by adding a mobile-sink support to the distance-vector based routing protocol. The sink is required to move between several anchor points at which the sink stops for a longer sojourn time. The movement of the sink between two anchor points is broken down into four states: pause, premove, move, and prepause. MintRoute is used for routing in the network. However, in the prepause state, after the mobile sink stops, the routes have to be quickly updated. These updates are done by MobiRoute by sending route update messages at a high rate. In Ref. [7], the authors present a simple yet effective routing protocol that uses restricted flooding to update the paths toward multiple mobile sinks in the network. This solution tries to find a compromise between the optimal routes and the number of messages needed to update these routes.

The rate of mobility is an important question that must be taken into account. If the rate is relatively low, optimal network performance can be reached by different solutions than in the case of high mobility. Considering low mobility (e.g., slowly varying environmental conditions and slow or rare motion of network nodes), it is affordable to rebuild all the routes in the network when the situation changes. Table-driven solutions become advantageous in this case, when optimal routes are found and maintained for efficient data communication. To avoid data-flow disruptions, a so-called multi path approach can help to prevent route breakages due to node mobility. If several node-disjoint paths are maintained between the source and the destination, it is possible to switch to the secondary route if the first one fails. For slowly moving nodes, it is also possible to repair the routes locally, even before the route breaks. This can be achieved with some mobility prediction, based on received signal strength, for example. The weakening received signal is an indication that the communication distance is increasing, and it could lead to a broken link soon. There are several routing algorithms specifically designed for WSNs with low mobility, such as GAF (Geographic Adaptive Fidelity) and TTDD (Two-Tier Data Dissemination), which attempt to estimate the node trajectories [19].

The problem can be quite different when the nodes move with high relative velocities. An example is the case of intervehicular networks, where despite the constraints on the movement of vehicles (i.e., they must stay on the lanes), the network will tend to experience very rapid changes in topology [19]. Mobility is considered high also in the case when the topology changes frequently, either because nodes move fast or because many of them keep moving at the same time. In these types of scenarios, it would be really hard to constantly maintain all routes, as this would result in unacceptable overhead of routing update messages. If we still want to set up routes in a highly mobile environment, this can be done on demand rather than being computed in advance. For mobile ad hoc networks, two such solutions were proposed and are widely used, namely, AODV (Ad-Hoc On-Demand Distance Vector) and TORA (Temporally Ordered Routing Algorithm). Both are examples of demand-driven solutions. The main advantage of the reactive solutions is that it would be wasteful to set up and maintain routes that are never used. However, both AODV and TORA have high energy costs during route discovery.

The most desirable routing protocol in mobile WSNs would perhaps be one that does not use preestablished paths at all for networkwide communication. Actually, this is not at all unrealistic. There are solutions specifically for sensor networks that use flooding, gossiping, or location-aware algorithms, to spread the information within the network. The most successful solutions are those that are based on distributed protocols using only locally available information. If the application requirements allow that, these types of solutions could support mobility in a seamless way.

We should also note that the effect of mobility is not so severe on solutions employing direct communication between the end points as it is on those using multi-hop paths. However, assuming direct communication between any two points within the network (or, equivalently, direct communication between any sensor and a central sink node) would require long communication distances with high power consumption for radio communication, which is just the opposite of the typical assumption on low-power sensor nodes with limited communication abilities.

14.6 Open Issues

Mobility in WSNs raises several questions and opens research areas when examined from different perspectives.

First, designing and building sensor motes that are capable of moving, but keeping the hardware itself small, cheap, and highly energy efficient requires a cutting-edge technological background and high professional skills. The ever-developing technology based on MEMS (microelectro mechanical systems) and nowadays NEMS (nanoelectro mechanical systems) devices assures a solid background for this. The open research issues here are technology driven, and newer areas open up rapidly as the time goes by. An interesting future direction besides nanotechnology would be toward the use of nano biotechnology in WSNs.

Second, from the networking point of view, handling mobility is a challenging area as well. The protocols employed must remain simple, efficient, and distributed on one hand, to satisfy the "classical" requirements of sensor networking, but must be highly intelligent as well to adaptively cope with the increased complexity resulting from mobility. Designing self-configuring and self-managing WSNs with mobility support would be the next overall target. Designing routing solutions that hide the underlying topology changes and handle node mobility seems to be a persistent hot topic. It should be noted, however, that there will be no unified routing solution to support mobility in all kinds of WSN application areas. Different mobility solutions and application requirements lead to numerous application-tailored solutions that are efficient for that given target only.

Third, the mobility management of the mobile equipments in a sensor network should attract attention as well. It is by far not trivial what kind of strategies have to be defined and followed while moving the mobile nodes or sinks to fulfill the application's requirements in the best way. In the framework of WSNs with mobile nodes, motion planning is an important issue that has not been sufficiently addressed yet [19]. The main task of motion planning is typically the computation, possibly in a distributed way, of the paths to be followed by the mobile nodes. Typically, the computational complexity of the algorithms for motion planning and related map building are usually beyond the capabilities of sensor nodes. If we add to this the communication constraints that a WSN usually has, the problem becomes again by far not trivial to solve. Recently, a new research field has been proposed that considers the information provided by a static WSN to guide a mobile robot in a given environment [19]. The idea behind is to consider the sensor nodes as an extension of the sensorial capabilities on board of the mobile robot (even when there are no sensors on board at all).

14.7 Conclusion

This chapter presented a thorough analysis on the possible advantages of employing mobile elements, sensors and sinks, in a WSN. The presence of mobile nodes can help in prolonging network lifetime, balancing energy consumption, reducing communication paths, and extending network coverage. Nevertheless, using mobile nodes has several consequences as well, e.g., regarding routing or MAC, which have to be handled carefully to preserve the gained advantages.

References

1. J. N. Al-Karaki and A. E. Kamal. Routing techniques in wireless sensor networks: A survey. *IEEE Wireless Communications*, 11(6):6–28, 2004.
2. P. Baruah, R. Urgaonkar, and B. Krishnamachari. Learning-enforced time domain routing to mobile sinks in wireless sensor fields. In *Proc., 29th Annual IEEE International Conference on Local Computer Networks (LCN)*, pp. 525–532, Washington, DC, 2004.
3. Z. J. Butler and D. Rus. Controlling mobile sensors for monitoring events with coverage constraints, In *Proc., IEEE International Conference of Robotics and Automation (ICRA)*, pp. 1568–1573, New Orleans, LA, April 2004.
4. A. Cerpa and D. Estrin. ASCENT: Adaptive self-configuring sensor networks topologies, *IEEE Transactions on Mobile Computing*, 3(3):272–285, 2004.
5. A. Chakrabarti, A. Sabharwal, and B. Aazhang. Using predictable observer mobility for power efficient design of sensor networks. In *Proc., 2nd International Workshop on Information Processing in Sensor Networks (IPSN)*, pp. 129–145, Palo Alto, CA, April 2003. Also in *Lecture Notes in Computer Science*, 2634.
6. K. Fall, A delay-tolerant network architecture for challenged internets. In *Proc., International Conference on Applications, Technologies, Architectures, and Protocols for Computer Communications (SIGCOMM)*, pp. 27–34, New York, 2003. ACM.
7. K. Fodor and A. Vidács. Efficient routing to mobile sinks in wireless sensor networks, In *Proc., 2nd International Workshop on Performance Control in Wireless Sensor Networks (PWSN)*, Austin, TX, October 2007.
8. M. Gerla and K. Xu. Multimedia streaming in large-scale sensor networks with mobile swarms. *ACM SIGMOD Record: Special Section on Sensor Network Technology and Sensor Data Managment*, 32(4):72–76, 2003.
9. P. P. Jayaraman, A. Zaslavsky, and J. Delsing. Sensor data collection using heterogeneous mobile devices. In *Proc., IEEE International Conference on Pervasive Services*, pp. 161–164, Istanbul, Turkey, June 2007.

10. D. Jea, A. A. Somasundara, and M. B. Srivastava. Multiple controlled mobile elements (data mules) for data collection in sensor networks. In *Proc., IEEE/ACM International Conference on Distributed Computing in Sensor Systems (DCOSS)*, Marina del Rey, CA, June 2005.

11. A. Kansal, M. Rahimi, W. J. Kaiser, M. B. Srivastava, G. J. Pottie, and D. Estrin. Controlled mobility for sustainable wireless networks. In *Proc., IEEE Sensor and Ad Hoc Communications and Networks (SECON)*, Santa Clara, CA, October 2004.

12. A. Kansal, A. Somasundara, D. Jea, M. B. Srivastava, and D. Estrin. Intelligent fluid infrastructure for embedded networks. In *Proc., ACM 2nd International Conference on Mobile Systems, Applications, and Services (MOBISYS)*, pp. 111–124, Boston, MA, June 2004.

13. H. S. Kim, T. F. Abdelzaher, and W. H. Kwon. Minimum-energy asynchronous dissemination to mobile sinks in wireless sensor networks. In *Proc., 1st International Conference on Embedded Networked Sensor Systems (SenSys)*, pp. 193–204, New York, 2003. ACM.

14. K. M. Krishna, H. Hexmoor, P. S. Rao, and S. Chellapa. A surveillance system based on multiple mobile sensors. In *Proc., 17th International FLAIRS Conference: Special Track on AI Techniques in Multi-Sensor Fusion*, May 2004.

15. B. Liu, P. Brass, O. Dousse, P. Nain, and D. Towsley. Mobility improves coverage of sensor networks. In *Proc., 6th ACM International Symposium on Mobile Ad Hoc Networking and Computing (MobiHoc)*, pp. 300–308, New York, 2005. ACM.

16. H. Luo, F. Ye, J. Cheng, S. Lu, and L. Zhang. TTDD: Two-tier data dissemination in large-scale wireless sensor networks. In *Wireless Networks*, 11(1–2):161–175, 2005.

17. J. Luo and J-P. Hubaux. Joint mobility and routing for lifetime elongation in wireless sensor networks. In *Proc., IEEE INFOCOM*, pp. 1735–1746, Miami, FL, 2005.

18. J. Luo, J. Panchard, M. Piorkowski, M. Grossglauser, and J-P. Hubaux. MobiRoute: Routing towards a mobile sink for improving lifetime in sensor networks. In *Proc., 2nd IEEE/ACM International Conference on Distributed Computing in Sensor Systems (DCOSS)*, San Francisco, CA, 2006.

19. P. J. Marron. Research directions of cooperating objects with mobile nodes. In *Proc., NSF Workshop on Data Management for Mobile Sensor Networks (MobiSensors)*, Pittsburgh, PA, January 2007. Position Paper.

20. S. Nittel, N. Trigoni, and N. Pettigrew. Data management in mobile ad-hoc ocean sensor networks. In *Proc., NSF Workshop on Data Management for Mobile Sensor Networks (MobiSensors)*, Pittsburgh, PA, January 2007. Position Paper.

21. T. S. Rappaport, *Wireless Commnications: Principle and Practice*, Prentice Hall, Englewood Cliffs, NJ, 2002.

22. R. C. Shah, S. Roy, S. Jain, and W. Brunette. Data MULEs: Modeling a three-tier architecture for sparse sensor networks. In *Proc., IEEE Workshop on Sensor Network Protocols and Applications (SNPA)*, pp. 30–41, Anchorage, AK, May 2003.

23. K. Sohrabi, J. Gao, V. Ailawadhi, and G. J. Pottie. Protocols for self-organization of a wireless sensor network. *Personal Communications, IEEE [see also IEEE Wireless Communications]*, 7(5):16–27, 2000.

24. A. A. Somasundara. Controllably mobile infrastructure for low energy embedded networks. *IEEE Transactions on Mobile Computing*, 5(8):958–973, 2006.
25. L. Tong, Q. Zhao, and S. Adireddy. Sensor networks with mobile agents. In *Proc., IEEE MILCOM*, vol. 22, pp. 688–693, Boston, MA, October 2003.
26. G. Trajcevski, P. Scheuermann, and H. Brönnimann. Mission-critical management of mobile sensors: Or, how to guide a flock of sensors. In *Proc., 1st International Workshop on Data Management for Sensor Networks (DMSN)*, pp. 111–118, New York, 2004. ACM.
27. Y-C. Tseng, Y-C. Wang, and K-Y. Cheng. An integrated mobile surveillance and wireless sensor (iMouse) system and its detection delay analysis. In *Proc., 8th ACM International Symposium on Modeling, Analysis and Simulation of Wireless and Mobile Systems (MSWiM)*, pp. 178–181, New York, 2005.
28. A. Vidács and J. T. Virtamo. Minimum transmission energy trajectories for a linear pursuit problem. In *Proc., 1st EuroFGI International Conference on Network Control and Optimization (NET-COOP)*, pp. 286–295, 2007.
29. Z. Vincze, D. Vass, R. Vida, and A. Vidács. Adaptive sink mobility in event-driven clustered single-hop wireless sensor networks. In *Proc., 6th Int. Network Conference (INC)*, pp. 315–322, Nice, France, 2006.
30. Z. Vincze, D. Vass, R. Vida, A. Vidács, and A. Telcs. Sink mobility in event-driven multi-hop wireless sensor networks. In *Proc., 1st Int. Conf. on Integrated Internet Ad hoc and Sensor Networks (InterSense)*, page Article No. 13, Nice, France, 2006.
31. Z. Vincze, D. Vass, R. Vida, A. Vidacs, and A. Telcs. Adaptive sink mobility in event-driven densely deployed wireless sensor networks. *International Journal on Ad Hoc and Sensor Wireless Networks*, 3(2–3):255–284, 2007.
32. K-C. Wang and P. Ramanathan. Collaborative sensing using sensors of uncoordinated mobility. In *Proc., IEEE/ACM International Conference on Distributed Computing in Sensor Systems (DCOSS)*, vol. 3560 of *Lecture Notes in Computer Science*, pp. 293–306. Springer, Marina del Rey, CA, 2005.
33. Z. M. Wang, S. Basagni, E. Melachrinoudis, and C. Petrioli. Exploiting sink mobility for maximizing sensor networks lifetime. In *Proc., 38th Hawaii International Conference on System Sciences*, Big Island, HI, January 2005.
34. A. Woo, T. Tong, and D. Culler. Taming the underlying challenges of reliable multihop routing in sensor networks. In *Proc., 1st International Conference on Embedded Networked Sensor Systems (SenSys)*, pp. 14–27, New York, 2003. ACM.
35. V. I. Zadorozhny. Mobility-aware query optimization in data intensive mobile sensor networks. In *Proc., NSF Workshop on Data Management for Mobile Sensor Networks (MobiSensors)*, Pittsburgh, PA, January 2007. Position Paper.

Chapter 15

Security in Wireless Sensor Networks

Sk. Md. Mizanur Rahman, Nidal Nasser, and Tarek El Salti

CONTENTS

Wireless sensor networks (WSNs) are networks of tiny sensing devices spread over a large geographic area, and can be used to collect and process environmental data like temperature, humidity, light conditions, seismic activities, and images of the environment. This data can be used to detect certain events and to trigger activities.

As sensor networks move closer toward widespread deployment of mission critical, security issues become a central concern. So far, much research has focused on making sensor networks feasible and useful, and has not concentrated on security. On the other hand, due to inherent resource and computing constraints as well as a lack of infrastructure, security in sensor networks poses different challenges than traditional network security.

These limitations represent major obstacles to the implementation of traditional computer security techniques in a WSN. The unreliable communication channel and unattended operation make the security defenses even harder. Fortunately, the new problems also inspire new researches and represent an opportunity to properly address sensor network security from the start. Many researchers have begun to address the challenges of maximizing the processing capabilities and energy reserves of wireless sensor nodes while securing them against attackers. The security issues we are going to discuss include traditional ones like secure and efficient routing, data aggregation, key management, and intrusion detection, as well as those particular to WSN, like building a sensor trust model and defense against physical attacks.

In this chapter, we are going to cover the major topics in WSN security, present the obstacles and the requirements in this field, classify many of the current attacks, and finally list their corresponding defending measures.

The chapter is organized as follows. First, we give an introduction and a background about security problems in sensor networks. We also talk about the difficulties of security in WSNs. Second, we list the security requirements of a WSN. Third, we categorize the major attacks on sensor networks, and outline the corresponding defending measures. Finally, we conclude the chapter by discussing some possible solutions to the security problems.

15.1 Introduction

Sensor networks must arrange several types of data packets, including packets of routing protocols and packets of key-management protocols, the latter being for security concern. The key establishment technique employed in a given sensor network should meet several requirements to be efficient. These requirements may include supporting in-network processing and facilitating self-organization of data, among others. However, the techniques for a secure application must minimally incorporate the following security goals: authenticity, confidentiality, integrity, privacy, scalability, and flexibility [1].

15.1.1 Security Goals

■ Authenticity: Authenticity enables a node to ensure the identity of the peer node it is communicating with, without which an attacker would impersonate a node, thus gaining unauthorized access to the resource and sensitive information and interfering with the operation of other nodes. The secure technique should guarantee that the communication nodes in the network have a way for verifying the authenticity of the other nodes involved in a communication, i.e., the receiver node should recognize the assigned ID of the sender node.

■ Confidentiality: Confidentiality ensures certain information is never disclosed to unauthorized entities. The secure technique should protect the disclosure of data from unauthorized parties. An adversary may try to attack a sensor network by acquiring the secret keys to obtain data. A better key technique controls the compromised nodes to keep data from being further revealed.

■ Privacy: Privacy is one of the key primitives for securing a sensor network in terms of disclosure of identity. By knowing the identities of the nodes in the network, outsider parties can set up some severe target-oriented attacks. As a result, the scare sources of the sensor nodes are drastically exhausted and network communication disrupted. To ensure privacy of the nodes, an anonymous communication protocol can be an effective tool for communication in the network without disclosing the IDs of the nodes.

■ Integrity: Integrity means no data falsification during transmissions. That is message being transmitted is never corrupted. In terms of key establishment techniques, the meaning is explained as follows. Only the nodes in the network should have access to the keys and only an assigned base station should have the privilege to change the keys. This would effectively prevent unauthorized nodes from obtaining knowledge about the keys used, and preclude updates from external sources.

■ Scalability: Efficiency demands that sensor networks utilize a scalable secure technique to allow for the variations in size typical of such a network. Techniques employed should provide high-security features for small networks, but also maintain these characteristics when applied to larger ones.

■ Flexibility: Secure techniques should be able to function well in any kind of environments and support dynamic deployment of nodes, i.e., the techniques should be useful in multiple applications and allow for adding nodes at any time. One of the challenges in developing sensor networks is to provide high-security features with limited resources. Sensor networks cannot be costly made as there is always a great chance that they will be deployed in hostile environments and captured for key information or simply destroyed by an adversary, which, in turn, can cause huge losses. Part of these cost-limitation constraints includes an inability to make sensor networks totally tamper-proof. Other sensor node constraints that must be kept in mind while developing a secure technique include battery life, transmission range, bandwidth, memory, and prior deployment knowledge.

■ Battery life: Sensor nodes have a limited battery life, which can make impractical the use of asymmetric-key techniques, like public-key cryptography, as they use much more energy for their integral complex mathematical calculations. This constraint is mitigated by making use of more efficient symmetric techniques that involve fewer computational procedures and require less energy to function.

■ Transmission range: Limited energy supply also restricts the transmission range. Sensor nodes can only transmit messages up to specified short distances because increasing the range may lead to power drain. Techniques like in-network processing can help to achieve better performance by aggregating and transmitting only processed information by only a few nodes. In this way the dissipated energy can be saved.

■ Bandwidth: It is not efficient to transfer large blocks of data with the limited bandwidth capacity of typical sensor nodes, such as the transmitter of the UC Berkeley Mica platform that has only a bandwidth of 10 kbps. To compensate, a secure protocol should only allow small chunks of data to be transferred at a time.

■ Memory: The memory availability of sensor nodes is usually 6–8 kbps, half of which is occupied by a typical sensor network operating system, like TinyOS. Secure protocols must use the remaining limited storage space efficiently by storing keys in memory, buffering stored messages, etc.

■ Availability: Availability ensures survivability despite Denial of Service (DoS) attacks. On the physical and media access control layer, the attacker can use jamming techniques to interfere with the communication on the physical channel. On the network layer, the attacker can disrupt the routing protocol. On higher layers, the attacker could bring down high-level services, e.g., key-management service.

■ Nonrepudiation: Nonrepudiation ensures that the origin of a message cannot deny having sent the message.

■ Prior deployment knowledge: As the nodes in sensor networks are deployed randomly and dynamically, it is not possible to maintain the knowledge of every placement. A secure protocol should not, therefore, be aware of where nodes are deployed when initializing keys in the network. A secure protocol is not judged based solely upon its ability to provide secrecy of transferred messages, but must also meet certain other criteria for efficiency in light of vulnerability to adversaries, including the three Rs of sensor networks: resistance, revocation, and resilience.

■ Resistance: An adversary might attack the network by compromising a few nodes in the network and then replicating those nodes back into the network. Using this attack the adversary can populate the whole network with his replicated nodes and thereby gain control of the entire network. A secure protocol must resist node replication to guard against such attacks.

■ Revocation: If a sensor network is invaded by an adversary, the secure protocol should provide an efficient way to revoke compromised nodes, a lightweight method that does not use much of the network's already limited capacity for communication.

■ Resilience: If a node within a sensor network is captured, the secure protocol should ensure that the secret information about other nodes is not revealed. A scheme's resilience is calculated using the total number of nodes compromised and the total fraction of communications compromised in the network. Resilience also means conveniently making new inserted sensors to join secure communications.

15.1.2 Challenges

Use of wireless links renders a sensor network susceptible to link attacks ranging from passive eavesdropping to active impersonation, message replay, and message distortion. Eavesdropping might give an attacker access to secret information, thus violating confidentiality. Active attacks could range from deleting messages to injecting erroneous messages, impersonating a node, etc., thus violating availability, integrity, authentication, and nonrepudiation. Nodes roaming freely in a hostile environment with relatively poor physical protection have a nonnegligible probability of being compromised. Hence, it is needed to consider malicious attacks not only from outside but also from within the network from compromised nodes. For high survivability, sensor networks should have a distributed architecture with no central entities; centrality increases vulnerability. Sensor networks can be dynamic due to frequent changes in topology, for example, an ad hoc sensor vehicular network. Even the trust relationships among individual nodes also change, especially when some nodes are found to be compromised. Security mechanisms need to be on the fly (dynamic) and not static, and should be scalable for hundreds of thousands of nodes.

15.1.3 Key Management

Cryptographic schemes, such as digital signatures, are often employed to protect both routing info as well as data. Public-key systems are generally espoused because of their upper hand in key distribution. In the public-key infrastructure, each node has a public-key/private-key pair. Public keys are distributed to other nodes, while private keys are kept to nodes themselves, and that too confidentially. The third party (trusted) called certification authority (CA) is used for key management. The CA has a public-/private-key pair, with its public key known to every node, and signs certificates binding public keys to nodes. The trusted CA has to stay online to reflect the current bindings, because the bindings could change overtime. The public key should be revoked if the owner node is no longer trusted or is out of network. A single key-management service for sensor networks is probably not a good idea, because it is likely to become the Achilles' heel of the network. If a CA is down/unavailable, the nodes cannot get the current public keys of other nodes to establish a secure connection. Also, if a CA is compromised, the attacker can sign any erroneous certificates with the private key. A naive replication of a CA can make the network more vulnerable, because compromising of a single replica can cause the system to fail. Hence, it is more prudent to distribute the trust to a set of nodes by letting these nodes share the key-management responsibility.

15.1.4   Secure Routing

The contemporary routing protocols for sensor networks cope well with message forwarding and processing but are not designed to accommodate defense against malicious attackers. No single standard protocol captures common security threats and provides guidelines to secure routing protocols. Nodes exchange network control information informally to establish routes between nodes; this is one of the potential targets for malicious attackers who intend to bring down the network. External attackers perform attacks injecting erroneous routing info, replaying old routing info, or distorting routing info to partition a network, or overload a network with retransmissions and inefficient routing. Internal compromised nodes are more severe for detection and correction. Routing info signed by each node will not work because compromised nodes can generate valid signatures using their private keys. Detection of compromised nodes through routing information is also difficult for the dynamic topology of some specific sensor networks' applications. Routing protocols must handle outdated routing information to accommodate dynamic changing topology. False routing information generated by compromised nodes can also be regarded as outdated routing information. As long as there are sufficient number of valid nodes, the routing protocol should be able to bypass the compromised nodes, this however needs the existence of multiple, possibly disjoint routes between nodes. A routing protocol should be able to make use of an alternate route if the existing one appears to have faulted.

15.2   Preliminaries

Public-key cryptography has been considered too expensive for small sensor nodes, because traditional public-key algorithms (such as Rivest Shamir Adleman algorithm (RSA)) require extensive computations and are not suitable for tiny sensors. However, the recent progress on elliptic-curve cryptography (ECC) [2] provides new opportunities to utilize public-key cryptography in sensor networks. The recent implementation of a 160 bit ECC on Atmel ATmega128, a CPU of 8 Hz and 8 bit, shows that an ECC point multiplication takes less than 1 second [3], which demonstrates that the ECC public-key cryptography is feasible for sensor networks.

The ECC can be combined with the Diffie–Hellman approach to provide the key-exchange scheme for two communication parties. ECC can also be utilized for generating digital signatures. The Elliptic Curve Digital Signature Algorithm (ECDSA) utilizes ECC to generate digital signatures for authentication and other security purposes [4,5]. Several approaches for encryption and decryption using ECC have been proposed [2,4]. Please refer to references [2,4,5] for details. Based on ECC, in Ref. [6], Du et al. proposed a key-management scheme for heterogeneous sensor networks, and compared it with other existing schemes and had shown better performance. This protocol is described in Section 15.4.1.3. Another protocol based on identity-based encryption (IBE) and pairing-based cryptography (PBC), discussed in Section 15.4.1.4, ensures better performance than the previous.

In the branch of cryptography, identity-based cryptography (IBC) [7] is an exception where an information that uniquely identifies users (e.g., IP or email addresses) can be used for both exchanging keys and encrypting data, and thus PKI is unnecessary. It only has become truly practical with the advent of PBC [8]. The bilinear pairing such as Weil pairing or Tate Pairing on elliptic and hyperelliptic curves has recently been found for applications in the design of cryptographic protocols. The first known implementation of pairings for sensor nodes based on the 8 bit/7.3828 MHz ATmega128L microcontroller (e.g., MICA2 and MICAz motes) has been investigated in Ref. [9], and it concludes that cryptography from pairings is indeed viable in resource-constrained nodes.

In Refs. [10,11], Leonardo et al. argue that IBE is the ideal cryptographic scheme for WSNs, in fact, because WSNs meet the strong needs of an IBE scheme. They further argue that WSNs are ideal scenarios for using IBE as well. They also discuss the uses and implementations of IBE in resource-constrained nodes and present some results. Specifically, they evaluate pairings, the most significant operation of IBE, over the MICAz—the new generation of MICA nodes [12]. The time-consuming part while evaluating IBE is the pairing computation. They described implementation issues and presented results of computing pairings over MICAz. MICAz is powered with the ATmega128 microcontroller (8 bit/7.38 MHz processor, 4 kB SRAM, 128 kB flash memory). Pairings are measured on a MICAz node running TinyOS [13]. The average execution time to compute a pairing is 30.21 s. The costs concerning random access memory (RAM) and read-only memory (ROM) (flash) memory are 1,831 and 18,384 bytes, respectively. Thus, the above result shows that pairing computation is feasible for sensor network nodes.

15.2.1 Elliptic Curves

Elliptic curves are considered interesting primarily as an alternative group structure. When, it comes to the implementation of common cryptographic protocols, certain advantages comes with the elliptic curve families, $E(F_q)$: $y^2 = x^3 + Ax + B$. The main advantage is that much smaller keys can be used, as there is no known polynomial-time algorithm for the discrete logarithm (DL) problem for the great majority of such curves. Given a point P on a curve E defined over a finite field F_q, where $q = p^m$, and p is a large prime; this is the problem of determining "a" for given "aP." In most circumstances, the points on such a curve form a simple cyclic group. Each point on the curve has an order. This order is the smallest positive integer, r, such that $rP = O$, where O is the identity point of the group, the so-called point at infinity. The number of points on the curve, the order of the curve, is referred to as #E. Every valid r divides #E. We also need to know the important relationship #$E = q + 1 - t$, where t is the "trace of the Frobenius," and t is relatively small—a constant for each curve. We note also the "twisted" curve, $E^t(Fq)$: $y^2 = x^3 + d^2Ax + d^3B$, where d is any quadratic nonresidue mod q. This curve has #$E^t = q + 1 + t$ points on it [14].

15.2.2 Elliptic Curve Groups and the Discrete Logarithm Problem

At the foundation of every cryptosystem is a hard mathematical problem that is computationally infeasible to solve. The DL problem is the basis for the security of many cryptosystems, including the elliptic curve cryptosystem. More specifically, the ECC relies upon the difficulty of the elliptic curve discrete logarithm problem (ECDLP) [15].

Recall that we examined two geometrically defined operations over certain elliptic curve groups. These two operations were point addition and point doubling. By selecting a point in an elliptic curve group, one can double it to obtain the point $2P$. After that, one can add the point P to the point $2P$ to obtain the point $3P$. The determination of a point nP in this manner is referred to as scalar multiplication of a point. The ECDLP is based upon the intractability of scalar multiplication products.

Although it is customary to use the additive notation to describe an elliptic curve group, some insight is provided by using the multiplicative notation. Specifically, consider the operation called "scalar multiplication" under the additive notation, that is, compute kP by adding together k copies of the point P. The multiplicative notation operation consists of multiplying together k copies of the point P, yielding the point $P^* P^* P^* P \ldots$, i.e., k times $P = Pk$.

In the multiplicative group Zp^*, the DL problem is: given elements r and q of the group, and a prime p, find a number k such that $r = qk$ mod p. If the elliptic curve group is described using the multiplicative notation, then the elliptic curve DL problem is: given points P and Q in the group, find a number such that $Pk = Q$; k is called the DL of Q to the base P. When the elliptic curve group is described using the additive notation, the ECDLP is: given points P and Q in the group, find a number k such that $Pk = Q$. In a real application, k would be large enough such that it would be infeasible to determine k in this manner.

Although most DL-based protocols were originally defined on the multiplicative group of a finite field F_q^*, the DL problem (and Diffie–Hellman problems) can of course be defined on any group. The protocols can thus be translated in terms of groups that possibly allow better security or more efficient arithmetic. In some groups, however, taking DLs is easy, for example, in the additive group of a finite field. Obviously, such groups are not suitable for cryptographic purposes, as the security is based on the hardness of the DL. Fortunately, there also exist groups in which solving the DL problem is believed to be harder than in F_q^*, where the index calculus method provides a subexponential algorithm. The group of points on an elliptic curve is an example of such a group.

15.2.3 Bilinear Pairings

Let G_1 be an additive group and G_2 be a multiplicative group of the same prime order, q. Let P be an arbitrary generator of G_1 ("aP denotes P added to itself a times"). Assume that the DL problem is hard in both G_1 and G_2. We can think G_1 as a group of points on an elliptic curve over F_q, and G_2 as a subgroup of the multiplicative group of a finite field F_{q^k} for some $k \in Z_q^*$. A mapping $\tilde{e}: G_1 \times G_1 \to G_2$, satisfying the following properties, is called a cryptographic bilinear map.

- *Bilinearity*: $\tilde{e}(aP, bQ) = \tilde{e}(P, Q)^{ab}$ for all P, $Q \in G_1$ and a, $b \in Z_q^*$. This can be restated in the following way: for P, Q, $R \in G_1$, $\tilde{e}(P + Q, R) = \tilde{e}(P, R)\tilde{e}(Q, R)$ and $\tilde{e}(P, Q + R) = \tilde{e}(P, Q)\tilde{e}(P, R)$.
- *Nondegeneracy*: If P is a generator of G_1, then $\tilde{e}(P, P)$ is a generator of G_2. In other words, $\tilde{e}(P, P) \neq 1$.
- *Computability*: A mapping is efficiently computable if $\tilde{e}(P, P)$ can be computed in the polynomial time for all P, $Q \in G_1$.

Modified Weil Pairing [16] and Tate Pairing [17,18] are examples of cryptographic bilinear maps.

15.2.4 Diffie–Hellman Problems

In this section, we recall the properties of Diffie–Hellman gap families. First of all, we assume the following terms to define Diffie–Hellman gap families.

- Let P be a point on an elliptic curve E, given by $y^2 = X^3 + \alpha X + \beta \bmod T$, where T is a prime
- $\langle P \rangle$ is a subgroup of E generated by P
- $|\langle P \rangle| = q$
- $a, b \in Z_q^*$

Hence, we can think G_1 as a group of points on the elliptic curve E. With this group we can define the following hard cryptographic problems.

- *Computational Diffie–Hellman (CDH) Problem*: Given a triple $(P, aP, bP) \in G_1$ for a, $b \in Z_q^*$, find the element abP
- *Decision Diffie–Hellman (DDH) problem*: Given a quadruple $(P, aP, bP, cP) \in G_1$ for a, b, $c \in Z_q^*$, decide whether $c = ab \bmod q$ or not
- *Gap Diffie–Hellman (GDH) Problem*: It is a class of problems where the CDH problem is hard but the DDH problem is easy
- *Bilinear Diffie–Hellman (BDH) Problem*: Given a quadruple $(P, aP, bP, cP) \in G_1$ for some a, b, $c \in Z_q^*$, compute $\tilde{e}(P, P)^{abc}$

Groups where the CDH problem is hard but the DDH problem is easy are called GDH groups. Details about GDH groups can be found in Refs. [19–21]. There are also some other Diffie–Hellman problems, which are not associated with our proposed protocol. Details about other Deffie–Hellman problems can be found in Ref. [22].

15.3   Types of Attacks

In this section, we describe the different kinds of attacks that are effective in sensor networks.

15.3.1   Passive Attacks

Passive attacks typically involve unauthorized "listening" to the routing packets or silently refusing execution of the function requested. This type of attack might be an attempt to gain routing information from which the attacker could extrapolate data about the positions of each node in relation to the others. Such an attack is usually impossible to detect, because the attacker does not disrupt the operation of a routing protocol but only attempts to discover valuable information by listening to the routed traffic.

15.3.2   Active Attacks

Active attacks are meant to degrade or prevent message flow between nodes. They can cause degradation or a complete halt in communications between nodes. Normally, such attacks involve actions performed by adversaries, e.g., replication, modification, and deletion of exchanged data.

15.3.3   DoS Attacks

DoS attacks occur when an attacker overloads nodes with useless traffic so that legitimate requests cannot be processed and resources cannot be accessed. The packets sent to the target will have randomly selected return addresses and often spoofed source addresses, so the target has difficulty finding the exact location of the attack.

Multiple adversaries such as I_1, I_2, I_3, and I_4 in cooperation can set a specific node, N, as a target to exhaust the resource of that node, as in Figure 15.1. One adversary I, with enough power, as in Figure 15.2, can set specific nodes N_1, N_2, N_3, N_4 as targets to exhaust the resources of the nodes. The main concept is to identify the node and make a target to the specific node.

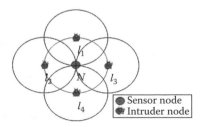

Figure 15.1   DoS, according to the target; multiple-to-one.

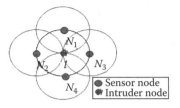

Figure 15.2   DoS, according to the target; one to multiple.

15.3.4   Wormhole Attacks

In wormhole attacks, an attacker records a packet in one location of the network and sends it to another location through a high quality out-of-band link, namely, a tunnel [23], made between the attacker's nodes in the network. Figure 15.3 shows a basic wormhole attack. The attacker replays packets received by I_1 at node I_2, and vice versa. If it would normally take several hops for a packet to traverse from a location near I_1 to a location near I_2, packets transmitted near I_1 traveling through the wormhole will arrive at I_2 before packets traveling through multiple hops in the network. The attacker can make S and D believe they are neighbors by forwarding routing messages, and then selectively drop data messages to disrupt communications between S and D. For most routing protocols, the attack has an impact on nodes beyond the wormhole endpoints' neighborhoods also. Node S will advertise a one-hop path to D so that C will direct packets toward D through S. In almost all on-demand routing protocols, the wormhole attack can be mounted by tunneling ROUTE REQUEST messages directly to nodes near the destination node. Because the ROUTE REQUEST message is tunneled through a high quality channel, it arrives earlier than other requests. According to the protocol, other ROUTE REQUEST messages received for the same route discovery will be discarded. This attack thus prevents other routes from being discovered, and the wormhole will have full control of the route. The attacker can discard all messages to create a DoS

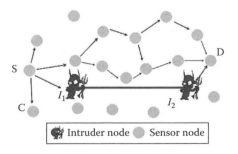

Figure 15.3   Wormhole attack: the adversary controls nodes I_1 and I_2 and connects them through a low-latency link.

attack, or more subtly, selectively discard certain messages to alter the function of the network [24]. An attacker with a suitable wormhole can easily create a sinkhole that attracts (but does not forward) packets to many destinations. An intelligent attacker may be able to selectively forward messages to enable other attacks and also be able to place wormhole endpoints at particular locations. Strategically placed wormhole endpoints can disrupt nearly all communications to or from a certain node and all other nodes in the network [24].

The neighbor-discovery mechanisms of periodic (proactive) routing protocols rely heavily on the reception of broadcast packets as a means for neighbor detection, and are also extremely vulnerable to this attack.

15.3.5 Rushing Attacks

Existing on-demand routing protocols forward a request packet that arrives first in each route discovery. In the rushing attack, the attacker exploits this property of the route-discovery operation, as shown in Figure 15.4. The initiator node initiates a route discovery for the target node. If the ROUTE REQUESTs for this discovery forwarded by the attacker are the first to reach each neighbor of the target (shown in gray in the figure), then any route discovered by this route discovery will include a hop through the attacker [25]. That is, when a neighbor of the target receives the rushed REQUEST from the attacker, it forwards that REQUEST, but will not forward any further REQUESTs from this route discovery. When nonattacking REQUESTs arrive later at these nodes, they will discard those legitimate REQUESTs. As a result, the initiator will be unable to discover any usable routes (i.e., routes that do not include the attacker) containing at least two hops (three nodes).

In general, an attacker can forward a route request more quickly than legitimate nodes can, so it can enter in a route. Such a route cannot be easily detected.

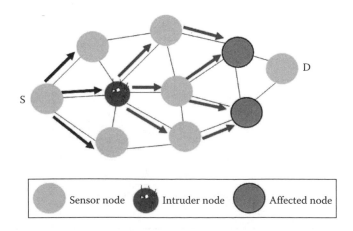

Figure 15.4   Network illustrating the rushing attack.

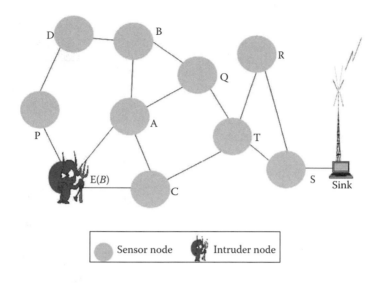

Figure 15.5   Network illustrating the masquerade attack.

15.3.6   Masquerade Attacks

This is a type of attack in which one system entity illegitimately poses as another entity to gain access to confidential systems, that is, one system assumes the identity of another. Suppose that a node A sends out a reference beacon to its two neighbors B and C. An attacker E can pretend to be B and exchange wrong time information with C, disrupting the time synchronization process between B and C, as shown in Figure 15.5.

15.3.7   Replay Attacks

A replay attack is a form of network attack in which a valid data transmission is maliciously or fraudulently repeated or delayed. Suppose Alice wants to prove her identity to Bob. Bob requests her password as proof of identity, which Alice dutifully provides (possibly after some transformation like a hash function); meanwhile, Eve is eavesdropping the conversation and keeps the password. After the interchange is over, Eve connects to Bob posing as Alice; when asked for a proof of identity, Eve sends Alice's password read from the last session, which Bob must accept.

15.3.8   Message Manipulation Attacks

The most direct attack against a routing protocol is to target the routing information exchanged between nodes. By spoofing, altering, or replaying routing information, adversaries may be able to create routing loops, extend or shorten source routes as shown in Figure 15.6. Thus adversaries attract or repel network traffic, generate false error messages, partition the network, increase end-to-end latency, etc. In this attack, an attacker

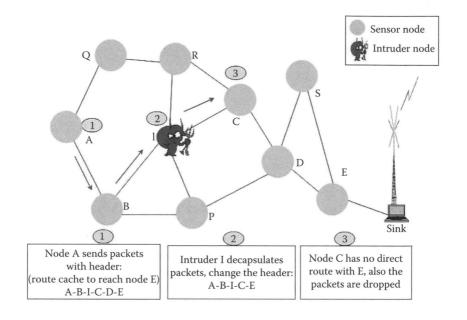

Figure 15.6   Network illustrating the message manipulation attack.

may drop, modify, or even forge the exchanged timing messages to interrupt the time synchronization process [26].

15.3.9   Delay Attacks

The attacker intentionally delays some of the time messages, e.g., the beacon message in the RBS scheme, so as to fail the time synchronization process. In Figure 15.7, the intruder node E receives the beacon message at time t_1 and sends it after a long period of time, t_n.

15.3.10   Sybil Attacks

In a Sybil attack, a single node presents multiple identities to other nodes in the network. The Sybil attack can significantly reduce the effectiveness of fault-tolerant schemes such as distributed storage, disparity, multi-path routing, and topology maintenance. Replicas, storage partitions, or routes believed to be using disjoint nodes could in actuality be using a single adversary presenting multiple identities.

Sybil attacks also pose a significant threat to geographic-routing protocols. Location-aware routing often requires nodes to exchange coordinate information with their neighbors to efficiently route geographically addressed packets. It is only reasonable to expect a node to accept but a single set of coordinates from each of its neighbors, but by using the Sybil attack an adversary can "be in more than one place at once."

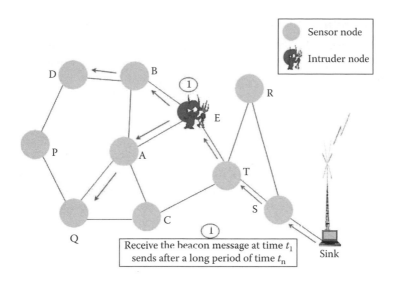

Figure 15.7   Network illustrating the delay attack.

15.4   Countermeasure

Here, we describe the countermeasure techniques that can prevent the attacks available in the sensor network. The combined effect of the techniques can prevent the attacks and all the techniques fulfill a part of the security requirements. In many ways the attacks can be prevented, but only one technique cannot prevent all the attacks; a combined effect is needed. Some of the techniques can prevent passive attacks and some can prevent active attacks. Among all the techniques in this section, we consider key management, anonymous communication, and intrusion detection.

15.4.1   Key Establishment and Management

In sensor networks, sensor nodes are resource constrained, which we already discussed. This resource constrain makes security applications a challenging problem for sensor networks. Efficient key distribution and management mechanisms are needed besides lightweight ciphers. Many key establishment techniques have been designed to address the trade-off between limited memory and security, but which scheme is the most effective is still debatable. Here, we describe some basic as well as some current proposed protocols.

15.4.1.1   Single Networkwide Key, Pairwise Key Establishment, Trusted Base Station, and Authentication

Single networkwide key, pairwise key establishment, and trusted base station are described in Ref. [1]. An authentication scheme is introduced; although it is not a key-management scheme, but it is used by many schemes.

15.4.1.1.1   Single Networkwide Key

Using a single networkwide key is by far the simplest key-establishment technique. In the initialization phase of this technique, a single key is preloaded into all the nodes of the network. After deployment, every node in the network can use this key to encrypt and decrypt messages. Some of the advantages offered by this technique include minimal storage requirements and avoidance of complex protocols. Only a single key is to be stored in the nodes' memory, and once deployed in the network, there is no need for a node to perform key discovery or key exchange, because all the nodes in the communication range can transfer messages using the key that they already share.

Though a single networkwide key may seem advantageous, the main drawback is that compromise of a single node causes the compromise of the entire network through the shared key. This scheme counters several constraints with less computation and reduced memory use, but it fails in providing the basic requirements of a sensor network by making it easy for an adversary trying to attack.

15.4.1.1.2   Pairwise Key-Establishment Scheme

The pairwise key establishment scheme, however, is one of the most efficient key establishment schemes in WSNs, because it does offer many additional features compared to other schemes, including node-to-node authentication and resilience to node replication.

For a network of n nodes in the pairwise scheme, the key predistribution is done by assigning each node a unique pairwise key with all the other nodes in the network, i.e., $n-1$ pairwise keys, which are retained in each node's memory, so that each node can communicate with all the nodes in its communication range. With each node sharing a unique key with every other node in the network, this scheme offers node-to-node authentication. Each node can verify the identity of the node it is communicating with. This scheme also offers increased resilience to network capture, as a compromised node does not reveal information about other nodes that are not directly communicating with the captured node. Through increased resilience, the scheme minimizes the chance for node replication. The drawback with the pairwise scheme is the additional overhead needed for each node to establish $n-1$ unique keys with all the other nodes in the network, and maintain those keys in its memory. Utilizing such a scheme makes a network size prohibitive because, as the number of nodes in the network increases, so do the number keys that must be stored in each node's memory. If there is a network of 10,000 nodes, then each node must store 9,999 keys in its memory. Because sensor nodes are resource constrained, this significant overhead limits the scheme's applicability, but it can be effectively used for smaller networks.

15.4.1.1.3   Trusted Base Station

The main problem of using the pairwise key establishment scheme is that every node in the network has to store $n-1$ key pairs. This can be eradicated when we use a trusted base station to send the session keys for the communication between any two nodes. This scheme is also called the centralized key distribution center (KDC) approach. The scheme has small memory requirement and a perfectly controlled node replication. It is resilient to node capture and makes it possible to revoke key pairs. The drawbacks are that it is not scalable and the base station becomes the target of attacks.

15.4.1.1.4   Authentication: µTESLA

Perrig et al. [27] at University of California, Berkeley presented a suite of security protocols optimized for sensor networks that they called "SPINS." The suite is built upon two secure building blocks, each performing individual required work: SNEP and µTESLA. SNEP offers data confidentiality, authentication, integrity, and freshness, while µTESLA offers broadcast data authentication. The µTESLA protocol, used on regular networks, is modified as SPINS for use in resource-constrained WSNs. SPINS incorporates TinyOS (operating system) in each node, all of which communicate with a base station. Most WSN communications pass through the base station and involve three communication types: node to base station, base station to node, and base station to all nodes.

The main goal of the SPINS protocol is to design a key-establishment technique based on SNEP and µTESLA to prevent an adversary from spreading to other nodes in the network through a compromised node. Each node in this scheme shares a secret key with the base station that is initialized before deployment. The following are some of the representations in this scheme used to illustrate how it works:

- Node A and node B are two communicating nodes in the network
- N_a is generated by node A
- X_{ab} is the master key shared between nodes A and B
- K_{ab} and K_{ba} are the encryption keys shared between node A and node B, which are derived from the master key, Xab
- K'_{ab} and K'_{ba} are the secret Message Authentication Code (MAC) keys shared between node A and node B, which are derived from the master key, Xab
- $\{M\}K_{ab}$ denotes the encryption of message M with key K_{ab}
- MAC(K'_{ab}, M) denotes the computation of MAC for message M with the MAC key, K'_{ab}

(1) *SNEP: Data confidentiality/authentication/freshness*: A combination of two schemes forms SNEP including a counter for semantic security and a bootstrapping scheme. Using this combination, SNEP is able to offer a number of advantages and only adds 8 bytes per message by reducing the communication overhead of the network. It uses a counter, like many other protocols, to offer authentication and freshness, but does so using means that also provide semantic security. Note that semantic security is nothing new and it is a common technique in cryptography, such as using the traditional counter (CTR) mode. Two counters are shared between nodes attempting to communicate with each other for which some of the source node's cryptographic techniques send the shared counters with a message to the destination node. General encryption can be used as a simple form of confidentiality, but is not sufficient to protect messages, whereas, semantic security offers far greater security by making it harder for an adversary to derive the original data even after obtaining one or more encrypted messages. In WSNs, sending messages with a counter can cause overhead, but, the energy can be saved by sharing the counter between both nodes and incrementing it each time the destination node

receives a message. As with other schemes, for better security, the same keys should not be used again and again. In SNEP, independent keys are used for encryption and MAC operations. The secret key shared between the source node A and the destination node B is used for deriving the encryption and MAC keys for each direction. The encrypted data has the form $E = D(K, C)$, where D is the data, K is the encryption key, and C is the counter. The MAC is $M = MAC (K', C||E)$.

In SNEP, the total message that node A sends to node B is: A $\rightarrow$ B: $D(K_{ab}, C_a)$, MAC($K'_{ab}, C_a||D(K_{ab}, C_a)$).

The semantic security property is satisfied as each time the message is encrypted, the counter value is incremented to a different value; thus, though the same message is encrypted, an adversary would not be able to decode the message. This is exactly the same as the traditional CTR mode in cryptography.

With SNEP, an adversary does have a chance of performing a DoS attack by constantly sending requests for counter synchronization, but this can be prevented either by sending the counter value with each encrypted message or by attaching a short MAC to the message that does not depend on the counter. Data authentication is done using the MAC. The counter value in the message prevents an adversary from replaying old messages, which would cause confusion and overhead in a WSN. As the counter value is kept at both ends of communication and the ID is not transferred with every message, communication overhead is negligible. The counter scheme also allows achieving weak freshness. If the counter value is verified correctly, it reveals the sequence of the messages, but only guarantees the sequence of messages, not that the reply from node B is caused by the message from node A. To achieve strong freshness that includes delay estimation, a nonce must be included with messages. To achieve strong freshness, node A sends a nonce N_a along with a reply message to node B, which resends the nonce with a reply message. This process can be optimized by implicitly using the nonce in the MAC computation; therefore, the entire SNEP protocol with strong freshness is

$$A \rightarrow B : N_a, R_a \quad \text{and} \quad B \rightarrow A : \{R_b\}_{(K_{ba}, C_b)}, MAC(K'_{ba}, N_a || C_b || \{R_b\}_{(K_{ba}, C_b)})$$

If the MAC correctly verifies, node A will know that the reply from B is a reply to its message. In this method, it is assumed that both communicating parties know the counter value so that it need not be sent with every message; though, in reality, messages might get lost or tampered and cause inconsistencies in the counter value. Protocols needed to synchronize the counter value include bootstrapping the counter value in the following manner: A $\rightarrow$ B : C_a with B $\rightarrow$ A : C_b, MAC($K'_{ba}, C_a||C_b$) and A $\rightarrow$ B : MAC($K'_{ab}, C_a||C_b$).

The counter value need not be encrypted because the protocol needs strong freshness for which both communicating parties use the counter as nonce. Also, MAC need not include the names A and B, as the keys they use, K_{ab}, state which nodes are participating in the communication. If node A realizes that the counter C_b of node B is not synchronized, it may request the counter of B with a message including N_a for strong freshness, or A $\rightarrow$ B : N_a and B $\rightarrow$ A : C_b, MAC($K'_{ba}, N_a||C_b$).

(2) μ*TESLA*: *Authenticated broadcasts*: Authenticating broadcasted data is a critical issue in WSNs, but previous solutions to this problem suffer from too much

communication and computation overheads, and therefore, are not so useful in resource-constrained WSNs. TESLA, one of these solutions, provides an inefficient scheme for broadcasting data with authentication by using the digital signatures technique, which adds 24 bytes of overhead to each message that is typically only allotted a packet size of 30 bytes. Thus, using TESLA can cause almost all of the packet size to be occupied for the code only. Also, TESLA discloses the key with every message packet it sends and receives, which can use a great deal of a WSNs energy. Finally, TESLA authenticates keys using a one-way key chain, which is not possible to be stored in each sensor node. Perrig et al. modified TESLA for authenticating broadcasted data in a way that involves no significant overhead, called μTESLA; this method reduces the energy needed for authenticating data using asymmetric mechanisms. Also unlike TESLA, which discloses the key every time a packet is sent or received, μTESLA does so only once in an epoch. The only limit with μTESLA is that it restricts the number of authenticated senders, as it is expensive to store the one-way key chain in a sensor node.

μTESLA is able to provide the asymmetric cryptographic type of authenticated broadcast through a delayed disclosure of symmetric keys. For broadcasting authenticated information between the base station and nodes of a WSN, μTESLA requires that the base station and nodes are loosely time synchronized and that each node knows an upper bound on the maximum synchronization error. When the base station wants to send a packet to all the nodes in a given network, it computes a MAC on the packet beforehand. Because all the nodes in the network are sure that only a base station can compute the MAC, the MAC key is not disclosed at this point in time so they will not be vulnerable to attacks from an adversary. The packets sent to the nodes are stored in their buffers until the base station discloses the corresponding keys. Once disclosed, the keys can be authenticated by the nodes using the one-way function, F. If a key is correct, a node can use it to authenticate the packet stored in its buffer.

Each MAC key is a sequence of keys generated by the function F. The sender chooses the last key, K_n, of the chain randomly and then generates the one-way key chain by repeatedly applying F. Supposing that the base station has sent packets P_1 and P_2 in the time interval t_1, P_3 and P_4 in t_2, P_5 in t_3, and P_6 in t_4, the nodes receiving the packets cannot verify their authentication immediately, so the nodes store them in buffers. Packets sent in a particular time interval are authenticated using the key that corresponds to that time interval. Let the difference of the time interval be two in this case, and the receiver node is loosely time synchronized with the base station and knows key K_0. Assuming that all the messages sending the key information about packets P_1–P_5 are lost and only the message that carries the key information about packet P_6 arrives, the receiver node can still authenticate the keys of the other packets by deriving the key information supplied for P_6. Thus, though some of the packets may have been lost, the nodes can still authenticate them using the keys received. To do this, μTESLA has multiple phases that perform a particular job each, including Sender Setup, Broadcasting Authenticated Packets, Bootstrapping New Receivers, and Authenticating Broadcast Packets.

(3) *Considerations for SPINS*: SPINS uses less of a sensor node's memory, i.e., while crypto routines occupy 20 percent of the space, μTESLA occupies 574 bytes, and 2 kB is the acceptable total used memory [27]. The scheme's performance is also efficient, as

the bandwidth of the WSNs is adequate for the cryptographic primitives that SPINS uses. Additionally, most of the SPINS designs may be used in other networks of low-end devices. Finally, the communication costs for SPINS are small, with security properties like data freshness, authentication, and confidentiality only adding an overhead of 6 bytes in a 30 bytes packet, which allows for inclusion in each packet [27]. SPINS can offer even greater advantages when restrictions on bandwidth and memory are slightly relieved.

Broadcasting and authenticating data are not that easy for individual nodes, as storing a one-way key in a node's memory is not possible, computation of the keys using a function generates much network overhead, and each node does not share a common key with every other node in the network. However, there are two solutions for this problem. Firstly, the base station is used by a node to transmit all data that has to be broadcasted to other nodes. Secondly, the node broadcasts the data to the base station while the base station generates the authenticating keys using the one-way key chain. It is efficient to implement the cryptographic primitives in a single block cipher as WSNs are resource-constrained and, therefore, cannot afford additional overhead for security. Yet, a strong cryptographic base is necessary for SPINS.

- Block cipher: Using RC5 can be very efficient in WSNs because of its small size and high efficiency. Moreover, as an algorithm it has been subject to scrutiny under many attacks. Using TEA could also work for block ciphers, but it is not subject to cryptanalysis scrutiny. DES and other algorithms are not usable for block ciphers due to their large size and high computation requirements that cannot be met in WSNs.

- Encryption function: The counter (CTR) mode of block ciphers can use the same function for both encryption and decryption, and the size of the cipher text is the same as the data in this mode. These two properties make this mode very useful while working in the encryption function of SPINS. Also, CTR mode offers semantic security, which is a strong cryptographic property already discussed. To use the CTR mode, both the sender and receiver nodes must maintain counters in their memory and possess an efficient way to synchronize the counters if needed. One advantage of maintaining a counter at both ends is that the messages now will not have an overhead of carrying the counter with them.

- Freshness: Using a counter and incrementing it every time a message is sent automatically provides weak freshness. For strong freshness, the sender must create a nonce and should include it in the request message to the receiver. SPINS uses a MAC function for generating random numbers and a counter is created to keep track of those created.

- Message authentication: Not only is a good encryption function necessary for data, but also a secure MAC is needed. As the block cipher is used more than once, CBC–MAC is used for MAC. An efficient way of message construction must be used to achieve authentication and message integrity. The construction $\{M\}_k$, $\mathrm{MAC}(k', \{M\}_k)$, in which M is the data, K is the encryption key, and K' is the MAC key is secure and protects the nodes from decrypting erroneous ciphered text.

Advantages of this scheme include that it is one of the best memory-efficient schemes discussed, provides strong security features with less complexity, has a universal design that allows its use in many low-end devices, incurs low communication cost, and offers authentication and strong data freshness with a minimum overhead. Disadvantages of this scheme include the μTESLA overhead from the release of keys after a certain delay, possibly the message delay.

15.4.1.2 Public-Key Schemes

A MICA2 mote developed by the University of California at Berkeley has an 8 bit, 7.3 MHz processor with 4 kB RAM and 128 kB of programmable ROM [28]. WSNs have mostly been using symmetric-key and other nonpublic-key encryption schemes [29]. A drawback to these schemes is that they are not as flexible as public-key schemes, but they are computationally faster. With limited memory, computing and communication capacity, and power supply, sensor nodes cannot employ sophisticated cryptographic technologies such as typical public-key cryptographs. The use of public-key cryptography on WSNs has not been tested enough to rule it out completely. Through the use of the MICA2 mote and TinyOS, public-key schemes are tested to determine their performance. ECC has a faster computation time, smaller keys, and uses less memory and bandwidth than RSA [3]. Both ECC and RSA can be accelerated with dedicated coprocessors. Recently, ECC has been used [3,29] for WSNs. The authors in Ref. [3] have implemented a way to execute public-key schemes on WSNs with 8 MHz processors, using elliptic curves in computing encryption keys. The MICA2 mote is also able to use ECC on its 8 bit processor. On these sensor nodes, public keys can be computed in less than 34 seconds using 1 kB of RAM and 34 kB of ROM [28]. In the following section, RSA and ECC schemes are described.

15.4.1.2.1 RSA and ECC Schemes

Both RSA and ECC have been researched for many years. It was developed in 1977 and is still one of the most popular public-key encryption technologies currently available. RSA relies on its strength due to the complication of factoring very large numbers. ECC was developed in 1985 independently by Koblitz and Miller. Its approach to public-key cryptography is based on the mathematics of elliptic curves. ECC can obtain the same security level as RSA while using a smaller key. A 160 bit ECC key has the same security as a 1024 bit RSA key [3]. A 224 bit ECC key compares to the 2048 bit RSA key [3]. This is due to the fact that it takes exponential algorithms to solve the elliptic-curve DL problem as opposed to small runtime algorithms to solve the large number factorization in RSA [3].

The RSA scheme generally introduces keys of size 512–2048 bit. Its takes a message, M, and composes a cipher text, C, using the key K. A method called the Chinese remainder theorem can be used to accelerate RSA. Two prime numbers, q and p, are multiplied together to get the modulus n [3]. Computing these modular multiplications, CRT can lower computation time by almost three-fourths [3]. Other factors like the

Montgomery multiplication and optimized squaring can reduce the RSA complexity by 25 percent.

ECC is computed by point multiplication on elliptic curves over prime integer fields or binary polynomial fields [3]. The implementation of ECC on WSNs is primarily interested in prime integer fields because binary polynomial field mathematics is poorly supported by slow processors. Operations of ECC scale linearly. This gives ECC an advantage over RSA on processors with small word sizes. Also, ECC grows in advantages as the key size grows.

CRT and modular multiplications for ECC and large integer mathematics for RSA are the most important operations in these cryptography schemes. Large numbers of multiplication operations need high memory reads and writes due to the small word size of the processor. The computation time is therefore reduced by optimizing the number of memory operations [3].

ECC was implemented on two 8 bit platforms. Performance optimizations were applied due to limited resources. RSA-1024 and RSA-2048 were also implemented for comparison [3]. ECC-160 resulted with a private key faster than RSA-1024. The performance was even more favorable when comparing ECC-224 to RSA-2048 [3]. ECC, on both platforms, outperformed RSA-1024 private-key operation. ECC also improves its performance over RSA as the word size of the processor decreases.

15.4.1.3 Routing-Driven Elliptic Curve Cryptography-Based Key-Management Scheme

In Ref. [6], the ECC-based key-management scheme is described for heterogeneous sensor networks. Here, the authors divide the key-management schemes in two phases: centralized key management and distributed key management.

15.4.1.3.1 Centralized Key Establishment

In the centralized ECC-based key-management scheme, a server is used to generate pairs of ECC public and private keys, one pair for each L-sensor and H-sensor. L-sensors are the leaf sensors and H-sensors are the header sensors in the hierarchical cluster network. The server selects an elliptic curve, E, over a large finite field, F, and a point P on that curve. Each L-sensor (denoted as u) is preloaded with its private key (denoted as $K_u^R = I_u$). An H-sensor has a large storage space and is preloaded with public keys of all L-sensors (such as $K_u^U = I_u P$). Each H-sensor (denoted as H) also stores the association between every L-sensor and its private key. An alternative approach is to preload each L-sensor with its public key, and then let every L-sensor send the public key to H after deployment. However, this scheme introduces large communication overhead, and furthermore, security problems, because an adversary may modify the public key during its route to H.

The preloaded keys in H-sensors are protected by tamper-resistant hardware. Even if an adversary captures an H-sensor, he could not obtain the key materials. Given the protection from tamper-resistant hardware, the same ECC public-/private-key pair can be used by all H-sensors, which reduces the storage overhead of the key-management scheme. Each H-sensor is preloaded with a pair of a common ECC public key (denoted as $K_H^R = I_H P$) and a private key (denoted as $K_H^R = I_H$). The public key of H-sensors, K_H^U, is also loaded in each L-sensor, and the key is used to authenticate broadcasts from H-sensors. The ECDSA algorithm [5] is used for authenticating broadcasts from H-sensors. When H broadcasts the routing structure information (e.g., the MST) to L-sensors, a digital signature is calculated over the message using H's private key. Each L-sensor can verify the digital signature by using H's public key, and thus authenticate the broadcast. In addition, each H-sensor is preloaded with a special key KH that is used by a symmetric encryption algorithm for verifying newly deployed sensors and for secure communications among H-sensors.

After selecting a cluster head H, each L-sensor, u, sends to H a clear (unencrypted) key-request message, which includes the L-sensor ID—u, and u's location. A greedy geographic-routing protocol (e.g., [40]) may be used to forward the key-request message to H. Note that the location of the cluster head is known to all L-sensors during cluster formation. An L-sensor sends the message to the neighboring L-sensor that has the shortest distance to the cluster head, and the next node performs a similar operation until the packet arrives at the cluster head.

After a certain time, the cluster head, H, should receive key-request messages from all (or most) L-sensors in its cluster, and then H uses a centralized MST (or SPT) algorithm to determine the tree structure in the cluster. Next, H generates shared keys for each L-sensor and its c-neighbors. For an L-sensor, u, and its c-neighbor, v, H generates a new key, $K_{u,v}$. Recall that H is preloaded with the public keys of all L-sensors. H encrypts $K_{u,v}$ by using u's public key and an ECC encryption scheme (e.g., [30]), and then H unicasts the message to u. L-sensor, u, decrypts the message and obtains the shared key between itself and v. After all L-sensors obtain the shared keys, they can communicate securely with their c-neighbors.

15.4.1.3.2 Distributed Key Establishment

The key setup can also be done in a distributed way. In the distributed key establishment, each L-sensor is preloaded with a pair of ECC keys—a private key and a public key. When an L-sensor (denoted as u) sends its location's information to its cluster head, H, u computes a MAC over the message by using u's private key, and the MAC is appended to the message. When H receives the message, H can verify the MAC, and then authenticate u's identity by using u's public key. Then, H generates a certificate (denoted as CAu) for u's public key by using H's private key. After determining the routing-tree structure in a cluster, the cluster head, H, disseminates the tree structure (i.e., parent–child relationship) and the corresponding public-key certificate to each L-sensor. The public-key certificates are signed by H's private key, and can be verified by every L-sensor, because each L-sensor is preloaded with H's public key. A public-key certificate proves the authenticity of a public key and further proves the identity of one L-sensor to another L-sensor.

If two L-sensors are a parent and its child in the routing tree, then they are c-neighbors of each other, and they will set up a shared key by themselves. For each pair of c-neighbors, the sensor with a smaller node ID initiates the key-establishment process. For example, suppose that L-sensor, u, and v are c-neighbors, and u has a smaller ID than v. The process is presented as follows:

1. Node u sends its public key $K_u^U = I_u P$ to v
2. Node v sends its public key $K_v^U = I_v P$ to u
3. Node u generates the shared key by multiplying its private key, I_u, with v's public key, K_v^U, i.e., $K_{u,v} = K_u^R K_v^u = I_u I_v P$; similarly, v generates the shared key $K_{u,v} = K_v^R K_u^u = I_u I_v P$

After the above process, nodes u and v share a common key and they can start secure communications. To reduce the computation overhead, symmetric encryption algorithms are used among L-sensors. Note that in the distributed key-establishment scheme, the assumption of having tamper-resistant hardware in H-sensors can be removed.

15.4.1.3.3 Performance Evaluation

Key space storage for this scheme can be evaluated as follows. Assume that the number of H-sensors and L-sensors in a heterogeneous sensor networks (HSNs) is M and N, respectively. Typically, assume that $M \ll N$. In the centralized ECC key-management scheme, each L-sensor is preloaded with its private key and the public key of H-sensors. Each H-sensor is preloaded with public keys of all L-sensors, plus a pair of private/public keys for itself, and a key K_H for newly deployed sensors. Thus, an H-sensor is preloaded with $N + 3$ keys. The total number of preloaded keys is

$$M \times (N + 3) + 2 \times N = (M + 2)N + 3M$$

According to the clarification of the deployment of nodes in the network, before or at deployment, all the L-sensors need to know their H-sensors (L-sensor is defined as a low-end sensor and H-sensor is defined as a high-end sensor) as well as the tree structure of the network. Hence, nodes have to be deployed according to the predefined tree structure, otherwise there is a mismatch between the private-key/public-key pair of the L-sensors and their corresponding H-sensors. This is a cumbersome procedure for deployment. Actually, to reduce the key space of the L-sensors in the network, this scheme needs to predefine the tree structure of the network. In that case, assume that there are M number of H-sensors and N number of L-sensors in the network, as each H-sensor is preloaded with public keys of all L-sensors, plus a pair of private/public key for itself, and an extra key for the newly deployed sensor. Each L-sensor is preloaded with its private key and the public key of its H-sensor (L-sensor knows its corresponding H-sensor, because nodes are deployed according to a predefined tree structure). Thus, an H-sensor is preloaded with $N + 3$ keys and an L-sensor is preloaded with two keys. Therefore, the total number of preloaded keys is

$$M \times (N + 3) + (2 \times N) = (M + 2)N + (3M)$$

which is already discussed.

If the scheme does not care about the predefined tree structure then each L-sensor needs to be preloaded with $M + 1$ keys. Thus, the total number of preloaded keys is

$$M \times (N + 3) + (M + 1) \times N = (2 \times M \times N) + (3 \times M) + N$$

Otherwise, the scheme introduces large communication overhead and furthers more security problems. This scheme is absolutely static because of the preloaded keys of the other nodes. Every node needs to maintain a relative connection with the other nodes.

In the distributed ECC key-management scheme, each L-sensor is preloaded with its public/private key. Each H-sensor is preloaded with a public/private key and key K_H. Thus, the total number of preloaded keys is

$$3M + 2N$$

In Ref. [54], Eschenauer and Gligor present a key management scheme based on probabilistic key pre-distribution for sensor networks. In this scheme if every sensor node is preloaded with m keys. The total number of preloaded keys in a network with $M + N$ sensors is

$$m(M + N)$$

This scheme is absolutely static because of the preloaded keys of the other nodes. Every node needs to maintain a relative connection with the other nodes.

15.4.1.4 Identity- and Pairing-Based Secure Key-Management Scheme

In Ref. [31], identity- and pairing-based key-management scheme is proposed; the protocol is described in the following sections.

15.4.1.4.1 Protocol Design

The base station (sink or system administrator) has the following extra tasks during the bootstrap of the network.

- Determining two groups, G_1 and G_2, of the same prime order, q. We view G_1 as an additive group and G_2 as a multiplicative group, as discussed in the preliminary chapter.
- Determining the bilinear map g: $G_1 \times G_1 \rightarrow G_2$, collision-resistant cryptographic hash functions H_1 and H_2, where H_1:$\{0, 1\}^* \rightarrow G_1$, a mapping from arbitrary-length strings to points in G_1 and H_2:$\{0, 1\}^* \rightarrow \{0, 1\}^\mu$, a mapping from arbitrary-length strings to a μ-bit fixed-length output.
- Generating the system's secret $\acute{\omega} \in Z_q^*$, where $Z_q^* = \{y | 1 \leq y \leq q - 1\}$. Anyone in the network does not know $\acute{\omega}$ except the base station (sink). The base station also uses this secret to generate the secret point of the nonadversary nodes.

Thus, the system parameters $\langle G_1, G_2, g, H_1, H_2 \rangle$ are known to the nonadversary nodes. The base station also provides the following parameters for nodes, regarding their IDs and secret points.

- Providing each node (L-nodes and H-nodes) a different, unique, and real ID individually, let us say

$$\text{ID}_{R1} = H_1(\text{ID}_{R1}^\ell), \ \text{ID}_{R2} = H_1(\text{ID}_{R2}^\ell), \ldots, \text{ID}_{RN} = H_1(\text{ID}_{RN}^\ell) \in G_1$$

for N number of nodes and the corresponding secret points, $SP_{R1}, SP_{R2}, \ldots,$ $SP_{RN} \in G_1$, which are defined as $SP_{Ri} = \acute{\omega}ID_{Ri} = \acute{\omega}H_1(ID_{Ri}^{\ell})$, where $i = 1, 2, \ldots, N$. Also provide each node a different random number $R_{Ni} \in Z_q^*$. If an L-node (leaf node) is not directly connected to an H-node (header node), then it uses its secret point and its communication neighbor's ID to generate the secret sharing key between itself and its communication neighbor, also to authenticate each other (the key generation and authentication techniques are described in the next section). If an L-node is directly connected to an H-node then it uses its secret point to generate the secret sharing key between itself and its H-node, also to authenticate each other (the key generation and authentication techniques are described in the next section). For a given set of $\langle ID_R, SP_R \rangle$, no one can determine the system's secret $\acute{\omega}$, as discussed in the preliminary chapter.

■ Providing each H-node, the IDs of all L-nodes, and a corresponding random number $R_N \in Z_q^*$. This random number is used to authenticate the L-node to its corresponding H-node. Also, the base station changes this random number in some time interval and sends it to the H-nodes, which is described in the next section.

With the above information any node can generate its own secret sharing key. Let us check for a node K; K has received its ID, ID_K, and the corresponding secret point, $SP_K = \acute{\omega}ID_K = \acute{\omega}H_1(ID_K^{\ell})$, from the base station. Hence, K can generate its own secret sharing key with its communication neighbor; let us say its communication neighbor is M and it has also received its ID, ID_M, and the corresponding secret point, $SP_M = \acute{\omega}ID_M = H_1(ID_M^{\ell})$. As a result, K and M both can generate their own secret sharing key without sharing their secret point as follows. K computes $K_{KM} = g(SP_K, ID_M) = g(ID_K, ID_M)^{\acute{\omega}}$ and M computes $K_{MK} = g(SP_M, ID_K) = g(ID_M, ID_K)^{\acute{\omega}}$. According to the properties in Section 2.1, the two values, K_{KM} and K_{MK}, are same; thus, without sharing the secrets, both K and M can generate their same secret sharing key. These equations also hold the previous cited property mentioned in Sections 15.2.3 and 15.2.4. Hence, no one can determine the system's secret $\acute{\omega}$ for a given set of IDs and the corresponding secret point $\langle ID_K, SP_K \rangle$. As a result, a node can generate its own secret key with its communication neighbor without sharing their corresponding secret points.

15.4.1.4.2 Key Management and Authentication

The base station works as a system administration and is responsible for generating IDs and corresponding secret points for all the nodes in the network. At the bootstrap of the network, any node knows its ID, secret point, and random number. Furthermore, the H-node knows IDs of all L-nodes as well as their random numbers. The random numbers of the nodes are changed by the base station in a time period and are distributed to L-nodes through their corresponding H-nodes. Thus, at the beginning, each node has a random number and it is changed after some period. This random number is used for authentication between an L-node and its communication neighbor (the neighboring L-node) or between an L-node and its H-node.

The key establishment and authentication processes are described in the following subsection.

15.4.1.4.2.1 Shared-Key Establishment and Authentication After the
bootstrap and cluster formation of the network, all L-nodes know their corresponding H-nodes connected either directly or through their communication neighbors (other
L-nodes). Remember that all the H-nodes are connected to each other either directly
or through other H-nodes. Furthermore, H-nodes are connected to the base station
either directly or through other H-nodes. Thus, nodes are connected in a hierarchical
fashion.

Let us consider an H-node with its ID, ID_{RH1}, corresponding secret point,
SP_{RH1}, and random R_{H1}; an L-node with its ID, ID_{RL1}, corresponding secret point,
SP_{RL1}, and random R_{L1}. Let us consider again that the H-node wants to authenticate to its neighbor L-node, ID_{RL1}. Hence, H-node generates its secret sharing
key, $K_{H1L1} = g(SP_{RH1}, ID_{RL1}) = g(ID_{RH1}, ID_{RL1})^{\omega}$, and an authentication code,
$Aut_0 = H_2(K_{H1L1}||ID_{RH1}||ID_{RL1}||0)$, and sends to ID_{RL1}. On the other hand, L-node
generates its secret sharing key, $K_{L1H1} = g(SP_{RL1}, ID_{RH1}) = g(ID_{RL1}, ID_{RH1})^{\omega}$,
and a verification code, $Ver_0 = H_2(K_{L1H1}||ID_{RL1}||ID_{RH1}||0)$, and compares it
with Aut_0; if Ver_0 matches with Aut_0 then it generates another authentication code,
$Aut_1 = H_2(K_{L1H1}||ID_{RL1}||ID_{RH1}||R_{L1}||1)$, and sends it to the H-node. Finally, the
H-node computes $Ver_1 = H_2(K_{H1L1}||ID_{RH1}||ID_{RL1}||R_{L1}||1)$ and compares it with
Aut_1; if it matches then the authentication is successful, otherwise a failure. Thus, all the
nodes (H-nodes and L-nodes) are authenticated to each other, either with their communication neighbors or H-nodes. Therefore, nodes in the network can communicate
securely with each other.

When the base station changes the random of the L-nodes, it encrypts the random
with the secret sharing key of the base station and H-nodes. As the L-nodes are authenticated to their corresponding H-nodes, H-nodes encrypt the random with the secret
sharing key of H-nodes and the corresponding L-nodes and send it to the L-nodes. If
an L-node is not directly connected with its H-node, rather it is connected via other
L-nodes, then the H-node performs double encryption. In the first encryption, it uses
the secret sharing key of the destination L-node. In the second encryption, it uses the
secret sharing key of its nearby L-node and sends the encrypted packet to this nearby
L-node. Consequently, the L-node decrypts the packet and looks at the destination ID,
and again encrypts with its communication neighbor's (L-node) secret sharing key and
sends to that L-node. Thus, the random eventually reaches to the destination L-node.
Notice that the intermediate L-nodes cannot get any information about the random
because it is double encrypted by the H-node.

15.4.1.4.3 Security Analysis and Performance Evaluation

In this section, cryptographic implementation and cryptographic analysis of the protocol
are described. At first, the implementation point of view, and later, the security analysis
and key space are described.

15.4.1.4.3.1 Key Space Saving In the previous protocol, nodes are static and
they cannot move at all. Hence, nodes are fixed in a certain position. In this protocol,
nodes can move or they can be fixed in a position; for both the cases the proposed

protocol works. As in the existing protocol nodes are fixed, comparison between the key spaces is given when nodes are fixed in a certain position.

Two cases of the scenario have been considered for comparison with the previous protocol. The first scenario is the case that is already mentioned in Ref. [6] as well as discussed in the previous section; in this case, the total number of preloaded keys is $M \times (N + 3) + 2 \times N = (M + 2)N + 3M$. The second scenario is the same as the first scenario, and the total number of preloaded keys is $M \times (N + 3) + (M + 1) \times N = 2 \times M \times N + 3 \times M + N$.

On the other hand, for this protocol there is no preloaded key at all, after cluster formation the key space could be N/M, to reduce the computational cost; considering nodes do not move as in the existing scheme. This can be clarified thus: at the bootstrap of the network, nodes store only their own secret points; later the nodes compute the secret sharing keys of their neighbor nodes. For equally distributed clusters of the nodes in the network, there are N/M number of nodes in each cluster, for M number of H-nodes. To reduce the computational cost, after formation of the clusters, nodes store only their neighbors' shared keys; thus, for the worse case, the total number of shared keys to store is N/M.

15.4.1.4.3.2 Prevention of Attacks The following well-known attacks are effective in sensor networks and this protocol prevents these attacks effectively.

Masquerade attack: In this scheme, all the nodes are authenticated networkwide to each other on their routes. Therefore, the intruder node/s cannot pretend as a valid node/s and cannot exchange wrong information between the valid nodes. Thus, the masquerade attack is not valid for our proposed protocol.

Reply attack: In this scheme, at first, an attacker node cannot pass the authentication process. Furthermore, the old verification/Authentication code is no more valid after a certain time period. Because the base station changes the random number of the nodes in the network, which is used to generate the verification code, nodes also perform encryption during communication.

Message manipulation attack: To perform this attack, an attacker needs to take part in the message communication. To this end, it is necessary to be a valid node in the network. In this scheme, an attacker cannot forge the path or packet. Thus, this attack is not effective in this protocol.

Delay attack: In this protocol, nodes can calculate the distance from the source to themselves, and thus, also can estimate the traveling time of the packet. Hence, if any intruder node wants intentionally to delay a packet, then it cannot be effective.

15.4.2 Anonymous Communication

In this section, we are describing some of the anonymous protocols for WSNs. Anonymous protocols are effective for preventing some target-oriented attacks, especially passive attacks.

15.4.2.1 Hierarchical Anonymous Communication Protocol

The hierarchical anonymous communication protocol (HACP) is proposed in Ref. [32]. HACP provides two different mechanisms to achieve anonymity—one is based on introducing dummy messages for anonymity within a cluster and the other is based on a ring-based approach for anonymous communication within cluster heads.

15.4.2.1.1 Network Model

Here, clustered sensor networks have been considered because clustering allows for scalability of MAC and routing. Cluster heads also serve as fusion points for aggregation of data, so that the amount of data that is actually transmitted to the base station is reduced. Clustering sensors into groups, so that sensors communicate information only to cluster heads and then the cluster heads communicate the aggregated information to the processing center, may save energy. Many clustering algorithms in various contexts have been proposed [33–36]. These algorithms aim at generating the minimum number of clusters such that any node in any cluster is at most d hops away from the cluster head.

The communication graph G(VCH, E) is used to represent the network in terms of cluster heads. VCH is the set of cluster heads and E is the set of communication edges (might be paths involving intermediate noncluster heads) connecting the cluster heads. We assume that G is connected.

A spanning tree is fixed initially in the graph. Next, using an Euler tour (that is a DFS tour) of the spanning tree in the graph, a ring is defined. Also, the ring formation can use the underlying routing protocol to achieve energy efficiency and load balancing.

This protocol is based on symmetric-key cryptographic techniques because of infeasibility of implementation of public-key protocols in sensor networks [27]. There exist a number of key predistribution schemes for sensor networks to set up secret keys among sensors [37]. In this protocol, it is assumed that each sensor shares a secret key with its cluster head. Also, each cluster head shares a symmetric key with its neighboring cluster heads in the ring. The following notations are used to describe the protocol: $E(M, K_{ij})$ to represent encryption of message M with K_{ij}, and $D(M, K_{ij})$ to represent decryption of message M with K_{ij}, where K_{ij} is the secret key shared by nodes i and j.

Tokens and frames. At any time there can be only one frame traversing through the ring. The nodes use a token-passing access mechanism to access a frame passing through the network. A node wishing to send data should first receive permission. When it gets the control of the token, it may transmit data in that frame. Each frame is of fixed length and contains the status of the token itself. A token can be either in free status or occupied status. The format of the frame is as follows:

$$\langle E((Token\|E(Message_{Header}, K_{sd})\|E(Message_{Data}, K_{sd})), K_{si})\rangle \qquad (15.1)$$

where
  K_{si} is the secret key shared between the source node s and node i that is the upstream neighbor of sender s
  K_{sd} is the secret key shared between the source node s and destination node d

The format of the token is as follows: $\langle Redundancy predicate\|Status\rangle$
  Redundancy predicate is used for checking the validity of the frame. For the frame to be verified successfully by node i, upon decryption the *Redundancy predicate* must be

fulfilled. "Status" specifies if the token is "occupied" or "free." If a token is free, a node can send data through that frame, else it cannot.

The format of the *Frame$_{Header}$* is as follows:

⟨*Redundancy predicate*||*Source Address*||*Destination Address*⟩

The format of *FrameData* is as follows: ⟨*Data length*||*Data Padding*⟩

Data length specifies the length of the total data in the packet. This is crucial when the amount of data needed to be sent is not enough to fill the whole frame. In that case, data to be sent is padded with some random number to meet the constraint that the size of the frame is of fixed length.

15.4.2.1.1.1 Anonymous Communication within a Cluster

Inserting dummy traffic in a network is a technique that hides the traffic patterns inside the network, making traffic analysis more difficult [38]. The generation of dummy traffic increases the anonymity of the messages sent through the mix network.

A dummy message is a fake message created by a sensor node. The final destination is its cluster head; the dummy message is discarded by the cluster head. Observers of the network and other nodes cannot distinguish the dummy from a real message.

In HACP, each sensor (including the cluster head) transmits messages at a Poisson rate, r_t. Thus, on an average, each sensor would send a message every $1/r_t$ seconds. Let r_s denote the sensing rate of each sensor. Thus, whenever there is sensed data to be sent, the sensor encrypts the data message with the secret key it shares with the cluster head and transmits it. Else the sensor sends dummy messages. Hence, the dummy messages are sent at a rate of $(r_t - r_s)$. Whenever a cluster head has a message to be sent to one of its cluster nodes, the cluster head simply encrypts the message with the secret key it shares with that sensor and sends. Whenever a sensor senses a packet transmission, it receives the packet and decrypts it with its key and checks if it is a valid packet.

15.4.2.1.1.2 Anonymous Communication between Cluster Heads

Whenever a node *i* receives a frame, it decrypts the frame using the key shared with its downstream node in the ring, and verifies the redundancy predicate. Once the *Redundancy predicate* is fulfilled, the following algorithm is executed.

1. If the node has no data to send, it just encrypts the resultant plain frame with the common key shared with its upstream node and retransmits the packet onto the ring.
2. If the status of the token is free and the node has some data to send to another node *D*, then *i* constructs the frame as follows:

   a. Node *i* constructs *Frame$_{Header}$* and *FrameData* as explained earlier using the key shared with the destination.
   b. Node *i* sets the *status* field in the token to occupied.
   c. Computes Equation 15.1 using its shared key with the upstream node and transmits the packet onto the ring.

3. If the status of the token is set to occupied, the node checks if the data in the frame is destined to itself by decrypting $E(Frame_{Header}, K_{sd})$ with the shared key and checking if the *Redundancy predicate* is fulfilled.

a. If the node is able to check the validity of the frame header, then it is addressed to node i, which makes a copy of it. It encrypts the whole frame with the key shared with its upstream node and transmits the frame onto the ring.

b. Else, if the node i is not able to check the validity of the frame header, then it is not the destination and the node just encrypts the whole frame with the key shared with its upstream node and transmits the frame onto the ring.

Once the frame returns to the source, the source repeats the procedure as long as it has data to send. When it has no more data to send it sets the status field of the token to free and assigns the whole frame to some randomly generated data. Then it encrypts the whole frame with its shared key with the upstream node and transmits the frame onto the ring.

15.4.2.1.2 Multiple Rings

In a network consisting of n nodes, the ring size is n. Thus, a message needs to be transmitted along the whole ring, and hence, each message is transmitted n times. To reduce the communication overhead (complexity), the protocol divides the graph into subgraphs and constructs rings within each subgraph. The same partition mechanism as presented in Ref. [27] is chosen. Once the partitions to subgraphs are available, then one ring in each subgraph is also available, which is formed by an Euler tour on the spanning tree of the subgraphs. The nodes that are part of more than one ring are called junction nodes. There are at most x nodes in each subgraph; thus, the time complexity is no more than x within a subgraph.

To enable communication with a node outside a subgraph, each ring is assigned a unique identifier, *RID*. Also, each node knows the *RID* of the ring to which the destination belongs. We introduce a new header—$E(Frame_{RID}, K_{sJ})$—in the frame to identify the destination's *RID*, where K_{sJ} is the common key shared by the source with the junction node that is also part of a ring that has to be traversed to reach the destination. The modified format of the frame is as follows:

$$\langle E((Token||E(Frame_{RID}, K_{sJ})||E(Frame_{Header}, K_{sd})||E(Message_{Data}, K_{sd})), K_{si})\rangle$$

The format of $Frame_{RID}$ is $\langle RIP||RID_D\rangle$. *RIP* is the redundancy predicate that has to be fulfilled so as to indicate successful decryption. RID_D is the ring identifier of the destination's ring. The sender encrypts $Frame_{RID}$ with the key shared with the junction node that is part of the ring that is on the way to the destination's ring.

When a node in one ring has data to send to a node in another ring, then the frame needs to be transferred from one ring to another until it reaches the ring of the destination. For this, each junction node maintains a forwarding routing table that specifies to the ring a frame addressed to a particular destination ring to which the frame has to be transferred. A junction node upon successful decryption of $E(Frame_{RID}, K_{sJ})$ stores a copy of the frame and then retransmits the frame. The junction node based on the *RID* of the destination node decides to which ring the frame has to be transferred. Then, it waits for a free token on the other ring to which it has to transmit the copied frame, encrypts the frame with the common key it shares with the next junction node on the way to the destination's ring, and transmits the frame. The process continues till the frame reaches the destination's ring, where the junction node of RID_D that receives the frame just assigns some random string to $E(Frame_{RID}, K_{sJ})$ and transmits the frame onto the ring RID_D.

This mechanism prevents local traffic from traversing the whole network. Even if an adversary were able to compromise a junction node, he would just be able to know the ring to which the frame was destined to and no more. The attacker could not even figure out the originating ring of the frame. Thus, this mechanism does not reduce the anonymity provided by the protocol. In some situations, only some nodes might have a need for anonymity in which case a ring has to be established only among those nodes. In such cases, the neighbors in a ring need not be physical neighbors in the network and these nodes can communicate using the shortest path available.

15.4.2.1.3 Performance of HACP

In this section, the performance of HACP is presented in terms of the overhead imposed and the anonymity provided.

15.4.2.1.3.1 Communication Overhead In HACP, whenever a node has data
to send, it captures a free token and sends data in that frame. Else, it just forwards the idle frame. The term communication overhead is used to represent the number of transmissions that correspond to idle frames. It should also be noted that the power consumption of a sensor can be derived from the average current drain [39] given by $I_{avg} = T_{on}{}^* I_{on} + (1 - T_{on})^* I_{stby}$, where T_{on} denotes the fraction of time receiver or transmitter is on. I_{on} is the current drain from the battery when the receiver or transmitter is on and I_{stby} is the current drain from the battery when both the transmitter and receiver are off. Therefore, higher the communication overhead higher the T_{on}, which implies higher is the power consumption.

Consider a ring with N number of nodes out of which N_a nodes have data to send at a rate of R packets per unit time. Let us say a frame can traverse the ring at a maximum of t times in one unit of time. The value of t depends on ring latency, which in turn depends on the transmission time of the frame, (T_{tr}), ring traverse time delay, (T_t), and processing delay at a node, (T_{proc}). Here, the delay incurred at a node to process the frame before forwarding it is ignored. Therefore, $t = 1/(N^* T_{tr} + T_{proc} + T_t)$.

If n tokens are present in the ring, then a maximum of $n^* t$ frames can be transmitted across the ring. Thus, ideally, it is likely to have the following condition satisfied, so that no idle frame is transmitted: $(N_a/N)^* R = n^*, t$.

Thus, the fraction of idle frames being transmitted over the ring is $1 - \{(N_a{}^* R)/(N^* n^* t)\}$. Thus, the communication overhead, i.e., number of transmissions corresponding to idle frames, is given by number of idle frames × number of nodes in the ring $= N - \{(N_a{}^* R)/(n^* t)\}$. The communication overhead increases almost linearly as number of nodes in the ring increases.

15.4.2.1.3.2 Security Analysis The data-centric behavior of sensor networks leaves them vulnerable to traffic analysis, and identification of event locations and active areas. Therefore, ensuring data anonymity is a crucial research area. Using traffic analysis, the attacker can compromise the network functionality by correlating data-flow patterns to event locations/active areas. The above-discussed protocol is a HACP that hides the location of nodes and obscures the correlation between event zones and data flow from snooping adversaries. Hiding the nodes from snooping adversaries the protocol prevents message-manipulation attack, delay attack, replay attack, as well as it prevents the nodes from passive attack. The protocol offers flexible trade-offs between degree of anonymity and communication delay overhead.

15.4.2.2 Finding Routes in Anonymous Sensor Networks

A problem and the corresponding algorithm for anonymous route discovery is discussed in Ref. [40]. The summary of the algorithm and simulation results are given in the following sections.

15.4.2.2.1 Problem and an Algorithm

It is considered that n sensors are placed arbitrarily in a two-dimensional space and assume the existence of one single sink. Sensors are assumed to have no identifications, not even their coordinates in space. It is assumed that the sink broadcasts one single question to all sensors and that n^* of the n sensors are the ones to answer. We call each of these n^* sensors a source and assume that sources are distributed arbitrarily amid the n sensors. The problem here addressed is that of finding routes from all sources to the sink. Because sources can only broadcast at low power, their answers are likely not to reach the sink directly, but need instead to be routed through the other sensors. All n sensors, even though $n - n^*$ of them do not have an answer for the sink, may have a part to play in aggregating and relaying the sources' answers.

The key premise underlying this approach is that each sensor is capable of measuring the amount of power it perceives in the sink's transmission. For sensor i, it denoted this measure by P_i. Clearly, if for sensors i and j, $P_i > P_j$, then i is closer to the sink than j is, provided that the sink's broadcast reaches all sensors isotropically (i.e., at the same power level for the same distance from the sink); although this is obviously no means of telling sensors apart from one another, because it only differentiates sensors radially with respect to the sink. By using such a property it is possible to provide routing from all sources to the sink.

At first, a simple distributed algorithm for execution by the sink and the sensors is introduced. Assume that an upper bound R on the greatest distance from the sink to a sensor is known to the sink. If B_0 is the power level at which the sink broadcasts its question, then sensor i, upon being reached by this broadcast and measuring P_i, can calculate its distance to the sink, denoted by R_i, and also its radial distance to the circle of radius R centered on the sink (that is, $R - R_i$); all it takes is that the sink broadcasts, along with its question, the values of B_0 and R. Let T and T_i denote the propagation times of an electromagnetic wave over the distances R and $R - R_i$, respectively (these can be computed easily given the wave's speed in the medium under consideration). The initial broadcast is by the sink, the action to be taken by a sensor upon receiving this message, and also how the sink or a sensor reacts upon receiving a message from a sensor. In the description of the algorithm, S_0 and S_i are used to denote the data structures used, respectively, by the sink and the sensor i to aggregate all information they receive. If sensor i is a source, then initially, S_i is assumed to contain its answer to the sink's question.

The description that follows is given in terms of Actions 1 and 2, respectively, for the sink and for a generic sensor i. Action 2, in particular, is dependent upon the product fr of the two parameters f and r. Each of these is, respectively, a number in the interval [0, 1] and the *radius* that a broadcast by a sensor is desired to reach. Once the value of r is known, it is assumed that sensors broadcast at a power level that is the same for all sensors, such that the locations at which the message can be received are exactly those that are no farther apart from the sensor than r.

Action 1. The sink broadcasts *Question* (B_0, R) and sets a timer to go off $2T$ time units later. In the meantime, upon receiving a message *Answer* $(*, S)$, the sink incorporates S into S_0. When the timer goes off, the source's answers to the sink are all summarized in S_0.

Action 2. Upon receiving the message *Question* (B_0, R), sensor i broadcasts *Answer* (P_i, S_i) if it is a source, and regardless of being a source or not sets a timer to go off $2T_i$ time units later. In the meantime, upon receiving a message *Answer* (P, S), sensor i incorporates S, suitably tagged with P, into S_i. When the timer goes off, sensor i checks whether S_i has had any information incorporated into it from an *Answer message*. In the affirmative case, it selects from S_i the entry whose P tag is greatest among all entries that have a P tag such that $P < P_i$. If the selection is successful (i.e., there is at least one candidate entry), then let P_j be this greatest P tag; sensor i then calculates R_j from P_j. If it is unsuccessful, then sensor i lets $R_j = \infty$. It then broadcasts *Answer* (Pi, Si) if $R_j - R_i > fr$.

15.4.2.2.2 Simulation Results

For each simulation, the n sensors were placed uniformly at random inside the circle and then the n^* sources were selected also at random. Every broadcast by a sensor during a simulation is assumed to reach exactly those sensors that lie within a circle of radius r centered on the emitting sensor. The value of r is determined so that the expected sensor density inside the circle is the same as that in the larger circle of radius R. If n_r denotes the expected number of sensors inside the circle of radius r, then $n_r/\pi r^2 = n/\pi R^2$, so it follows that

$$r = \sqrt{\{n_r/n\}} * R \qquad (15.2)$$

The parameter r is then a function of n_r, so in the experiments the two parameters that vary are f and n_r. Results for $n = 2000$, $n^*/n = a \times 10^{-b}$ with $a = 1, 2, 5$ and $b = 1, 2, 3$, $f = 0.1, 0.3, 0.5$, and $n_r = 9, 11, 13, 15$ (by Equation 15.2, these values of n_r correspond respectively to $r \approx 0.067R, 0.074R, 0.081R, 0.087R$).

Simulation results suggest that the radially decaying power perceived by the sensors when moved farther away from the emitting sink is capable of sustaining the construction of routes from randomly placed sources back to the sink.

15.4.2.2.3 Security Analysis

In this protocol, networks of anonymous sensors are considered and the problem of constructing routes for the delivery of information from a group of sensors in response to a query by a sink is addressed. To circumvent the restrictions imposed by anonymity, the protocol relies on using the power level perceived by the sensors in the query from the sink. Hence, the disguise of the IDs of the node from the outsider node thus prevents the passive attack.

15.4.3 Intrusion Detection

Intrusion detection systems (IDS) are important security tools in computer networks. There exist many approaches to the problem of intrusion detection in traditional computer networks and wireless ad hoc networks [41], but literature on this topic with regard to sensor networks is scarce. The goal of failure recovery is to extend the lifetime of a sensor network by restarting or reprogramming failed or misbehaving nodes. In combination, these two measures raise the cost for a potential attacker. Even if an attacker manages to capture a node and abuses it for his own purposes, there is a chance that the aberrant behavior of this node will be detected and the node be recovered, thus nullifying the attack.

When trying to protect a system from malicious use, it is important to define the goals and capabilities of potential attackers. Attackers that try to capture nodes by taking control of the code they are executing are considered. This would allow an attacker to take part in the network's ordinary operation and thus exercise a certain influence on the outcome of its operation, and to exploit the resources of the captured nodes. DoS attacks are not considered.

There are many possible ways for an attacker to inject a malicious code into a node, including the exploitation of weaknesses in its application code or in protocols used for application management, or physical vulnerabilities. The impact of software vulnerabilities can be minimized by using quality assurance tools like code verification and others. Defending against physical attempts at rewriting the application code requires barriers that make access to the physical features of the node's hardware as difficult as possible [42]. However, all defense mechanisms increase the cost of a sensor network. Therefore, it may be sensible to devote resources to intrusion detection and recovery to mitigate the effects of attacks.

Active measures against physical manipulations are also possible. The sensors already built into sensor nodes could help detect physical manipulations. For example, if a node is relocated, acceleration sensors can trigger off the key material, rendering the node inoperable within the network. In principle, all defense mechanisms can be circumvented, but the required effort should be prohibitively high. Generally, we would like to avoid that attacking a single node becomes cheaper if many nodes have already been attacked.

Here, an emotional-ant-based approach is discussed to identify possible preattack activities, and subsequently correspond with a centralized intrusion detection mechanism. Security monitoring in the sensor network is achieved by the foraging behavior of natural ant colonies. Ants may be positioned at relevant locations in the interconnected sensor networks, and for some of the related vocabularies to be described in this section, please refer [43,44]. An important advantage of the proposed approach is that the intruder-traversed trails could be easily available.

15.4.3.1 Intrusion Detection on Sensor Networks Using Emotional Ants

From the above discussion, we can understand there are many techniques for intrusion detection. In Ref. [45], intrusion detection on sensor networks using emotional ants (IDEA) has been proposed and the protocol is as follows.

15.4.3.1.1 Ant Colony Approach

Data-mining approaches for intrusion detection were first implemented in mining audit data for automated models for intrusion detection. Several data-mining algorithms are applied to audit data to compute models that accurately capture the actual behavior of intrusions as well as normal activities [46]. Audit-data analysis and mining combine the association rules and classification algorithm to discover attacks in audit data. Other approaches include fuzzy-rule-based classifiers [44], genetic programming techniques [43], support vector machines, and decision trees [Ant7]. A hierarchical distributed IDS architecture is analyzed in Ref. [47]. In Ref. [48], a self-organized ant-colony-based clustering technique is introduced (ANTIDS) to detect intrusions. Different adaptive and self-organized techniques have been already envisaged in designing the IDS. Here, some of the related works contemplate in a different form, where the ant agent would create a framework that allows users to define the characteristics of a given interaction due to intrusion. The ant system, given in Refs. [49,50], and the emotional-ant approach for IDS are presented as follows.

15.4.3.1.2 Ant Colonies System

An agent x can be described by the tuple $\langle \beta, S \rangle$, where β is the belief of an agent and s is the state of an agent. The set of agents A is actually organized into families to facilitate interagent collaboration and communication. It is needless to mention that all agents are sharing a common goal to find the intrusion in an optimized manner. Ant agents are typically involved in this distributed system by virtue of their pheromone concentration differences. A pheromone is a chemical substance an ant can drop in the environment. The pheromone then propagates (by Brownian motion) through the environment as well as evaporates over time.

The relatively simple mechanism of the pheromone demonstrates a simple yet effective distributed decision-making mechanism.

The basic algorithm is described as follows:
  Initialize pheromone values (τ)
    while termination condition not met do
      for $j = 1$ to k do
        $S^j \leftarrow$ ***construct solution*** (τ)
      end for
    Apply online delayed pheromone update ($\tau, s^1, \ldots, s^k$)
    end while

The *initialize pheromone values* step basically initializes all the pheromone values to the same positive constant value and adheres to the following conditions:

■ Whether or not the node has already been visited by any ant, a memory (called tabu list) is maintained. It expands within a particular traversal and is then emptied between visits.
■ The inverse of the distance $\eta_{ij} = 1/d_{ij}$ is called visibility. Visibility is based strictly on local information and represents the hemistich desirability of choosing node *j* when an ant is in node *i*. Visibility is used to direct the searching capabilities of ants, although a constructive method based on its sole use could produce a very low-quality solution.
■ The amount of virtual pheromone trail, $\tau_{ij}(t)$, is on the edge and is updated online.

The *apply online delayed pheromone update* $(\tau, s^1, \ldots, s^k)$ is used to store the track and edge details in the tabu list with the following pheromone update rule:

$$\tau_j \leftarrow (1 - \rho)\tau_j + \sum_{j=1}^{k} \Delta s^j \tau_j \qquad (15.3)$$

where
$\Delta s^j \tau_j = f(s^j)$ if s^j contributes to τ_j, otherwise it is 0
$\Delta s^j \tau_j$ is the combination of a solution s_j to the update for pheromone value τ_j
k is the number of solutions used for updating the pheromones
ρ is the evaporation rate
f is a function which usually maps the quality of a solution to its inverse

15.4.3.1.3 Emotional-Ants System

The basic idea is to identify the affected path of intrusion in the sensor network by investigating the pheromone concentration. The work also emphasizes the emotional aspects of agents, where they can communicate the characteristics of a particular path among themselves through the pheromone update. Therefore, in a sensor network, if the ants (here called emotional ants) are placed, they could keep track of the changes in the network path, following a certain knowledge base of rules depicting the probable possibilities of attack. Once the particular path among nodes is detected by the spy emotional ant, it can communicate the characteristics of the path through pheromone balancing to the other ants; thereafter, the network administrator could be alerted. The entire model has been inspired by several contemporary works. In pure cognitive form, the approach finally incorporates two basic parts:

■ Emotion and its utility in decision making
■ Transformation of the ant agent into an emotional ant

First, the emotion model of generic agents has been discussed. The structure of any emotion model is primarily based on certain thematic reactions exhibited by the agents with the real world. They encompass affinity, satisfaction, dejection, and approach [Ant20].

15.4.3.1.3.1 Emotional-Agent Definition Here, the agents have to be conceived and created (characteristics should be defined, e.g., name and class). An agent can be of the type of an object, where its function is related to the environment, or of the type no object, where its functions are related to the actions that the agent must carry out.

15.4.3.1.3.2 Basis of Rules Rules for possible attacks scenarios, as suggested by the network administrator, are cumulated. Again, certain primitives could be proposed to formulate these rules of sensor networks.

15.4.3.1.3.3 Emotion Templates These templates stand for individuals and represent different emotion states, i.e., if the parameters match with existing agents. For a clear picture of the emotion-exchange model some mathematical concepts are formulated in the forthcoming section.

15.4.3.1.3.4 Emotional Model of Ants This work closely adopts the strategy followed by the ant colony system [49,50]. Here, in the ant colony system, only the ant that generated the best tour since the beginning of the trail is allowed to globally update the pheromone concentration on the branches.

15.4.3.1.4 Security Analysis

Popular ways to secure a sensor network are by including cryptographic techniques or by safeguarding sensitive information from unauthorized access/manipulation, and by implementing efficient intrusion detection mechanisms. The discussed method is an ant-colony-based intrusion detection mechanism, which could also keep track of the intruder trials. This technique could work in conjunction with the conventional machine-learning-based intrusion detection techniques to secure the sensor networks. The discussed model of emotional ants presents collaborative distributed intelligence as a distributed coordination problem in the face of uncertainty, incomplete information with soft time, and resource constraints. An important feature of the IDEA model framework is the ability to perceive behavioral patterns, deliberate, and act based on self-organizational principle initiated with probability values. Thus, it detects the misbehaving nodes in the network. Hence, it prevents all kinds of attacks that are based on message injection, such as message manipulation attack, replay attack, delay attack, etc.

15.4.3.2 Applying Intrusion Detection Systems to Wireless Sensor Networks

An intrusion can be defined as a set of actions that can lead to an unauthorized access or alteration of a certain system. The task of IDS is to monitor computer networks and systems, detect possible intrusions in the network, and alert users after intrusions have been detected, reconfiguring the network if this is possible. A wireless network consists of a collection of nodes that are able to maintain a wireless communication channel between each other without relying on any fixed infrastructure. This and other factors make wireless networks substantially different from wired networks. Also, IDS for sensor networks must send the alerts to the base station to warn the human user. Finally, the IDS must be simple and highly specialized for reacting against specific sensor network threats and the specific protocols used over the network. A general IDS architecture of Ref. [51] is described in the following sections.

15.4.3.2.1 General IDS Architecture for Sensor Networks

15.4.3.2.1.1 Detection Entities

The constraints inherent to sensor networks, such as sparse resources and limited battery life, impose a cautious planning on how the detection tasks are performed. As in ad hoc networks, IDS agents must be located in every node. However, for the sake of performance, the architecture of these agents must be divided into two parts: local agents and global agents.

- Local agents should monitor the local activities and the information sent and received by the sensor. This is only carried out when the sensor is active, and the sensor only manages its own communications. Thus, the overheads imposed on the sensor node are low.
- Global agents should watch over the communications of their neighbors, and can also behave as watchdogs. However, not all nodes can perform this operation at the same time, because this operation would require sensors to analyze the contents of all packets in their radio range. Therefore, only a certain subset of the nodes must watch over the network communications at a time.

Once any agent, global or local, discovers a possible breach of security in the network, it must create and send an alert to the user. The only way the user can be reachable is through the base station. Hence, all alerts must be sent to the base station. The mechanism for sending the alerts to the base station depends on the underlying architecture of the sensor network, but it must assure that all alerts reach their destinations safely (using mechanisms such as μ*Tesla*).

15.4.3.2.1.2 Data Structures

Every agent, and hence every node, must store information about its surroundings to work properly. This information can be divided into two categories: knowledge about the security (an alert database that contains information about alerts and suspicious nodes), and knowledge about the environment (a list of the neighbors of the immediate neighbors of the node, which can be updated over the lifetime of the node using the received messages).

Every node has an internal alert database, which is used for storing the security information generated by the node agents. The format and size of that database are implementation-dependant (i.e., they depend on the protocols used in the sensor network). Nevertheless, they must contain the following fields: time of creation, classification, and the source of the alert.

This neighbor list can be obtained a priori, if any deployment information is available, or after the deployment, using the same assumption of the LEAP [52] protocol (the network will be secure in the first t seconds after the deployment). One problem of this list is its memory footprint. The list grows as a quadratic function, $((n^2) + n)$, of the number of one-hop neighbors; hence, it is not scalable for high density networks. However, the size of the list can be reduced using Bloom filters [53], storing the neighbors' list of every neighbor as a bit array. For a configuration of $k = 1$ (hash functions) and $m/n = 2$ (number of bits doubles the number of neighbors), the size of the list is reduced by 75 percent, at the cost of introducing a probability between 16 percent and 40 percent of false positives.

15.4.3.2.1.3 Local Agents

The task of local agents is to discover any attack or threat that can affect the normal behavior of the sensor nodes by analyzing only the local sources of information. These sources are the actual status of the node, i.e., packets received and sent by the node, the measurements made to the environment, and all the available information about its neighbors.

What kind of attacks should be detected by the local agent? First, attacks against the physical or logical safety of sensor nodes can be discovered if the nodes are able to know whether or not they are being manipulated. Sensor nodes are able, for example, to detect whether they are being reprogrammed, so they can raise an alarm before allowing the execution of any new code.

Node measurements are also vulnerable. Because the primary task of the sensor nodes is to analyze environmental data, any adversary can try to influence this process for his own benefit. Nevertheless, all data being read by the nodes comes from the real world, and follows certain patterns and limits. Therefore, anomaly-detection techniques can be used for monitoring these measurements. For example, if the node is going to be deployed in a static place, any variation in the accelerometer means that the node is being taken by an unknown source, so it will raise an alarm. Finally, the local agent also monitors packets that are addressed directly to the node. However, there are some issues related to locally processed packets that are protocol independent, e.g., the incorporation of a new node into the network and jamming of the signal.

In static scenarios, where few nodes are added after the initial deployment, every node can take advantage of the list of known neighbors. Every time a node receives a packet from a new neighbor, it will add it to the list, and raise an alarm. If human users of the network know that they did not include any node into the network, they will be aware that the new node belongs to an adversary. Also, if a node tries many times to send a packet but the channel is not available, misuse techniques can be employed to detect if this is an abnormal situation that must raise an alarm.

15.4.3.2.1.4 Global Agents

As stated before, global agents must be in charge of analyzing packets that their immediate neighbors send and receive. They can also behave as watchdogs, receiving and processing the packets relayed by next-hop nodes using protocol-dependant techniques. Because global agents are able to receive packets from both the neighbors that are relayed by the next-hop (due to the broadcast nature of the communications), they can be prepared to detect whether a certain node is dropping or modifying packets by analyzing those packets.

However, if all global agents are activated and listening to their neighborhoods at the same time, analyzing the network would be a costly operation in terms of energy. As a result, only a certain subset of nodes that cover all communications in the sensor network should activate their global agents. How this task is done depends on the underlying architecture of the sensor network. There are two basic architectures that specify how the sensors route the information over the network and how sensors group themselves. These two architectures are called hierarchical and flat.

In hierarchical configurations, sensors are grouped into clusters. One of the members of the cluster behaves as a server, or a "cluster head" (which can or cannot be more powerful than the other nodes), with management and routing tasks. On the other side, in flat configurations, information is routed sensor by sensor (every sensor of the network participates in the routing protocol), and almost all sensors have the same computational capabilities and constraints.

In hierarchical architectures, global agents are activated in every cluster head, because the combination of all cluster heads covers (in most cases) the entire sensor network. Consequently, total network coverage is assured. This approach helps to preserve the overall energy of the system because cluster heads are either more powerful than other nodes or are rotated periodically.

This clustering solution adds some complexity to the network in the creation and maintenance of clusters with maximum global coverage, adding a possible point of attack and the overhead of periodical control messages. However, there is an alternative distributed solution that can be applied to flat architectures without organizing them into clusters or adding more powerful nodes, which are called spontaneous watchdogs.

15.4.3.2.1.5 Spontaneous Watchdogs
The spontaneous-watchdog technique relies on the broadcast nature of sensor communications, and takes advantage of the high density of sensors being deployed in the field. For every packet circulating in the network, there are a set of nodes that are able to receive both that packet and the relayed packet by the next-hop. Hence, all these nodes have a chance to activate their global agents to monitor those packets. The main goal is to activate only one global agent per packet circulating in the network. The process is as follows:

- Every active node will receive all packets sent inside its neighborhood, due to the broadcast nature of communications.
- The node will check if it is the destination of the packet. If not, it will not drop the packet instantly. Instead, it will check if the destination of the packet is in its neighborhood (thus it could receive any packet forwarded by the destination). This check can be done because all nodes store a list of neighbors for every node in its neighborhood.
- If true, the node can be a spontaneous watchdog. Consequently, it will calculate how many nodes in the network are in its same situation.
- If the number of nodes that fulfill the requirements are n, a single node will select itself as a global agent for this packet with a probability of $1/n$. This process can resemble as n people with 1 dice of m sides each, where $n = m$, trying to obtain a 1 in the dice to activate the global agent.

15.4.3.2.2 Security Analysis
In the above section, a technique is discussed to optimally watch over the communications of the sensors to detect anomalies and monitor the packets. Thus, any doubtful behavior of a node can be detected and sent to the base station; finally, there is action to

prevent this misbehavior. Thus, the outsider nodes cannot take part in the reprogramming and cannot disguise themselves. Hence, message manipulation attack, sybil attack, masquerade attack, and wormhole attack can be prevented.

15.5 Conclusions

In sensor networks, sensor nodes are resource-constrained; thus, security applications are the challenging problem for sensor networks. In this chapter, we have focused on the attacks regarding sensor network security and their countermeasure techniques. Only one technique cannot prevent all the attacks. Hence, a combined effect is needed. According the resource constraints and applications, the security protocol should be selected.

Besides that, to understand the protocols, we also discussed the mathematical background and some other related preliminaries. We mainly focus on the latest proposed protocol and in some cases the basic protocol. All the individual protocols' security analysis is given separately in their corresponding subsections. There is no protocol existing that prevents all the attacks. And still there is no solution existing for the DoS attack in sensor network security.

References

1. Y. Xiao, V.K. Rayi, B. Sun, X. Du, F. Hu, and M. Galloway, A survey of key management schemes in wireless sensor networks, *Computer Communications Journal*, 30(11–12), 2314–2341, 2007, Elsevier, Science Direct.
2. N. Koblitz, Elliptic curve cryptosystems, *Mathematics of Computation* 48, 203–209, 1987.
3. N. Gura, A. Patel, A. Wander, H. Eberle, and S.C. Shantz, Comparing elliptic curve cryptography and RSA on 8-bit CPUs, in *Proceedings of the 6th International Workshop on Cryptographic Hardware and Embedded Systems*, Boston, MA, August 2004.
4. N. Koblitz, *A Course in Number Theory and Cryptography*, 2nd edn., *Graduate Texts in Mathematics*, vol. 114, Springer-Verlag, New York, 1994.
5. I. Blake, G. Seroussi, and N. Smart, *Elliptic Curves in Cryptography*, London Mathematical Society, Lecture Note Series 265, Cambridge University Press, Cambridge, U.K., 1999.
6. X. Du, M. Guizani, Y. Xiao, S. Ci, and H.H. Chen, A Routing-driven elliptic curve cryptography based key management scheme for heterogeneous sensor networks, *IEEE Transactions on Wireless Communications*, Accepted for publication, Apr. 2007 (to appear). Extended version of, A routing-driven key management scheme for heterogeneous sensor networks, *IEEE International Conference on Communications,* 2007 (ICC'07), ISBN: 1-4244-0353-7, 24–28 June 2007, Glasgow, DOI: 10.1109/ICC.2007.564, pp. 3407–3412.
7. A. Shamir, Identity-based cryptosystems and signature schemes, in *CRYPTO'84: On Advances in Cryptology*, pp. 47–53, Springer-Verlag, 1984.

8. R. Sakai, K. Ohgishi, and M. Kasahara, Cryptosystems based on pairing, in *Symposium on Cryptography and Information Security (SCIS2000)*, Okinawa, pp. 26–28, January 2000.
9. L.B. Oliveira, D.F. Aranha, E. Morais, F. Daguano, J. Lopez, and D. Ricardo, TinyTate: Computing the tate pairing in resource-constrained sensor nodes, in *6th IEEE International Symposium on Network Computing and Applications (NCA 2007)*, Cambridge, MA, pp. 318–323, July 12–14, 2007.
10. L.B. Oliveira, D. Ricardo, J. Lopez, F. Daguano, and Loureiro, Identity-based encryption for sensor networks, in *5th IEEE International Conference on Pervasive Computing and Communications Workshops (PerComW'07)*, White Plains, NY, pp. 290–294, March 2007.
11. L.B. Oliveira, D. Aranha, E. Morais, F. Daguano, J. L'opez, and R. Dahab, TinyTate: Identity-Based Encryption for Sensor Networks, available at http://eprint.iacr.org/2007/020.pdf.
12. J.L. Hill and D.E. Culler, Mica: A wireless platform for deeply embedded networks, *IEEE Micro*, 22(6), 12–24, 2002.
13. P. Levis, S. Madden, J. Polastre, R. Szewczyk, K. Whitehouse, A. Woo, D. Gay, J. Hill, M. Welsh, E. Brewer, and D. Culler, TinyOS: An operating system for wireless sensor networks, in W. Weber, J. Rabaey, and E. Aarts, eds., *Ambient Intelligence*, Springer-Verlag, New York, 2004.
14. The Tate Pairing, available at http://www.computing.dcu.ie/~mike/tate.html.
15. Elliptic Curve Cryptography Tutorial, http://www.certicom.com/index.php/ecc
16. D. Boneh and M. Franklin, Identity based encryption from the weil pairing, *SIAM Computing*, 32(3), Extended Abstract in Crypto 2001, 586–615, 2003.
17. S.L.M. Berreto, H.Y. Kim, and M. Scott, Efficient algorithms for pairing-based cryptosystems, in *Advances in Cryptology-Crypto'2002, LNCS 2442*, Springer-Verlag, Berlin, Germany, pp. 354–368, 2002.
18. S. Galbraith, K. Harrison, and D. Soldera, Implementing the tate pairing, in *Algorithm Number Theory Symposium-ANTS V, LNCS 2369*, Springer-Verlag, Berlin, Germany, pp. 324–337, 2002.
19. D. Boneh and M. Franklin, Identity-based encryption from the weil pairing, in *Advances in Cryptology—CRYPTO'01, Lecture Notes in Comput Science*, vol. 2139, Springer-Verlag, Berlin, Germany, pp. 213–229, 2001.
20. D. Boneh, B. Lynn, and H. Shachum, Short signatures from the weil pairing, in *Advances in cryptology—ASIACRYPT'01, Lecture Notes in Comput Science*, vol. 2248, Springer-Verlag, Berlin, Germany, pp. 514–532, 2001.
21. A. Joux and K. Nguyen, Separating decision Diffie–Hellman from Diffie–Hellman in cryptographic groups, Cryptology ePrint Archive, Report 2001/03, available at http://eprint.iacr.org/2001/03/.
22. R. Dutta, R. Barua, and P. Sarkar, Pairing-based cryptographic protocols: A survey, Cryptology ePrint Archive, Report 2004/064, available at http://eprint.iacr.org/2004/064.
23. Y.C. Hu, A. Perrig, and D.B. Johnson, Packet leashes: A defense against wormhole attacks in wireless ad hoc networks, in *Proceedings of the 22nd Annual Joint*

Conference of the IEEE Computer and Communications Societies (*INFOCOM 2003*), San Franciso, CA, 2003.

24. Y.C. Hu, D. Johnson, and A. Perrig, SEAD, secure efficient distance vector routing for mobile wireless ad hoc networks, in *IEEE Workshop on Mobile Computing Systems and Applications*, Callicoon, NY, June 2002.

25. H. Y-Chun, A. Perrig, and D.B. Johnson, Rushing attacks and defense in wireless ad hoc network routing protocols, in *WiSe '03: Proceedings of the 2003 ACM Workshop on Wireless Security*, San Diego, CA, ISBN: 1581137699, pp. 30–40, September 19, 2003.

26. X. Du, M. Guizani, Y. Xiao, and H.-H. Chen, Secure and efficient time synchronization in heterogeneous sensor networks, *IEEE Transactions on Vehicular Technology*, 57(4), 2387–2394, July, 2008.

27. A. Perrig, R. Szewczyk, V. Wen, D. Culler, and D. Tygar, SPINS: Security protocols for sensor networks, in *Proceedings of 7th Annual International Conference on Mobile Computing and Networks MOBICOM 2001*, Rome, Italy, July 2001.

28. D. Malan, M. Welsh, and M.D. Smith, A public-key infrastructure for key distribution in TinyOS based on elliptic curve cryptography, in *Proceedings of 1st IEEE International Conference Communications and Networks* (*SECON*), Santa Clara, CA, October 2004.

29. A.S. Wander, N. Gura, H. Eberle et al., Energy analysis of public-key cryptography for wireless sensor networks, in *Proceedings of the 3rd IEEE International Conference on Pervasive Computing and Communications* (*PERCOM*), Kauai, HI, March 2005.

30. B. Karp and H. T. Kung, GPSR: Greedy perimeter stateless routing for wireless networks, in *Proceedings of the 6th Annual International Conference on Mobile Computing and Networking*, pp. 243–254, 2000.

31. Sk. Md. M. Rahman, N. Nasser, and K. Saleh, Identity and pairing-based secure key management scheme for heterogeneous sensor networks, in *Proceedings of 4th IEEE International Conference on Wireless and Mobile Computing, Networking and Communications (WiMob 2008)*, Avignon, France, October 2008.

32. A. Durresi, V. Paruchuri, M. Durresi, and L. Barolli, A hierarchical anonymous communication protocol for sensor networks, in *Proceedings of 2005 IFIP International Conference on Embedded and Ubiquitous Computing (EUC-05)*, Nagasaki, Japan, LNCS 3824, Springer-Verlag, pp. 1123–1132, December 2005.

33. S. Bandyopadhyay and E.J. Coyle, An energy efficient hierarchical clustering algorithm for wireless sensor networks, in *Proceedings of the 22nd Annual Joint Conference of the IEEE Computer and Communications Societies* (*INFOCOM 2003*), San Francisco, CA, March–April, 2003.

34. C.F. Chiasserini, I. Chlamtac, P. Monti, and A. Nucci, Energy efficient design of wireless ad hoc networks, in *Proceedings of European Wireless*, February 2002.

35. O. Younis and S. Fahmy, Distributed clustering in ad-hoc sensor networks: A hybrid, energy-efficient approach, in *Proceedings of IEEE INFOCOM'04*, Hong Kong, March 2004.

36. O. Younis and S. Fahmy, HEED: A hybrid, energy-efficient, distributed clustering approach for ad-hoc sensor networks, *IEEE Transactions on Mobile Computing*, 3(4), 366–379, October 2004.

37. W. Du, J. Deng, Y.S. Han, S. Chen, and P. Varshney, A key management scheme for wireless sensor networks using deployment knowledge, in *Proceedings of IEEE INFOCOM'04*, March 2004.

38. C. Daz and B. Preneel, Reasoning about the anonymity provided by pool mixes that generate dummy traffic, in *6th International Workshop, Information Hiding (IH'04)*, Revised selected papers, *Lecture Notes in Computer Science* (LNCS-3200), Springer-Verlag, Toronto, Canada, May 2004.

39. E.H. Callaway, Jr., *Wireless Sensor Networks: Architectures and Protocols*, Auerbach Publications (an imprint of CRC Press), New York, 2003.

40. R.C. Dutta and V.C. Barnosa, Finding routes in anonymous sensor networks, arXiV:cs.NI/0507021, vol. 1, Jul 7, 2005, EUC 2005, *LNCS 3824*, pp. 1123–1132, 2005.

41. Y. Zhang and W. Lee, Intrusion detection in wireless ad-hoc networks, in *MOBICOM 2000*. ACM Press, 2000.

42. R. Anderson and M. Kuhn, Low cost attacks on tamper resistant devices, in *5th International Workshop on Security Protocols*, Paris, France, LNCS 1361, Springer-Verlag, pp. 125–136, April 1997.

43. A. Abraham, Evolutionary computation in intelligent web management, evolutionary computing in data mining, in *Studies in Fuzziness and Soft Computing*, A. Ghosh and L. Jain, eds., Springer Verlag, Germany, Chapter 8, pp. 189–210, 2004.

44. A. Abraham, R. Jain, S. Sanyal, and S.Y. Han, SCIDS: A soft computing intrusion detection system, in *6th International Workshop on Distributed Computing (IWDC 2004)*, A. Sen et al. eds. Springer Verlag, Germany, *Lecture Notes in Computer Science*, vol. 3326, pp. 252–257, 2004.

45. S. Banerjee, C. Grosan, A. Abraham, and P.K. Mahanti, Intrusion detection on sensor networks using emotional ants, *International Journal of Applied Science and Computations*, 2005.

46. D. Barbará, J. Couto, S. Jajodia, and N. Wu, ADAM: A testbed for exploring the use of data mining in intrusion detection, in *ACM SIGMOD Record*, 30(4), 15–24, December 2001.

47. S. Marsella, W.L. Johnson, and C. LaBore, Interactive pedagogical drama, in *Proceedings of 4th International Conference on Autonomous Agents (ICMAS)*, 2000.

48. V. Ramos and A. Abraham, ANTIDS: Self-organized ant-based clustering model for intrusion detection system, in *4th IEEE International Workshop on Soft Computing as Transdisciplinary Science and Technology*, Muroran, Japan, May 2005.

49. J. Deng, R. Han, and S. Mishra, INSENS: Intrusion-tolerant routing in wireless sensor networks, Technical Report CU-CS-939-02, Department of Computer Science, University of Colorado, November 2002.

50. M. Dorigo and L.M. Gambardella, Ant colony system: A cooperative learning approach to the travelling salesman problem, *IEEE Transactions. Evolutionary Computation*, 1, 53–66, 1997.

51. R. Roman, Z. Jianying, and J. Lopez, Applying intrusion detection systems to wireless sensor networks, in *3rd IEEE Consumer Communications and Networking Conference (CCNC 2006)*, vol. 1, pp. 640–644, January 2006.

52. S. Zhu, S. Setia, and S. Jajodia, LEAP: Efficient security mechanisms for large-scale distributed sensor networks, in *10th ACM Conference on Computer and Communications Security (CCS'03)*, Washington, DC, October 2003.
53. B. Bloom, Space/time trade-offs in hash coding with allowable errors, *Communications of the ACM*, 13(7), 422–426, July 1970.
54. L. Eschenauer and V. D. Gligor, A key management scheme for distributed sensor networks, in *Proceedings of the 9th ACM Conference on Computer and Communication Security*, Washington, DC, November 2002.

Chapter 16

Network Management in Wireless Sensor Networks

Hamid Mukhtar, Ali Hammad Akbar, Shafique Ahmad Chaudhry, Ki-Hyung Kim, and Seung-Wha Yoo

CONTENTS

Wireless sensor networks (WSNs) have drawn a lot of attention from the research community in recent years owing to the fact that these networks extend the usability and business potential of various networked devices across the globe and into our daily lives, while mandating an all-the-time availability. To keep such networks always operational, robust network management is needed. The classically perceived purview of network management of monitoring, and maybe controlling a network, tends to be lacking in the sense that WSNs present newness not only in their applications but also in their vulnerabilities to both intrinsic and induced failures. In this chapter, we investigate various network-management aspects that relate to the off-beat aspects of WSNs, particularly in architecture, design, and applications. We substantiate newer dimensions to the classical network-management framework and underpin their significance for emerging application scenarios.

16.1   Introduction

Wireless sensor networks (WSNs), which encompass a large number of intelligent sensors, are designed to transform the previous paradigm of collecting data from the environment by offering a new way of monitoring and controlling the environment. The emerging range of WSN applications includes medical, military, home automation, environmental monitoring, and industrial monitoring and control.

WSNs differ from traditional networks in many ways, presenting some unique characteristics. Sensor nodes are generally projected with small dimensions (cm^3 or mm^3), and this size limitation results into severe resource constraints, such as limited battery power, low computational and memory resources, scarce wireless bandwidth, and limited communication capability. However, the overall network performance and network gain could be substantial, if the networkwide resources of WSNs are well managed.

The processes of monitoring, managing, and controlling the operation of a network can be referred to as "network management." In more concrete terms, such management functions coordinate configuration, security, operation, and maintenance, of all elements of a network. The network management, however, requires different approaches for each network. As such, a wide range of traditional management solutions are not directly applicable to WSNs due to their distinctive attributes. The applications in WSNs are designed to work under their inherent energy, bandwidth, and resource limitations, as compared to traditional networks, where they are designed to achieve better performance and response time. In fact, faults commonly occur in large-scale WSNs, which have hundreds of thousands of nodes and severe resource constraints. In this case, therefore, maintenance of components or recharging the nodes is not feasible, and sensor nodes are usually considered disposable. Moreover, many findings of [3] demonstrate that the environment interference and configuration errors can even cause the loss of an entire WSN prior to the start of its operation.

Moreover, sensor nodes tend to undergo harsh conditions, which increase the likelihood of dramatic changes in node configurations of WSNs. Therefore, a sensor network management system must allow a network's self-formation, self-organization, and self-configuration, in the absence of prior topological knowledge. In case of node deployments in a perfunctory manner for self-managed WSNs, the nodes bootstrap and perform a self-test [12]. Afterward, they discover the locality and observe the energy levels, the usage state, and the administrative situation. All these activities are conducted at the network element level by the self-management functions. If the nodes discover their location, they are able to divide themselves into groups. Although sensor network management is exceedingly significant, there is no existing generalized solution for WSN management [1].

Furthermore, traditional or temporary networks are intended for a large number of user applications. This means that the network elements are installed and configured to support various types of services. On the other hand, WSNs are application oriented in general. In contrast with the traditional or temporary networks, wireless sensor nodes operate a common application cooperatively.

To address such behavioral characteristics of WSNs, it is necessary to devise a network management system that possesses new types of management functions. In this chapter, such a management framework for monitoring and controlling of WSNs will be explored and suggested.

The remainder of this chapter is organized as follows. Section 16.2 reviews design goals of WSN management followed by management dimensions in Section 16.3. Section 16.4 describes the design alternatives for management architectures. Section 16.5 discusses some contemporary research efforts in the area of WSN management. Section 16.6 describes IP-USN as an integrating technology which is followed by Section 16.7 on WSN management as an FCAPS (fault, configuration, accounting, performance, and security) model. Section 16.8 concludes the chapter.

16.2   Design Goals of WSN Management

Management applications, when designed for traditional networks, may have restrictions in terms of throughput and latency, as compared to the hardware limitations such as memory, energy, and processing power, when designed for sensor networks. The goals of network management for sensor networks need to set up a clear and direct relationship with the application-oriented design of sensor networks. The goals are defined as follows:

16.2.1   Scalability

The nodes in a WSN are assumed to be deployed in large numbers, and new nodes may be added to the network dynamically. Network management architectures should be able to work with dense deployment of nodes as well as high volumes of data generated in the network.

16.2.2   Limited Energy Consumption

Sensor devices mostly run on batteries. The management operations should be lightweight on a node's resources to prolong its lifetime; in this manner, contributing to the network lifetime as a whole. The management operations should be efficient and should consume the least possible energy.

16.2.3   Memory and Processing Limitations

Sensor devices are supposed to have a constrained memory and limited processing power. The management applications should impose less storage overhead for the management information. The information bases and managed agents should be optimized to cater for the memory and processing limitations; they should have a limited processing overhead and code footprint to optimally use the sensor resources.

16.2.4 Limited Bandwidth Consumption

Sensor networks may have a bandwidth limitation in the presence of high channel impairments. The designed management applications should keep this consideration in mind. In such cases, the bulky management queries may have to be optimized to cater for the link characteristics. Moreover, the energy cost associated with communication is usually more than that of sensing and processing. Therefore, management queries should be optimized to consider the bandwidth limitations and the energy cost associated with bulky transmissions.

16.2.5 Adaptability to Network Dynamics

The management applications should be adjustable to network dynamics and changes in the network topology. They should also be able to gather the state of the network and the topological changes that occur in the network. The features like addition of new nodes, failure of the current nodes, and node mobility within and outside the network can also be supported in these networks according to the application requirements.

16.2.6 Fault Tolerance

The faults in sensor networks are different from those in traditional networks. A fault may be caused by a node that may be in the sleep mode to conserve energy. A fault may also be caused by a node that ran out of energy or is disconnected from the network because of mobility or network partitioning. The management system and applications should exhibit self-healing characteristics and they should be aware of such network dynamics.

16.2.7 Network Responsiveness

The events in the network should be quickly reported to the network management system. Network dynamics, such as movement of nodes, failure of a node, and network partitioning of a subnetwork, must be quickly detected by the management system. The management system should be able to react and adapt to these changes, and adjust accordingly through management policies.

16.2.8 Cost of Equipment

The management system should minimize the deployment cost by requiring less equipment to be deployed in the network for management applications. The management systems are supposed to reuse the current network hardware and efficiently utilize its resources for management operations.

16.3 Management Dimensions

The classical paradigm of network management is defined as two planes, i.e., management functionalities and management levels. Management functionalities cover fault management, configuration management, accounting management, performance management, and security management. Together they are called the FCAPS model. Management levels define the abstraction of the purview of network management.

16.3.1 Management Functionalities

Traditionally, WSNs have been considered as mission oriented and application centric, but some recent proposals have defined new roles, e.g., running multiple applications on a WSN and service-centric nature. The execution of multiple applications, with the service perspective, gives altogether a different outlook to the applications of WSNs. For example, one WSN deployed in a region of interest can perform various military applications, including target tracking, mine detection, and recognizing a friend or enemy.

All these conditions demand that management functions must be performed autonomously on the sensor nodes. An autonomic network-management paradigm is, therefore, needed at large for WSN management. Here, we briefly describe the self-based autonomicity of network management.

16.3.1.1 Self-Management

Self-management can be described as "management of the network through managed objects." Self-management functionality transforms components into composite entities to optimally provide required functions. It makes the systems manage themselves according to an administrator's goals with least human intervention. The realization of self-management can be better understood with the configuration, optimization, healing, accounting, and security model, as depicted in Figure 16.1. It may also be noted that the need of self-management solutions makes configuration management, in particular, significant. The fault, performance, account, and security management, all emanate and converge into the configuration management.

16.3.1.2 Self-Configuration

Self-configuration of an element determines its operational and maintenance characteristics as well as application execution, data communication, and data forwarding. The self-configuration module makes managed components to configure themselves automatically in accordance with high-level policies.

16.3.1.3 Self-Healing

Self-healing describes the property that each node has the ability to perceive that it is not operating correctly and, without human intervention, makes the necessary adjustments to restore itself to normal operation.

16.3.1.4   Self-Accounting

Self-accounting enables the measurement of the use of network and other local resources.

16.3.1.5   Self-Security

Self-security should minimize unauthorized or accidental access to all the elements as well as the network resources. Security-management functions deal with ensuring legitimate use, maintaining confidentialities, etc.

16.3.1.6   Self-Optimization

Self-optimization means to continually seek the ways to improve the network performance. Nodes keep identifying and seizing opportunities to make the entire system's performance more efficient.

16.3.2   Management Levels

In terms of logical architectural design, management functionalities depend on the respective management level they belong to. The approaches to define the management levels and resulting frameworks may be categorized into "top-down" and "bottom-up" approaches. The former facilitates more clarity and granularity because the analysis of upper layers helps defining the necessities of lower levels. On the contrary, the later approach is wieldier. The management levels are categorized into business, service, network, and network-element levels.

16.3.2.1   Business

The business level represents the business objectives that mainly deal with service development and cost-benefit analysis. These objectives form the main source of determining the basic skeleton and feature the blueprint of a WSN. The level also outlines the sensor network with a special focus on network setup, maintenance, sensing, processing, and communication costs.

16.3.2.2   Service

A WSN service is the functionality that is associated with application objectives. A common objective in designing these services is to minimize the energy consumption. The examples of WSN services are data gathering, processing, and communication. The network-management services' level defines the set of operations/processes that must be implemented to carry out certain functionality. The other aspect is to define the scope, conditions, and parameters for these functions. A given service is then guaranteed on the execution of these functions. Management functions are the representations of detailed granular functional portions of a management service, as perceived by users. The management framework, therefore, must include the functionality blueprint of WSN applications, and users.

16.3.2.3 Network

This level deals with the network-level considerations, e.g., nodes collaboration, network coverage, connectivity, data aggregation, and communication. The relationships between sensor nodes are defined and maintained. As an objective function, the network gain is optimized in the presence of nodes' sensing, processing, and communication constraints.

16.3.2.4 Network-Element Management

This level covers the power management, mobility management, state management, and task management of wireless sensor nodes. If the sensor nodes are heterogeneous in design and roles, the management system must consider such differences in designing a management system for a WSN.

16.3.2.5 Element-Level Management

The network-element level deals with the network elements that need to be managed or execute some management function. Based on the scope of an application being executed, a management element could be a single node or a group of nodes. In case the application requires a large number of nodes, the network element may be associated with a cluster of nodes.

16.4 Design Alternatives of Management Architectures

The architectural design of a management architecture lies at the heart of any management system. As per the requirements and goals of WSNs, the architecture should be lightweight, scalable, and adaptable. Moreover, it should ensure fault tolerance, responsiveness, and low production cost. In this section, we discuss the commonly proposed and adopted architectural design approaches for WSNs.

16.4.1 Policy-Based Approach

Policy-based management is an approach to determine operating rules for the management, administration, and control of network resources. Policies define the behavior of the components in different operating situations, which helps in increasing the adaptability and the fault tolerance of the network. Policies for fault, configuration, accounting, performance, and security management are defined to enhance the autonomy and efficacy of the network. For example, MANNA (A Management Architecture for WSNs) [1] uses policies to define the interaction between the network managers and agents for decision making in the operating environment.

16.4.2   Management by Delegation

WSNs are resource-constrained networks and energy is usually the scarcest resource in such restricted environments. Putting too much sensing, processing, or communication overhead on a constrained device will drain its energy. Moreover, the cost associated with communication is usually dominating the cost of sensing and processing. Therefore, delegation of management responsibilities to the less-constrained nodes and aggregation of management information on the way to the manager can help in saving the network's energy and can considerably increase the network lifetime. The management responsibility in this case is delegated to the nodes closer to the end nodes, and localized processing is used to increase the network lifetime. Guerilla management [6] and LoWPAN Network management Protocol (LNMP) [7] employed such an approach where the management tasks were distributed to nodes with higher power.

16.4.3   Decentralized Management

Centralized management has always been a performance bottleneck in traditional systems as well as in WSNs. The centralized approach of management concentrates the traffic toward the central manager causing congestion on the active links in the network. The WSN's traffic flow toward the central manager can drain the energy of the nodes on that particular route. Moreover, centralized management is not recommended in WSNs due to the network-partitioning problem where a subpart of the network becomes disconnected for some time. Therefore, distributed management can be used to ensure reliability and energy efficiency. Furthermore, in decentralized management, each manager can have its own domain, can ensure localized processing of management information, and can help in decreasing the network latency and the traffic in the network. The managers can communicate with each other and can share management information, thus ensuring more reliability. MANNA [1] uses decentralized management where different locations for managers and agents can be selected according to the application goals.

16.4.4   Layered Management

In a layered approach, the management tasks are divided among different layers, and each layer performs its task independently and provides interfaces to communicate with other layers, if necessary. Different schemes can coexist for management tasks for different layers. Furthermore, the task of modification is simpler, which helps in maintenance. In Ref. [13], the authors employ a cluster-based middleware framework that makes use of the layered approach and divides the tasks between the cluster- and the resource-management layers. In Ref. [14], the authors use the layered approach to distribute the tasks among the node, the network, and the application-specific layers.

16.4.5   Level-Based Management

Complex systems can be divided into small manageable system components. Management is traditionally level-based, as mentioned in Section 16.4.2 (management levels). Different approaches can be defined for each level of management. For example, for business and service management, service-oriented architectures (SOA) can be used where the design of a large system can be divided into small manageable components and each service can be made available as a web service. The lower levels, i.e., the network and the network-element levels, can exist as functional units that are combined to provide these services. MANNA uses such an approach where the lower-level information is reported as functional units and the higher-level services are at the abstract layer where they make use of the functional units. Bridge of the SensorS (BOSS) [11] uses such an architecture where the management information is gathered from the sensor network and is made available as a service, which can be discovered by the users.

16.4.6   Mobile or Intelligent Agent-Based Approaches

In the mobile agent-based approach for network management, the processing task in the form of a mobile code is moved to the locations where the management data is present. The code is executed locally and only the results are sent back to the manager. This local processing of data results in less consumption of network bandwidth. Additionally, reliability can be ensured in the case of network partitioning where the data is processed at the local agent and the results can be sent when the connection is restored. Agilla [15] is an example of such architectures where the middleware helps in the deployment of mobile agents. Mobile agent-based policy management for WSNs is proposed in Ref. [16]. In this architecture, operating policies and rules are enforced by the help of management agents.

16.5   Contemporary Research Efforts

16.5.1   MANNA

MANNA [1] is a policy-based architecture for WSN management. MANNA presents the technical basis to how management can be performed in WSNs. It presents a generic architecture for management of WSNs highlighting its inherent dependency on the application for which it is being developed. It extends the traditional two-dimensional management, i.e., "Management levels" and "Management functional areas," by adding "WSN functionalities" as a new dimension to use considered network functionalities of WSNs. The new WSN functionalities' dimension covers configuration, maintenance, sensing, processing, and communication issues of WSNs.

Here, the management information is mapped as WSN models. Based on the information obtained with these models, management functions and services are executed according to the application-specific policies. MANNA describes many different network models, e.g., sensing and communication coverage area maps, behavioral and dependence models, network topology and energy usage status, and the cost of equipment. The management functions can consist of information obtained from these models and some examples of management functions are topology-map-discovery function, energy-level-discovery function, and coverage-area-supervision function. These services use the data that is obtained from the management models. Finally, a set of these functions can be combined to offer services according to the policies defined for the applications.

MANNA architecture is divided into three architectures: functional, physical, and informational. The functional architecture illustrates the delegation of management responsibility among different management entities, i.e., manager, agent, and the management information base (MIB). It defines the optimum location for placement of managers and agents in the sensor network. The physical architecture defines how to implement the functional architecture. It defines protocol profiles that may be satisfactory for each application type and proposes that the interfaces between the management entities should use lightweight protocol stacks. The informational architecture provides the information that should be supported in the network. The model divides the information into different types of support and managed classes, according to their functionality. The managed classes include attributes related to network, managed element, equipment, system, phenomenon, connection, and environment.

16.5.1.1 WSN Functionalities' Dimension of MANNA

In addition to the two traditional dimensions of network management, as mentioned in Section 16.3, MANNA architecture proposes a new dimension, WSN functionalities, in consideration of the special characteristics of WSNs. This covers five main WSN functionalities: configuration, sensing, processing, communication, and maintenance. These are as follows.

16.5.1.1.1 Configuration

WSNs can be characterized by their composition of nodes and the organization of the network. The composition can be homogeneous where all nodes in the network have the same hardware capabilities, or it can be heterogeneous and can be composed of nodes with different capabilities. Moreover, the topology and the organization of the network also influence the configuration. The nodes can be organized hierarchically in a tree-based topology or the network can have a flat topology where all the nodes are at the same level. At the element level, the nodes in a sensor network can be characterized either as sink nodes, cluster heads, or source nodes.

16.5.1.1.2 Sensing

It covers the data-gathering mechanisms for the sensor nodes. The collection can be continuous where sensor nodes collect data from the environment continuously or it can be reactive where the sensor responds to a query from a manager or starts the collection in reaction to some event in the environment.

16.5.1.1.3 Communication

It covers the monitoring mechanisms for WSNs. The monitoring can be continuous where the data is sent periodically to the management system. It can be On Demand where it answers user's requests. It can also be Event-Driven communication where the sensor node sends the information in reaction to an event of interest. Lastly, it can be programmed where the sensor sends data to the manager according to the policies defined by the application.

16.5.1.1.4 Processing

Processing can be basic processing where simple mapping is done or the threshold-based filtering is used. Contrariwise, it can be a correlation where decisions are taken on the basis of the gathered data.

16.5.1.1.5 Maintenance

This functionality describes the necessity of autonomic reconfiguration, protection, optimization, and healing in WSNs.

16.5.2 BOSS

BOSS [11] defines an architecture to connect Universal Plug and Play (UPnP)-based devices to non-UPnP-based sensor devices. UPnP allows devices to connect seamlessly and to simplify the implementation of networks. BOSS exploits automatic device- and service-discovery features of UPnP networks for the service-discovery management of WSNs. In this architecture, a service agent listens to the service advertisements from the sensor network. Upon reception, it makes the services available through UPnP. Additionally, BOSS also defines a UPnP device-management service where data related to device information, context, localization, synchronization, power, security, and discovered services is reported. Thus, BOSS provides a network-management and service-discovery architecture for WSNs. The advantage in BOSS is that UPnP requires zero configuration and it can be easily integrated with many technologies. The disadvantage is the overhead of the amount of data that is reported from the sensor network.

16.5.3 SNMS

Sensor Network Management System (SNMS) [17] presents an application-cooperative management system for WSNs. It is composed of three core services. First, it consists of a query system to enable rapid, user-initiated acquisition of network health and performance data. Second, it consists of a logging system to enable recording and retrieval of system-generated events. Finally, it comprises a lightweight network layer that can operate in parallel with the application's network layer [17]. The system supports two traffic patterns: Collection and Dissemination. Collection is required to obtain health data from the network, and Dissemination is required to distribute management commands and queries. For the collection pattern, a collection tree protocol is defined that minimizes state requirement by being independent of the neighbor table and minimizes traffic by construction of a tree after the collection of messages. The constructed tree is adaptive and adjusts according to the network topology. For dissemination, the protocol ensures reliable transport-level dissemination of messages. It proposes Drip, a protocol for reliable dissemination of all messages to a managed device in the network. Each component wishing to use Drip registers a specific identifier, which represents a reliable dissemination channel. Messages received on that channel will be delivered directly to the component. It requires each node to periodically check the subscribed channels and cache the extracted data from the most recent message received on the channel and it requires them to return the cached data in response to the periodic messages. This helps in ensuring reliability even when the application fails. The main advantage of SNMS is its low memory consumption and reduced traffic burden. SNMS also uses aggregation techniques to bundle the query replies together to save network energy.

16.5.4 Mobile Agent-Based Policy Management

Mobile agent-based policy management [16] introduces mobile agents into the WSN management to support autonomy of the tasks in the network. They employ a hierarchical architecture for mobile agent-based policy management of WSNs. The architecture is composed of policy manager as the higher layer, cluster policy agent as the middle layer, and local policy agent as the lower layer. A policy manager manages multiple cluster policy agents and it acts as a global policy decision point (PDP). Cluster policy agents act as intermediate PDPs and local policy agents act as policy enforcement points (PEPs). Policies are disseminated hierarchically from Policy Managers to the Local Policy Agents. Any node can dynamically take over the functionality of any other node to ensure survivability. This architecture also defines a policy information base, which is defined in a data structure similar to Simple Network Management Protocol (SNMP).

16.6 IP-USN as an Integrating Technology

Internet Protocol–based ubiquitous sensor networks (IP-USNs) are formed by the merger of WSNs with the IP network. These networks are distinctive and challenging at the same time, in the sense that these networks are very diverse as compared to other wireless technologies (see Figure 16.1).

	IP-based	Proprietary
Technology	IP-USN	MANET, WSN
Connectivity scope	Access, edge, core, internet	Access

Figure 16.1 IP-USNs encompass all the scopes of connectivity.

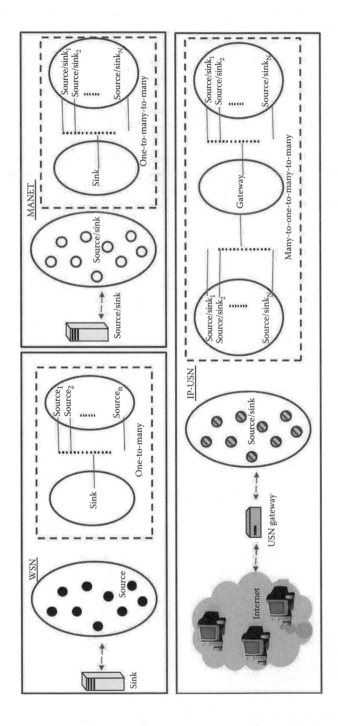

Figure 16.2   Architectural presentations of WSNs, MANETs, and IP-USNs (inset: Communication paradigm).

The management considerations for these networks greatly depend upon their unique characteristics that arise from their unique design. A management architecture for a USN must consider the differences that USNs have from WSNs and mobile ad hoc networks (MANETs). Figure 16.2 illustrates the architectural, operational, and functional differences between WSNs, MANETs, and USNs. Architecturally, a USN varies from these networks mainly in the network entities and the communication model. The operational differences lie within the device roles of USNs, especially due to the user and application heterogeneity. The functional differences arise from their application-specific form factors and code footprints. These unique characteristics of IP-USNs pose specific requirements for management systems, as shown in Table 16.1.

These requirements directly impact the design of a management system for IP-USNs, as shown in Figure 16.3. The first requirement implies the support of query generation and processing for different types and formats of queries at different network entities, such as devices in the Internet domain, the gateway, and devices within the WSN domain. The second requirement points to the fact that the network entities should be intelligent enough to handle the end-to-end query and response transmission under the restriction that channel behavior may vary under wireless-cum-wired and multi-hop paths because the intermediate devices may fail to deliver such transmission. The third requirement states the importance of distributing the network intelligence by the optimal placement of management roles, namely manager and managed agents, across the devices in the network. The differing form factors of devices in the Internet, the gateway, and the sensor nodes bring up the fourth requirement, i.e., the inclusion of proxies and translation gateways (wherever required) to map complex syntax and bulky queries into

Table 16.1 Considerations and Requirements for IP-USN Management System

Considerations for IP-USNs	Requirements for IP-USN Management System (IP-USN NMS)
C.1 IP-USNs exhibit user heterogeneity	R.1 Queries should be supported across user domains
C.2 Communication is across networks	R.2 Network-management framework must cognize and act to network and channel behaviors
C.3 Network elements are many and heterogeneous	R.3 Elements of IP-USN NMS must be distributed across networks, optimally
C.4 Syntax and semantics vary across networks	R.4 Translators and proxies should be embedded in IP-USN NMS, wherever necessary
C.5 Querying types and scopes vary across networks	R.5 Consistent query types and specific Management Information Base (MIB) must be defined

NMS design elements; requirements and purview

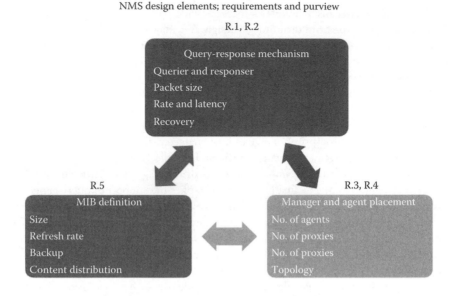

Figure 16.3   Relationship between the IP-USN management requirements (overlay) and design elements (inlay).

lighter and manageable queries, while keeping the same semantics. The last requirement identifies the need to define the MIBs according to the characteristics of the underlying network, to enable the information retrieval of the required management information.

The management purview leads to the goals for the network management station (NMS), including its protocol design. These goals that form a basis of the USN-NMS design are presented in the next section.

16.6.1   Goals for IP-USN NMS

The goals of the management system for USNs are being outlined here with an avid emphasis on the orthogonality of WSNs and IP networks.

G.1 The management system for USNs must provide backward compatibility with legacy management protocol in IP, such as SNMP and its variants.

G.1.1 Network elements that can implement SNMP readily must be identified.

G.1.2 Network elements that cannot implement SNMP must connect to the management system through an SNMP parser and a proxy.

G.2 The management system for USNs must pose least on the communication of the USN.

G.2.1 A resource-discovery mechanism must exist to circumvent futile communication with the unavailable nodes.

G.2.2 A mechanism must exist that classifies network elements as available, alive, sleeping, or expired.

G.2.3 A fragmentation mechanism must exist to allow a large-sized management packet to be split into the least number of fragments.

G.3 The management system must place managers and agents onto the network elements optimally.

G.3.1 A mechanism must exist that identifies the need of the number of manager and managed agents.

G.3.2 A mechanism must exist that assigns a manager and managed roles to appropriate network elements.

G.4 The management system must distribute and utilize MIB to ensure information availability.

G.4.1 A mechanism must exist that determines the constituents of MIB for all the network elements.

G.4.2 A mechanism must exist that ensures correctness and availability of MIB.

G.4.3 A mechanism must exist that provides resilience to network-element failures,

G.4.4 A mechanism must exist that distributes new MIB definitions or incremental information with relevant network elements.

16.6.2   LNMP as an Example Architecture

The LNMP [7] was proposed to provide a robust and adaptable management architecture through the identification and consideration of the special nature of 6LoWPANs (IPv6 over low-power personal area networks). 6LoWPANs are considered a realization of the IP-based sensor networks, which would facilitate ubiquity of sensors by enabling the IP on them. The LNMP encompasses an operational and informational architecture for these networks. The management operations are carried out in two steps. First, network discovery is performed to get a snapshot of the available network elements through hierarchical device state monitoring. After the device discovery, the second step is the actual management of the available devices. In the network discovery phase, at the initialization phase all the end devices report their status to their parent neighboring coordinator (also known as 6LoWPAN forwarder), and the coordinator eventually reports the states to the gateway through its ancestor coordinators. After network initialization, only the changes to the subordinate states are reported to the gateway. The architecture also supports device monitoring through enabling the use of the SNMP on the devices. As the bandwidth available for 6LoWPAN networks is severely limited, and only 33 bytes of data can be sent over to the User Datagram Protocol (UDP) in the worst case, a proxy is used on the 6LoWPAN gateway to reduce the amount of control overhead of SNMP packets. The incoming SNMP packets are parsed and converted into lightweight local management messages, which are sent to the destined devices, and the replies are translated back to the SNMP at the gateway and sent to the managers. The informational architecture consists of MIBs, which are designed to consider the special characteristics of these networks. The informational architecture focuses on reusing the existing standard MIBs, and defines an information base for the special nature of 6LoWPAN networks.

16.7 Network Management as FCAPS Model: A Fresh Perspective

The "ubiquity" in USNs is perhaps the most significant phenomenon that gives a new meaning to WSNs. It implies that sensor networks may be created and disintegrated at the user's prerogative spontaneously, and may interact at large with other sensor networks either directly in the proximity, or through distant sensor networks via the intermediaries. The classical definition of sensor networks is therefore modified from application centricity to user centricity, from static to mobile, and from source–sink to source/sink–sink/source paragon. With such a role of USNs, each sensor network purports associativity with more than one application through a temporally varying users' group management system. The application of the widely modeled FCAPS management aspects to WSNs has been investigated in numerous studies in varying details. However, the granularities of FCAPS implementation take an equally extended perspective when USNs are considered as the natural transformation of WSNs. In this section, we encompass such "value-added" FCAPS management perspectives, with only newer aspects explained (Figure 16.4).

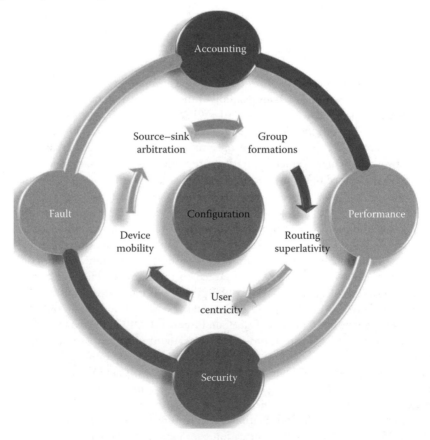

Figure 16.4 FCAPS management perspectives.

16.7.1 User Centricity

The provisioning of multiple sensors on a single hardware platform makes the USNs more usable than ever before. The connectivity of USNs to heterogeneous networks such as the IP networks, the telecom networks, and other proprietary networks brings about a large user base to USNs. Each user may initiate a differing application onto the same sensor network. Therefore, the application specificity of sensor networks morphs into user specificity. It is the user-centric nature of USNs that triggers the phenomena of group formation and user mobility. With regards to user centricity, the FCAPS model may handle the following management issues (Table 16.2).

16.7.2 Group Formations

It is turning out to be a viable proposition to execute multiple applications concomitantly over a singular sensor network through the abstraction of each application in the form of a logical group, where each group may have a one-to-one correspondence to a sensor type. Although each group may be independent of each other metaphorically, in all practical purposes, these groups may depend upon a shared set of sensor nodes. Therefore, preemptive binding may be incorporated for resource reservation, or equivalently, conflict resolution may be utilized for late binding (Table 16.3).

Table 16.2   User Centricity for Different Management Areas

Management Aspect	User Centricity
Fault	User failure; device failure; connectivity failure
Configuration	User preferences; RAM and device occupancy
Accounting	Access control; user billing
Performance	SLA agreement and QoS violations
Security	User authentication; encryption

Table 16.3   Group Formation for Different Management Areas

Management Aspect	Group Formation
Fault	Group failure; violation; resource starvation
Configuration	Group definition; node bid; selection; acknowledgement
Accounting	User-to-group log; nodes per group statistics
Performance	Group formation success rate
Security	Group policy breaches; group AAA management

16.7.3   Source–Sink Arbitration

The classical operation of a sensor network could be described simply as a multiple source-single-sink paradigm. With the notion of multiple groups operating at free will in USNs, both the nodes originating the data and the nodes to which such data terminates at, may lie within the same logical group let alone the same sensor network. A sensor node may turn out to be a source for a logical group, while receiving data from another logical group. Such arbitration of sources and sinks gives a new dimension to the complexity of routing processes, and consequently, to the FCAPS management (Table 16.4).

16.7.4   Routing Superlativity

With the eccentricity of group formation and the resulting arbitration of sources and sinks in USNs, routing seems to emerge as the dominant activity. Unless effective means are formulated to collate optimal routing paths for nondisjoint routes, USN energy resources may wither out very quickly. A superlative routing architecture must therefore be brought about to handle redundant data transmissions by identifying spatial and temporal overlap between routable data of arbitrary sources and destinations. The relationship between such a routing mechanism and the FCAPS is summarized in the table (Table 16.5).

Table 16.4   Source–Sink Arbitration for Different Management Areas

Management Aspect	Source–Sink Arbitration
Fault	Source–sink conflict; source overload; sink overload
Configuration	Send and receive buffer size; scheduling
Accounting	No. of active sources and sinks; network longevity
Performance	Goodput per sink; throughput per source
Security	Data aggregation and privacy management

Table 16.5   A Superlative Routing Mechanism for Different Management Areas

Management Aspect	Routing Superlativity Mechanism
Fault	Route request broadcast failure; route error
Configuration	Routing kernel settings; routing metric
Accounting	No. of broadcasts; no of active, stale, and invalidated routes
Performance	Goodput per route; throughput per source
Security	Routing kernels broadcast legitimacy; black-hole attempts

Table 16.6 Device Mobility for Different Management Areas

Management Aspect	Device Mobility
Fault	No. of sessions lost; no. of disappearing devices
Configuration	HA and FA assignment; session-migration statistics
Accounting	SLA for mobility-aware sessions; pay per packet
Performance	Migration success rate; migration time
Security	Session-privacy management; authorization control at FA

16.7.5 Device Mobility

In USNs, sensor nodes may be classifiable as stationary and mobile. The agility of the FCAPS management system to changing network dynamics would greatly depend upon its diagnosis and response to the mobility behavior of the sensor nodes. Particularly of interest is the ability of a management system to preempt the void that is created amidst device mobility—how to migrate the active sessions on a mobile sensor node, and how would the mobile node adjust to the alien environment. The table summarizes the management issues that spring up from mobility (Table 16.6).

16.8 Conclusion

In this chapter, network management for WSNs is presented in breadth. Network management in WSNs is identified and appreciated to be unique, and specific goals are defined as design considerations. Then management purview, including its dimensions, functionalities, and levels, is presented. Contemporary design approaches are presented followed by various practical realizations both in the form of research initiatives and industrial implementations. IP-USN, which is the extended version of WSNs, is presented as the integration technology to identify a thorough set of challenges, requirements, and goals for USNs. LNMP is then presented as an example of IP-USNs. The newer and evolving aspects of FCAPS are presented to identify emerging phenomena that will determine how future management architectures shall be formed.

References

1. L.B. Ruiz, *MANNA: A Management Architecture for Wireless Sensor Networks*, PhD. dissertation, Federal Univ. of Minas Gerais, Belo Horizonte, MG, Brazil, Dec. 2003.
2. L.B. Ruiz, I.G. Siqueira, L.B. e Oliveira, H.C. Wong, J.M.S. Nogueira, and A.A.F. Loureiro, Fault management in event-driven wireless sensor networks, in *Proc. ACM MSWiM Conf.*, Atlanta, GA, Oct. 2004.

3. L.B. Ruiz, F.B. Silva, T.R.M. Braga, J.M.S. Nogueira, and A.A.F. Loureiro, On impact of management in wireless sensor networks, in *Proc. IEEE/IFIP NOMS*, Seoul, Korea, Apr. 2004.

4. L.B. Ruiz, T.R.M. Braga, F.A. Silva, H.P. Assuncao, J.M.S. Nogueira, and A.A.F. Loureiro, On the design of a self-managed wireless sensor network, *IEEE Communications Magazine*, 43(8), 95–102, 2005.

5. W. Chen, N. Jain, and S. Singh, ANMP: Ad hoc network management protocol, *IEEE Journal on Selected Areas in Communications*, 17(8), 1506–1531, Aug. 1999.

6. C.-C. Shen, C. Srisathapornphat, and C. Jaikaeo, An adaptive management architecture for ad hoc networks, *IEEE Communications Magazine*, 41(2), 108–115, Feb. 2003.

7. H. Mukhtar, S.A. Chaudhry, K. Kang-Myo, A.H. Akbar, K. Ki-Hyung, and S.W. Yoo, LNMP–Management architecture for IPv6 based low-power wireless personal area networks (6LoWPAN), in *Proc. IEEE/IFIP NOMS*, Salvador, Brazil, pp. 417–424, 2008.

8. I.F. Akyildiz, W. Su, Y. Sankarasubramaniam, and E. Cayirci, Wireless sensor networks: A survey, *Computer Networks*, 38(4), 393–422, 2002.

9. N. Kushalnagar, G. Montenegro, G, J. Hui, J, and D. Culler, 6LoWPAN: Transmission of IPv6 packets over IEEE 802.15.4 Networks, RFC 4944, Sep. 2007.

10. M. Welsh and G. Mainland, Programming sensor networks using abstract regions, in *Proc. USENIX NSDI Conf.*, San Francisco, Mar. 2004.

11. H. Song, D. Kim, K. Lee, and J. Sung, Upnp-based sensor network management architecture, in *Proc. ICMU Conf.*, Osaka, Japan, Apr. 2005.

12. S. Meguerdichian, S. Slijepcevic. V. Karaya, and M. Potkonjak, Localized algorithms in wireless ad-hoc networks: Location discovery and sensor exposure. in *MobiHOC—Symposium on Mobile Ah Hoc Networking and Computing*, Long Beach, CA, pp. 106–116, Oct. 2001.

13. Y. Yu, B. Krishnamachari, and V.K. Prasanna, Issues in designing middleware for wireless sensor networks, *IEEE Network Magazine Special Issue*, 18(1), 15–21, Jan. 2004.

14. Z. Li, X. Zhou, S. Li, G. Liu, and K. Du, *Issues of Wireless Sensor Network Management*, ICESS, LNCS, Springer, Berlin/Heidelberg, 2005.

15. C. Fok, G. Roman, and C. Lu, Mobile agent middleware for sensor networks: An application case study, in *Proc. IEEE ICDCS Conf.*, Columbus, Ohio, June 2005.

16. Z. Ying and X. Debao, Mobile agent-based policy management for wireless sensor networks, in *Proc. IEEE WCNM Conf.*, New Orleans, LA, Sep. 2005.

17. G. Tolle and D. Culler, Design of an application-cooperative management system for wireless sensor networks, in *Proc. EWSN*, Istanbul, Turkey, Feb. 2005.

18. W. Louis Lee, A. Datta, and R. Cardell-Oliver, WinMS: Wireless sensor network-management system, An adaptive policy-based management for wireless sensor networks, *Tech. Rep. UWA-CSSE-06-001*, The University of Western Australia, Perth, Australia, June 2006.

Chapter 17

Deployment in Wireless Sensor Networks

Nadjib Aitsaadi, Nadjib Achir, Khaled Boussetta, and Guy Pujolle

CONTENTS

In this chapter, we provide a deep overview of the different wireless sensor networks' (WSNs') uniform and nonuniform deployment strategies proposed in the literature. Generally, the observed area is characterized by the geographical irregularity of the sensed events. We expose the sensor deterministic and probabilistic event detection models, and the different metrics for WSN evaluation. Thereafter, we formalize the generalized WSN deployment problem, which is a multi-objectives problem and known to be NP-hard. We describe all the exact and heuristic algorithms found in the literature. We implement all the WSN deployment strategies and we compare them in terms of (1) number of sensors deployed, (2) event detection confidence, (3) network connectivity, and (4) computation complexity.

17.1   Introduction

Wireless sensor networks (WSNs) are collections of small resource-constrained devices, which consist of computer processors, memory devices, and wireless channels for transmitting and receiving information. As a result, WSNs are capable of observing and monitoring their environment. When interesting events are detected, information is routed from one node to another and eventually gathered in gateway nodes or base stations. Recently, sensor networks have generated considerable community enthusiasm to apply themselves to a wide range of applications such as environmental monitoring, military-target tracking, weather forecast, home automation, intrusion detection, etc.

Although the challenges in sensor networks are diverse, researches have mainly focused on fundamental networking challenges, which include routing protocols, energy minimization, sensor localization, and data gathering [1,2]. Unfortunately, few studies have been achieved regarding the sensors deployment process.

In this chapter, we provide a deep overview of the different strategies proposed in the literature to provide sensor deployment over uniform and nonuniform events distributed within the monitored area. The rest of this chapter is structured as follows: the next section provides an overview on sensor detection models including deterministic and probabilistic detections models. Thereafter, Section 17.3 introduces the different metrics for WSN evaluation. Section 17.4 is devoted to the description of the different sensor network deployment strategies proposed in the literature, including the generalized problem formalization, uniform deployment strategies, and nonuniform deployment strategies. Finally, Section 17.5 summarizes the chapter and we present open issues in WSN deployment.

17.2   Event Detection Models

An important factor in WSN deployment issue is the sensor's detection capabilities. This detection capability highly depends on physical and environmental characteristics, such as obstacles, environment noise, event speed, hardware sensor reliability, distance between event and sensor, etc. Hence, numerous theoretical detection models with different complexity levels can be proposed. They are based on applications needs, devices features, and environment characteristics.

Two main mechanisms can be implemented in a sensor to decide if an event appeared in the deployment field. In the first mechanism, the event detection decision is made in each sensor that detects the event independently of remainder sensors, which means that neither communication nor sensors' collaboration are necessary to make a decision if the event appears or not. In the first mechanism, we have a local decision. On the opposite with the second mechanism, the decision is collaborative between sensors that detect the event. We have a distributed and collaborative decision. Considering collaborative-sensing mechanism means that sensors collaborate and converse between each other to make a decision of event detection. The collaboration needs a specific data-link protocol to materialize a distributed event detection decision. This collaboration between sensors has many advantages such as increasing the detection probability, reducing energy consumption [3,4], less uncertainty, and false alarms in event detection.

Generally, in the literature we can distinguish among three sensor detection models families. The first family, which is the more simplistic one, is called the binary detection model. The second one, which is more realistic, is called the distance-related probabilistic sensor detection model. The last family, called the tracking detection model, is an extension of the second model, which includes the event duration at each point.

17.2.1   Binary Model

The binary detection model is considered the most simplistic event detection model. This model takes into account only the distance separating the event from a sensor and makes an abstraction of other physical environment parameters. The binary model supposes that an event is surely detected if and only if it appears within a sensor's sensing range. In other words, the event detection probability is equal to 1 if the distance between the event and a sensor is less than a sensing range. Otherwise, if an event appears outside the sensing range, the event is not detected. Generally, a sensing-range radius, R_{max}, depends on the sensor hardware characteristics.

If we assume that a sensor, s, was positioned at the coordinates (x, y) and an event, e, occurs at the coordinates (x', y'), then the event detection probability, $P(s, e)$, is defined as

$$P(s, e) = \begin{cases} 1 & \text{if } \|se\| \leq R_{max} \\ 0 & \text{else} \end{cases} \tag{17.1}$$

where R_{max} is the sensing range, and $\|se\|$ is the Euclidean distance between the sensor s and the event e ($\|se\| = \sqrt{(x - x')^2 + (y - y')^2}$).

This model was mainly considered in works addressing area-coverage problems, such as target detection or the k-coverage problem [5,6]. The model aims to simplify the analysis, but unfortunately, it does not realistically reflect sensing capabilities of sensors.

17.2.2 Probabilistic Detection Model

A new sensor detection model was proposed seeking for more realism, namely, distance-related probabilistic sensor detection model, rather than a binary event detection model. With this model, the event detection probability is inversely proportional to the distance separating the sensor and the event. More precisely, confidant and maximum sensor monitoring circles are defined. If any event (event (1) in Figure 17.1) occurs within the confident circle, then event detection probability is considered as equal to 1. If an event (event (2) in Figure 17.1) occurs outside the confidant circle but within the maximum circle, then the detection probability decreases with the distance. Finally, when the distance is larger than the radius of the maximum circle, then the event (event (3) in Figure 17.1) is no longer detected.

We find, in the literature, two expressions of the general detection probability according to the distance between the sensor and the event. These models were mainly inspired by the radio signal propagation theory [7]. If we assume that a sensor s was positioned at the coordinates (x, y) and an event e occurs at the coordinates (x', y'), then the event detection probability $P(s, e)$ for the two models is defined as the following [8–10]:

$$P(s, e) = \begin{cases} 1 & \text{if } \|se\| \leq 1 \\ \dfrac{\alpha}{\|se\|^{\beta}} & \text{if } 1 \leq \|se\| \leq R_{\max} \\ 0 & \text{if } R_{\max} < \|se\| \end{cases} \qquad (17.2)$$

$$P(s, e) = \begin{cases} 1 & \text{if } \|se\| \leq 1 \\ e^{-\alpha \|se\|} & \text{if } 1 \leq \|se\| \leq R_{\max} \\ 0 & \text{if } R_{\max} < \|se\| \end{cases} \qquad (17.3)$$

where
$R_{\max}$ is the radius of the maximum circle
$\|se\|$ is the Euclidean distance between the sensor s and the event e

Finally, α, β are sensor-technological and event-characteristic parameters. More precisely, α represents the energy distortion factor due to obstacles. Depending on the environment, β is typically less than 5 [11].

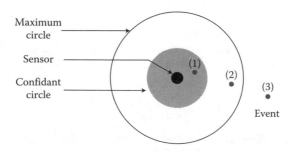

Figure 17.1 Sensor detection model.

17.2.3 Tracking Detection Model

In Ref. [12], the authors extend the event detection probability models described above. This new model takes into account the event duration. The main idea is to increase the event detection probability when the event stays for a long time at the same point. So the event detection probability depends on variable t that quantifies the time duration of an event at a certain point. In this case, the event detection probability is

$$P(s, e, t) = \begin{cases} P(s, e) & \text{if } 0 < t \leq T \\ 1 - [1 - p(s, e)]^{\lfloor t/T \rfloor} & \text{if } t > T \end{cases} \tag{17.4}$$

T is the time period during which the sensing algorithm is executed and a decision is made if an event appeared. This model can be used in the tracking application.

17.3 Deployment Criteria

Sensor deployment must be done keeping in mind the optimization of one or more metrics related to the application needs. Hereafter, we introduce the most important metrics defined in the literature.

17.3.1 Number of Deployed Sensors

Until now, the majority of the works consider low-cost sensors. This assumption deals with the fact that the sensing devices, embedded within sensors, assumed to monitor very simplistic phenomena such as temperature, pressure, moisture, light, sound, and magnetism. However, if we consider a deployment of hundreds or thousands of sensors we need to take into consideration the global cost of the network. Thus, the number of deployed sensors is one of the important metrics that needs to be dealt within WSN deployment process.

On the other hand, in some specific applications, considering low-cost sensors is not realistic. Indeed, sensor cost is highly depending on the application target and the monitoring environment. For example, in the case of oceanographic applications such as pollution monitoring, offshore exploration, or disaster prevention, the sensor cost is very high. In this environment, the sensing part can be much more sophisticated to detect some specific phenomena and to resist to the environment influence.

17.3.2 Coverage and K-Coverage

One of the fundamental issues in a WSN is the coverage problem. Coverage can be defined as the percentage of the deployment area where the event is monitored by at least one sensor. This is called 1-coverage. Obviously, in an ideal deployment, the coverage should be equal to 100 percent. Unfortunately, sensors are subject to failure, measurement errors, damages, or energy exhaustion. In this case, the more general k-coverage ($k \geq 1$) was defined, where each point of the monitored area must be monitored by at least k sensors.

17.3.3 Connectivity

The main task of a sensor is to detect an event within its coverage field. However, this event must be transmitted by the sensor to a specific node, called sink, having in charge the notification of the event occurrence to the applications users. If the sink is within the communication range of a sensor, the event can be easily transmitted to the sink. However, if the sink is not within the communication range of the node detecting the event, the event notification must be transmitted from one sensor to another in a multi-hop manner until reaching the sink node. In this case, the sensor network topology must be a connected graph to permit the transmission of an event notification from any sensor to the sink.

Connectivity and coverage are related, because they are affected by the sensors position. In [6,13], the authors formulate sufficient conditions to guarantee the coverage and connectivity in WSNs. The sufficient conditions are impacted by the sensor positions, communication range (R_c), and sensing range (R_s). For example, in Ref. [6], the authors prove that if $R_c \geq \sqrt{3}R_s$ and the area is fully covered, then the communication graph is connected.

Similar to coverage criteria, some studies focused on the k-connectivity paradigm, and proposed algorithms to build K different paths between each sensor and sink. In this case, we have more reliability on data transmission. If one path falls, we have $K - 1$ paths maintaining the connectivity. Moreover, building K different paths makes easy to conceive and realize distributed mechanisms as load-balancing of traffic. Consequently, we can reduce the energy consumption and extend the WSN lifetime.

17.3.4 Detection Probability

As shown in Section 17.2.2, the detection probability is not necessarily equal to 1 when an event appears inside the coverage circle. To measure the event detection probability, we generally use models described in Equations 17.2 and 17.3. In real deployment, the event detection probability required in each point can be less than 1. It depends on many parameters: region importance, event appearance frequency, and event propagation.

Moreover, in many sensor network applications (such as fire detection alarms, water quality monitoring, etc.), the supervised area can request different detection levels, depending on the event's location. For exemple, in the case of a fire detection system, high detection probabilities (close to 1) can be required for risky areas (e.g., those close to chemical deposits or to habitats). However, for low fire-risky places, relative low detection probabilities are sufficient.

Using the detection probability as a metric we can compute the deployment satisfaction rate. This metric represents the percentage of the deployment area where the generated detection probabilities are greater than the required detection probabilities. In other terms, the satisfaction rate represents the similarity between the required and the generated detection probability distributions. Ideally, the two distributions should be identical. In real deployment, it is very difficult to obtain identical distributions.

17.3.5   Network Lifetime

A sensor is equipped with a battery storing power to supply to different sensor hardware devices. This battery has small capacity, so the WSN lifetime strongly depends on energy consumption. To ensure a certain WSN working duration, we have to deploy the right number of sensors at the right positions. Sensors redundancy is necessary to reach a requested lifetime by a customer.

Deployment has a direct impact on the WSN lifetime. When we deploy few sensors we obtain a weak lifetime. But when we deploy many sensors, the lifetime does not increase automatically. In WSNs, when the number of sensors increases then the traffic size rises. Consequently, the energy consumption increases and lifetime decreases. Thus, deploying more sensors does not solve the problem. To increase the lifetime, we have to put more sensors at the right place by taking into consideration, for example, the sinks positions, event frequency, and eventually any specific mechanism implemented in sensors to build the paths to the sinks (routing protocol).

17.4   Sensor Network Deployment Strategies

In this section, we formalize the deployment problem. Thereafter, we will present the different deployment strategies found in the literature.

17.4.1   Problem Statement

We consider a sensor field area, denoted $\mathcal{A}$. To reduce the computational complexity of the problem, the area is discretized. We suppose that $\mathcal{A}$ is a square, with a side equal to n units. A unit is defined as a normalized physical distance (e.g., 1 or 10 m). To simplify, we will refer, in the rest of this chapter, to each square unit of $\mathcal{A}$ by its baricenter point. In other words, when we say that a sensor is located in the point $p_{(i,j)} \in \mathcal{A}$ then this means that the sensor is placed in the baricenter of the corresponding square unit. Similarly, the event detection probability of a unit square is computed considering the detection probability of its baricenter. Finally, we consider that any event occurring inside a unit square is detected with probability 1 by a sensor that would be placed in its baricenter point.

In an efficient monitoring, the event detection is formalized as a probabilistic detection model. Each point, $p_{(i,j)}$, in $\mathcal{A}$ is associated with a required minimum probability detection threshold, denoted $r_{(i,j)}$.

Ideally, a good WSN deployment algorithm should lead to obtain that $\forall p_{(i,j)} \in \mathcal{A}$, the measured detection probability of that point is greater than $r_{(i,j)}$, and the sensor graph connectivity G is connected. The detection probability in a point $p_{(i,j)}$ is estimated by all the sensors in its vicinity, but the event detection model is not collaborative. The detection probability of the point $p_{(i,j)}$, denoted $\mathcal{P}_{(i,j)}$, is estimated by all the sensors available in the monitored area as

$$\mathcal{P}_{(i,j)} = 1 - \prod_{(x,y)\, \in\, \text{Grid}} [1 - p((i,j),(x,y))]^{D(x,y)} \qquad (17.5)$$

where $D(x, y)$ denotes the deployment bivalent variable. If $D(x, y)$ equals to 0, it means that no sensor is deployed at grid point $p(i, j)$. If $D(x, y)$ equals to 1, it means that a sensor is deployed at grid point $p(x, y)$.

Obviously, if a sufficiently large number of sensors are deployed, it is possible to satisfy the objective: $\mathcal{P}_{(i,j)} \geq r_{(i,j)}$, $\forall p_{(i,j)} \in \mathcal{A}$. Nevertheless, taking into account cost considerations, the number of sensors is also a critical metric. In addition to the satisfaction of the requirement on the minimum detection probability thresholds, a second objective in a deployment problem is to minimize the sensors number. Formally, the aim is to find a WSN deployment solution to the following multi-objectives optimization problem:

1. Minimize the number of deployed sensors needed to satisfy the following two constraints (2 and 3).
2. For each point $p_{(i,j)} \in \mathcal{A}$, minimize the difference between the required detection probability threshold, $r_{(i,j)}$, and the after-deployment resulting detection probability, $\mathcal{P}_{(i,j)}$.
3. Ensure the network connectivity.

The multi-objectives optimization problem described above is an NP-hard problem. The size of the solution space is finite but very large (equal to 2^{n^2}). To resolve this optimization problem, we can choose an exact method as Branch & Bound. Unfortunately, we cannot apply it to large areas because of its exponential complexity. The second alternative is to resolve the problem by using heuristics. Inopportunely, theses methods cannot guarantee to obtain the optimal solution. However, the main advantage is the polynomial complexity and the running time, which can be reasonable.

In what follows, we develop the exact and heuristic methods found in the literature. We classify them into two families. The first family regroups the methods where the required detection probabilities are homogeneous on all the deployment area. The second family regroups all methods where the required detection probabilities are not identical in all the deployment area.

17.4.2 Uniform Deployment Strategies

We talk about uniform deployment strategies when the monitored event has the same importance in all the deployment area. That means, all required detection probabilities, $r_{(i,j)}$, are identical at each point, $p_{(i,j)}$. We can find in the literature two uniform deployment processes, namely Random and Regular.

17.4.2.1 Uniform Random Deployment

Random deployment consists of droping sensors randomly on the deployment area. The sensor positions should follow uniform distribution. The choice of uniform distribution is motivated by the homogeneous nature of the monitored event.

The main advantage of random deployment is its feasibility. When the deployment area is not reachable, we cannot place the sensors by hand on the area. Consequently, only the random deployment, using for example air plan, can solve the problem. Typical applications are volcano supervision, battlefield application, etc.

Random deployment has many drawbacks. To ensure a full coverage and network connectivity, we have to deploy many sensors, so the network is oversized and the deployment cost can be expensive. Also, the lifetime of the network is a critical challenge if the number of sensors is large, which can increase the data traffic and energy consumption.

17.4.2.2 Regular Deployment

In a regular deployment, sensor positions follow a regular structure as triangle, square, polygon, etc. Because the monitored event is homogeneous, a regular deployment ensures the same density of sensors in all the deployment area. Regular deployment guarantees a full coverage and a network connectivity. But it imposes that the monitored area is accessible to place sensors.

Many papers study the regular deployment and comparison with a uniform random deployment is made. In Ref. [14], the authors proved that the quality of monitoring (QoM) in regular deterministic deployment is better than the uniform random deployment case. A comparison is made with different regular topologies as triangle, square, and hexagon. In Ref. [15], an incremental deployment process is proposed. The main idea is to deploy a new sensor at point $p_{(x,y)}$ if no sensors are placed inside the sensor coverage circle that would be deployed. In Ref. [?], the authors propose to deploy the sensors according to the grid topology. They take also into account obstacles in the deployment area. In Ref. [16], the deployment process builds a hexagon topology. They show that hexagon topology gives better performance, in terms of required minimal number of sensors, than triangular and square topologies.

17.4.3 Nonuniform Deployment Strategies

Nonuniform deployment strategies consist of computing the number and positions of sensors to deploy, and takes into account event characteristics. As shown in Section 1.4.4, the event detection is not homogeneous in the deployment area. Consequently, the area is characterized by the geographical irregularity of sensed events. Hereafter, we explain and develop the different deployment strategies found in the literature.

17.4.3.1 Optimal Solution

A nonuniform deployment problem can be formulated as follows:

$$\begin{cases} \min\left[\sum_{i=0}^{n-1}\sum_{j=0}^{n-1}D(i,j)\right] \\ \forall i,j \in \{0, 1, ..., (n-1)\} : \mathcal{P}_{(i,j)} \geq r_{(i,j)} \end{cases} \tag{17.6}$$

where $D(i,j)$, $\mathcal{P}_{(i,j)}$, and $r_{(i,j)}$ represent deployment bivalent variable, generated detection probability, and requested detection probability, respectively, of point (i,j).

The problem formulated in Equation 17.6 is a typical Binary Integer Programming, which has been proved to be NP-hard [16]. To compute the optimal solution, we apply the Branch & Bound algorithm [17]. It is a general algorithm for finding optimal solutions of various optimization problems, especially in discrete and combinatorial optimization. Branch & Bound is an intelligent process to visit a set of admissible solutions. Large subsets of candidates are discarded when the cost is surely (mathematically proven) greater than a current solution.

The main problem of Branch & Bound is the running time and the complexity, which is exponential in our case. To give an idea on the Branch & Bound running time, we launch three simulation scenarios with different area sizes: 5×5, 6×6, and 7×7. The required detection probabilities are shown in Figures 17.2 through 17.5. For the area 5×5, the optimal solution is found after 12.02 seconds. For the area 6×6, the optimal solution is computed after $1,771.93$ seconds. Finally, for the last scenario 7×7, Branch & Bound converges after $332,49$ hours. We notice that the space solutions size increases very quickly. For the first scenario the size space is equal to 2^{25}, for the second scenario the size space is equal to 2^{36}, and finally, for the last scenario the size space is equal to 2^{49}.

17.4.3.2 Distribution-Based Random Deployment

The distribution-based random deployment strategy is similar to random deployment in uniform events. The difference consists on the choice of the distribution. In the case of uniform event distribution, the sensor positions follow uniform distribution. However, when an event is not uniformly distributed on the deployment area, we likely need

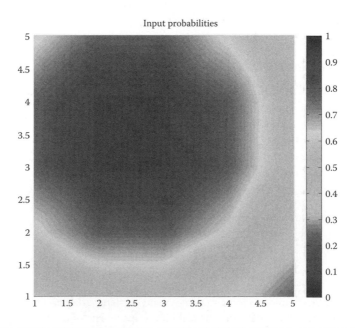

Figure 17.2   Required detection probabilities, area 5*5.

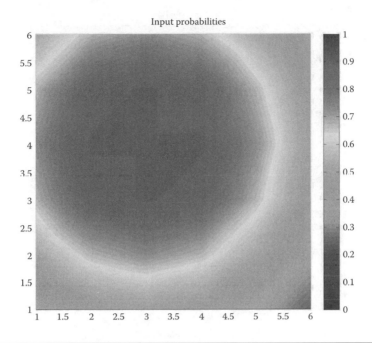

Figure 17.3   Required detection probabilities, area 6*6.

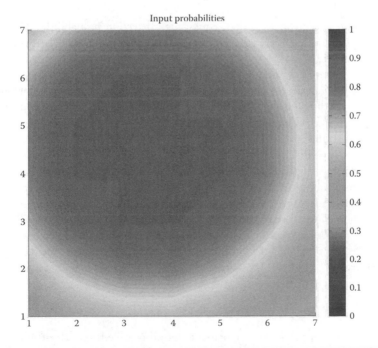

Figure 17.4   Required detection probabilities, area 7*7.

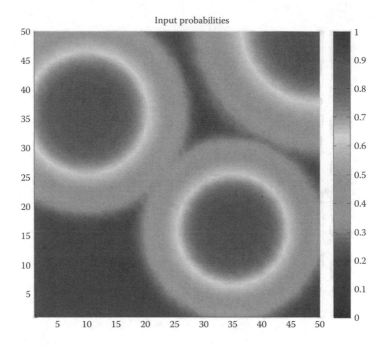

Figure 17.5   Area 50*50 with its desired detection probability values.

to deploy more sensors in areas that request high detection probabilities. Under these constraints, we have to choose a sensor distribution that maximizes the satisfaction rate after deployment.

The main drawbacks and advantages in uniform random deployment persist in nonuniform random deployment. However, in this case, the additional problem is the choice of sensors distribution. The difficulty is to determine the right sensor distribution that should be in accordance with the distribution of requested probabilities.

In Ref. [18], the authors study the random deployment process by using the probabilistic detection model. They propose a new deployment algorithm named probabilistic coverage algorithm (PCA). In Ref. [18], simulation results show that the event detection probabilities obtained by PCA are better than a binary model. In Ref. [19], a comparison is made between three deployment strategies: (1) Poisson process, (2) uniform random distribution, and (3) regular grids. The authors prove that a grid deployment renders asymptotically lower node density than random deployment. Consequently, regular topologies deploy fewer sensors than random deployment for the same performance.

17.4.3.3 Max-Avg-Coverage

In [20], the authors propose a deployment algorithm called *Max-Avg-Cov*. The main idea of this deployment strategy is to maximize the average coverage of the deployment

area. The authors define three mains variables *MAT-P*, *MAT-M*, and *VEC-M**.
MAT-P is a sensor detection matrix, $MAT\text{-}P = [P((i,j),(x,y))]$. We remark that for
an $n \times n$ deployment area, the total grid points are equal to n^2. Consequently, *MAT-P*
is a square matrix, contains n^2 rows and n^2 columns. Each matrix element, *MAT-
P(u,v)*, represents the detection probability generated at point v by a sensor deployed
at point u. We transform coordinates from two dimensions to one dimensions, so
$u = y \times n + x$ and $v = j \times n + i$. A row, u, in *MAT-P* determines the detection probability generated in all points of the deployment grid when a sensor is placed at point
$(\lfloor u/n \rfloor, u\%n)$. $MAT\text{-}M = [m_{ij}]$ is a miss probability matrix. This matrix is equal to
$\mathbf{1}\text{-}MAT\text{-}P$; $\mathbf{1}$ is the square matrix where all elements are equal to 1. In other terms,
$MAT\text{-}M = [1 - P((i,j),(x,y))]$. *VEC-M** is a vector equal to $[M_1, M_2, \cdots, M_{n^2}]$; it
determines a miss probability in each point after the deployment process is finished. At the
beginning, no sensor is deployed, so *VEC-M** is initialized to the all-1 vector, $VEC\text{-}M^* =
[1, 1, \ldots, 1]$.

We remind that the aims are to minimize the number of sensors to deploy and
maximize the satisfaction rate. *Max-Avg-Cov* is an iterative algorithm; in each step one
sensor is positioned. *Max-Avg-Cov* deploys a sensor in point u where $\sum_v MAT\text{-}M(u,v)$
is minimal, Which means a sensor is deployed at a point that minimizes the global
impact by summing the changes in the miss probabilities for all the points in the
grid. The deployment process is stopped when the generated detection probabilities are
greater than the requested detection probabilities, or the number of sensors deployed
exceed the allocated budget of sensors. *Max-Avg-Cov* pseudo code is illustrated in
Algorithm 17.1.

Algorithm 17.1: *Max-Avg-Cov* pseudo code

Num_Sensor = 1;
repeat
> **for** *i=1 to n^2* **do**
> > $\tau_i = \sum_j^{n^2} MAT\text{-}M(i,j)$;
>
> Select grid point k such that τ_k is minimum;
> Deploy a sensor at point $(\lfloor k/n \rfloor, k\%n)$;
> *%Update miss probabilities due to sensor on grid point k%*;
> **for** *i:=1 to n^2* **do**
> > $VEC\text{-}M^*(i) = VEC\text{-}M^*(i) \times MAT\text{-}M(k,i)$;
>
> Delete kth row and column from the matrix *MAT-M*;
> Num_Sensor++;

until *(Num_Sensor > Sensor_Max)* **or** $\left(\forall i,j : \mathcal{P}_{(i,j)} \geq r_{(i,j)}\right)$;

The computational complexity for *Max-Avg-Cov* algorithm is equal to
$O(Sensor\_Max \times n^2)$. The maximum value of *Sensor_Max* is n^2. Consequently, the
computational complexity is equal to $O(n^4)$. Finally, the process creates a matrix that
contains n^4 elements, so the algorithm memory size to reserve is heavy.

17.4.3.4 Max-Min-Coverage

In Ref. [20], the authors propose another deployment algorithm called *Max-Min-Cov*. The main idea of this process is to deploy a sensor at the grid point with minimum coverage. That means, we deploy a sensor at a point with the highest value of miss detection probability.

Max-Min-Cov uses identical data structure as *Max-Avg-Cov*. Thus, we have *MAT-P*, *MAT-M*, and *VEC-M** variables that contain the same data. *Max-Min-Cov* pseudo code is illustrated in Algorithm 17.2.

Max-Min-Cov complexity is equal to $O(n^4)$ and it is identical to *Max-Avg-Cov* complexity. In addition, *Max-Min-Cov* creates a matrix with n^4 elements, so the memory size to run the process is considerable when the deployment area is large.

Algorithm 17.2: *Max-Min-Cov* pseudo code

Place first sensor randomly;
Num_Sensor = 1;
repeat
  %Update miss probabilities due to sensor on grid point k%;
  for $i:=1$ to n^2 **do**
  | $VEC\text{-}M^*(i) = VEC\text{-}M^*(i) \times MAT\text{-}M(k, i)$;
  end
  Select grid point k such that $VEC\text{-}M^*(k)$ is maximum;
  Deploy a sensor at point $(\lfloor k/n \rfloor, k\%n)$;
  Delete kth row and column from the matrix $MAT\text{-}M$;
  Num_Sensor++;
until $(Num\_Sensor > Sensor\_Max)$ **or** $\left(\forall i, j : \mathcal{P}_{(i,j)} \geq r_{(i,j)}\right)$;

17.4.3.5 Min-Miss

In Ref. [21], the authors propose a new deployment process called *Min-Miss*. This strategy is an iterative algorithm, one sensor is deployed at a step. The authors define for each grid point a new metric called over miss probability, denoted $\tilde{m}(x, y)$. The later quantifies benefit in coverage when a new sensor is added at point (x, y). The main idea of *Min-Miss* is the following. At first, all possible free grid points (no sensors are deployed in these points) are selected to form a set, $\mathcal{AP}$. Thereafter, for each point in $\mathcal{AP}$ an over miss probability is calculated. Finally, a new sensor is deployed at a point that minimizes over miss probability, which means a sensor is deployed in a position that maximizes the detection probabilities in the area.

Min-Miss uses *VEC-M** and *MAT-OMP$_{xy}$* variables. *VEC-M** is an identical variable as defined in *Max-Avg-Cov*. *MAT-OMP$_{xy}$* is a square matrix $(n \times n)$ associated with grid point (x, y). It contains the miss probability introduced in all grid points when a sensor is placed at point (x, y). So *MAT-OMP$_{xy}$*$(i, j) = 1 - P((i, j), (x, y))$. Over miss probability of point (x, y) is equal to

$$\tilde{m}(x, y) = \sum_{i,j} MAT\text{-}OMP_{xy}(i, j) \tag{17.7}$$

Min-Miss deploys a new sensor at point (u, v) that minimizes over miss probability,

$$\tilde{m}(u, v) = \min_{x,y} \sum_{i,j} MAT\text{-}OMP_{xy}(i, j) \tag{17.8}$$

When a sensor is deployed, *Min-Miss* updates the miss probabilities for *VEC-M**. The process should be stopped when all requested detections probabilities are satisfied or all available sensors are deployed. *Min-Miss* pseudo code is illustrated in Algorthim 17.3.

Algorithm 17.3: *Min-Miss* pseudo code

Num_Sensor = 1;
$\mathcal{AP}$ initialized with all grid deployment points, $\mathcal{AP} = \{(x_i, y_i)\}, i, j \in \{1, 2, \dots, n\}$;
repeat
  %simulate sensor deployment in all points of $\mathcal{AP}$%;
  for *i:=1 to $\|\mathcal{AP}\|$* **do**
    Compute $MAT\text{-}OMP_{x_i y_i}$;
    Compute over miss probability $\tilde{m}_{x_i y_i}$;
  Select a point (u, v) to deploy a new sensor;
  $\tilde{m}_{uv} = min_{x_i, y_j \in \mathcal{AP}} (\tilde{m}_{x_i y_i})$;
  Deploy a new sensor at point (u, v);
  %Update miss probabilities due to sensor on grid point (u, v)%;
  for *i:=1 to n^2* **do**
    $VEC\text{-}M^*(i) = VEC\text{-}M^*(i) * MAT\text{-}OMP_{uv}(\lfloor i/n \rfloor, i\%n)$;
  %Update $\mathcal{AP}$%;
  $\mathcal{AP} = \mathcal{AP} \backslash \{(u, v)\}$;
  Num_Sensor++;
until (*Num_Sensor > Sensor_Max*) **or** $(\forall i, j : \mathcal{P}_{(i,j)} \geq r_{(i,j)})$;

We observe that *Min-Miss* makes anticipation in deployment to decide where to place a new sensor. The anticipation is materialized by simulating a sensor deployment in all candidates points (free grid points). Afterward, the best solution is selected. Unfortunately, this anticipation is expensive in running time; the computational complexity is equal to $O(n^6)$. If we choose the solution that maximizes over miss probability, it consists of deploying a sensor in Worst Coverage case. This strategy is named *Max-Miss*, which is not interesting in our case, because its only advantage is to give an idea on lower bound of coverage.

The authors also propose a new detection probability model. It is based on the models described in Section 17.2.2. It includes the uncertainty in sensors placement to the predetermined locations computed by any deployment algorithms. The error in sensor locations is modeled by a Gaussian probability distribution. Computed coordinates (x, y) serve as the mean values with standard deviation σ_x and σ_y in the x and y directions, respectively. The error probability that a sensor will be deployed at point (x_1, y_1) and the predetermined coordinates are (x, y) is equal to

$$EP_{xy}(x_1, y_1) = \frac{\exp\left[-\frac{(x_1-x)^2}{2\sigma_x^2} - \frac{(y_1-y)^2}{2\sigma_y^2}\right]}{2\pi\sigma_x\sigma_y} \tag{17.9}$$

By applying total probability theorem, a sensor should be deployed at point (x, y) and detect event appeared at point (i, j) with a probability P^*:

$$P^*((i,j),(x,y)) = \frac{\sum_{(x_1,y_1)\in A}\left[P((i,j),(x_1,y_1))EP_{xy}(x_1,y_1)\right]}{\sum_{(x_1,y_1)\in A}EP_{xy}(x_1,y_1)} \tag{17.10}$$

17.4.3.6 Diff-Deploy

In Ref. [12], the authors propose a new deployment strategy called *Diff-Deploy*. The authors formalize a deployment process as control theory system. A deployment matrix D (input) is transformed to matrix I (output). I is defined as logarithmic miss probability matrix, where $I(x,y) = \ln(VEC\text{-}M^*(y*n+x))$. We have

$$VEC\text{-}M^*(i) = \prod_{x,y}(1 - P((\lfloor i/n\rfloor, i\%n),(x,y)))^{D(x,y)} \tag{17.11}$$

We apply a logarithmic function to the above equation:

$$I(i,j) = \ln\left(VEC\text{-}M^*(j*n+i)\right) = \sum_{x,y}\left(D(x,y)\ln\left[1 - P((i,j),(x,y))\right]\right) \tag{17.12}$$

It has been proved that the system is Linear Shift Invariant (LSI). $g(x,y)$ is the impulse response that characterizes how an input D is transformed to I. $g(x,y)$ is defined as

$$g(x,y) = \ln\left[1 - P((x,y),(0,0))\right] \tag{17.13}$$

If we use the model described in Equation 17.2, then

$$g(x,y) = \begin{cases} \infty & \text{if } \sqrt{x^2+y^2} \leq 1 \\ \ln\left(1 - \frac{\alpha}{\sqrt{x^2+y^2}^\beta}\right) & \text{if } 1 < \sqrt{x^2+y^2} \leq R_{\max} \\ 0 & \text{else} \end{cases} \tag{17.14}$$

Because the system is an LSI, the output system I can be obtained by the convolution result of the input D and the impulse response g. Equation 17.15 shows the two-dimensional convolution:

$$I(x,y) = D(x,y)*g(x,y) = \sum_i\sum_j\left(D(i,j)\times g(x-i,y-j)\right) \tag{17.15}$$

The two-dimensional convolution can be transformed into the matrices multiplication by constructing specific matrices. This technique is developed in Refs. [22,23]. Consequently, the system is represented by

$$I_p = G_p \cdot D_p \tag{17.16}$$

where
I_p and D_p are vectors with n^2 elements
G_p is a square matrix with n^4 elements
I_p and G_p represent the required detection probabilities and the detection models used, respectively.

The aim is to find D_p, which contains the deployment topology. So,

$$D_p = G_p^{-1} \cdot I_p \tag{17.17}$$

The elements of D_p computed in Equation 17.17 can be different from 0/1 values. The deployment topology D_p is not realistic but it gives an idea where to place sensors. The main idea of *Diff-Deploy* consists of deploying a new sensor at the grid point corresponding to the greatest value of D_p. After placement of a new sensor an update of I_p is made. The same process is repeated until all requested detection probabilities are satisfied or all available sensors are deployed. The pseudo code of *Diff-Deploy* is illustrated in Algorithm 17.4.

Algorithm 17.4: *Diff-Deploy* pseudo code

%*Compute the inverse of G_p* %;
G_p^{-1} = *inverse*(G_p);
remainI$_p$ = I_p ;
%*Initialize all elements of D_p to zero (no sensors are deployed)* %;
$\forall i : D_p(i) = 0$;
repeat

    nextD$_p$ = G_p^{-1}.*remainI*$_p$;
    Find the maximum value *nextD*$_p(k)$ in *nextD*$_p$ which satisfies the following constraints:

      ■ *remainI*$_p(k) < 0$
      ■ $D_p(k) == 0$

    $D_p(k) + +$;
    Num_Sensor $+ +$;
    %*update remainI*$_p$%;
    remainI$_p = I_p - G_p.D_p$;
    All positives values in *remainI*$_p$ receive zero;

until (*Num_Sensor* > *Sensor_Max*) **or** $\left(\sum_u^{n^2} remainI_p(u) \geq 0 \right)$;

To compute the complexity of *Diff-Deploy*, we have to specify the exact algorithm to obtain G_p^{-1}. Many inverse matrix algorithms exist in the literature, for example, Gaussian elimination, LU-decomposition, Cholsky decomposition, and QR-decomposition. The main difference consists in complexity. If we use the LU-decomposition algorithm, the computational complexity of *Diff-Deploy* is equal to $O(\frac{4}{3}n^6)$. Consequently, *Diff-Deploy* is not scalable and it is not expensive in memory usage(G_p contains n^4 elements).

17.4.3.7 Mesh

In Ref. [24], the authors propose a new deployment algorithm inspired from image processing and three-dimensional modeling, namely, mesh representation. Generally, mesh representation allows convenient modeling of arbitrary surfaces, where meshes serve as basic primitives to approximate a surface [25]. In image processing, mesh representation is used as a reduced representation of the image by polygonal elements. Several types of meshes representations exist. The representation is said regular when all meshes have the same shape (i.e., triangle, rectangle, hexagon, etc.) and the same dimension. Irregular mesh representation is composed of heterogeneous mesh shapes. Finally, hierarchical mesh representation is obtained using a particular mesh shape (e.g., triangle) but with nonuniform edge sizes.

In wireless sensor deployment context, meshes nodes represent sensors positions and each arc is the Euclidean distance between two sensors. The hierarchical mesh representation is suitable in the differentiated deployment problem. Different shapes could be considered, as triangles, rectangles, etc.

The algorithm proposed is named Differentiated Deployment Algorithm (DDA). The basic idea is to permit progressive meshes division of $\mathcal{A}$ as long as it is considered beneficial. Mesh division can be considered as beneficial when cost function $\mathcal{CF}$ decreases (or the profit function $\mathcal{PF}$ increases). Otherwise, the division is not permitted, and the mesh is marked.

The algorithm starts considering an initial unmarked mesh, which depends on the chosen shape. In the rectangle case, it is defined by the four corners delimiting the deployment surface. In the triangle case, the initial unmarked meshes consist of two triangles. Each corner of the surface is a vertex of at least one of these two initial unmarked meshes. Then, the algorithm evaluates the cost function, $\mathcal{CF}$, obtained with the current deployment (sensors located at the nodes placement). Thereafter, for each unmarked mesh a division test is performed. Here, a key point is to choose the new sensor location. Several strategies are possible. For instance, the new node could be placed in an equidistant location to the vertices of the divided mesh. Another alternative would be to place the new node in the mid of one given arc of the mesh. The algorithm evaluates once again the cost function, $\mathcal{CF}$, obtained with the new proposed mesh division. The mesh division is refused and the mesh is marked if the difference between the resulting cost function value after and before division is bigger than a constant threshold. So, the DDA can accept some divisions that do not decrease the cost function to skip local minima. Otherwise, the proposed division is retained as an acceptable operation in the list of candidates that could be performed in the next step. Once the test division of all unmarked meshes is finished, the division proposition (in the list of candidates), which leads to the smallest cost function value, is chosen. The elected mesh is divided and the algorithm continues checking for a new mesh division. The algorithm ends when all meshes are marked or all sensors are deployed.

The pseudo code of the DDA method is illustrated in Algorithm 17.5.

Algorithm 17.5: *DDA* pseudo code

Build initial meshes;
repeat
    Calculate the current cost function value, $C.F_a$ before division;
    Put the queue F to $\emptyset$;
    for *Each mesh M not marked* **do**
        Simulate the division of the mesh M;
        Calculate the new cost function value, $C\mathcal{F}_b$, after division;
        if $C\mathcal{F}_b - C\mathcal{F}_a \leq$ *Threshold* **then**
            Put the mesh M in the queue F;

        else
            Mark the mesh M;

        if $F \neq \emptyset$ **then**
           Choose the best solution, M_{best}, in the queue F;
           Mark the mesh M_{best};
           Accept division for the mesh M_{best};
           Add the generated meshes;
           Num_Sensor $++$;

until (*Num_Sensor* > *Sensor_Max*) **or** (*All meshes are marked*);

The computational complexity of the DDA is equal to $O(n^6)$, so the running time increases according to the number of unmarked meshes or the size of the deployment area. However, the DDA is not expensive in memory usage.

17.4.3.8 Differentiated Sensor Deployment Based on Tabu Search Approach

In Refs. [26,27], the authors propose a pseudo random deployment algorithm based on the meta-heuristic Tabu Search. Tabu search is a local search optimization technique that tries to minimize a cost function $F(x)$, where x represents a parameter vector, by iteratively moving from a solution x to a solution x' in the neighborhood of x (according to a neighborhood function $H(x)$) until a stopping criterion is satisfied or a predetermined number N of iterations is reached. The Tabu Search algorithm is independent of the detection model. This model provides input parameters to the method, though some other detections models can be used.

A Tabu Search algorithm is adapted to the differentiated deployment problem described in Section 17.4.1. The initialization of the method, the neighborhood function, the cost function, and the new specific steps are detailed hereafter.

1. *Initialization*

   The convergence of the Tabu Search method depends on the judicious choice of the initial solution (s_0). Ideally, the first solution has to be close to the optimal one; otherwise, because the maximum number of iterations is fixed, the algorithm may stop before reaching the optimal solution.

   The authors consider that the decision, $D(x, y)$, of deploying a sensor in a point $p(x, y)$ is a random variable, which follows a Bernoulli distribution with parameter $\alpha_{(x,y)}$. The binary form of the decision rule motivates the choice of the Bernoulli law. Precisely, $D(x, y)$ can assume a value of 1 with a probability of $\alpha_{(x,y)}$ and the value of 0 with a probability of $(1 - \alpha_{(x,y)})$.

   The parameter $\alpha_{(x,y)}$, associated to a point $p_{(x,y)}$, is chosen as the percentage of the points located in the vicinity of $p_{(x,y)}$ and is not receiving the required probabilities of detection. The vicinity, denoted $E_{(x,y)}$, is defined as the set of neighbor points located inside the maximum monitoring circle of a sensor that would be placed in $p_{(x,y)}$. Formally,

   $$\alpha_{(x,y)} = \frac{1}{\|E_{(x,y)}\|} \sum_{(i,j) \in E_{(x,y)}} 1_{\{r_{(i,j)} > \mathcal{P}_{(i,j)}\}} \tag{17.18}$$

Here, $1_{\{cond\}}$ is the indicating function, which is equal to 1 if the condition cond is true and 0 otherwise. The initialization stage of our Tabu Search approach follows these steps:

■ Step 1: The initial Tabu Search solution is started assuming zero deployed sensors. Thus, $\forall p_{(x,y)} \in \mathcal{A}$, Bernoulli parameters are computed using Equation 17.18 with $\mathcal{P}_{(x,y)} = 0$.

■ Step 2: Generate a list, L_{init}, including all points of $\mathcal{A}$. L_{init} is a decreasingly sorted list of points according to their Bernoulli parameters.

■ Step 3: In L_{init}, select the point $p_{(x,y)}$ with the highest Bernoulli parameter and remove it from the list. If the actual detection probability, $\mathcal{P}_{(x,y)}$, associated to $p_{(x,y)}$ is lower than $r_{(x,y)}$, then a decision to deploy a sensor in $p_{(x,y)}$ is randomly generated through a Bernoulli decision rule with parameter $\alpha_{(x,y)}$.

■ Step 4: If the Bernoulli decision is to deploy a sensor ($D(x, y) = 1$), then (1) the probabilities of detection for all points in the vicinity of $p_{(x,y)}$ (the set $E_{(x,y)}$) are recomputed and (2) L_{init} is updated (sorted).

■ Step 5: If L_{init} is not empty, go back to **Step 3**.

When the stop criterion of Step 5 is satisfied, the resulting positions of deployed sensors are considered as the initial solution s_0. The latter one is saved in the algorithm's memory, called the Tabu List. In the remaining part of the chapter, we will refer to this list as T. The goal of the Tabu list is not to block the method on a local minimum of the cost function.

2. *Neighborhood exploration function*

After the initialization stage, a Tabu Search method executes N times the neighborhood exploration stage. Here, N is a chosen fixed parameter, which must be set to limit the number of Tabu Search iterations.

During the nth iteration of the neighborhood exploration stage, a given number V of possible neighbors of the solution selected in the previous iteration, noted s_{n-1}, are generated and evaluated. Neighboring solutions are possible solutions that can be reached from s_{n-1} by a basic transformation. Solutions that are present in the Tabu List T are considered unreachable neighbors.

The authors propose two neighboring generation methods, namely, Suppression-oriented stage (H_{supp}) and Additional-oriented stage (H_{add}). These two methods alternate in the successive iterations of our Tabu Search approach to determine the set V. Both stages are detailed hereafter.

■ *Suppression-oriented stage* (H_{supp}): The aim of this stage is to suppress some sensors among those deployed in over-covered areas. The method proceeds with the following steps:

Step 1′: Compute the Bernoulli parameters for all $p_{(x,y)} \in \mathcal{A}$ using Equation 17.19 and assuming the deployment obtained in the last Tabu Search iteration.

$$\beta_{(x,y)} = \frac{1}{\|E_{(x,y)}\|} \sum_{(i,j) \in E_{(x,y)}} \left[\left(1 - \frac{r_{(i,j)}}{\mathcal{P}_{(i,j)}} \right) \times 1_{\{r_{(i,j)} < \mathcal{P}_{(i,j)}\}} \right] \quad (17.19)$$

Step 2′: Generate a list, L_{supp}, including all the points of $\mathcal{A}$ where a sensor ($D(x,y) = 1$) is deployed. The list is then decreasingly sorted according to the resulting Bernoulli parameters.

Step 3′: In L_{supp}, select and remove the point $p_{(x,y)}$ with the highest Bernoulli parameter, and randomly generate the decision to suppress the sensor in $p_{(x,y)}$ through a Bernoulli decision rule with parameter $\beta_{(x,y)}$.

Step 4′: If the Bernoulli decision is to suppress the sensor ($D(x,y) = 0$), then the probabilities of detection for all points in the vicinity of $p_{(x,y)}$ are recomputed. In this case, the list L_{supp} is updated (sorted) with the new values of the Bernoulli parameters associated to each point in L_{supp}.

Step 5′: If L_{supp} is not empty, go back to Step 3′.

Once the stop criteria on Step 5′ is satisfied, the next Tabu Search iteration alternates toward an additional stage.

■ *Additional-oriented stage* (H_{add}):

The aim of this stage is to add more sensors to the actual deployed ones in under-covered areas. The execution is very similar to the initialization stage, except **Step 1** and **Step 2** are replaced by the following steps:

Step 1″: Compute the Bernoulli parameters for all $p_{(x,y)} \in \mathcal{A}$ using Equation 17.18 and assuming the deployment obtained in the last Tabu Search iteration.

Step 2″: Generate a list L_{add} including all the points of $\mathcal{A}$ where a sensor is not deployed. List one is then decreasingly sorted according to the resulting Bernoulli parameters.

Steps 3 through 5, as detailed in the initialization stage, are repeatedly executed until list L_{add} is empty.

3. *Cost function*

   After the neighborhood exploration (a suppression- or an additional-oriented stage) during the nth iteration, an elected solution s_n must be chosen among the V explored candidates. This solution (which cannot be in the Tabu list T) is the one which is provided by minimizing a given cost function, F. The cost function reflects two objectives of the optimization problem described in Section 17.4.1: minimize the number of deployed sensors and maximize the satisfaction rate in requested detection probabilities. The first objective could be quantified by counting the number of deployed sensors. Formally, as defined in Section 17.4.1 $D(x, y) = 1$, if a sensor is deployed in the point $p_{(x,y)}$. Otherwise, $D(x, y) = 0$. To minimize the cost function, F includes the following term:

$$\sum_{(x,y) \in \mathcal{A}} D(x, y) \tag{17.20}$$

The second objective is integrated into the cost function through the following penalty function:

$$\text{Penalty} = \sum_{(x,y) \in \mathcal{A}} \frac{\left[r_{(x,y)} - \mathcal{P}_{(x,y)}\right]^+}{r_{(x,y)}} \tag{17.21}$$

Here, $\left[r_{(x,y)} - \mathcal{P}_{(x,y)}\right]^+$ denotes the projection of $r_{(x,y)} - \mathcal{P}_{(x,y)}$ in $\mathcal{R}^+$. Formally,

$$\left[r_{(x,y)} - \mathcal{P}_{(x,y)}\right]^+ = (r_{(x,y)} - \mathcal{P}_{(x,y)}) \times 1_{\{r_{(x,y)} > \mathcal{P}_{(x,y)}\}} \tag{17.22}$$

According to the expression of the penalty function, a good deployment solution should lead to obtain detection probabilities higher than (or ideally, equal to) the required detection thresholds. If this is not satisfied, the penalty function value translates how far is the solution to the required thresholds. This is exactly the second objective of the optimization problem.

From the above objective expressions, the authors define two cost functions, F_{supp} and F_{add}. The function F_{supp} is used to choose the best next-iteration solution in the case of a suppression-oriented stage. The function F_{supp} is formulated using only Equation 17.21. On the other hand, the cost function F_{add} is used in the case of the additional-oriented stage. In this case, both terms associated to each objective of the optimization problem are integrated into the cost function through the following additive expression:

$$F_{\text{add}} = \sum_{(x,y) \in \mathcal{A}} \left[D(x,y) \right] + \text{Penalty} \qquad (17.23)$$

The pseudo code of the Tabu search deployment process is illustrated in Algorithm 17.6.

Algorithm 17.6: *Tabu search* pseudo code

Compute initial solution s_0;

$s_{out} = s_0$;

bool sup-sensors = true;

Tabu-List = $\{s_0\}$, set of T last solutions visited;

for $i=0$ *to* N **do**

    neighborhood = $\emptyset$;

    if *sup-sensors* == *true* **then**

        for $j=1$ *to* V **do**

            $s_i^j = H_{supp}(s_i)$;

            neighborhood = neighborhood + $\{s_i^j\}$;

        Cost-Function $= F_{supp}$;

    else

        for $j=1$ *to* V **do**

            $s_i^j = H_{add}(s_i)$;

            neighborhood = neighborhood + $\{s_i^j\}$;

        Cost-Function $= F_{add}$;

    for $j=1$ *to* V **do**

        if $s_i^j \in$ *Tabu-List* **then**

            neighborhood = neighborhood - $\{s_i^j\}$;

    Select the best solution s_i^{best} in neighborhood, s_i^{best} minimizes the cost function;

    Cost-Function $\left(s_i^{best} \right) = min_{s_i^j \in neighborhood} \left[Cost\text{-}Function \left(s_i^j \right) \right]$;

    $s_{i+1} = s_i^{best}$;

    %update output solution s_{out} %;

    if *Cost-Function* $\left(s_i^{best} \right) <$ *Cost-Function* (s_{out}) **then**

        $s_{out} = s_i^{best}$;

    %update Tabu-List%;

    Tabu-List = Tabu-List + $\{s_i^{best}\}$;

    %Alternate between H_{supp} and H_{add} %;

    sup-sensors = not(sup-sensors);

The computational complexity of Tabu search is equal to $O(NVm^2n^2)$. N is the number of iterations of the Tabu search process and V is the size of the neighborhood. These two parameters are chosen by the designer; they are calibrated according to the specific deployment scenario. A coverage circle of a sensor covers a set of cells in the area; m represents the number of cells in length or width of the subarea covered by a sensor. m depends on a sensor coverage range R_c, and m is equal to $\lceil \frac{R_c}{\sqrt{2}} \rceil$. We remark that the Tabu search complexity depends on deployment area dimensions, detection characteristics of sensor (R_c), and parameters of the Tabu search process (N, V). If N, V and m are not large, the product, NVm^2, is equal to constant C_1, so we can say that we have a quadratic complexity $O(C_1 n^2)$.

In Ref. [28], the authors extend the Tabu search deployment process presented in Ref. [27]. The extension consists of guaranteeing the network connectivity, so all constraints of the optimisation problem (number of sensors, satisfaction rate, and connectivity) described in Section 17.4.1 are included. The authors adapt the initialization stage and the selection process of solutions in the neighborhood presented in Ref. [27] to build network connectivity. The remainder steps and cost functions are identical to Ref. [27].

17.4.4 Deployment Strategies Comparison

To evaluate and compare the performance of different deployment strategies exposed above, we implemented all methods, namely, Random, Grid, MIN_MISS, MAX_MIN_COV, MAX_AVG_COV, Mesh-DAA, Diff-Deploy, and Tabu Search approach. The programming is made with C++.

The comparison is based on different metrics as number of sensors deployed, satisfaction rate in detection probabilities θ (percentage of area units receiving a detection probability larger than the required detection probability threshold), computational complexity, memory consumption, and network connectivity. We compare the deployment methods in two stages. Initially, we focus the comparison on all the metrics introduced above except network connectivity. Afterward, we focus the comparison including network connectivity.

We fixed the sensor parameters values, α, β, R_s, and R_c, to 1, 1, 5, and 3, respectively. We chose R_c less than $\sqrt{3}R_s$ to illustrate how deployment algorithms can ensure and build the connectivity. We recalled that if R_c is greater than $\sqrt{3}R_s$, the sufficient condition to guarantee the connectivity is the full coverage of the deployment area. We considered an area with $50 * 50$ units. The required detection probabilities thresholds are illustrated in Figure 17.5.

For a regular deployment we chose a grid topology, so the shape is rectangle. In the Mesh method, we chose the triangle shape for meshes. To divide a mesh, we placed a new sensor in the middle of one given arc. The cost function is the miss detection probabilities rate, and is equal to $(1 - \theta)$.

We calibrated the Tabu search process by fixing the number of iterations, the size of the Tabu list, and the size of the neighborhood explored to 100, 10, and 15, respectively. For the random deployment, we selected the best deployment topology among 1500

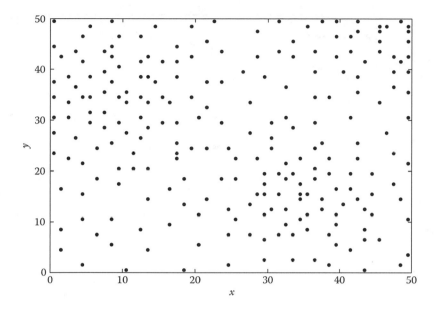

Figure 17.6   Sensors positions in the case of Tabu Search approach.

(10×15) random deployment topologies generated. The number of random deployment topologies is equal to the number of all solutions explored by Tabu Search strategy.

Figure 17.6 shows the deployed sensors positions obtained by the Tabu Search approach. The number of deployed sensors is equal to 233 for a satisfaction rate θ equal to 96.52 percent. For the same number of sensors, the satisfaction rates obtained when using the Random, Grid, MIN_MISS, MAX_MIN_COV, MAX_AVG_COV, Diff-Deploy, and Mesh-DDA approaches are equal to 77.12, 83.40, 85.52, 83.96, 82.56, 96.2, and 93 percent, respectively.

However, to reach the satisfaction rate obtained by the Tabu Search apporach, the Random, Grid, MIN_MISS, MAX_MIN_COV, MAX_AVG_COV, Diff-Deploy, and Mesh-DDA approaches must deploy 450, 552, 337, 373, 518, 234, and 249 sensors, respectively.

We can notice from the above results Tabu Search and Diff-Deploy reduce highly the number of deployed sensors while improving the satisfaction rate. But the computational complexity of Diff-Deploy, which is equal to $O(\frac{4}{3}n^6)$, is larger than the computational complexity of Tabu Search, which is equal to $O(c_1 n^2)$. Also, the memory consumption in Diff-Deploy is more important than the memory consumed in Tabu Search, because Diff-Deploy manipulates square matrix with n^4 elements. The Mesh-DDA method also gives a good result but is more complex than Tabu Search.

In Figure 17.6, we plot the sensors positions using the Tabu Search approach. Compared to Figure 17.5, we can notice a clear concentration of the deployed sensors in

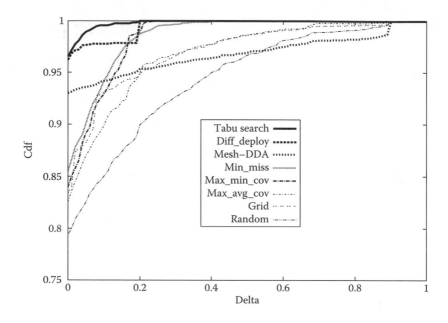

Figure 17.7 Cdf between desired and obtained detection probabilities.

the areas requiring high detection probability thresholds (top right, top left, and down right). Figure 17.7 illustrates the cumulative distribution function (cdf) obtained by all deployment strategies of X, where X is a random variable, which can take values in the set of $\{\vartheta_{i,j} | \forall i, j \in \mathcal{A}\}$. Formally,

$$\vartheta_{i,j} = (r_{(i,j)} - p_{(i,j)}) \times 1_{\{r_{(i,j)} > p_{(i,j)}\}} \tag{17.24}$$

ϑ_i can take values in $[0, 1]$. Each curve in Figure 17.7 indicates for each δ value in the x axis the probability $P(X \leq \delta)$. From the above figure, we can notice that the Tabu Search, Diff-Deploy, and Mesh-DDA approaches provide the best performances compared to all other deployment algorithms. Moreover, we can observe that the Tabu Search approach satisfaction rate reaches very quickly 100 percent of the area units compared to the other approaches. This means the not satisfied units receive a detection probability very close to the required detection probability thresholds.

To compare the network connectivity between the different approaches, we run the second version of Tabu Search where the connectivity is ensured by construction. Tabu Search method deploys 276 sensors and the satisfaction rate, θ, is equal to 98.68 percent. With the same number of sensors, the satisfaction rates, θ, of Random, Grid, MIN_MISS, MAX_MIN_COV, MAX_AVG_COV, Diff-Deploy, and Mesh-DDA approaches are equal to 85.84, 75.28, 91.16, 86.36, 84.96, 100, and 100 percent, respectively. We can observe that Diff-Deploy and Mesh-DDA reach 100 percent in satisfaction rate. Figure 17.8 shows the cdf of random variable X defined in Equation 17.24.

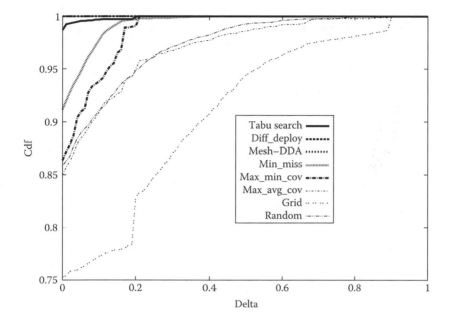

Figure 17.8   Cdf between desired and obtained detection probabilities considering connectivity.

We calculate the number of connected components in all deployment topologies generated by the different deployment strategies. The network is connected if the graph of connectivity contains only one connected component. Figures 17.9 and 17.10 illustrate the graphs of connectivity of Tabu Search and Diff-Deploy. We can see clearly that only the Tabu Search approach provides a connected graph.

Figure 17.11 represents the connected components related to the connectivity graphs resulting from the various deployment methods. We give the number of nodes for each connected component. We plot only the 10 largest connected components related to each graph. The largest connected component is the one that contains the greatest number of nodes. We note that only the graphs resulting from the Tabu Search and the Grid approaches have one connected component. In the case of Grid, we certainly guarantee the connectivity, but the performances in terms of detection probabilities are not satisfactory (see Figure 17.8). The number of sensors deployed in Grid is slightly higher than that of Tabu Search. The cause is due to the construction of the grid shape. The other methods contain more than one connected component, which means that the network is not connected. For example, Diff-Deploy and Mesh-DDA have 134 and 36 connected components, respectively. The Tabu Search method is the only method that ensures the connectivity and produces a better satisfaction rate compared to other methods, while minimizing the number of required sensors.

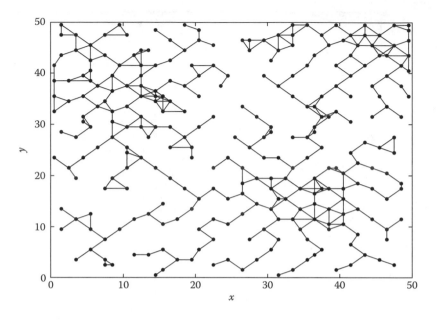

Figure 17.9   Tabu Search, graph of connectivity.

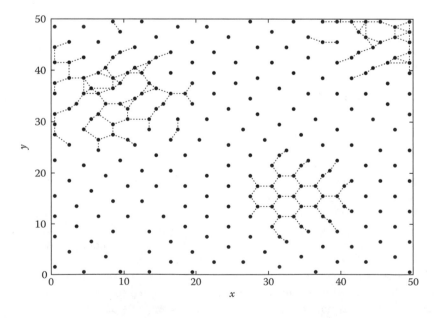

Figure 17.10   Diff-deploy, graph of connectivity.

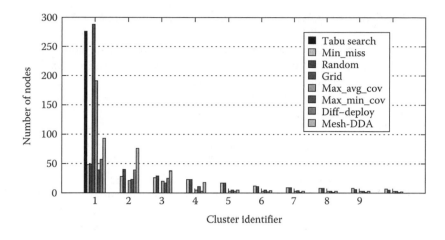

Figure 17.11   Connected components of the graphs of connectivity.

17.5   Conclusion and Open Issues

We provide in this chapter a deep overview on the deployment strategies for WSNs. We started the chapter by introducing the main sensor detection models proposed in the literature. Then, we highlighted the principal metrics used to evaluate WSN deployment. Thereafter, we introduced the principal WSN deployment algorithms found in the literature. Finally, we provided a performance comparison between these algorithms. However, there are still many issues that we have not discussed in this chapter due to the lack of space, such as, energy consumption, reliability, and data aggregation. We believe that the deployment issue discussed in this chapter can be considered as an important first step to establish a WSN.

Acknowledgment

We would like to thank Melissa Mecheri for revising text grammar.

References

1. I.F. Akyildiz, W. Su, Y. Sankarasubramaniam, and E. Cayirci, Wireless sensor networks: A survey, *Computer Networks: The International Journal of Computer and Telecommunications Networking*, 38(4), 393–422, 15 March 2002.
2. J.N. Al-Karaki and A.E. Kamal, Routing techniques in wireless sensor networks: A survey, *IEEE Wireless Communications*, 11(6), 6–28, 2004.
3. P. Kumar, A. Mallikarjuna, and D. Janakiram, Distributed collaboration for event detection in wireless sensor networks, *Proceedings of the 3rd International Workshop*

on Middleware for Pervasive and Ad-Hoc Computing, Grenoble, France, pp. 1–8, 2005.

4. Z. Xue-Yu and C. Yang, Collaborative detection probability of mobile target for coverage in large-scale WSN, *International Conference on Wireless Communications Networking and Mobile Computing WiCom*, Shanghai, China, pp. 2710–2714, 2007.

5. C. Huang and Y. Tseng, The coverage problem in a wireless sensor network, *Book: Wireless Sensor Networks and Applications*, San Deigo, CA, pp. 115–121, 2003.

6. X. Bai, S. Kuma, D. Xua, Z. Yun, and T. Lai, Deploying wireless sensors to achieve both coverage and connectivity, *MobiHoc '06: Proceedings of the 7th ACM International Symposium on Mobile Ad-Hoc Networking and Computing*, Florence, Italy, pp. 131–142, 2006.

7. S. Kuo, Y. Tseng, F. Wu, and C. Lin, A probabilistic signal-strength-based evaluation methodology for sensor network deployment, *AINA '05: Proceedings of the 19th International Conference on Advanced Information Networking and Applications*, Taipei, Taiwan, 2005.

8. X. Li, P. Wan, Y. Wang, and O. Frieder, Coverage in wireless ad-hoc sensor networks, *ICC'02, IEEE International Conference on Communications*, New York, 2002.

9. C. Tai-Lin, R. Parameswaran, and S. Kewal K., Optimal sensor distribution for maximum exposure in a region with obstacles, *Global Telecommunications Conference IEEE GLOBECOM*, California, 2006.

10. A. Elfes, Occupancy grids: A stochastic spatial representation for active robot perception, *Book : Atonomous Mobile Robots : Perception, Mapping, and Navigation*, Vol. 1, pp. 60–70, IEEE Computer Society Press, 1991.

11. M. Hata, Empirical formula for propagation loss in land mobile radio services, *IEEE Transactions on Vehicular Technology*, 29(3), 317–325, 1980.

12. J. Zhang, T. Yan, and S.H. Son, Deployment strategies for differentiated detection in wireless sensor networks, *SECON '06: Proceedings of the 3th Annual IEEE Communication Society Conference on Sensor and Ad Hoc Communications and Networks*, Virginia, Reston, VA, Vol. 1, pp. 316–325, 2006.

13. Y. Wang, C. Hu, and Y. Tseng, Efficient deployment algorithms for ensuring coverage and connectivity of wireless sensor networks, *WICON '05: Proceedings of the First International Conference on Wireless Internet*, Budapest, Hungary, pp. 114–121, 2005.

14. S. Megerian, F. Koushanfar, G. Qu, G. Veltri, and M. Potkonjak, Exposure in wireless sensor networks: Theory and practical solutions, *Journal of Wireless Networks ACM Kluwer Academic Publishers*, 8(5), 443–454, 2002.

15. L.F.M. Vieira, M.A.M. Vieira, L.R. Beatriz, A.A.F. Loureiro, D.C. da Silva Jr., and A.O. Fernandes, Efficient incremental sensor network deployment algorithm, *SBRC Brazilian Symposium on Computer Networks*, Brazil, 2004.

16. M.M. Iqbal, I. Gondal, and Dooley L., Dynamic symmetrical topology models for pervasive sensor networks, *Proceedings of INMIC*, Lahore, Pakistan, 2004.

17. A. H. Land and A. G. Doig, An automatic method of solving discrete programming problems, *Econometrica*, 28(3), 497–520, 1960.

18. N. Ahmed, S. S. Kanhere, and S. Jha, Probabilistic coverage in wireless sensor networks, *Proceedings of the The IEEE Conference on Local Computer Networks 30th Anniversary*, Sydney, Australia, 2005.
19. Zhang H, and Hou J. C, Is deterministic deployment worse than random deployment for wireless sensor networks?, *INFOCOM*, Barcelona, Spain, 2006.
20. S. Dhillon and K. Chakrabarty, Sensor placement for effective coverage and surveillance in distributed sensor networks, *IEEE Wireless Communications and Networking Conference*, 3(20), 1609–1614, 2003.
21. Y. Zou and K. Chakrabarty, Uncertainty-aware and coverage-oriented deployment for sensor networks, *Journal of Parallel and Distributed Computing*, 64(7), 788–798, 2004.
22. H. C. Andrews and B. Hunt, *Digital Image Restoration*, Prentice Hall, Englewood Cliffs, NJ, 1977.
23. B. Hunt, A matrix theory proof of the discrete convolution theorem, *IEEE Transaction on Audio Electroacustic*, AU-19, 285–288, 1971.
24. N. Aitsaadi, N. Achir, K. Boussetta, and G. Pujolle, Differentiated underwater sensor network deployment, *IEEE/OES OCEANS'07*, Aberdeen, U.K., pp. 1–6, 2007.
25. L.B. Jordan, Progressive geometrical compression of arbitrary shaped video objects, PhD thesis, EPFL, 1998.
26. N. Aitsaadi, N. Achir, K. Boussetta, and G. Pujolle, Déploiement différencié des réseaux de capteurs, *8eme Colloque Francophone de Gestion de Réseaux et de Services GRES*, Hammamet, Tunisia, 2007.
27. N. Aitsaadi, N. Achir, K. Boussetta, and G. Pujolle, A tabu search approach for differentiated sensor network deployment, *Fifth IEEE Consumer Communications and Networking Conference IEEE CCNC*, Las Vegas, NV, pp. 163–167, 2008.
28. N. Aitsaadi, N. Achir, K. Boussetta, and G. Pujolle, Heuristic deployment to achieve both differentiated detection and connectivity in WSN, *IEEE 67th Vehicular Technology Conference—VTC-Spring*, Marina Bay, Singapore, pp. 123–127, 2008.

INTEGRATED RFID AND SENSOR NETWORKS

Chapter 18

Integrated RFID and Sensor Networks: Architectures and Applications

Aikaterini Mitrokotsa and Christos Douligeris

CONTENTS

Radio frequency identification (RFID) systems and wireless sensor networks (WSNs) represent two key technologies for ubiquitous computing that have attracted considerable attention in recent years because their use revolutionizes diverse application areas. However, these two technologies have separate research and development areas.

The integration of RFID and sensor networks can increase their utilities to other scientific and engineering fields by exploiting the advantages of both technologies. In this chapter we investigate why the integration of RFID systems and WSNs is important and we identify the key requirements to achieve efficient and effective integration. We present possible architectures for integrating RFID and WSNs and provide a detailed list of real-world application examples.

18.1   Introduction

Radio frequency identification (RFID) systems and wireless sensor networks (WSNs) are emerging as the most ubiquitous computing technologies in history due to their important advantages and their broad applicability. RFID communication is fast, convenient, and its application can substantially save time, improve services, reduce labor cost, thwart product counterfeiting and theft, increase productivity gains, and maintain quality standards. Common applications range from highway toll collection, supply chain management, public transportation, controlling building access, animal tracking, developing smart home appliances, and remote keyless entry for automobiles to locating children.

RFID systems are mainly used to identify objects or to track their location without providing any indication about the physical condition of the object. WSNs on the other hand, are networks of small, cost-effective devices that can cooperate to gather and provide information by sensing environmental conditions such as temperature, light, humidity, pressure, vibration, and sound. WSNs provide cost-effective monitoring of critical applications including industrial control, border monitoring, environmental monitoring, military, home and healthcare applications.

RFID technology has received great attention and it has been deployed extensively in industrial applications. On the other hand, sensor networks have been the focus of great research activity but they have been around mainly as a proof of concept with the main exception of their adoption in military applications. The evolution of RFID and WSNs has followed separate research and development paths and has led to distinct technologies. Nevertheless, there are many applications where the identity or the location of an object is not sufficient and extra information that can be retrieved through sensing environmental conditions is important. Although sensor networks may be used in these environments as well, the location and identity of an object remain critical information that can be retrieved through RFID systems. The optimal solution in these cases is the integration of both technologies because they complement each other.

In this chapter, we first investigate why the integration of RFID and WSNs is significant. Furthermore, we list the key requirements for integrating RFID and WSNs in an efficient and flexible way. Subsequently, we present possible integration architectures and discuss the existing and suggested integration scenarios in real applications. A variety of possible architectures for integrating RFIDs and WSNs have also been studied before by Mitsugi et al. [49], Liu et al. [45], and Zhang and Wang [75]. In this chapter, we present the different possible integrated architectures based on the proposal by Liu et al. Nevertheless, we provide a detailed and comprehensive study and we update each type of integration strategy with current commercial as well as research academic integrating approaches.

18.2   Why Integrate RFIDs and WSNs

The integration of the promising technologies of RFID and WSNs will maximize their effectiveness, give new perspectives to a broad range of useful applications, and bridge the gap between the real and the research/academic world. This is because the resulting integrated technology will have extended capabilities, scalability, and portability as well as reduced unnecessary costs [46].

Extension of capabilities and functionalities: Considering the fact that RFID networks can provide critical information, such as the identity and the location of an object, by merging RFIDs with WSNs additional information can be retrieved, while the potential for exploiting this information is multiplied. For instance, in supply chain management we are able not only to track food products but also to monitor their environmental conditions and detect when perishables go off.

Scalability–portability: RFID systems integrated with WSNs enjoy the advantages of wireless communication. The transmission and processing of critical data and information is facilitated without the burden and inconvenience of wired transactions while saving valuable time. Portable RFID readers can further speed the collection of data and ease procedures in varying applications. For instance, healthcare applications, including monitoring everyday medication of elderly or monitoring patients for diagnosing deceases, can be extremely facilitated without rendering patients immobile through cumbersome data wirings.

Reduce unnecessary costs: Reducing the cost of employed services is a critical factor in many applications including industrial ones. The requirement is to achieve the desired goal with the minimum possible cost by supporting backup solutions in case of undesired circumstances. For instance, perishable goods can be monitored so that in case they are not preserved properly their transport can be terminated, thus avoiding unnecessary additional transport costs.

18.3   Requirements for Integrating RFID Networks and WSNs

The integration of RFIDs and WSNs should be performed in such a way that specific requirements are met to have an efficient and effective solution. Some of the most important requirements that should be taken in consideration are the following [13,15]:

Accurate and reliable communication: In traditional client server networks large data streams are transferred from servers to clients. However, in integrated RFID and WSNs the data flow is mainly transferred from a large number of devices (clients) to a few servers. Subsequently, servers are expected to process all the received information from RFIDs and sensors in a reliable way and to allow the appropriate action to be taken within a short period of time. Reliability and accuracy are also expected for the data transferred to the applications (or users) of the integrated system within a tolerable latency. Of substantial importance is the capability of the integrated RFID-sensor network to deliver data to the required destination with reliability and to provide a confirmation for the successful completion of a task. The reliability and accuracy of an integrated RFID-sensor network is also dependent on the criticality of the specific application. In not so critical applications a lower degree of reliability is required.

Energy efficiency: Considering the fact that both sensor nodes and active RFID tags present scarce resources, the integrated RFID-sensor network should take into account this limitation. The integrated system should be energy efficient to make sure that accurate and reliable communication will be achieved with the minimum possible energy consumption.

Network maintenance survivability: Considering the large number of devices that can be employed in an integrated RFID-sensor network, among the most important requirements for such a network is the ability to perform remote device configuration and remote device software updates. Thus, we can achieve a high survivability and an efficient maintenance of the network with an acceptable cost. Furthermore, it is important that the integrated network is able to recover in case of possible denial-of-service (DoS) attacks. A way to achieve that would be the adoption of intrusion tolerance and mitigation mechanisms such as replicating critical network devices.

18.4   Possible Architectures of Integrated RFIDs and WSNs

18.4.1   Integrated RFID Tags with Sensors

Integrated RFID tags with sensors or sensor-tags, as we are going to refer to them from now on, can be discriminated into two main categories: integrated sensor-tags that are able to communicate only with RFID readers and integrated sensor-tags that are able to communicate with each other and form a cooperative ad hoc network. In this section, we will provide the main features of these two categories of integrated sensor-tags and we will also present an overview of the available research and commercial proposals for each category.

18.4.1.1   Integrated Sensor-Tags with Limited Communicating Capabilities

One of the simplest ways of integrating RFID networks and WSNs is the integration of sensing capabilities in RFID tags. Many RFID tags have incorporated sensors in

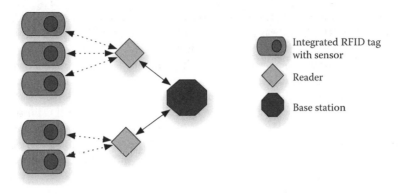

Figure 18.1 Integration architecture of sensor-tags, RFID reader, and base station.

their design and, thus, they are able to take sensor readings and to transmit them later to a reader. Nevertheless, when RFID tags are given sensing capabilities the line between RFID networks and sensor networks becomes blurred because the sensor-tags use the same protocols and mechanisms for reading tag IDs and for collecting sensed data. In this architecture, integrated sensor-tags function as normal RFID tags and they are equipped with a unique identification, while integrated sensors are used to collect sensing information related to the environment, the existing conditions, and the associated objects (Figure 18.1). The integration of RFID tags with sensor nodes is based on converting the sensors' analog signal by the A/D module while the resulting data is forwarded by the readers to the base station [45].

There are many commercial and academic proposals for integration of RFID sensor-tags. However, there is a discrimination between active, semi-active, and passive integrated RFID sensor-tags.

18.4.1.1.1 Active Sensor-Tags

Active sensor-tags use batteries to power their communication circuitry, sensors and micro-controllers. Thus, they have a rather long range (approximately 30 m) and they are able to achieve high data and sensor activity rates. Nevertheless, because a battery is used, the device cost and the weight is increased while the lifetime of the RFID sensor-tag is limited.

An active RFID sensor-tag created through the integration of sensors with UHF RFID tags and printed on low-cost environmental paper for frequencies up to 950 MHz was proposed by Ferrer-Vidal et al. [28]. The proposed integrated sensor-tag uses embedded rechargeable thin film batteries that increase the nodes' lifetime. Considering that paper is one of the cheapest organic materials, the proposed sensor-tags present an attractive advantage that will trigger large-scale adoption of integrated sensor-tags. Rida et al. [64] have also proposed an RFID prototype including sensing capabilities and a battery source on a low-cost paper substrate.

The design of a sensor-embedded radio frequency identification (SE-RFID) system based on active RFID tags was proposed by Deng et al. [21]. The main advantage of the proposed system is that the sensors sample the external data independently and periodically with or without the presence of the reader in the tag activation zone. Deng et al. [21] have proposed two different architectures for the SE-RFID tags. In the first proposed architecture multiple sensors may be embedded in a single RFID tag, while in the second architecture each sensor is embedded in a single RFID. Furthermore, they have evaluated one of the proposed architectures by developing a real-time health monitoring system (HEMS) using a SE-RFID. In HEMS, the goal was to develop a constant monitoring system, which would be able to continually monitor, reevaluate, and diagnose diseases.

A commercially available active vibration sensor-tag (24TAG02V) [11] and an active temperature sensor-tag (24TAG02T) [10] was developed by Bisa Technologies. Both sensor-tags operate at 2.4 GHz, their range is 100 m and they employ an anticollision mechanism. Hundred tags can be read simultaneously and their battery life is four years. The temperature sensor-tag collects real-time temperature from items and transmits them to a reader for logging. When an item's temperature reach a specified intolerable temperature an alarm can be triggered. Its temperature measurements ranges from $-50°C$ to $150°C$ with an accuracy of $1°C$. The vibration sensor-tag detects and records either continuous or impulsive vibrations or impacts and it has minimum sensitivity 200 mV/g, resonance sensitivity 4 V/g, and resonance frequency 90 Hz.

Other commercial active sensor-tags include TELID 310 [39], an active temperature sensor-tag by Microsensys, and Callistro and Elara, active temperature and humidity sensor-tags respectively by Adage Solutions [1].

18.4.1.1.2 Passive Sensor-Tags

Passive sensor-tags receive operating power from RFID readers. Thus, a battery does not limit their lifetime. They offer several advantages such as smaller size, lower cost, and longer life cycle. The feature of unlimited lifetime can be exploited in applications where neither batteries nor wired connections are feasible because of the relative cost, weight, or because of other reasons. The main limitation of the passive sensor tags is that they have to be close to an RFID reader to function properly.

A passive sensor-tag with incorporated temperature and photosensors that can be used for environmental monitoring was proposed by Cho et al. [14]. The proposed sensor-tag is powered by an external ISM band RF signal and it senses ambient temperature and light. Zhou and Wu [76] have proposed a passive UHF RFID tag with an embedded magnetic sensor. Their system includes a 900 MHz RFID front-end circuit and a mirror-based magnetic sensor in standard CMOS process. Among its main advantages, one can point its high sensitivity and its low-power consumption.

The design of a long range passive RFID-tag for sensor networks was proposed by Kitayoshi and Sawaya [43]. The proposed tag has a range longer than 10 m, operates at 2.45 GHz and 915 MHz ISM bands and it is composed of a divided microstrip antenna and a passive voltage multiplying circuit. In order to demonstrate the validity of the proposed tags, the authors have also fabricated passive sensor-tags for temperature monitoring for a range larger than 9 m.

The design of a wireless identification and sensing platform (WISP) was proposed by Sample et al. [66]. WISP is a battery-free RFID sensor device and as all the passive RFID tags, is powered via the RF energy transmitted by an RFID reader. WISP is implemented as a printed circuit board (PCB) and its range is approximately 4.5 m. WISP is the first micro controller integrated as part of a passive UHF RFID tag.

An integrated passive sensor-tag (ICT Tag Sensor) was proposed by Instrumentel [31]. In this passive sensor-tag, power is provided by inductive coupling, enabling the tag to operate without batteries. Thus, the sensor-tag is ideal for applications where the weight and the size of a battery would interfere with the sensing capabilities. These passive sensor-tags can be tailored to support multiple sensors and to interface with many communication protocols. Among their most important features, the onboard microcontroller and differential amplifier as well as their multiple sensor capabilities are included.

Microchip [55] in cooperation with Digital Angel have designed and developed a passive implantable sensor-tag which can be used to determine glucose levels in the bodies of animals and humans, without the need for diabetics to draw blood glucose levels. The RFID sensor-tag is passive, powered by the scanner signal, avoiding the need for a battery on the sensor-tag. Many measurement sensors require an accurate reference voltage; To achieve that, the used glucose sensor has a specific circuit architecture, which provides precise and stable measurements of physiological parameters, allowing for accurate measurement of glucose concentration. The patent was granted in October 2006 and is titled "Embedded Bio-Sensor System" [55].

A passive sensor-tag that can be used to measure the body temperature of animals was also developed by Digital Angel. The passive sensor-tag called Bio-thermo [9] is syringe implantable in a glass-tube form, operates at 134.2 kHz carrier frequency and follows the ISO 11785 standard [32]. It allows the noninvasive monitoring of temperatures in pets and it is able to detect infections and diseases at an early stage. However, in Bio-thermo there is no available memory space to store memory data.*

Commercially available passive RFID tags with integrated sensors were also developed by three Japanese Companies OKI, NYK Logistics Japanese and IIILLS [63] as well as by Microsensys (TELID 210) [38] and Alien Technology (ALB-2484) [3].

18.4.1.1.3 Semi-Active Sensor-Tags

Semi-passive sensor-tags function as passive RFID tags when the generated RF power is sufficient to operate, otherwise they operate in a semi-active mode using batteries. An integrated passive and battery-powered semi-passive UHF RFID tag which supports the EPC Gen 2 protocol [25] was proposed by Kim et al. [40]. The proposed sensor-tag functions as a passive RFID tag, when the generated RF power is sufficient to operate. In other cases, the sensor-tag functions in a semi-active mode using battery power. The sensor-tag is also employed with a rewritable nonvolatile memory bank formed by Ferroelectric RAM (FeRAM) and an on-chip temperature sensor.

---

*Bio-thermo is already commercially available in the United Kingdom, Japan, and Philippines.

There are also commercial semi-active sensor-tags that have been released. The German firm KSW-Microtec has produced the first semi-active RFID sensor-tag with an integrated sensor called VarioSens [44]. The VarioSens sensor-tag operates at 13.56 MHz and it is compliant with the ISO 15693 standard [33]. VarioSens is an upgraded version of an active sensor-tag produced by KSW-Microtec called Tempsens. It has 1024 bytes memory and, thus, it can hold 720 temperature readings versus 292 bytes and 64 readings in TempSens. While TempSens only offers password protection, VarioSens provides three levels of security. In VarioSens permissions for reading, writing, and erasing data on the tag can be defined. Another important feature that VarioSens provides is the ability to monitor the battery levels and tell the amount of power left. Its operating temperature ranges from $-20°C$ to $50°C$, while its temperature accuracy is $1°C$.

Phase IV engineering Inc. has produced a CMOS device called SensIC RFID ASIC [56], which is able to measure and transmit temperature as well as the value of an external capacitive MEMS sensor. SemsIC RFID ASIC may operate both in passive or active mode. It operates at 134.2 kHz frequency and is compatible with ISO 14223 standard and ISO 11784/5. It is able to measure temperatures that range from $-40°C$ to $125°C$ while its temperature accuracy is $\pm0.2°C$. ThermAssureRF is another commercial semi-active temperature sensor-tag that was proposed by Evidencia [26].

18.4.1.2 Integrated Sensor-Tags with Extended Communicating Capabilities

Integrated sensor-tags that are able to communicate only with RFID readers can be considered as RFID tags with some additional sensing capabilities but with limited communicating capabilities. However, it is possible to integrate sensor nodes with RFID tags so that the integrated sensor-tags will be able to communicate with each other as well as with other wireless devices. Thus, this category includes the integrated sensor-tags that exceed the limitations of possible communication only with an RFID reader and are able to communicate with each other through a cooperative ad hoc network (Figure 18.2).

A research approach which integrates RFID tags with sensor nodes was proposed by Ruzzelli et al. [57]. The main goal of the approach was to add an on-demand wakeup capability on sensor nodes to reduce energy consumption and to eliminate idle listening in WSNs. The proposed approach is called RFID impulse and it is achieved by using an RFID tag attached to each sensor node that is also provided with an RFID reader capability. By attaching RFID tags to sensor nodes, it is possible to remotely wake up the microprocessor and to radio the receiving sensor nodes on demand. The RFID impulse technique can use either a passive or an active RFID tag.

A characteristic example of integrated sensor-tags that are able to communicate with each other is the commercially available iRFID tag [37], an active, intelligent Radio Frequency Identification Device, produced by Machine Talker, a constructor of RFID tags designed to serve as wireless network nodes. iRFID tags are active tags with integrated onboard sensors for measuring environmental conditions, such as temperature, light, vibration, or even battery levels. The tags operate at 900 MHz and communicate via a proprietary air-interface protocol. When iRFID tags are activated and they are in

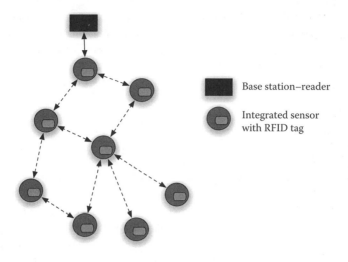

Figure 18.2   Integrated sensor-tags that form a cooperative ad hoc network.

proximity with each other, they automatically form a wireless mesh network and transfer sensor data among themselves. iRFID tags are able to communicate in a range of 200 m. Depending on how the devices are implemented they may communicate with other data systems via WiFi or wired networking protocols. iRFID tags have already been tested at several large oil refineries [8].

Integrated RFID sensor-tags that can communicate with each other were also developed and used in the CoBIs project [17], a European project which focuses on connecting business process management with the physical world. The integrated sensor-tags can collect data related to the conditions around them and transmit and share these data with each other. Their communication is performed using a proprietary peer-to-peer protocol. Each tag is employed with a movement sensor, a wireless transceiver as well as components for storing and processing rules related to business management. The data transmitted between the sensor-tags include apart from their unique ID, information sensed from the environmental conditions. The transmission range of each CoBIs RFID tag is 3 m. A trial of this project has been performed in BP by placing the integrated CoBIs RFID tags on 20 to 40 of BP's containers. Integrated sensor-tags are able, depending on the used business rules, to monitor the containers and trigger an alarm if the total volume of stored chemicals allowed in specific locations is violated or if potentially reactive chemicals are stored close to each other. The network of the integrated sensor-tags is able to communicate with a wider network via base stations.

Sensitech has also released an RFID-sensor integrated device called Temp Tale RF-enabled (TTRF) [68] temperature monitoring device. TTRF is built into an active RFID tag and it is composed of a temperature sensor, a radio chip, and an antenna. The sensor periodically records and stores temperatures, while the active tags transmit the sensed data to RFID readers. This data is collected centrally and it can be used to trigger alarms when, for instance, there is a danger of perishable goods to go off because of too high or too low temperatures. This integrated sensor-tag has battery power and a microprocessor and it is able to operate within an RF mesh networking environment. It operates in the 915 or the 868 MHz ISM band and the temperature measurements range from −30°C to 70°C.

Another type of integrated RFID sensor-tags were developed by Aeroscout [2]. More precisely, Aeroscout has produced a WiFi-based active RFID tag that is employed with a motion sensor and it has an optional built-in temperature sensor which is able to sense environmental temperature and trigger alarms based on reaching a configurable threshold. The tags operate at 2.45 GHz and transmit standard WiFi messages that can be transmitted to wireless access points (802.11 b/g). Aeroscout T3 tags (the latest version) have a ten year battery life, a read range of 100 m, 1 byte memory, and the temperature measurements range from 0°C to 100°C.

An enhanced type of RFID tags, referred as "multi-hop" tags, have been developed in NTT lab [62]. These tags are able not only to transmit but also to relay and read data. They can be configured as "reader" or replay devices. They are battery powered, operate in the 429 MHz band, and their range is less than 1 km. This special type of tags were initially developed for repelling monkeys or other intruders and preventing them from vandalizing farms and disturbing domestic animals. To achieve this, RFID tags are attached to monkeys or other disturbers. When these disturbers attempt to approach the farm, they are detected by RFID readers and residents are notified by e-mails. At the same time, light or sound alarms may be used to scare the disturbers.

18.4.2 Integrating RFID Readers with Wireless Sensor Nodes

Another possible strategy of integrating RFID systems with WSNs is by integrating RFID readers with sensor nodes. In this integration scenario, the existence of three types of devices is assumed: the integrated RFID readers/sensor nodes, simple RFID tags, and the sink or base station. This type of integration was first introduced by Zhang and Wang [75]. They called the integrated RFID reader/sensor node "a smart node." The integrated smart nodes can be considered as sensor nodes that can be used as RFID readers extending their sensing capabilities. Smart nodes are able to relay information and to be configured as relay nodes of a WSN. They are able to communicate with each other by creating an ad hoc communication network. The integrated RFID reader/sensor node is able to function as a router and to pass messages to the right destination. The smart nodes are responsible for collecting data from simple RFID tags in their range and communicate with each other to relay data to the sink/base station where all the data is collected and processed by a human. The architecture of this integrated network, illustrated in Figure 18.3, is similar to the hierarchical clustering-based two-tiered WSN.

This type of integration strategy gives new perspectives in likely applications. The limitations of traditional RFID readers including their passive operation, their serious mobility issues because of their big volume, and the position of their antennas limit their potential applications. The integrated smart node is smaller, less expensive, and easier to be deployed. However, this strategy of integration presents also some important

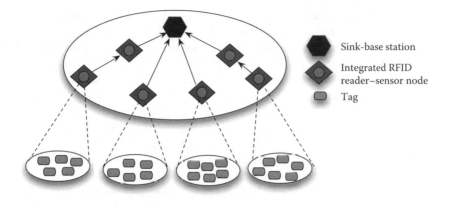

Figure 18.3   Integrating RFID readers with wireless sensor nodes.

disadvantages, because it is characterized by many-to-one traffic patterns and it presents some problems related to energy imbalance among smart nodes. Yang et al. [74] identified that in this type of integration because smart nodes have a fixed transmission range, the amount of traffic that is required to be forwarded will increase considerably as the distance to the base station becomes shorter. Subsequently, smart nodes that are closer to the base station will run out of battery early and areas of the network will remain unmonitored. Yang et al. [74] studied this type of integration and proposed a strategy that can be used to balance the energy consumption of the network and lengthen its lifetime.

Yang et al.'s proposal for balancing the load among the readers is based on adding more readers in the area near the sink. However, by adding more readers in the network, the cost of the network would be increased and more collisions would be caused. These disadvantages are outweighed by the increase of the network's survivability. More precisely, Yang et al. [74] studied how many nodes should be added in the neighborhood of the sink to get the best trade-off. Furthermore, they proposed a node distribution strategy to achieve a balanced energy reduction and to maximize the lifetime of sensor nodes. They showed that this strategy will substantially improve the network's lifetime.

Another approach for integrating an RFID system into a sensor network was proposed by Englund and Wallin [24]. The proposed system is able to collect data from RFID tags spread over a large area. More precisely, they have focused on the deployment of a system where RFID tags could be read from distances that overcome the range of ordinary RFID readers. This is achieved by connecting each RFID reader with an RF transceiver. Thus, information can be forwarded to and from the reader in distances of 100–200 m. Thus, a whole network of nodes is created, which are able to function as routers and forward messages to the right destination. Each node consists of an RF reader and an RF transceiver. To make the nodes functional a microcontroller is used to coordinate the different components in each node. More precisely, each node in the network consists of a microcontroller, an RF transceiver, an RF antenna, an RFID reader, an RFID antenna, and a battery. The MICA2 platform [20] from Crossbow technologies was used for the deployment of this project.

Integrated RFID readers/sensor nodes have also been produced commercially. The SkyeRead M1-mini [69] is an RFID reader produced by SkyeTek. It has a diameter of 1 in. and a thickness of 0.1 in. [18]. The reader's small dimensions make it suitable for a range of size-sensitive mobile RFID applications. The M1-mini has battery life that can last over two weeks of operation and offers a read rate of 20 tags per second. It operates at 13.56 MHz frequency and it can read and write to EPC tags and smart labels as well as to tags complying with the ISO 15693, ISO 14443, and ISO 18000 standards. Furthermore, the SkyeRead M1-Mini RFID reader can be connected directly with the Crossbow Mica2Dot [48] sensor mode resulting in an integrated RFID reader/sensor node. Another commercial RFID reader developed by Alien Technology is ALR-9770 [4] which is "equipped" with up to four antenna sets and it is able to communicate via the 802.11b/g standard.

A third commercial solution for integrating RFID readers with sensor networks was proposed by Gentag [27]. Gentag, an IP development company, issued a patent for adding sensor networks to RFID readers in mobile phones, laptops, PDAs, and other wireless devices. The developed patent is the base technology that will allow consumers to use cell phones to read almost any type of RFID tag.

18.4.3 Mix Architecture

In the mix architecture RFID tags and sensor nodes are physically distinct devices but they coexist in an an integrated network and they work independently. The main advantage of such a mixed architecture is the fact that there is no need to design a hardware integrated device. However, there is the possibility of communication interference between the RFID tags/readers and sensor nodes because in that case they are all physically disctinct devices. The procedures that should be followed to avoid this interference may cause additional overhead.

Initially, the mix architecture was discussed by Zhang and Wang [75]. According to Zhang and Wang [75] an integrated RFID-sensor network that follows the mix architecture consists of three types of devices: the smart stations, the normal RFID tags, and the normal sensor nodes (Figure 18.4). A smart station is a special device which is composed of an RFID reader, a microprocessor, and a network interface. Smart stations do not present power constraints and they are able to aggregate information from RFID tags and sensor nodes and to transmit them to a local host or to a remote LAN.

Information from RFID tags and sensor nodes can be transmitted to the base station. Because smart stations do not face power constraints, the traditional Internet protocol architecture can also be deployed. Thus, smart stations are able to perform not only data processing but also routing protocols and transport protocols such as TCP. A communication protocol that can be used in such a heterogeneous environment is the 802.11/WiFi technology.

An integration framework of RFID and WSNs that follows the mix architecture is SARIF [13] proposed by Cho et al. According to this framework the integrated system is composed of an integration server, RFID networks, and a WSN. The integration server

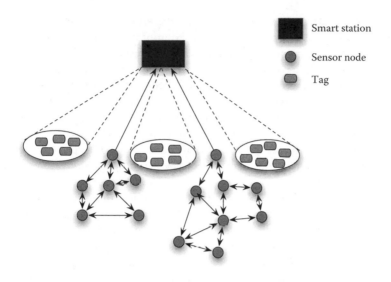

Figure 18.4 Mix architecture of RFID tags and sensor nodes.

is an important component that manages the main tasks of the WSN and the RFID network. The RFID network is composed of an information server, RFID readers, and tags while the sensor network is composed of a gateway and sensor nodes. The information server of the RFID network communicates with the integration server and it transmits information related to the RFID tags. The integration server, depending on the information it has received from the information server, initiates a task in the sensor network. The integration server may also access the RFID network and assign tasks to it. The authors have also evaluated the proposed integration framework by developing a prototype and demonstrated that SARIF can achieve energy efficiency through load balancing.

Some commercial solutions able to support the integration of RFID and WSNs according to the mix architecture have already been proposed. For instance, RFID Anywhere [58] is a commercial platform which includes rich features, broad hardware, standard and protocol support, and architecture flexibility that developers and integrators need to produce integrated RFID/sensor applications. More precisely, RFID Anywhere allows the efficient integration of RFID tags/readers, environmental sensors, barcodes, and mobile devices.

18.5 Integration Scenarios for RFIDs and WSNs in Various Applications

18.5.1 Healthcare Applications

RFID and WSNs have separately been used in medical healthcare applications such as emergency care, stroke rehabilitation [52], dental implants [34], and patient and staff tracking in hospitals. However, by integrating these two technologies their potentials will be extended. The treatment quality can be improved, because patient conditions would be monitored continuously and doctors can be easier notified in cases of a sudden deterioration of a patient's condition. The patient's location can be tracked using RFIDs, while their condition can be monitored using sensors. Furthermore, medication errors such as outdated treatments orders, inaccurate medical records, and increased costs can be avoided with the use of an integrated RFID-sensor network [49]. Possible applications of integrated RFID and WSNs in the healthcare sector include sensing temperature, measuring blood pressure, heartbeat rate, or pH value. Among the proposed approaches of integrating RFID and sensor networks in healthcare applications we present some indicative applications.

IMEC-Netherland (IMEC-NL) a Dutch research institute has built prototypes of human-monitoring systems using active RFID tags, which are integrated with sensors [71]. The integrated sensor-tags are used to record and transmit data about a patient's vital signs. The human monitoring system is mainly used for investigating conditions such as epilepsy and sleep apnea. Until now, a hospitalized patient suffering epilepsy is monitored via electrodes, which are attached to his face and scalp and they are connected with wires into a box. The box collects data about the patients' facial and brain activity. This data is analyzed and used to track a patient's condition. Similarly, a patient that suffers from apnea is monitored via electrodes attached to his face to measure eye and jaw muscle movement and brain activity. The IMEC research institute investigated the possibility of providing the patients the ability to move and even the option to be monitored while being at home by making these procedures wireless. The researchers of the institute use sensors to monitor the patients' brain activity. If the sensors detect unexpected brain activity they transmit an alarm to an RF interrogator 10 m away. The prototype has been tested at the Universitaire Ziekenhuizen Leuven (UZ) in Belgium.

University hospital of Ghent in Belgium implemented an RFID-based real time locating system (RTLS) to provide nurses and other caregivers with a patient's location in the event of an emergency [6]. The implemented integrated RFID-sensor network detects when a patient is having cardiac distress and sends to the caregivers an alert indicating the patient's location. In the proposed prototype Aero Scout T2 active Wi-Fi tags are used, which transmit the tags' unique IDs to the hospitals Wi-Fi network.

An integrated scheme of RFID and WSNs for in-home medication monitoring and elder healthcare was developed by Ho et al. [30]. The goal of the system is to monitor the quantity of medicine required by elder people and to assist them in taking the accurate amount of medicine as an extension of the Caregiver's prototype by Intel labs. The system is composed of an HF RFID reader, a UHF RFID reader, a weight scale, a base station, and three modes. HF RFID tags are placed on each medicine bottle in order to identify each bottle, while the HF RFID reader is used to monitor the location of the medicine bottles within its range. The movement or replacement of a bottle is detected by regular reads on a readers' range. By using a weight scale, in combination with RFID tags embedded on medicine bottles, it is possible to determine which medicine and how much of the medicine was used by the patient. Each patient also bears an RFID tag and the patient is notified by the associated RFID reader to take the required medicine (via sound or light alarm). The proposed prototype is an extension of the approach proposed by Intel Labs [29], where an HF RFID reader and some tags are used along with two sensor nodes (motes) to monitor a patient's intake of medicines.

Intel Research Seattle and University of Washington also proposed a smart home prototype system, called "Caregiver's Assistant and CareNet Display" [22,67] that is able to detect, monitor, and record the daily living activities of elder people by collecting data through postage stamp-size wireless RFID tags attached to household objects. Information regarding the objects and the time that they were touched, are collected and transmitted using a WSN. Statistical methods are used on this data to detect high-level activities and to fill out entries from a state-mandated activities of daily living form (ADL). The goal of the proposed prototype system is to help elder people manage their everyday activities without the need for an around-the-clock caregiver. Thus, caregivers may focus on the quality of care of elder people rather than on performing tedious tasks.

Kim et al. [41] have deployed integrated sensor-tags deployed on a blood monitoring and management system in hospitals. The proposed system can be used to continuously monitor the temperature of the blood bank refrigerator as well as to track the location of a blood bag. Moreover, the proposed system can be used to prevent patient-blood mismatching as well as to perform temperature monitoring of blood bags. This is achieved using a sensor network in blood banks and RFID sensor-tags that are attached to blood bags. Crossbow Technology MTS420CA sensors were used to record the temperature of the refrigerator, blood bank room, and the temperature of the blood bag. The TempSens KSW RFID sensor-tags were used to tag the blood bags, which have enough memory to save 64 measurements of temperature data.

A passive wireless RFID sensor, which can be implanted in the patient's esophagus wall to detect impedance changes associated with gastroeophageal reflux was developed by Ativanichayaphong et al. [5]. Integrated sensor-tags have also been used to monitor patients' dental health. More precisely Instrumentel [31] in cooperation with a U.K. dental school has attached pH sensor-tags in dentures to monitor the levels of food's acidity or alkalinity in a patient's mouth [16].

18.5.2 Integrated RFID and WSNs in Supply Chain Management

RFID systems and WSNs have extensively been used in the supply chain for inventory control, product tracking, and asset monitoring, while sensor networks are used for space and environment monitoring. Nevertheless, the integration of RFID systems with sensor networks open new directions. RFID systems are able to accurately identify objects, but often sometimes provide unreliable information concerning the location of an object. Sensors on the other hand, present many advantages in recognizing the location of an object but they are unable to identify it. The efficient integration of RFID and WSNs offers great advantages in accurate location tracking [49]. Furthermore, the integration of RFID and WSNs allows the condition monitoring of products and the detection of "dangerous" environmental conditions, such as a high temperature or humidity for fragile, sensitive or valuable products. Furthermore, integrated RFID and WSN technologies allow for the automatic condition and tamper detection from a distance without direct and manual inspection in a convenient, inexpensive and less error prone way. There is a great range of possible integration approaches of RFID and WSNs in the supply chain management. We list some of the most important ones.

An integrated scheme of RFID and WSNs for automated asset tracking was proposed by McKelvin et al. [47]. The proposed scheme integrates wireless sensor nodes and RFID readers. More precisely, a wireless sensor node is connected to a host (i.e., ordinary PC) in which an inventory of tagged products is maintained in a database. Another wireless sensor node is integrated in an RFID reader (reader node). A user of the host node is able to perform queries on the database which are later relayed to the reader node via the WSN. The query is then transmitted to the RFID reader and the required data is retrieved. The communication is bidirectional. Thus, data can be sent from the reader to the host device using the same interface.

PROMISE [53] was a European project that focused on monitoring and tracking a product through its life cycle using smart embedded devices. One of the goals of this project was the use of intergraded RFID tags and sensors to retrieve information and subsequently to improve the way that products are made, used, and recycled. Furthermore, some companies such as InfraTab [35] and CliniSense [19] have focused on developing integrated sensor-tags to monitor the freshness of perishable goods. Their focus was the combination of temperature and time duration to estimate possible bacterial growth.

18.5.3 Other Applications

Apart from applications in the field of healthcare and supply chain management, numerous other real-world scenarios of integrated RFID and WSNs exist. In this section, we present application scenarios in diverse areas of science and engineering including fire detection, monitoring shipping containers, the condition of weapons in battlefields as well as managing cattle. In all cases integrated RFID and WSNs facilitate procedures and provide efficiency. However, in all cases different technical challenges existed that had to be overcome to achieve smooth operation of the integrated system.

An application of integrated RFID and sensor networks to notify firefighters [7] in case of a fire's ignition, was introduced by a wireless communication company in Melbourne, called Telexpath. The proposed scheme is based on the integration of RFID chips that operate at 433 MHz and use a proprietary air interface protocol and wireless thermal sensors. In case a thermal sensor senses temperatures within 2° of a predetermined setting, it sends its unique ID number to an interrogator. The next step is the cross check of the tag's ID number and a notification is sent to a person's cell phone. Thus, the firefighters are notified very quickly and they are able to respond to a fire incident faster and more efficiently. This approach can also be extended for alerting people not only in case of fires but also in case of other emergencies or equipment failures.

A similar approach that can be applied for early fire detections was proposed by researchers at the University of California in Berkeley [23]. This approach is based on the development of a GPS-enabled wireless sensor, called Firebug, which collects real-time data from sensors about approaching fires and transmits these data via an RF tag which contains a Chipcon mote. Thus, firefighters have access to information related to the speed and the intensity of an attack and are able to respond appropriately.

Siemens IT Solutions and Services [73] have conducted a thorough research on the use of integrated RFID and sensor network technologies used to monitor shipping containers. They have proved that it is possible to achieve continuous monitoring of shipping containers from their departure until they reach their final destination. The proposed approach uses active RFID transponders and sensors. The collected data via RFID are transmitted over GSM and GPRS telecommunication networks to a satellite telecommunication service. Information that is transmitted may include the temperatures or the location of the containers. The advantages of continuous monitoring of shipping containers provide significant advantages regarding the placement of order. For instance, in case an alarm is triggered, while the goods are still on the sea, the customer can be notified and perform a new order.

BP has also adopted the integration of RFID and WSNs to manage efficiently the chemical inventory and increase the stock visibility at a petrochemical plant in the United Kingdom. This project involves the communication and cooperation of integrated RFID sensor-tags and data sharing between each other. In the integrated RFID-sensor network each RFID sensor-tag collects data and transmits them to any other node in the network. The employed RFID sensor-tags are designed to monitor the conditions around them and to provide alerts according to predetermined rules. Each tag is employed with a movement sensor, a wireless transceiver, and it uses a proprietary peer-to-peer protocol to communicate with each other. Each sensor-tag transmits except from its unique ID, details about environmental conditions to all other nodes within 3 m range.

A mixed heterogenous architecture of RFID systems and sensors to track multiple people was proposed by Mori et al. [50]. The proposed system integrates distributed floor pressure sensors and a complete RFID system. By using floor pressure sensors or pyroelectric sensors it is possible to detect an area that indicate the presence of a person. RFIDs are used to identify people, because sensors fail to track each person when their existing areas overlap. Experimental results prove that the proposed approach is able to track multiple people in daily life situations effectively.

Integrated RFID-sensor networks have also been embraced by the U.S. Navy [65] to monitor the condition of valuable aircraft parts in storage. The U.S. Navy in cooperation with the George Institute of Technology in Atlanta employed an RFID system that doesn't need RFID readers to scan each tag, but instead it uses battery-powered sensor-tags that are able to communicate with each other and transmit information between themselves. The data is transmitted from the final transponder to a single reader. The RFID sensor-tags are able to measure temperature, humidity, and air pressure and to communicate with other RFID sensor-tags. The RFID sensor-tags are self-powered and they have a two year battery life. The tags transmit information only after being interrogated by a base station, which needs to send a specific security code as well. Thus, the possible leakage of information to unauthorized enemies, that could reveal sensitive information concerning the location of the ship or other sensitive inventory information, is prevented.

HP also uses integrated RFID-sensor networks in two prototype applications, namely the smartLOCUS and the smartRack [51]. SmartLOCUS controls and monitors a sensor network of cameras and readers. A sensor overlay network connects RFID readers to inexpensive video cameras via an 802.11b network. The cameras provide information concerning, for instance, the movement of items in a warehouse.

In the smartRACK project thermal sensors and high-frequency RFID readers are used to monitor the temperature of servers' cabinets. Each shelf in the rack is equipped with an RFID reader designed to read high-frequency signals from servers with special chips storing the machine's unique ID number. An RFID reader is employed on each server cabinet and 14 antennas are employed (configured) on the server door to read the 13.56 MHz RFID tag attached to each of the servers in the cabinet. Five to six thermal sensors are wired to the cabinet door and linked to the reader. The thermal sensors monitor the temperature of the servers located in the servers' cabinets. HF RFID readers and sensors are networked and the collected data is used to produce two-dimensional graphics in real time that represent the temperature profile of each cabinet. This application is essential for companies that use a large number of servers. Abnormally high temperatures would send an alarm and notify the personnel about the possible problem.

Another application of integrated RFID and WSNs is their use in battlefields [36]. Recently, a prototype was produced that can be embedded on weapons to track how many times war fighters fire their weapons during a battle. Thus, the Army can assess when each weapon has reached the end of its lifetime. The prototype is based on integrating RFID tags with sensors. This is achieved by placing on each weapon a piezo-electric sensor, a tiny processor, and an RFID tag. The piezoelectric sensor is able to determine when the weapon has been fired by sensing the recoil. The tiny processor has limited memory and it is able to record and store the output from the sensor. Finally, the RFID tag is used to transfer the data to an RFID reader. The prototype is able to track the number and times of rounds fired as well as to deduce characteristics from the intensity of the firing, such as acceleration, heat, and resonant electromagnetic frequencies. All these characteristics can help on estimating the useful time of the weapon.

Integrated RFID and WSNs have also been used to sense the level of oil in tanks [8], to monitor the vibration levels of large cooling fans used in the oil-refining processes as well as to monitor and facilitate the mixing of different crude oil levels during the formation of gasoline. All three uses are pilot projects performed by Sense-Comm using the i-Sense Talker devices which are an enhanced version of an iRFID tag [37] containing sensors different from those normally used in iRFID tags. Integrated RFID sensor-tags will facilitate considerably the above-mentioned processes in oil refineries. For instance, traditionally to monitor the fluid levels in tanks a mechanical gauge is used which should be checked manually or by dipping a wired sensor into the tank. However, neither solution is optimal considering that these areas are hazardous and it is expensive to perform these procedures either with wires or with manual controls.

Another application of integrated RFID and WSNs is the easy management of cattle. Zigbeef [70] is marketing hardware that can be used by ranchers to manage their herds more easier. Zigbeef uses integrated RFID sensor-tags, which can sense animal movements and transmit these data to an RFID reader. These data could also be used in rodeos where the sensor-tags can be attached to rodeo bulls. Thus, using a mobile reader, the audience will be able to observe data regarding how much the bull was actually bucking the rider before throwing him. ZigBeef is also planning to extend the present system with mesh capabilities so that ZigBeef tags will be able to transmit data from one to another. Thus, the read range will be extended depending on how many cattle are spread throughout the area and, finally, the data will be sent to a reader.

Other applications of integrated RFID-sensor networks is the development of an RFID-based memory assistant [59], which is able to notify a person when leaving the house without carrying all the necessary items. All the essential items are tagged, while a pressure sensor detects her presence at the house's front door and activates an RFID reader if something is missing. Another project included the use of temperature sensor-tags attached in a cow's stomach [60] to predict its child birth. The system is based on measuring the cow's body temperature and the indication of an abrupt decent in a cow's body temperature 24 hours before its calf birth.

The Japanese Ministry of Internal Affairs and Communication (MIC) has also developed a system that is based on integrated sensor-tags for gathering information about a disaster area [61]. The sensor-tags are sprinkled from helicopters and important information is collected from the disaster area including the possible existence of human disaster victims.

An integration scenario of RFID and WSNs for a tour group system was proposed by Chen et al. [12]. According to this integration scenario the guide of each group holds a badge which emits 4 kHz signals. Each group member carries a ticket with a passive RFID tag in which is stored in the group's ID. Sensor nodes are attached to direction boards to display simple guiding directions. The signals from the leader's badge are recognized by the wireless sensor nodes and thus, the leader's location can be tracked. Some nodes in the WSN are selected as help centers and they are connected to a laptop and an RFID reader.

Integrated RFID and WSNs is also a research direction for Aerospace in Auto-ID Labs [54]. By using such systems on aircraft parts, the configuration of the aircraft will be known precisely. It is expected that by doing so, the aircraft's maintenance and repair will be considerably facilitated. Moreover, the delivery of customer-related services such as ticketing, baggage-processing, and meals are also being considered as research directions for using integrated RFID and WSNs.

18.6 Conclusions and Open Issues

Undoubtedly, the integration of RFID and WSNs is an imminent step that will lead to a high level of synergy and more technological advances. This integration will give us the advantage not only to reveal an item's location and identity but also its current state. These integrated networks will extend traditional RFID systems and will give us an important advantage in controlling environments and industrial processes. However, more effort is necessary to achieve efficient integration of RFIDs and WSNs.

To achieve a broader adoption of integrated RFID and WSNs, it is important to resolve some open issues and challenges such as the reduction of possible interference in large integrated RFID networks and WSNs, because the greater the number of wireless devices in a network, the larger the potential of a possible interference. Thus, it is important to define a good collision-free schedule for both WSNs and RFIDs.

An important step toward the wide integration of RFID and WSNs would be the deployment of tools, methods, and approaches that would be general enough to be used in a wide range of applications. However, it is important while deploying these tools methods and standards to take into consideration the restrictions concerning the available resources. Moreover, the developed approaches should be general enough so that their evolvement into standards shall be a possible option.

An interesting initiative toward this direction was the IntelliSense RFID project [72] part of the Nordic research program NORDITE. IntelliSense's goal was the development of multi-protocol RFID devices that are able to operate at different frequency bands with different communication protocols. Thus, dual-band sensor-tags can be developed, that could be used in multiple applications. For instance, a multi-band sensor-tag may be used by UHF RFID readers for a logistics application and by consumers to retrieve data stored in its memory through an HF RFID reader integrated in a cell phone.

Additionally, the creation of simulators that combine RFID and WSN technologies would be significant in the wide deployment and study of integrated RFID and WSNs. Furthermore, considering that the market is cost driven the success of integrating RFID and WSNs will highly depend on the choices of the lowest cost material and the simplest and most effective manufacturing processes.

Acknowledgments

We would like to thank the reviewers for providing valuable feedback and Christos Dimitrakakis for additional proofreading. This work was partially supported by the Netherlands Organization for Scientific Research (NWO) under the RUBICON "Intrusion Detection in Ubiquitous Computing Technologies" grant awarded to Aikaterini Mitrokotsa.

References

1. Adage Solutions, Products, http://www.adage.se/Adage/Page____240.aspx, Accessed July 2009.
2. Aeroscout Enterprise Visibility Solutions, AeroScout T3 tags, http://www.rfidglobal.org/product/2007_7/aeroscout_t3_tags.html, Accessed July 2009.
3. Alien Technology, Alien Technology ALB-2484 tag, http://www.rfidsolutions online.com/ecommcenters/alien.html, Accessed July 2009.
4. Alr-9770 Series Multi-Protocol RFID Readers, 2005, http://www.directouchpos.com/products/downloads/alr9774.pdf, Accessed July 2009.
5. T. Ativanichayaphong, J. Wang, W.-D. Huang, S. Rao, H.F. Tibbals, S.-J. Tang, S.J. Spechler, H. Stephanou, and J.-C. Chiao, Development of an implanted RFID impedance sensor for detecting gastroesophageal reflux, in *Proceedings of IEEE International Conference on RFID*, Grapevine, TX, pp. 127–133, March 26–28, 2007.
6. B. Bacheldor, Belgium hospital combines RFID, sensor to monitor heart patients, RFID Journal, March 2007, http://www.rfidjournal.com/article/articleview/3120/1/1, Accessed July 2009.

7. B. Bacheldor, Fighting fires with RFID and wireless sensors, *RFID Journal*, November 2006, http://www.rfidjournal.com/article/articleview/2799/1/1, Accessed July 2009.

8. B. Bacheldor, Oil refineries to test sensor tags, *RFID Journal*, January 2007, http://www.rfidjournal.com/article/articleview/3006/1/1, Accessed July 2009.

9. Life Chip. Bio-Thermo, http://lifechip.com.au/products.php?id=4, Accessed July 2009.

10. Bisa Technologies, 2.4 GHz temperature sensor tag (24TAG02T), http://bisatech.com/product.asp?pid=2&zid=3&do=view&id=49, Accessed July 2009.

11. Bisa Technologies, 2.4 GHz vibration sensor tag (24TAG02V), http://bisatech.com/product.asp?pid=2&zid=3&do=view&id=50, Accessed July 2009.

12. P.Y. Chen, W.T. Chen, C.H. Wu, Y.C. Tseng, and C.F. Huang, A group tour guide system with RFIDs and wireless sensor networks, in *Proceedings of the 6th International Conference on Information Processing in Sensor Networks*, Cambridge, MA, pp. 561–562, April 25–27, 2007.

13. J. Cho, Y. Shim, T. Kwon, and Y. Choi, SARIF: A novel framework for integrating wireless sensor and RFID networks, *IEEE Wireless Communications*, 14(6), 50–56, December 2007.

14. N. Cho, S.-J. Song, S. Kim, S. Kimn, and H.-J. Yoo, A 5.1– W UHF RFID tag chip integrated with sensors for wireless environmental monitoring, in *Proceedings of the 31st IEEE European Solid-State Circuits Conference (ESSCIRC'05)*, Grenoble, France, pp. 279–282, 2005.

15. R. Clauberg, RFID and sensor networks, in *Proceedings of RFID Workshop*, University of St. Gallen, Switzerland, September 2004.

16. J. Collins, Passive tag powers sensor networks, switches, *RFID Journal*, April 2005, http://www.rfidjournal.com/article/view/1520/1/1, Accessed July 2009.

17. J. Collins, BP tests RFID sensor network at U.K. plant, *RFID Journal*, June 2006, http://www.rfidjournal.com/article/view/2443/1/1, Accessed July 2009.

18. J. Collins, ShyeTek shrinks the RFID reader, *RFID Journal*, January 2004, http://www.rfidjournal.com/article/articleview/778/1/1, Accessed July 2009.

19. J. Collins, Sensing a product's shelf life, *RFID Journal*, April 2005, http://www.rfidjournal.com/article/view/1539/1/1, Accessed July 2009.

20. Crossbow, Mica2 wireless platform, http://www.xbow.com/Products/product details.aspx?sid=174, Accessed July 2009.

21. H. Deng, M. Varanasi, K. Swigger, O. Garcia, R. Ogan, and E. Kougianos, Design of Sensor-Embedded Radio Frequency Identification (SE-RFID) systems, in *Proceedings of IEEE International Conference on Mechatronics and automation*, Henan, China, pp. 792–796, June 2006.

22. E. Dishman, Inventing wellness systems for aging in place, *IEEE Computer Magazine*, 37(5), 34–41, May 2004.

23. D.M. Doolin and N. Sitar, Wireless sensors for wild re-monitoring, in *Proceedings of SPIE Symposium on Smart Structures & Materials NDE 2005*, San Diego, CA, March 6–10, 2005.

24. C. Englund and H. Wallin, RFID in wireless sensor networks, Master thesis, Communication Systems Group, Department of Signals and Systems, Chalmers University of Technology, Goteborg, Sweden, April 2004.

25. EPC Global, Class 1 Generation 2 UHF air interface protocol standard "Gen 2" http://www.epcglobalinc.org/standards/uhfc1g2/uhfc1g2_1_1_0-standard-20071017.pdf, Accessed July 2009.

26. Evidencia, ThermAssureRF wireless temperature recorder, http://www.evidencia.biz/products/acti-tag.htm, Accessed July 2009.

27. R.B. Ferguson, Gentag patent adds RFID sensor network feature to mobile devices, December 2006, http://www.eweek.com/c/a/Mobile-and-Wireless/Gentag-Patent-Adds-RFID-Sensor-Network-Feature-to-Mobile-Devices, Accessed July 2009.

28. A. Ferrer-Vidal, A. Rida, S. Basat, L. Yang, and M.M. Tenzeris, Integration of sensors and RFID's on ultra-low-cost paper-based substrates for wireless sensor networks applications, in *Proceedings of the 2nd IEEE Workshop on Wireless Mesh Networks (WiMesh 2006)*, Reston, VA, pp. 126–128, September 25, 2006.

29. K. Fishky and M. Wang, A flexible, low-overhead ubiquitous system for medication monitoring, Intel Research Technical Report IRS-TR-03-011, October 2003.

30. L. Ho, M. Moh, Z. Walker, T. Hamada, and C.-F. Su, A prototype on RFID and sensor networks for elder healthcare: Progress report, in *Proceedings of the 2005 ACM SIGCOMM Workshop on Experimental approaches to Wireless Network Design and Analysis, Applications, Technologies, Architectures, and Protocols for Computer Communication*, Philadelphia, PA, pp. 70–73, 2005.

31. Instrumentel Telemetric Technologies, Tag sensor: ICT tag sensor, http://www.instrumentel.com/specs/Tag%20INTRA%20SENSE.pdf, Accessed July 2009.

32. International Organization for Standardization (ISO), ISO 11785: 1996—Radio frequency identification of animals, http://www.iso.org/iso/catalogue_detail?csnumber=19982, Accessed July 2009.

33. International Organization for Standardization (ISO), ISO 15693-2: 2006, Identification cards—Contactless integrated circuit cards—Vicinity cards—Part 2: Air interface and initialization, http://www.iso.org/iso/iso_catalogue/catalogue_tc/catalogue_detail.htm?csnumber=39695, Accessed July 2009.

34. C. Ilic, Using tags to make teeth, *RFID Journal*, October 2004, http://www.rfidjournal.com/article/view/1206/1/1, Accessed July 2009.

35. InfraTab, Monitor, track and trace perishables, http://www.infratab.com, Accessed July 2009

36. J. Jackson, Ready, aim, record: Army's prototype system uses RFID tags to track weapons use, *GCN government Computer News*, May 2008, http://www.gcn.com/Articles/2008/05/01/Ready-aim-record.aspx, Accessed July 2009.

37. Machine Talker, iRFID—Intelligent radio frequency identification device, *Q3 2006, Preliminary Specification*, http://www.machinetalker.com/pdf/iRFID-Q3-2006-Spec.pdf, Accessed July 2009.

38. Microsensys, Telid 210 passive sensor TAGs, http://www.microsensys.de/products/sensors/TELID210/TELID210.html, Accessed July 2009.

39. Microsensys, Telid 310 passive sensor TAGs, http://www.microsensys.de/products/sensors/TELID310/TELID310.html, Accessed July 2009.

40. S. Kim, J.-H. Cho, H.-S. Kim, H. Kim, H.-B. Kang, and S.-K. Hong, An EPC Gen 2 compatible passive/semi-active UHF RFID transponder with embedded feram and temperature sensor, in *Proceedings of IEEE Asian Solid-State Circuits Conference (ASSCC'07)*, Jeju, Korea, pp. 135–138, November 12–14, 2007.

41. S. J. Kim, S. K. Yoo, H. O. Kim, H. S. Bae, J. J. Park, K. J. Seo, and B. C. Chang, Smart blood bag management system in a hospital environment, *Personal Wireless Communications*, Springer, Berlin/Heidelberg, Vol. 4217/2006, pp. 506–517, September 2006.

42. D. Kiritsis, Ubiquitous product life-cycle management using product embedded information services, in *Proceedings of International Conference in Intelligent Maintenance Systems (IMS'2004)*, Arles, France, July 2004.

43. H. Kitayoshi and K. Sawaya, Long range passive RFID-tag for sensor networks, in *Proceedings of 62nd IEEE Vehicular Technology Conference (VTC'05)*, Dallas, TX, Vol. 4, pp. 2696–2706, September 25–28, 2005.

44. KSW Microtec, Active RFID—VarioSens, http://www.ksw-microtec.de/index.php? ILNK=Active_RFID_VarioSens&iL=2.

45. H. Liu, M. Bolic, A. Nayak, and I. Stojmenovie, Integration of RFID and wireless sensor networks, in *Proceedings of Sense IP 2007 Workshop at can SenSys*, Sydney, Australia, November 6–9, 2007.

46. A. Mason, A. Shaw, A.I. Al-Shamma'a, and T. Welsby, RFID and wireless sensor integration for intelligent tracking systems, in *Proceedings of 2nd GERI Annual Research Symposium GARS-2006*, Liverpool, U.K., June 2006.

47. M. L. McKelvin, M. L. Williams, and N. B. Berry, Integrated radio frequency identification and wireless sensor network architecture for automated inventory management and tracking applications, in *Proceedings of the 2005 Conference on Diversity in Computing*, Albuquerque, NM, pp. 44–47, 2005.

48. Mica2dot Series, http://www.willow.co.uk/html/mpr5x0-_mica2dot_series.html, Accessed July 2009.

49. J. Mitsugi, T. Inaba, B. Patkai, L. Theodorou, J. Sung, T. Sanchez Lopez, D. Kim, D. McFarlane, H. Hada, Y. Kawakita, K. Osaka, and O. Nakamura, Architecture development for sensor integration in the EPCGlobal network, *White Paper WP-SWNET-018, Auto-ID Labs*, July 2007.

50. T. Mori, Y. Suemasu, H. Noguchi, and T. Sato, Multiple people tracking by integrating distributed floor pressure sensors and RFID system, in *Proceedings of 2004 IEEE International Conference on Systems, Man and Cybernetics*, Vol. 6, pp. 5271–5278, The Hague, the Netherlands, October 10–13, 2004.

51. M. C. O'Connor, HP kicks off us RFID demo center, *RFID Journal*, October 2004, http://www.rfidjournal.com/article/articleview/1211/1/50, Accessed July 2009.

52. School of engineering and Harvard University Applied Sciences, Wireless sensor networks for medical care, September 2006, http://fiji.eecs.harvard.edu/CodeBlue, Accessed July 2009.

53. PABADIS, PABADIS based product oriented manufacturing systems for re-configurable enterprises, http://www.uni-magdeburg.de/iaf/cvs/pabadispromise/beschreibung_english.pdf.

54. B. Patkai and D. McFarlane, RFID-based sensor integration in aerospace Technical Report AEROID-CAM-009, Auto-ID Lab, University of Cambridge, U.K., November 2006.
55. Patent storm, U.S. Patent 7125382 – Embedded Bio-Sensor system, October 2006, http://www.patentstorm.us/patents/7125382/fulltext.html, Accessed July 2009.
56. Phase IV Engineering Inc., SensIc RFID ASIC, http://www.phaseivengr.com/Application_PDFs/61-100005-00_Rev_1_0_SensIC_RFID_ASIC.pdf, Accessed July 2009.
57. A.G. Ruzzelli, R. Jurdak, and G.M.P. O'Hare, On the RFID wake-up impulse for multi-hop sensor networks, in *Proceedings of 1st ACM Workshop on Convergence of RFID and Wireless Sensor Networks and their Applications (SenseID) at the Fifth ACM Conference on Embedded Networked Sensor Systems (ACM SenSys 2007)*, Sydney, Australia, November, 2007.
58. RFID anywhere overview, 2005, http://www.sybase.com/content/1034553/rfidanywhereoverview.pdf.
59. RFID in Japan, RFID-based alert when you leave home, March 2006, http://rfidinjapan.wordpress.com/2006/03/24/rfid-based-alert-when-you-leave-home/, Accessed July 2009.
60. RFID in Japan, RFID tags in cow's stomach predict child birth, March 2006, http://rfidinjapan.wordpress.com/2006/03/23/rfid-tags-in-cows-stomach-predict-child-birth/, Accessed July 2009.
61. RFID in Japan, Sprinkling RFID sensor tags from the sky, March 2006, http://rfidinjapan.wordpress.com/2006/03/20/sprinkling-rfid-sensor-tags-from-the-sky/, Accessed July 2009.
62. RFID in Japan, RFID system for repelling monkeys, June 2006, http://rfidinjapan.wordpress.com/2006/06/27/rfid-system-forrepelling-monkeys/, Accessed July 2009.
63. RFID in Japan, Battery-less sensor tags, http://rfidinjapan.wordpress.com/2006/04/20/battery-less-sensor-tags/, Accessed July 2009.
64. A. Rida, R. Vyas, S. Basat, A. Ferrer-Vidal, L. Yang, S.K. Bhattacharya, and M.M. Tentzeris, Paper-based ultra-low-cost integrated RFID tags for sensing and tracking applications, in *Proceedings of the 57th Electronic Components and Technology Conference (ECTC'07)*, Reno, NV, pp. 1977–1980, 2007.
65. M. Roberti, Navy revs up RFID sensors, *RFID Journal*, June 2004, http://www.rfidjournal.com/article/articleview/990/1, Accessed July 2009.
66. A.P. Sample, D.J. Yeager, P.S. Powledge, and J.R. Smith, Design of a passively-powered, programmable sensing platform for UHF RFID systems, in *Proceedings of 2007 IEEE International Conference on RFID*, Gaylord Texan Resort, pp. 149–156, March 26–28, 2007.
67. M. Philipose, S. Consolvo, T. Choudhurg, K. Fishkin, M. Perkowitz, I. Smith, D. Fox, H. Kautz, and D. Paterson, *Demonstration in the Sixth International Conference on Ubiquitous Computing (UbiComp'04)*, 7–10 September 2004, Nottingham, England.

68. Sensitech, ColdStream infrastructure: Integrated RF-enabled temperature monitoring infrastructure, Sensitech Cold Chain Visibility, http://www.sensitech.com/PDFs/applications/ColdStream_InfraStructure.pdf, Accessed July 2009.
69. SkyeTek, Inc., SkyeTek RFID Readers SkyeRead M1-Min, Product Reference Guide, 2004, http://www.ece.osu.edu/~cglee/ECE682Y/Doc/SkyeModule_M1-mini_Reference_Guide.pdf, Accessed July 2009.
70. C. Swedberg, ZigBeef offers ranchers a long-distance cattle head count, *RFID Journal*, http://www.rfidjournal.com/article/articleview/3935/1/1, Accessed July 2009.
71. C. Swedberg, Dutch researchers focus on RFID-based sensors for monitoring apnea, epilepsy, *RFID Journal*, December 2007, http://www.rfidjournal.com/article/articleview/3780/1/1, Accessed July 2009.
72. O. Vermesan, N. Pesonen, C. Rusu, A. Oja, P. Enoksson, and H. Rustad, IntelliSense RFID—An RFID platform for ambient intelligence with sensors integration capabilities, *ERCIM News*, 67, 40–41, October 2006.
73. R. Wessel, Cargo-tracking system combines RFID, sensors, GSM and satellite, *RFID Journal*, January 2008, http://www.rfidjournal.com/article/articleview/3870, Accessed July 2009.
74. G. Yang, M. Xiao, and C. Chen, A simple energy-balancing method in RFID sensor networks, in *Proceedings of 2007 IEEE International Workshop on Anti-Counterfeiting, Security, Identification*, Xiamen, China, pp. 306–310, April 16–18, 2007.
75. L. Zhang and Z. Wang, Integration of RFID into wireless sensor networks: Architectures, Opportunities, in *Proceedings of the Fifth International Conference on Grid and Cooperative Computing Workshops (GCCW'06)*, Changsha, China, pp. 463–469 October 2006.
76. S.-H. Zhou and N.-J. Wu, UHF RFID Front-end with magnetic sensor, in *Proceedings of the 8th International Conference on Solid-State and Integrated Circuit Technology*, Shanghai, China, pp. 554–556, October 2006.

Chapter 19

Integrated RFID and Sensor Networks for Smart Homes

Falko Dressler, Abdalkarim Awad, Sebastian Dengler, and Reinhard German

CONTENTS

The quick evolution of technology has enabled us to bring many things to reality in short periods of time. Thus, the vision of smart homes will become more feasible in the near future. Technologies such as wireless sensor networks (WSNs) and radio frequency identification (RFID) have attractive characteristics that make them great candidates to be engaged in this environment and they can greatly benefit from each other. The term "smart home" comprises various approaches engaged in living and working now and in the future. The objectives of the various approaches range from enhancing comfort in daily life to enabling a more independent life for elderly and handicapped people. The term "ubiquitous computing", coined by Mark Weiser in his essay, *The Computer for the 21st Century*, describes the ubiquity of computer and information technology. The task of "smart objects", implanted into everyday items, is to sense the immediate environment using various types of sensors, and to process this information. This functionality assigns a kind of artificial intelligence to common, well-known objects and enables comprehensive information-processing and interconnection of almost any kind of everyday object. The (preferably) transparent and hidden technology ranges from "wearable computers" and "smart clothes" to "intelligent" artificial replacements. It supports the user in almost every part of his life by extending his cognition and information-processing capacity and tries to compensate for certain handicaps. The challenge regarding smart homes, especially for supporting the elderly and handicapped, is to compensate for handicaps and support the individuals to give them a more independent life for as long as possible. In this chapter, a common architecture for smart-home environments is developed, mapped to an experimental setup, and finally evaluated. This architecture primarily consists of a sensor network that is functionally enhanced by mobile robots and passive RFID tags, which in turn complement the functionality of the sensor network.

19.1   Introduction

Sensor and actor networks (SANETs) consist of sensor and actuator nodes—usually communicating using wireless technology—that perform distributed sensing and actuation tasks. In recent years, the needs and interests in the capabilities of SANETs has increased. Obviously, SANETs are heterogeneous networks having widely differing sensor and actuator node characteristics [4,15]. The basic idea is that sensor nodes provide a communication infrastructure while gathering and distributing information about the physical environment [12]. They can, in a certain way, also be used for localization and navigation issues as is partially shown in [5,24]; however, this is not the focus of this chapter. Actuators are meant to take decisions based on information from the distributed sensor nodes or local systems. For example, they can offer management and maintenance services, such as repair tasks or energy refreshment, to the sensor network. In addition, actuator nodes are supposed to provide mobility, e.g., with mobile sensors, so as to improve the network through controlled mobility. We believe that in many cases more powerful actuators, such as mobile robots, are used to satisfy the needs of more complex tasks. These robots are autonomous machines that are capable of movement in a given environment [6,20]. A common class of robots are wheeled robots. They can take a variety of different tasks, as already mentioned, to improve and support the sensor network. We consider that, assistance of the sensor network by mobile robots, e.g., for maintenance and optimized deployment, and assistance of the robots by the sensor network, e.g., to provide location information, must be distinguished.

Conventionally, items are often tagged with a bar code that is used to identify the items. An alternative to bar code is the RFID tag. Basically, an RFID tag represents a simple chip with limited storage and computational capabilities and a transmitter for radio communication. Whereas, in general, active and passive RFID tags need to be distinguished, we concentrate on passive tags that do not need to carry a dedicated battery. Instead, an external energy source is used to wirelessly support the RFID tag with a limited amount of energy—just enough to transmit a short message. This message contains a unique number called electronic product code (EPC) stored in an EEPROM inside the small computer to the external RFID reader. Some RFID applications in smart-home environments and the associated security threats are presented in Ref. [14]. The RFID can be utilized to identify products, places, pets, and even people. One of the main advantages of the RFID tag is that the reading process does not require a line-of-sight position—in comparison to the bar code, which requires line-of-sight alignment. The only requirement is that the RFID tag must be within a certain range from the reader. This range strongly depends on the antenna of the reader. Using RFID technology, hundreds of RFID tags can be read simultaneously. One of the largest fields that RFID technology is being used in is supply chain management [30].

Other application scenarios for RFIDs that also focus on the application in smart homes include the tracking of frequently lost objects (FLOs) such as keys, activity monitoring (e.g., identifying people), and support for SANETs (e.g., for localization and untethered information transfer). All these applications become easier if they are RFID-enabled. In the context of smart-home scenarios, it has already been shown that RFID technology provides a number of benefits. For example, self-sensing places based on RFIDs have been analyzed using the so-called smart plugs [18]. Similarly, activity recognition in the home using simple and ubiquitous sensors has been investigated to support more complex smart-home scenarios [28]. Trumler et al. developed a smart doorplate using RFID technology as a building block for pervasive environments [29]. Finally, a programmable pervasive space has been developed based on mixed active and passive RFIDs [23]. These examples clearly outline the benefits of RFID-supported pervasive environments. In all these scenarios, a central computer is important to connect and organize the local network, which, in the scenario presented in this chapter, consists of RFID readers, wireless sensor nodes, and mobile actuators such as robot systems.

The term "smart home" covers a variety of practical and theoretical approaches dealing with ideas of life, living, and working today and in the future [32]. One task is to automate and solve everyday-life problems in different areas, such as home entertainment and health care, and to integrate these single solutions into an overall network. Another task is to combine several single solutions found in the different sections within the scope of a global task. All obviously heterogeneous networks, and deployed devices should be interconnected within a large network covering the entire living area, which is in turn connected to a global outside network such as the Internet. The global task is to collect and analyze data, to respond in a almost self-organized and self-controlled way. So, routine tasks can by solved almost without user interaction, and the environment or especially the home seems to be "intelligent" in a certain way. At the beginning of the 1990s, various concepts and standards—like Konnex (KNX), European Installation Bus (EIB), European Home Systems (EHS), and many others—evolved from different approaches. But these concepts were based on the user's interaction. This interaction was represented by a set of rules; each of them assigned a certain way to respond. It was possible to control plugs, light, and other systems from inside or outside the building, but without inherent intelligence. With increasing progress in technology, the possibilities for concepts using more artificial intelligence instead of control mechanisms increased. In past years, several approaches were presented, each of them offering a single solution. So no date can be named for the first and original smart home. A small selection of relevant and well-known projects with different aspects is

- InHaus Duisburg (Fraunhofer Institutes) [27]
- LIVEfutura (Fraunhofer Institutes) [22]
- T-Com-Haus (Berlin—T-Com) [3]
- Futurelife (Huenenberg/Zug, Beisheim Group Metro) [8]
- Various projects in Munich (BW University/TUM/Microsoft) [7,21,25]
- OnStar at Home (Detroit, Internet Home Alliance) [11]
- TRON Intelligent House (Nishi Azabu, K. Sakamura) [26]
- Smart Medical Home Research Laboratory (University of Rochester) [1]
- SENTHA (Berlin—several German Universities and research groups) [19]
- INGA (Innovations network building automation) [2]

In the next step, research will head for artificial intelligence, self-learning, and self-organizing systems. The environment represented by intelligent objects is no longer managed by a central control. It is gathering and analyzing data while working almost autonomously. So it is possible to identify (behavior) patterns in all sections of our life, like consumption, sleeping, etc. The user must not feel reliant on technology, or watched and controlled, which requires transparent technology and hidden complexity, especially with respect to the wide range of different users—elderly/younger people, healthy/handicapped people, people interested in technical improvements/people who do not want to or are not able to get familiar with those details.

Another goal would be to make the rapidly evolving improvements affordable and usable for everybody. So the new technology has to be merged with existing living concepts to make it a common part of our daily life. Up to now there is no project that can be considered the best or most innovative, because all of them are following different approaches with various differing targets. The most common scenarios are covering comfort and help in everyday life, recreations and (home) entertainment, retaining independence when getting older or living with a handicap, and medical applications. The first category targets comfort or economic goals, making life easier and at the same time reducing costs. As far as improvements of work processes are concerned, it is a wide field of activity, which is not covered in this chapter.

The focus of this chapter is on helping elderly or handicapped people to live a more independent life as long as possible. Those people are an important target group concerning smart-home environments, so special requirements have to be taken into account. For example, a health monitoring and emergency help system has to be established.

The requirements made to control infrastructure and interfaces have to be easy and self-explanatory. The user should be integrated and feeling well in the new environment. When developing smart-home environments for this target group the main focus is on compensating handicaps and limitations. In our work, we established and tested a SANET in a realistic environment to assume certain tasks, such as monitoring and control of domestic systems, home entertainment, and health care. Based on the collected data, a central diagnosis center or station is able to analyze and monitor the patient's behavior. For example, unused active systems like lights can be identified, analyzed, and controlled. For health-care applications, the collected data can be correlated. Concerning the sensor network, the operation is subject to certain technical constraints. So the sensor nodes are only conditionally reliable. Furthermore, new functionality for maintenance, management, and diagnosis ought to be integrated. Thus, the deployment of mobile robots seems to be a well-suited approach. They may validate certain data measured by sensor nodes, allocate tasks, reprogram sensor nodes, and localize frequently lost objects (FLOs), such as keys, and several other cooperative tasks. In addition, the mobile robots can be used as base stations for the smart-home environment. Our work comprises to set up a test scenario showing the mentioned functionality in smart homes. The test environment is meant to be the networking laboratory. The development is based on sensor nodes (BTnodes),* RFID chips (various types), and mobile robots (Robertino).† The basic functionality like routing, sensor–robot communication, and robot control has been already partially developed in prior work [13].

19.2 Our Smart-Home Scenario

In the following, an application scenario for smart homes is introduced [13], assuming a usual apartment owned and inhabited by an elderly person. The subject may be handicapped in several ways, as senses and capacity to remember are not so good anymore. For an elderly or handicapped person it is more difficult to deal with common routine tasks, like control of heating and the air conditioning, or even a simple home entertainment system. The increasing age is one reason why the subject's vital signs and values should be checked continuously in short periods. Also the person is not able to see the physician anytime he needs or wants to, because of the fact that he is unable to make his way to the physician's practice. Hence a solution must be developed that enables the patient to talk to his physician in a comfortable way. The physician must have the possibility to remote-view vital signs and values, and to interact with the patient, as if being in a normal practice. Additionally, the family members may be spread over the whole world living hundreds or thousands of kilometers away. Often there are not even any caregivers in the neighborhood. Thus, in case of an emergency, a reliable emergency system must call the physician and the ambulance and notify the immediate family about the incident.

---

*http://btnode.ethz.ch/
†http://www.openrobertino.de/

19.2.1 Objectives

Of particular concern to an elderly or handicapped person, the higher-level goal is to compensate any limitations in any part of his life as far as possible, to enable the patient to live a more independent life as long as possible. Now keeping the global task in mind, it is possible to list a basic set of objectives for a smart-home scenario. So the first objective to be named is the development of a sensing and monitoring system, which takes over the part of the named subject's wasting senses and if necessary maintaining his capacity to remember. In detail, the architecture must take care of domestic systems, air conditioning, lights, and heating, as well as control the basic functions of home entertainment and security systems. To interact with the physician and to get the vital signs and values checked keeping the physician up to date, as an additional goal, the architecture must be developed to gather data, analyze data locally, make data accessible, initiate an emergency call, and provide video/audio communication. Additionally, mobile robots should be used for maintenance and management tasks concerning the network, for global-task allocation as mobile sensors, and for localizing FLOs.

These objectives require a remarkable amount of technology to be used. But the fact that this technology must be transparent and easy to use is one of the most important points. All improvements and technology are useless if they are not accepted by the user. Thus, the user–system interface must be as simple and powerful as possible. And the system must basically operate in a self-organized way.

19.2.2 Real-Life Requirements and Lab Limitations

To map the developed architecture to a realistic representation, certain requirements are applied to the components and assemblies as well as to the environment and the user. The mapping must incorporate the given budget and possibilities. All devices must provide the necessary interfaces to be interconnected in the overall network. Although their design has to be aligned with their predetermined domain. In apartments and family houses, the requirements differ to those of office and factory buildings. Small and nicely built devices are necessary for a certain comfort. The possibility for a high-speed Internet connection is required to satisfy the needs for several services, like audio/video telephony and interactive medical care. In terms of smart homes for elderly and handicapped people, a backup Internet connection is a must to ensure the reliability of the emergency system. Even if the network collapses, the inhabitant should be able to initiate an emergency call. With the research in (W)BAN systems going on very quickly, the possibilities are improved and so versatile and more exciting technology will be affordable and in a certain way invisible in the near future, like sensor plasters or implants. Up to now, the technical units and devices commonly used in (W)BANs may interfere with normal life sometimes, being present, visible, and tangible all the time. Arrangements are also made to the environment and the user, as if the apartment is considered to be suitable to set up a sensor network either wired or wireless. To deploy mobile robots the flooring too has to be adapted. The users must be willing to use and understand the technology on a certain lower level to handle it, even if efforts are made to simplify processes and usability as far as possible.

For developing a demo scenario in our lab, several limitations have been identified for using the existing hardware and software as a basis to implement a smart-home test environment, which maps the designed common architecture. Obviously, the environment is limited or rather simplified in certain terms. In contrast to an apartment or a family house, the laboratory is furnished with some chairs, tables, and experimental setups. Additionally, workstations and other technical equipment are arranged. So the concept of a usual living space must be abstracted. Within the budget it is not possible to set up an environment with all named devices, like domestic, home entertainment, security system, and air conditioning. Other systems, such as heating and light, defy control that is maintained by the university's maintenance services. The dedicated server provides proxy and Web services. Within the available hardware, no components to implement a BAN are included. The used RFID reader offers a maximum reading distance of 10 cm, which implies that the RFID tags ought to be placed on the ground. They should also be placed in a way so as not to interfere the robot's movement. Concerning these facts, the architecture has to be generalized and in a certain way simplified to get an abstraction for experimental purposes.

19.3 Common System Architecture

In the following section, a common system architecture based on the given objectives is developed. To show the evolving possibilities in smart-home environments, certain existing constraints are sometimes disregarded. In Figure 19.1, an overview for a common system architecture, especially referring to the needs of elderly and handicapped people, is depicted. The central point of interest, in our case, is an apartment that should be changed into a smart home. Therefore, the apartment is equipped with several technological systems. These systems have to be interconnected supporting the user in a transparent way, as far as possible. Wireless network technology should be used to link various well-known concepts, such as security or home entertainment systems, as well as so-called smart objects. These smart objects are common objects with small integrated microprocessors communicating via wireless networks. They are typically equipped with various sensors to observe the environment and it is possible to use them as actuators as well. Smart objects are usually mentioned in conjunction with the idea of ubiquitous computing, which characterizes the omnipresent information and computer technology implanted into everyday items. But also the immediate and distant surroundings, covering neighbors, physicians, or hospitals, and the family, existing networks and technology as the Internet, the public telephony network, and home automation systems, must be integrated into the idea of a smart-home environment.

A SANET, sometimes using smart objects such as RFIDs, provides the possibilities to control various systems in a self-organized and self-learning way. In terms of economy, common processes, concerning domestic systems, air conditioning, light, and heating, may be optimized making them work together in a more intelligent way. One well-known technique is to use sensors and actuators to set up feedback loops. So sensors and actuators may provide functionality based on predefined rules as well as behavior evolved from a learning process. The environment is thus controlled in a self-organized

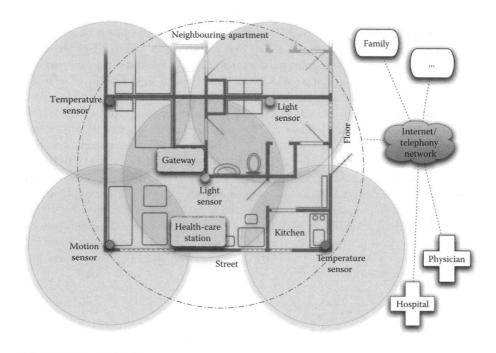

Figure 19.1 Overview—A common system architecture.

way. To give a few examples, the heating can be controlled economically as well as the light and the air conditioning. Furthermore, the fridge may keep a list of the contents to automatically order depleted foods directly from the supermarket, which are then delivered to the apartment. Also intelligent carpets using RFID technology could be part of a security system. However they may be part of a surveillance system to ensure that the patient is still alive, capturing motion and other behavior patterns as well.

Besides normal or implanted sensors or sensor plasters, like the ones developed at Fraunhofer Institute for Reliability and Microintegration (IZM) [17], "smart clothes", i.e., wearable computers and other smart objects, may be used to keep an eye on the patient while he lives in a common environment. If possible, neighbors may be involved for immediate help too. An emergency system may directly notify the neighbors, without additional costs, using a Bluetooth-enabled device, in case of unusual vital signs. A reliable emergency system must also notify the physician and the hospital in case of an emergency. For these purposes, existing networks, like the Internet or the telephony network, may be used. The system can rely on several kinds of third-party mechanisms. A pager or mobile device may be called over the public telephony network or a directly connected emergency notification system may be used via the Internet, which depends on the given interfaces provided by existing emergency notification systems.

As the apartment is now connected to the outside world, the spectrum of possibilities is enlarged drastically. A central health-care station in the apartment may gather and provide health information to the user, the medical helpers, and even the family. Also the communication system can be improved from normal telephony to audio/video calls. So the physician and the patient can talk to each other in a video conference call, instead of meeting in the practice.

Obviously, the various heterogeneous networks must be merged. Therefore, one or more gateways between the networks are needed. The sensor network may communicate using Bluetooth technology while the rest of the systems are WLAN-enabled. As it is possible to provide more than one gateway, the first may connect the Bluetooth sensor network to the internal WLAN while the second handles incoming and outgoing calls as well as Internet services. Mobile robots are another helpful extension to the smart-home environment, as they are able to provide a variety of services. The mentioned gateway functionality, between the sensor network and the WLAN, can be covered by a mobile robot for example. On one hand they may be used to support and maintain the network to encourage the reliability of the network and on the other hand to observe the environment and the inhabitants, while locating frequently lost objects (FLOs) using RFID technology.

The use of passive RFID tags eliminates battery constraints that represent one of the most challenging issues in sensor networks. It is not possible to find an object even if the battery is discharged—regardless of whether the object is equipped with a location system. As passive RFID tags are not dependent on an external power supply, they are usually very small and nearly every object can be equipped with these tags. Furthermore, navigation assistance and distributed coordination for mobile robots can be provided by writable RFID tags implanted into the floor of the environment. For example, the manufacturer Vorwerk is producing carpets with integrated RFID tags that can be used as a basis for building intelligent environments based on super-distributed RFID-tag infrastructures [10,31].

19.4 Implementation

As the test scenario was to be set up in our laboratory, the architecture's mapping specified had to be within the boundaries of possibility. In our lab, several BTnode sensor nodes and mobile robots called Robertino are available. With the available implementations and basic technology from prior work, considering the limitations pointed out in one of the following sections, the architecture is mapped to an experimental setup, as it is depicted in Figure 19.2, which represents the research environment. It may be roughly divided into the following subnetworks: Bluetooth, Wireless LAN, RFID, Internet (TCP/IP), and the telephone network.

With regard to the architecture, a sensor network is mentioned to interconnect the deployed BTnode sensor nodes and to diffuse sensor data values over the network, which is implemented by a Bluetooth network. A private WLAN has already been installed in the laboratory using a wireless access point. So it possible to access the mobile-robot system from a workstation or other systems within the network. The access point is connected to the Internet and to the university network. A gateway has been set up in terms of a dedicated server, which is used for several purposes by the ROSES project. It provides an Internet connection via the university network and a Web service. The RFID system is not a network in the common sense but it provides the possibility to transmit data from the RFID tags to the RFID reader.

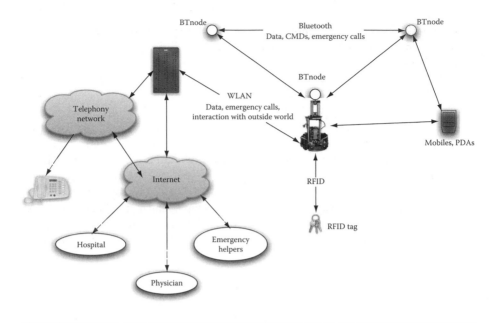

Figure 19.2 Architecture for the lab setup.

19.4.1 Wireless Sensor Network

The wireless sensor network is based on BTnodes interconnected via a common Bluetooth network. On top of the Bluetooth communication layer, a simple multi-hop protocol based on ad hoc on-demand distance vector (AODV) has also been implemented by the BTnut API. Within a multi-hop protocol it is possible to send messages to sensor nodes, even if they are not directly connected to each other. That fact eases the problem of covering the whole laboratory or the living space of a smart home with the desired sensor networks, respectively. This means that sensor nodes can be placed wherever there is the greatest need. So the problems of either wired solutions and even the constraints, concerning common Bluetooth communication ranges, are solved. In addition, using Bluetooth to set up the sensor network affords to integrate Bluetooth-enabled mobile devices.

As the BTnodes ought to operate as sensor and actuator nodes, a sensor may be connected to one of the I/O lines provided by the BTnodes. Two possible ways exist to connect sensors or sensor boards to the I/O lines using either the connector on the usbprog rev2 board, as shown in Figure 19.3b, or a direct connection using two other connectors placed on the BTnode itself, as depicted in Figure 19.3a and c. The sensors may be plugged directly to the connector pins. Here, a light sensor called TSL252R is used. In the first approach, the sensor has been installed on a simple sensor board, to develop a proper and reliable but simple solution. In Figure 19.3d, the circuit diagram is shown.

In the second approach, the recommended sensor board, BTsense v1.1a, from the BTnode developers' site and the according software has been used (see Figure 19.3). The

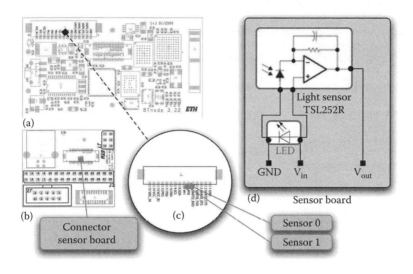

Figure 19.3   BTnode connectors assembly—Sensor board circuit diagram.

BTsense board is a simple sensor board that is connected by the J2 connector to the BTnode. It was designed at the Institute for Pervasive Computing, ETH Zurich, to support the education of students. The size of the board (2×4 cm) allows a fixing to the side of the BTnode [9].

It offers the possibility to connect

- TC74 temperature sensor (digital, I2C)
- TSL250R/TSL251R/TSL252R light sensor (analog)
- AMN1 passive infrared motion sensor (digital, logic level)
- 7BB-12-9 buzzer
- Optional I2C digital sensors
- One optional external analog sensor

In terms of the smart-home sensor network, a simple protocol, set on top of the multi-hop routing protocol, has been developed. It comprises three types of messages, which are divided into subtypes according to the needs. So the different message types may be handled with different priorities, if necessary. The main message types and their subdivisions are shown in Figure 19.4. If any uncommon parameter is recognized in the sensor network covering the environment, the BTnode, which detected the anomaly, initiates a broadcast "emergency call" to diffuse the information. So any device integrated into the Bluetooth sensor network is able to react in an appropriate manner, if it is using the same protocol. The emergency call messages are split into "light sensor messages," "temperature sensor messages," and "motion sensor messages," with the option to be extended. Here, only light, temperature, and motion sensors are available, so the given subdivision satisfies the demands.

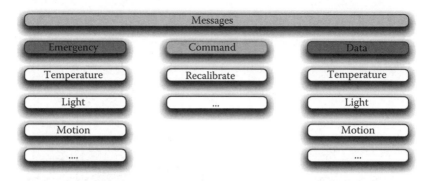

Figure 19.4   Smart-home sensor protocol—Message types.

An appropriate response to an emergency call, initiated by a light sensor, would be to switch on/off the light. Concerning the experimental setup, the actuator activity is simulated by a buzzer sound. If the initiator does not receive an acknowledgment within a certain time limit, it repeats sending the emergency call message. This continues until an actuator confirms the initiated emergency call with an acknowledgment, after finishing the action. In both directions, broadcast messages are used because an acknowledgment from one of the nodes ought to stop the reactions on other nodes or the mobile robot. So, in case of an emergency call the fastest solution wins and stops the others by confirming the request in the first place.

Command calls are initiated to start a remote process on a sensor node. In this case, a command for recalibration has been implemented, which simulates a recalibration process by blinking LEDs, as mentioned above. A command call is acknowledged by the receiving sensor node as well. In this case, a message is addressed to the initiator only because no other devices are involved.

Data calls are messages sent intermittently to maintain a chronicle of the environment status. The gathered data can be analyzed or just shown to users in and outside the smart home. The name "call" was chosen to keep uniformity of naming. Actually, it is a simple unacknowledged message.

19.4.2   Mobile Robots

We are employing the mobile-robot platform, Robertino, for which we already developed a number of hardware extensions (including an RFID reader board as described below). The basic software functionality is provided by the development framework, Robrain [16]. We developed Robrain in several student projects. This also includes several plug-ins, such as an RFID connector, "rfidReader"; a localization and mapping unit, "PathFinder"; and a video camera extension, "vdcUnit." The smart home plug-in merges the offered solutions to implement a behavior plug-in, which defines the mobile robot's behavior within the smart-home scenario. The Robertino has to be integrated into the sensor network and must be able to locate FLOs by passing them on its way, somewhere in the designed experimental environment. To integrate the mobile robot into the sensor network a BTnode is used, which is connected to a common USB port. It is possible to verify values measured by sensor nodes using the BTnode on top of the robot and the appropriate sensors. With a camera device, pictures and small video sequences can be stored for multiple purposes. Additionally, the Robertino embodies a base station for the sensor network and interface between the gateway, which represents the outside world, and the sensor network.

With regard to the sensor network, the mobile robots perform two functions. First, the robots build the data sink for neighboring sensor nodes. Thus, they work as a relay between the sensor nodes and the central gateway. Furthermore, the mobile robots ought to be responsible for maintenance and management tasks concerning validation and substitution as well as recalibration of sensor nodes and data storage. So, if the robot notices multiple unacknowledged emergency calls, it checks the sensor node. If a sensor node shows a suspicious behavior, the mobile robot instructs the BTnode to recalibrate by initiating a command call. In either case the Robertino takes a picture of the current situation and notifies the next-level instance—the user—via Bluetooth SMS messages or by using the Internet connection, if no Bluetooth-enabled mobile device is available.

The robot ought to observe the laboratory on predefined paths. To manage the sensor network and observe the environment, the robot's navigation within the environment has to be laid out to locate certain coordinates within the environment. Therefore, the PathFinder plug-in is used. Regarding the plug-in's abilities, it is predestined to find the optimal path from a given starting point to a certain target. If the robot decides to get in touch with the sensor node based on the above mentioned reasons, the plug-in finds the optimal path referring to an assigned map of the environment.

As no further localization approaches are deployed, the installed BTnodes must be marked on the environmental map. Therefore, a table of the installed BTnodes is set up to assign a sensor node's address with an area on the map. Additionally, the mobile robot may now use the dynamic mode offered by the PathFinder plug-in to update the given map. While crossing the room on a predefined path the robot is able to locate FLOs using the RFID reader system installed on the Robertino. If an FLO is found, the object's ID is assigned to the current position, in terms of coordinates on the map used in the PathFinder plug-in.

19.4.3 Radio Frequency Identification

Before discussing the implementation of the RFID module, we introduce some RFID basics as relevant in our scenario. As shown in Figure 19.5, the RFID system consists mainly of three parts: the RFID tag, the reader, and the information service host. The reader has an antenna that transmits a signal, which is also providing the necessary energy for the passive RFID tag to decipher the transmission and to respond to the query. This signal is initiated by the reader on request of the application. In our scenario, we periodically poll the environment, and when the RFID tag is within the transmission range, it transmits the EPC to the reader. The data obtained by the reader can then be transmitted to a central computer.

Several RFID tag types exist in the market: passive, active, and semi-active. The passive tags have no internal power source, instead they use the low power signal present

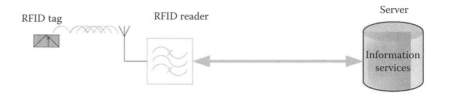

Figure 19.5   RFID transmission system.

in the antenna, which comes from the reader to power up and transmit a response. The passive tags can communicate with the reader in a range from few centimeters to few meters. On the other hand, active RFID tags have batteries (expected battery life is up to 10 years), which enable them to have a larger transmission range, and even more storage capabilities. But they are bigger and more expensive. The semi-active (or sometimes called semi-passive) tags use internal battery to power their circuits, but they rely on the reader for signal transmission.

For testing purposes, we attached an RFID reader module of type TLB-12-AA, depicted in Figure 19.6 (right), to the lower side of the Robertino robot. Several issues

Figure 19.6   Robertino robot with attached RFID reader (below the gripper arm) and some typical passive RFID tags (left) and the selected RFID reader module TLB-12-AA (right).

should be taken into consideration for the selection of an appropriate RFID reader. For our scenario, we decided to rely on passive RFID tags. Therefore, the range that we will deal with is about 10 cm. Because we will attach the reader to a robot, it should be a low-power device. Moreover, the size should be compact and the price should be reasonable. In Figure 19.6 (left), the modified Robertino as well as some typical RFID tags used in our smart-home scenario are shown.

Typical RFID tags available in the mass market are listed in Table 19.1. We bought several of these systems and tested them in our lab. The results of our tests showed that almost all RFID tags available can be used with no impact on the application. The main difference lies in the capabilities (storage), and the availability of collision detection and security measures.

The RFID reader is controlled by a special plug-in for our robot control system, Robrain, named "rfidReader." The output of the reader is stored in an XML-like encoded file. An example is shown in Figure 19.7. It contains the sample output of a single call of the reader. The OK in the first line indicates that the method call was performed successfully. The RETCODE in the second line lists the error code. The RFID tag that was found is of type UNIQUE and it contains the EPC of 1234567890.

Table 19.1 Comparison of Typical Passive RFID Tags as Available in the Mass Market

Transponder	Frequency	ID	EEPROM	Anticollision	Crypto
Unique	135 kHz	40 bit	N/A	N/A	N/A
Hitag-1	135 kHz	32 bit	2048 bit	Yes	Yes
Hitag-S	135 kHz	32 bit	32/256/ 2048 bit	Yes	Yes
Hitag-2	135 kHz	32 bit	256 bit	Yes	Yes
Mifair	13.56 MHz	32 bit	1/4 kByte	Yes	Yes
EPC Gen-2	860–960 MHz	64–96 bit	64–96 bit	Yes	Yes
µ-Chip	2.45 GHz	128 bit	N/A	N/A	N/A

```
<RETSTATUS>OK</RETSTATUS>
<RETCODE>200</RETCODE>
<NAME>get Unique Nonblocking</NAME>
<INFO>UNIQUE</INFO>
<OUTPUT>1234567890</OUTPUT>
```

Figure 19.7   XML-like encoded RFID reader output.

19.4.4 Gateway/Mobiles

The gateway is the bridge between the in-home network and the outside world. An apache Web server is used to make data, gathered by the sensor network and stored by the Robertino, available from outside the smart home. Following the architecture's design, an approach for audio/video calls was implemented. Due to the fact that several public and commercial approaches already evolved, it is not necessary to test these systems. So, such a system is not implemented in our experimental setup. The given approaches are quite well suited to satisfy the requirements, even if improvements, especially in terms of reliability, are desired.

To show the possibility of communication between the mobile robot and a mobile device, the Robertino ought to be able to send SMS messages to a mobile phone. In our approach, messages are transmitted via Bluetooth using the available Bluetooth module on the BTnodes. So, no further hardware is required, and additionally, no further costs arise.

19.5   Demonstration

A demonstration, which is described in this section, was performed to test and to evaluate the presented architecture. The demonstration scenario is set up to show the interaction

Figure 19.8   Experimental setup.

of the particular solutions. Figure 19.8 shows a picture of the experimental setup. First, an area has been defined to represent a simple smart-home environment. In our case, it embodies a small apartment including a living area and a kitchen. The area is framed by surrounding wooden plates. A sensor network connects three BTnodes, each equipped with the BTsense v1.1a sensor boards and according sensors. Each node is configured to use one sensor. So, a light, temperature, and motion sensor node is available. Each sensor node distributes the measured sensor data in time intervals of 30 seconds. The time intervals have been calculated to limit the demo to about 5–10 minutes.

If the sensor values are out of range the nodes initiate emergency calls. These values have been predefined to satisfy the needs in the lab. A value is out of range, if the light value falls below $300 * 10^{-6}$ W/cm^2 or exceeds $800 * 10^{-6}$ W/cm^2, also if the temperature falls below 15 °C or exceeds 35 °C, and if the motion sensor did not detect any motion during the last minute. Also, a fourth node is configured to behave as a simple actuator, to demonstrate a feedback loop. In our scenario, the BTnode acts as a switch if the light value is out of range. This is simulated by a beep sequence using the buzzer on the BTsense sensor board.

The motion sensor is used to show the mobile robots behavior. If the sensor sends more than five unacknowledged emergency calls the robot stops observing the living area, acknowledges the call, and moves directly to the node. It takes a picture of the actual situation and sends an SMS message via Bluetooth to a mobile phone. Then the Robertino returns to the position where it stopped before and continues observing the living area. The observation path, the robot's expected movement, and the BTnodes positions are depicted in Figure 19.9. The gateway is implemented by a host providing a Web service on port 80, which executes a CGI script. The host is connected to the laboratory's private WLAN as well as to the Robertino.

A dynamic generated Web page presents all gathered data and the according plots, as depicted in Figure 19.10. The Web page includes status information shown on the left-hand side. In particular, the latest data items that were downloaded from the mobile robots are listed. In fact, this list outlines the successful measurements in the sensor network. The sensor data is displayed in more detail on the bottom of the Web page. Here, all temperature, light, and motion detection values are listed. To provide further statistical information to the user, plots can be created for all these measures.

Furthermore, information collected from RFID tags is included in the Web page. In the middle, the RFID location table is depicted. As can be seen, a single RFID object has been located recently and the position can be drawn onto the map.

19.6   Implementation Experiences

Using the developed architecture, we expected to easily integrate two complementary technologies: sensor networks and RFID tags. From a software engineering point of view, both systems need a completely different approach because of the differences of the execution environments. Sensor nodes can usually be programmed in a traditional

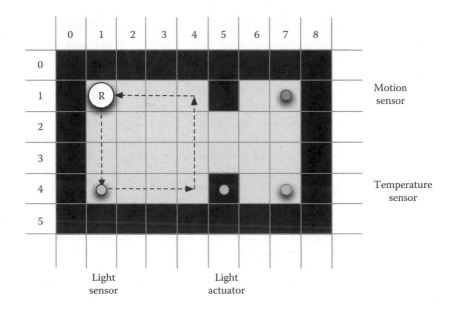

Figure 19.9   Snapshot of the Web page.

way, i.e., software applications are developed, installed, and run on the nodes. In contrast, RFIDs, at least in most cases, require external decision-taking and provide only limited computational capabilities despite their well-designed storage subsystems. Also, the transmission range of sensor networks is much larger compared to RFID systems.

To develop an integrated approach, the specific capabilities of both systems need to be identified and exploited. Furthermore, the application that is to be run in this environment needs to be clearly separated into modules for each part. If this process has been accomplished successfully, the integrated approach using sensor networks and RFIDs provides many advantages.

In our developed demonstrator, we coupled sensor networks and mobile robots with RFID technology. In the following, we summarize the main issues that turned out to be the limiting factors of the implementation process.

The BTnodes and the according API are well suited to quickly set up WSNs and to prototype sensor network applications. The nodes are equipped with radio interfaces and various I/O lines. Concerning their size, the nodes are offering an amazing computational power. Although, it is quite easy to program the BTnodes, as software is implemented in ANSI C and the API is constantly improved to extend their abilities. Furthermore, a very good and constantly extended programming tutorial is available, which shows how to implement and use the offered features. Unfortunately, well-known battery and memory constraints are still present in some ways. So, the nodes cannot be run for a longer time without external power supply. Concerning our implementation, no difficulties in terms of memory constraints evolved. Regarding the sensor hardware, the first approach was a very simple sensor board implemented by our own, whereas in the second approach

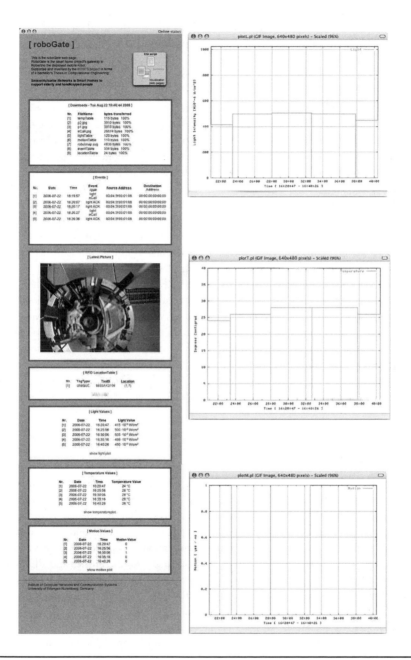

Figure 19.10 Experimental setup—Schematic.

the recommended BTsense v1.1a sensor boards were used. Both approaches solved the task of sensing the environment very well. However, the second solution is much more comfortable. The BTnut API provides methods to address the single sensors when using the BTsense sensor board and the sensor network protocol covers all necessary message types needed within the smart-home scenario. It is possible to respond to emergency situations as well as to gather usual environmental data, e.g., by exploiting the possibility to execute certain BTnode commands on command.

The Robertino is already well engineered and offers a wide range of possible applications. The skills are constantly increasing as new hardware and software is developed within the ROSES project. It is not difficult to integrate the Robertino into the used networks because of the fact that a WLAN card and a BTnode linked to the USB serial device are available. The plug-ins, either developed in prior work or implemented with regard to this work, enabled the robot to behave in the recommended way. The PathFinder plug-in has been developed to calculate the optimal path through the environment. We extended the plug-in to access the current position of the robot. The RFID reader and the camera solved their tasks without any major problems. Nevertheless, only card-like RFID tags could be used, because the coverage of the reader is limited to about 10 cm and the robot must be able to pass the tag without problems.

Regarding the combination of SANETs with RFID technology, it must be said that many minor problems such as the low reading distance of typical RFID readers (at least if passive RFID tags are concerned) requires a very precise movement control of the mobile robots. Also, nonmethodological issues such as the proper physical connection of the reader to the robot and other problems due to mechanical issues still dominate the development and deployment process in a smart-home demo setup. This can obviously be handled much easier if commercial solutions (which might be much more expensive) can be defined and used in a growing market for smart-home applications.

19.7 Conclusion

The increased interest in ubiquitous computing affords a wide spectrum of new, small and easy-to-use technology. The smart objects, sensor nodes, actuators represented by control units or mobile robots, and other technologies that are meant to be small but powerful helpers in nearly every part of our life have become more and more available and affordable. With the rapidly increasing research and development progress taken into account, soon these objects will become as common as mobile phones and PDAs today. Concerning health care, it becomes possible to observe and advise patients while living in their familiar environment, instead of spending months or even years in a hospital. Medical and senior care becomes easier and more secure in case of an emergency, because immediate help is available. All facts considered, it follows logic that sensor/actuator networks are well suited for supporting elderly and handicapped people in a smart-home environment.

The experimental smart-home setup represents a good approach for research in smart-home environments, keeping in mind that many things have been abstracted to a certain extent and a few problems are still waiting to be solved. Regarding the reliability of the developed systems interacting within a smart-home scenario, certain weaknesses have been observed, although the various parts are working very well on their own and also in combination with other systems. Another problem with respect to the system's reliability concerns the connectionless Bluetooth multi-hop network, as it is not providing any transport-layer mechanisms for reliable data transmission.

The challenge was to design a smart-home environment especially to support elderly and handicapped people. In either case, smart-home technology merging heterogeneous networks consisting of sensors and actuators, automated systems, mobile robots, and the Internet is able to compensate certain limitations and may help its users to live a longer, more independent life in many ways.

References

1. Center for Future Health—Smart Medical Home Research Laboratory, 2001–2005. Online; http://www.futurehealth.rochester.edu/smart_home/.
2. INGA e.V.—Innovationsnetzwerk Gebäudeautomation, 2003. [Online; http://www.inga.de/].
3. T-Com Haus—Intelligenter Wohnkomfort, 2006. Online; http://www.t-com-haus.de/.
4. I. F. Akyildiz and I. H. Kasimoglu. Wireless sensor and actor networks: Research challenges. *Elsevier Ad Hoc Network Journal*, 2:351–367, October 2004.
5. A. Awad, T. Frunzke, and F. Dressler. Adaptive distance estimation and localization in WSN using RSSI measures. In *10th EUROMICRO Conference on Digital System Design—Architectures, Methods and Tools (DSD 2007)*, pp. 471–478, Lübeck, Germany, August 2007, IEEE.
6. M. A. Batalin and G. S. Sukhatme. Coverage, exploration and deployment by a mobile robot and communication network. In *International Workshop on Information Processing in Sensor Networks*, pp. 376–391, Palo Alto, April 2003.
7. Bauland GmbH. Vision Wohnen, 2005. Online; http://www.visionwohnen.de/.
8. O. Beisheim. Futurelife—Smart home Beisheim-Gruppe metro. Online; www.futurelife.ch.
9. J. Beutel, M. Dyer, O. Kasten, M. Ringwald, and K. Römer. BTsense V1.1a, 2006. Online; http://www.btnode.ethz.ch/.
10. J. Bohn and F. Mattern. Super-distributed RFID tag infrastructures. In *2nd European Symposium on Ambient Intelligence (EUSAI 2004)*, pp. 1–12, Eindhoven, the Netherlands, 2004. Springer.
11. Continental Automated Buildings Association. OnStar, 2002. Online; http://www.internethomealliance.com/pilots_projects/family/onstar_at_home/.
12. D. Culler, D. Estrin, and M. B. Srivastava. Overview of sensor networks. *Computer*, 37(8):41–49, August 2004.
13. S. Dengler, A. Awad, and F. Dressler. Sensor/actuator networks in smart homes for supporting elderly and handicapped people. In *21st IEEE International Conference on Advanced Information Networking and Applications (AINA-07): First International Workshop on Smart Homes for Tele-Health (SmarTel'07)*, Vol. II, pp. 863–868, Niagara Falls, Canada, May 2007, IEEE.

14. M. K. Divyan and K. Kwangjo. Security for RFID-based applications in smart home environment. In *2007 Symposium on Cryptography and Information Security (SCIS 2007)*, Sasebo, Japan, January 2007.

15. F. Dressler, *Self-Organization in Sensor and Actor Networks*, John Wiley & Sons, New York, December 2007.

16. F. Dressler and M. Ipek. An extensible system architecture for cooperative mobile robots. Technical Report 06/06, University of Erlangen, Department of Computer Science 7, December 2006.

17. R. Dünkler, Body Area Network—Aufbau- und Verbindungstechnik (IZM), 2002. Online; accessed September 5, 2006.

18. H. Elzabadani, A. Helal, B. Abdulrazak, and E. Jansen. Self-sensing spaces: Smart plugs for smart environments. In *From Smart Homes to Smart Care: 3rd International Conference on Smart Homes and Health Telematics (ICOST 2005)*. IOS Press, Sherbrooke, Québc, Canada, 2005.

19. W. Friesdorf. SENTHA—Seniorengerechte Technik im Alltag/Technik im Haushalt zur Unterstützung der selbständigen Lebensführung älterer Menschen, September 1997. Online; http://www.sentha.tu-berlin.de/.

20. B. P. Gerkey, R. T. Vaughan, and A. Howard. The player/stage project: Tools for multi-robot and distributed sensor networks. In *International Conference on Advanced Robotics*, pp. 317–323, Coimbra, Portugal, July 2003.

21. S. Glubrecht, K. Greiner, D. Wichmann, and E. Steidl. Haus der Gegenwart-Wir wollen wissen, wie wir wohnen, 2006. Online; http://www.haus-der-gegenwart.de.

22. G. Goldacker. LIVEfutura. Online; http://www.livefutura.de/.

23. S. Helal, W. Mann, H. El-Zabadani, J. King, Y. Kaddoura, and E. Jansen. The gator tech smart house: A programmable pervasive space. *Computer*, 38(3):50–60, March 2005.

24. N. B. Priyantha, H. Balakrishnan, E. D. Demaine, and S. Teller. Mobile-assisted localization in wireless sensor networks. In *24th IEEE Conference on Computer Communications (IEEE INFOCOM 2005)*, Miami, FL, March 2005.

25. H. Ruser. Technische Universität der Bundeswehr München—Smart HOME, 2002. Online; http://smarthome.et.unibw-muenchen.de/de/.

26. K. Sakamura. TRON—Intelligent House, 1998. Online; http://tronweb.super-nova.co.jp/tronintlhouse.html.

27. K. Scherer. inHaus Duisburg—Innovationszentrum—Intelligente Raum- und Gebäudesysteme—Phase1, April 1995. Online; http://www.inhaus-duisburg.de/.

28. E. M. Tapia, S. Intille, and K. Larson. Activity recognition in the home using simple and ubiquitous sensors. In *Pervasive Computing*. Springer, Berlin, Heidelberg, 2004.

29. W. Trumler, F. Bagci, J. Petzold, and T. Ungerer. Smart doorplate. *Personal and Ubiquitous Computing*, 7(3):221–226, 2003.

30. VeriSign. The EPCglobal network: Enhancing the supply chain, 2005. White Paper.

31. Vorwerk. Vorwerk is presenting the first carpet containing integrated RFID technology, 2005. Online; http://www.vorwerk-teppich.de/sc/vorwerk/rfid_en.html.

32. Wikipedia. Intelligentes Haus—Wikipedia, The Free Encyclopedia, 2006. Online; http://de.wikipedia.org/w/index.php?title=Intelligentes_Haus&oldid=1917438.

Chapter 20

Integrated RFID and Sensor Networks for Health Care

Melody Moh, Teng-Sheng Moh, and Zachary Walker

CONTENTS

Radio frequency identification (RFID) and sensor networks are two wireless technologies that will soon become an integral part of everyday life. Each of them has many potential applications in ubiquitous and pervasive computing. On one hand, RFID has experienced a very fast growth in the industry, in both technology and applications. On the other hand, sensor network technology has attracted tremendous attention in the research community. This chapter attempts to bridge the gap between industry and academia focuses. We present a study of using RFID and sensor networks in the application of health care. This is an exceedingly important area as the growth of aging population is becoming a universal, worldwide phenomenon and an impeding social issue.

20.1   Introduction

Radio frequency identification (RFID) and sensor networks are both wireless technologies that provide unlimited future potentials. While the industry has witnessed a rapid growth in developing and applying RFID technology, and the network research community has devoted tremendous efforts in sensor networks, these two communities would benefit greatly by learning from each other.

Wireless sensor networks (WSNs) have recently presented numerous exciting challenges for scientists and researchers, military and government officials, and the business community. This is largely due to endless potential applications offered. Some well-known, existing applications include "smart dust" networks of very small wireless micro-electromechanical sensors (MEMS) that could track and report almost everything, from enemy movements in a military operation to cold areas in a room. In addition, advanced wireless sensors can also catch manufacturing defects, track fertilizer runoff from farms into lakes, or monitor patient movements in a hospital.

Even though a sensor network can sense many different types of objects, yet it has its limitation. It normally senses objects by their physical, chemical, or biological properties (such as temperature, light, sound, or movement). Without these sensible characteristics, an object becomes difficult to be sensed. By applying RFID as a special sensor to a sensor network, any RFID-tagged objects can be easily "sensed" (i.e., identified), thus overcoming a significant weakness of sensor networks. This integration can be viewed as the next advancement in RFID and in sensor networks—not only in terms of technology but also in terms of opening a door to other possible applications that never existed in the past.

Many recent studies have indicated that aging population is a worldwide phenomenon. The population of seniors aged 65 and older in the United States would grow from 10.6 million in 1975 to 18.2 in 2025, an increase of 72 percent, although the overall population increase is only about 60 percent. The aging trend is not only restricted to the United States but is in fact a global phenomenon. The worldwide population of seniors over age 65 would be more than double from 357 million in 1990 to 761 million in 2025. Longevity has caused expensive age-related costs such as those for disabilities, diseases, and therefore, health-care expenses.

In pursuing the enormous potential of integrating RFID into sensor network technology while addressing the aging population health-care needs, in this chapter we target our study in the area of health care. Note that we study proposals and projects in health-care research that utilizes both technologies. Because the focus is the integration of the two technologies, we do not include in this chapter those studies that use only one single technology (sensor networks or RFID).

This chapter is organized as follows: Section 20.2 describes three existing proposals in applying RFID and sensor networks for in-hospital applications, and Section 20.3 presents three that apply in the out-of-hospital health-care scenarios. A new application development platform for integrating RFID and sensor networks for health care is then proposed and illustrated in Section 20.4. Finally, Section 20.5 concludes the chapter.

20.2 Survey of Proposals Using RFID and Sensor Networks for Intelligent Hospitals

In this section, we describe first an application that analyzes and manages the supply and demand of hospital personnel [20]. This is followed by an application that maintains and tracks highly sensitive medical supplies [7]. Finally, a proposal of building a pervasive-sensing hospital is illustrated [18,19].

20.2.1 Analysis of Hospital-Personnel Flow for Supply and Demand Management

Making use of the unique feature of RFID that can accurately identify a person, Xiao et al. studied supply and demand management for hospital personnel such as doctors, nurses, and patients [20]. These people have RFID tags attached so that their flows are clearly identified, and the bottlenecks of supply and demand may be investigated and improved.

In the proposed application, RFID tags are worn as plastic bands strapped onto a person's wrist. Associated with the unique ID for a patient would be the patient's name, date of birth, gender, a detailed medical-record-billing information, etc., that may be stored in a remote database (DB). For doctors and nurses, the RFID tags may be embedded in their access ID's, which are frequently used for various levels of room access.

These unique IDs would be detected by RFID readers, which may be either fixed in each room or mobile in tablet-style personal computers (PCs) with wireless connections. These RFID readers are connected through wireless networks to a DB server, where these unique ID's and their associated information are stored.

Doctors' and nurses' activities may be analyzed in strategic places where resources are especially scarce and important, such as emergency rooms. This may be done by recording their flows, including the time period each spends on certain patients or waiting for certain supplies. Thus, their activities may be carefully analyzed and bottlenecks may be identified. This would improve the efficiencies of both personnel flow and resource management.

On the other hand, patient access and services may also be analyzed. This may be done by recording patient flows, such as the time a patient spends on certain stages of a treatment, including registration, various waiting rooms, tests, and laboratories. This would potentially advance patient care and treatments.

In such applications, the RFID used are mainly wearable RFID tags for people within a hospital, and the associated RFID readers. Sensor networks may be added to the wireless infrastructure to improve connectivity and data collection.

20.2.2 Tracking Vital and Highly Sensitive Medical/Life Supplies

Although RFID can accurately identify a person or an object, as shown in the study presented in the last section, advanced sensors can sense much more complex object properties. Kim et al. developed and demonstrated clinically a system that monitors the temperature of blood bags and tracks their locations [7]. The specific sensors and RFID devices used for the project have been carefully described.

Blood is an invaluable life supply used in hospitals. It is very sensitive to temperature and should be stored in stable-temperature environments. In fact, it must be kept at a temperature between 2 °C and 6 °C so as to maintain its quality and stability. If it is left to sit at room temperature, even as short as 30 minutes, the red corpuscle becomes hemolytic and the lifetime of the blood is seriously reduced.

A hospital blood bank is in charge of managing every blood bag—starting at the time when it is received from a blood donor until it is given to a blood recipient for transfusion. It is thus critical to maintain and monitor the temperature and location of a blood bag, so as to maintain its temperature and thus usability, and to promptly and correctly give it to the right patient for transfusion.

Using both RFID and sensor network technologies, the authors developed a 3T (time, temperature, tracking)-enhanced system to improve temperature management of blood samples and to prevent patient-blood mismatch. The system tracks the movement and location of blood bags in a designated time interval, and monitors its change in temperature. It also generates data reports shared by medical personnel.

The system used the Crossbow Technology MTS420CA sensors to record the temperature of blood bags, refrigerators, and blood banks. It is also used in conjunction with the MICAz mote to monitor related factors such as humidity and light acceleration. ZigBee, or IEEE 802.15.4 wireless technology, is used to transmit the temperature data to the sink node, a MIB510CA sensor, which is also attached to the MICAz platform mote.

For RFID temperature sensors, i.e., RFID tags and RFID readers, the TempSens from KSW Co. are used. They are active RFID tags using the 13.56 MHz band. Each has a built-in paper-type battery that lasts over 16 months, and SRAM memory for 64 temperature measurement data. The system also includes a location-tracking system (LTS), which includes a host computer, containers, stations, operators, controllers, and shift controllers.

The system process begins with the refrigerator in the blood bank. Inside, RFID and sensors would collect temperature data (blood bag ID, refrigerator temperature, etc.), and communicate with the outside sink node. The data is then sent to both the laptop computer and the RFID reader. This data is then stored in the DB server, which is connected within the hospital and to the hospital information system, allowing medical staff to access the information either via wireless local area networks (WLAN) or via a Web server.

When a blood bag is taken from the refrigerator, the LTS transports the blood bag to the final designated place. The LTS ensures that these blood bags are tracked and medical staffs are alerted to their arrivals. The DB server also allows medical personnel to query the time a blood bag is released and its exact location.

The entire system has been tested and its usability and efficiency have been verified at the Shin-chon Severance Hospital in Seoul, Korea. It has been proved that this automatic, intelligent system of blood bag transportation improved the efficiency of time and efforts of the medical staff. In addition, the system has provided a more accurate, analytical data history; it has therefore reduced the rate of human error, of the use of poor-quality blood, and of blood-discard rate.

20.2.3 Building a Pervasive-Sensing Hospital

Based on the foundation of the two applications presented above, it should become clear that RFID and sensor networks can be used to build a smart, pervasive hospital. Wu et al. proposed to build a smart hospital by leveraging RFID and sensor network technologies [18,19].

Similar to the two projects discussed in the previous sections, in this project, the authors observed that RFID and sensor networks improve efficiencies in operations through precise capturing of data. RFID tags can be used in the ID badges for both health-care professionals and patients, as well as in fixed assets of hospitals for various purposes. With the help of the RFID tags and sensor networks, physicians and nurses may be tracked and contacted quickly in case of emergencies.

In addition, RFID and sensor network technologies help in improved management of drugs within a hospital setting. For example, they can be used to track drug dispensation and validation of medical dosages. Their application in laboratory services include validation of patients' samples and improved accuracy of laboratory results, tracking of blood supplies and validation of blood-type matching (as explained clearly in the previous section), and controlled disposal of waste material.

A challenge following these applications is that RFID and sensor nodes generate a large amount of real-time or near real-time data. This data is very difficult to be organized, managed, and used in an efficient way. Addressing this challenge, the authors built a system, eWellness, which organized and facilitated sensor data based on the notion of events. eWellness included an edgeware paradigm of organizing hospital RFID sensor data, and a three-layer event representation and reasoning model to map raw sensory information to semantic events in the application domains [18].

The edgeware design is based on the notion of an event for modeling the inherent spatial–temporal properties of RFID and sensor data in the context of hospital applications. As an example, an event could be triggered when an infusion pump is tracked as leaving the emergency room without an authorization.

Expressing data in terms of events provided the ability of handling large volumes of heterogeneous sensor data in real-time pattern, and of maintaining relationships among discrete granular events across the hospital as well as correlating those events with temporal, spatial, and causal context.

A prototype system that tracks infusion pumps has been tested in a local hospital, with preliminary data showing an improvement in the management of available equipments. Associating with the prototype, a generalized event-ontology language, called EO, is later proposed [19]. It is developed to help constructing application context as well as reasoning with the events.

20.3 Survey of Proposals Using RFID and Sensor Networks for Health Care Outside of Hospitals

This section first explains a proposal that provides mobile telemedicine services to patients, aiming to improve remote diagnosis and to reduce critical response time during medical emergencies [20]. Next, a wireless health-monitoring system, Health Tracker 2000 [16], is discussed. Finally, a prototype of in-home medicine-intake-assisting system for elder patients [5,11] is described.

20.3.1 Mobile Telemedicine Services

In addition to improving the efficiency and management of patient care inside a hospital, RFID and sensor network technologies also provide unique opportunities for delivering quality health care to patients outside of hospitals. Xiao et al. proposed a real-time patient-monitoring system that can use RFID and smart sensors to collect a patient's vital signs, including electrocardiogram (ECG), pulse rate, basal temperature, oxygen level in blood, and acidity inside the esophagus [20]. In addition, intelligent location-tracking functions are also incorporated to locate patients especially in the case of emergencies.

In this system, microsensors are noninvasively attached to a patient for collecting vital signs including ECG, pulse rate, and basal temperature. RFID tags are used to identify and locate patients. The data collected by sensors and RFID tags is then sent to sensor heads and RFID readers, respectively, which is then transmitted to a PC, a personal digital assistant (PDA), or even a cell phone. Medical personnel can then monitor patients' vital signs and perform remote diagnosis anywhere and anytime. This system can thus improve the accuracy of remote diagnosis and reduce critical response time to medical emergencies. In addition, it can potentially provide a pervasive health-care system beyond hospitals.

20.3.2   A Wireless Health-Monitoring System

With extensive research in the medical field and the aging of baby boomers, Teaw et al. predicted that the average life expectancy in the United States will rapidly increase [16]. The swiftest surge in our senior population will take place between 2011 and 2030. During this 19 year interval, seniors will expand from 13% to 22% of our population. Keeping this phenomenon in mind, quite a few research groups have proposed using RFID and sensor networks for elder health care, including the projects to be presented in this and the next sections.

Teaw et al. designed a wireless health-monitoring system, the Health Tracker 2000 [16]. It combined wireless sensor networks, RFID, and the existing vital-sign-monitoring technology, which monitors vital signs while keeping track of users' locations.

The system included a patient wearing Patient Tracker, from which vital signs are sent to a base station (BS). Vital signs monitored included heart rate/pulse, blood pressure, respiration rate, and body temperature. The BS, upon collecting various data, may then send the aggregated data to the monitoring station that may reside with a family member or a health-care provider, or in a hospital. The patient-tracker device may be implemented using existing technologies. One or more patients may be monitored from a single BS. This system can be installed in all types of homes or facilities. The RF waves send vital signals and location information of patients to their relatives or medical personnel, and also send crucial alarm signals at life-threatening situations.

To accurately measure the vital signs of a patient, the authors carefully reported their selection of various sensors and RFID devices. Recall that vital signs include temperature, pulse rate, breath rate, and blood oxygen level. For thermal sensors, National Semiconductor LM 92 is chosen over LM 34 because it has a minimum supply voltage of 2.7 V (vs. 5 V in LM 34) and a resolution of 0.3°F (vs. 0.5°F in LM 34). For oximeter, the Nonin ipod is used as a digital oximeter sensor that goes over the user's finger. It would measure the blood oxygen level as well as the pulse rate. The sensor for respiration rate has been considerably more difficult to find, and the authors have not been able to obtain a suitable respiration-rate sensor for their system.

The Health Tracker 2000 system also uses RFID for the tracking component. Similar to other systems described above, an RFID tag is worn by each user, and an RFID reader is added to the BS. The use of wireless technologies offered by RFID and sensor networks made it possible to install the Health Tracker 2000 system in all types of homes and facilities; RF waves can travel through walls and fabrics, sending vital-sign and location information to the central monitoring computer via a miniature transmitter network. Such information can then be easily accessed from any location over the Internet. This is another fine example of demonstrating how RFID and wireless sensor network technology are instrumental in providing pervasive health care outside the hospital.

20.3.3 A Prototype for In-Home Elder Health Care

In this section, we describe another project that built a patient-care system based on integrating RFID and sensor network technologies, proposed by Ho et al. [5,11]. The authors observed that the population of seniors aged 65 and older has steadily increased in the United States as well as in other parts of the world. Longevity has caused expensive age-related disabilities, diseases, and therefore, health care. In addressing this aging population medication needs, the authors targeted a prototype for an in-home elder patient medication-assistant system.

The project included an initial learning phase and then a development phase. The learning phase investigated technology compatibility and capabilities through a sensor network interacting with a simulated RFID system. Simulating software modules were described as they provided excellent learning experiences, and were needed before hardware purchases were possible.

In the development phase, the strengths of both HF RFID (lower cost) and UHF RFID (long distance), along with sensor motes, were all utilized and applied onto a medicine monitor system. The system monitored the amount of medicine elder patients required and assisted them in taking the accurate amount of medicines.

The system consists of seven components—three motes, an HF RFID reader, a UHF RFID reader, a weight scale, and a BS. In the following, their roles in the medicine monitor system are described in detail. HF RFID tags are placed on each medicine bottle to identify each bottle. An HF RFID reader is used in conjunction to track all medicine bottles within the range of the reader. By performing readings of all the tags at a regular interval, the system is able to determine when and which bottle is removed or replaced by the patient. The short range of the HF reader is actually desirable for this aspect of the application. The weight scale monitors the amount of medicine on the scale. Combining changes in weight and in the HF tag event, the medicine bottle and the amount of medicine taken can be determined when the patients take their pills.

Next we describe tracking of the patient. A UHF RFID system, including a reader and one or more tags, is used to track the elder patient who needs medicines. This patient wears a UHF tag that may be detected by the associated RFID reader within 3–6 m. The applied wireless ID (AWID) UHF RFID reader is chosen for this subsystem. The Patient Mote communicates with the UHF Reader to monitor patient arrival at the door of a room or other areas where the system is installed. Thus, the system is able to determine that the patient is in the vicinity, and alerts the patient to take the required medicines via a buzzer.

To provide user interactions with the system, a GUI is required with a display. An embedded display may be used for this purpose. With limited resource, an emulated display within the Base Station PC software was developed as a replacement. The display emulates and provides a graphic user interphase (GUI) to assist the patient. The system uses large font sizes for various medication/vitamins and of different colors for pill quantity; this is to make it easier for elder patients. Alternatively, pictures of various medicine brands/bottles may be used to replace medicine names.

The authors have successfully built a prototype targeted for in-home medication monitoring for elder patients. Both HF and UHF RFID technologies, as well sensor networks are used. They noted that future works would extend the prototype from medication monitoring to a broader elder home-care system, from one room to an entire house, featuring more sensors and RFID components distributed at various strategic places and on various household items. Another extension would be adding the capability of notifying family members via email, and networking with an external health-care center monitor system for any assistance via the Internet.

20.4 A Development Platform for Sensor Networks and RFID for Health Care

In this section, a new application development platform for the integration of RFID and sensor networks targeting at health-care applications is described. A preliminary version of the work has been reported in Ref. [17]. After the introduction in Section 20.4.1, a brief overview on programming abstractions and related middleware projects will be presented in Section 20.4.2. The core components of the development platform will be described in Section 20.4.3, followed by protocol implementation illustration in Section 20.4.4, and a summary will be drawn in Section 20.4.5.

20.4.1 Introduction

Currently, most RFID and sensor network applications, including those presented above, are implemented as complex, low-level programs that specify the behavior of individual sensor nodes [12]. In order for RFID and sensor networks (RSN) to realize their full potential and to succeed commercially, however, it is necessary to enable users of RSN to develop their own applications that tailor to their specific needs. For example, in the case of a room designed to care for patients and elders [5], it is desirable that health-care experts can configure and control these RFID and sensors as to what and how often to sense, what event to log, and when to signal an alarm or take an action. Similarly, in the case of tracking hospital personnel or blood bags [7], hospital management may develop an application suitable for their specific purposes. In addition, it is necessary that they may change these configurations and control parameters as they need.

These users of RSN, such as health-care experts and hospital management mentioned above, are what we would call "domain specialists." They are not scientists who design RFID, sensors, or sensor networks, nor are they software engineers who can develop complex programs acting on these sensors. It will be beneficial that they are able to develop tailored applications without sophisticated scientific or programming background.

Toward this goal, this section attempts to propose an application development platform for RSN. The platform is to be used by domain specialists to develop applications for configuring and controlling RFID and sensors for their specific targets. It is highly desirable that the platform is simple to use, yet robust enough to support a good variety of complex applications. A preliminary version of this work has been presented in Ref. [17].

Six approaches taken in the design of middleware for wireless sensor networks (WSNs) have been identified in Ref. [9] (more discussions on WSN middleware are given in Section 20.4.2). Among them, the event-based approach is chosen for this work. It provides fast responses and offers less communication overhead. These features are advantageous to the RSN applying to health care.

In choosing an event-based middleware that is both easy to use and able to support complex application development, we have selected Java Agent-based DEvelopment framework, or JADE [1]. It is a Java-based middleware for the development and run-time execution of peer-to-peer applications. It is based on the agent paradigm and can seamlessly work and interoperate both in wired and wireless environment. It can greatly simplify the development of distributed applications composed of autonomous entities that need to communicate and collaborate. Furthermore, it can be seamlessly executed on every type of Java Virtual Machine (VM), except for Java Card. The advantage of inherent portability of Java also increases its usability for RSN, as it is simple to migrate applications among Java VM-supported platforms.

The proposed framework attempts to maintain the spirit of an RSN platform, with minimized communication overhead and reduced computational complexity, while acknowledging that the capabilities of both RFID and sensor nodes will steadily increase in the future.

The key technical contributions of the proposed platform include (1) the application of JADE in RSN environments, (2) abstraction of RFID, sensor, and event types, (3) a user interface for application development, and (4) the architecture used to monitor application states.

20.4.2 Programming Abstractions and Related Middleware Projects

In this section, we first discuss programming abstractions and related middleware projects in RSN; this is followed by a description of JADE with a brief introduction of Sun SPOT project.

20.4.2.1 Programming Abstractions

One important aspect in programming for RSN is programming abstraction, i.e., providing programmers with abstractions of sensors and sensor data. Three main programming abstractions used in system and application programming in sensor networks have been identified: data-based model, agent-based model, and macro-programming model [3]. The data-based model implies that the data is kept on the sensor itself and is retrieved through node-directed queries [3,9]. The agent-based model is correlated with the reacting-process operational paradigm, in which the reacting process determines its own course of action based on sensor input and messages from other reacting processes. The macro-programming abstraction refers to a system programmed as a whole, with global behavior in mind rather than on a node-to-node basis. This abstraction is very useful, allowing the application developer to express the system behavior as the creation, combination, and transformation of states [21].

Macro-programming is the approach taken in the proposed framework, as it allows domain specialists not familiar with programming to define system behaviors using familiar terms, namely, the states [21]. In addition, we also include some aspects of agent-based abstraction provided by JADE (see Section 20.4.2.3).

20.4.2.2   Middleware

As mentioned above, most RSN applications have been implemented as complex, low-level programs that specify the behavior of individual RFID or sensors. Clearly, a single hardware platform will most likely be insufficient for the wide range of possible applications. To avoid the development of application-specific hardware, it would therefore be desirable to have a (small) set of platforms with different capabilities to cover the design space [4,12].

Middleware is the standard approach for hiding the complexity of distributed systems. In RSN, a middleware is to collect large amounts of data from RFID tags and sensors, and perform aggregation and management of data in appropriate formats for the target-distributed applications [4,21].

Yoncki and Bacon have done an extensive survey on middleware for WSNs [21]. They have identified six approaches: data-driven, event-based, quality-of-services (QoS)-oriented, Internet-oriented, agent-based, and centralized; they have also summarized the systems that have taken each approach. Among them, the event-based approach has been chosen for our proposed framework. It can be generalized as a system that provides a publish/subscribe service for an RSN, where the subscribers are the applications and the publishers are the RSN nodes. Events occur in response to changes in the monitored phenomenon and are subsequently reported to the subscriber. This approach accrues less communication overhead; a single subscription message sent into the RSN can provide event information for the lifetime of the application. In RSN, when communication costs are at a premium, this approach is therefore highly desirable. Other middleware taking the event-based approach include DSWare [8], Mires [13], and Scope [14].

20.4.2.3   JADE

JADE is a Java-based middleware using agent paradigm. An agent is an autonomous, proactive software component used to program peer-to-peer activity in a distributed environment. An agent-based system is a middleware platform used to interconnect agents [1]. JADE is also a Foundation for Intelligent Physical Agents (FIPA)-compliant agent platform and allows for interoperability with other FIPA-compliant platforms. JADE also provides a homogenous set of application programming interface (APIs) that are independent from the underlying network and Java versions, thus allowing for uniformity and portability of design.

JADE includes FIPA-specified services such as white page and yellow page services. White pages service is a list of all agents by a global name and address. Yellow pages service, also called the directory facilitator (DF) in JADE, contains information about services that are provided by the agents. An agent registers its services along with semantic information describing the service. Agents in need of a service may then use semantic queries to lookup, through DF, services available within the agent platform [2].

Functionality of agents in JADE is programmed through behaviors. (Note that "behaviors" not "behaviors" is the proper spelling when using the JADE environment.) Each agent in JADE has its own thread of execution; the thread then sequentially executes programmer-supplied behaviors. JADE provides an extendible base class Behavior that is extended into specific behaviors.

As discussed in Section 20.4.1, it would be desirable to have a platform for RSN that not only is simple to use, but also able to support complex application developments. JADE can be seamlessly executed on most types of Java VM, including J2EE, J2SE, and J2ME [1]. Furthermore, it may be considered as a light-weight middleware: A miniature version of JADE, LEAP, has the same API as JADE and runs on J2ME (VM for resource-constrained computing environment). It is compatible with connected limited mobile device (CLDC)/mobile information device profile (MIDP) 1.0 environments that are common in resource-constrained devices (such as mobile phones and PDAs). The JADE run-time memory footprint, in an MIDP 1.0 environment, is around 100 kB, but can be further reduced to 50 kB using the "ROMizing" technique (i.e., compiling JADE together with the JVM) [1]. Furthermore, JADE provides redundant white page and yellow page services and the means for handoff and discovery in the case of failure. This adds to the robustness of the system, which is another feature needed by RSN applications.

Sun Labs have recently developed Sun SPOT (Small Programmable Object Technology), a hardware system that is specifically designed to support application development for wireless sensors and transducers (sensors combined with actuator mechanisms) [15]. It supports Java and common integrated development environments (IDE), and features the Squawk VM, a small J2ME VM. It also allows running these applications "on the metal" (directly on the central processing unit (CPU) without any underlying operating system (OS)). It is an example of hardware platforms well suited for the proposed work.

20.4.3 The Application Development Platform

The proposed framework is a publish/subscribe system built on top of the service discovery framework provided by the JADE platform. RFID and sensors, which run an instance of the JADE platform, register their presence with the JADE yellow pages service. Applications may then subscribe to their services.

This section first describes preliminaries and data structure, followed by a detailed presentation of the application development process, and concluded with some remarks on energy management.

20.4.3.1 Preliminaries and Data Structure

20.4.3.1.1 Sensor Registration and Service Discovery

The first step in the proposed platform is service discovery. All RFID and sensors must register their services. An RFID agent or sensor agent advertises its services in JADE yellow pages service and waits for subscriptions. When registering, it provides information that will be useful to application developers.

An RSN may be supplemented with a vast array of sensors that vary in both sensing capability and complexity. We have identified three classifications of sensors used as more general abstractions: quantitative-based, state-based, and identification-based (will be discussed in detail in Section 20.4.3.2). Table 20.1 lists suggested information that sensors provide during service registration. Note that an RFID/sensor may provide services related to more than one category. For example an RFID sensor will likely

Table 20.1 RFID and Sensor Information Reported during Service Advertisement

	RFID/ Sensor Category	RFID/ Sensor Type	Location	Units of Measurement	Precision of Measurement	Sample Rate	Set of States
Quantity-based	X	X	X	X	X	X	
State-based	X	X	X			X	X
Identity-based	X	X	X			X	

provide, in addition to identification-based service (identifying the objects detected), a quantitative service (reporting population size of detected objects).

20.4.3.1.2 Data Model

Each RFID or sensor maintains its own minidatabase of sensing data. The DB consists of data samples and their timestamps. The amount of data stored in the table varies by the requirements of applications currently using the RFID or sensor. The RFID or sensor may also maintain a description of its capabilities including the data described in Table 20.1.

20.4.3.1.3 Event Types

Each of the three types of RFID/sensors has a set of related events. Example events for each class of sensor are shown in Table 20.2. In addition, Table 20.3 shows an example

Table 20.2   A Minimal Set of Sensor-Event Types

Sensor Type	Event Type	Inputs to Event Generator	Data Attached to Event	Description
Quantity-based	Threshold	Threshold value	Data value, timestamp	Sensor value has increased/ decreased past threshold
	Sliding-window threshold	Threshold, window size	Sensor value, timestamp	Sensor value has increased/ decreased more than threshold in window
	Average-window threshold	Threshold, window size	Average over window, timestamp	The average value over the window has increased/ decreased past threshold
State-based	Enter state	State S	Current state, timestamp	Sensor has entered state S

Table 20.2 (continued)   A Minimal Set of Sensor-Event Types

Sensor Type	Event Type	Inputs to Event Generator	Data Attached to Event	Description
	Exit state	State S	Current state, timestamp	Sensor has exited state S
	Change of state	State S, state T	Current state, timestamp	Sensor has transitioned from state S to T
	Time in state threshold	State S, time t	Current state, time-entered state, timestamp	Sensor has been in state S for longer than time t
	Premature exit	State S, time t	State S, new state, timestamp	Sensor exits state S before time t has passed
Identification–based	Object detection	Object identifier	Object identifier, timestamp	Previously identified object detected
	New object detection	N/A	Object identifier, timestamp	New object detected
	Object removal	Object identifier	Object identifier, timestamp of last detection	Object detected in last sensor observation is no longer detected
	Time-referenced object removal	Object identifier, timeout interval	Object identifier, timestamp of last detection	Object detected in a previous sensor observation not detected for longer than timeout interval

Table 20.3   Event Set for an RFID Reader

	Event Type	*Inputs to Event Generator*	*Data Attached to Event*	*Description*
Identification-based sensor events	RFID tag detection	Tag ID	Tag ID, timestamp	Previously identified RFID tag detected
	New RFID tag detected	N/A	Tag ID, timestamp	New RFID tag detected
	RFID tag removed	Tag ID	Tag ID, timestamp of last detection of RFID tag	RFID tag detected in last RFID reader poll is no longer detected
	RFID tag removed	Tag ID, timeout interval	Tag, Timestamp of last detection of RFID tag	RFID Tag detected in a previous poll has not been detected for longer than timeout interval
Quantity-based sensor events	Population size threshold	Population threshold	Current population size, timestamp	Population size has increased/ decreased past threshold
	Sliding-window population size threshold	Population threshold, window size	Current population size, timestamp	Population size has increased/ decreased more than threshold in window
	Average-window population size threshold	Population threshold, window size	Average population size over window, timestamp	The average population size over the window has increased/ decreased past threshold

of an event set for an RFID reader. By identifying event types related to sensor categories, we are able to make an extendable system where a complete set of sensor types need not be identified during the initial design. Each of the event types in Table 20.2 is represented in the system by a behavior (behavior in JADE, referring to Section 20.4.4.1

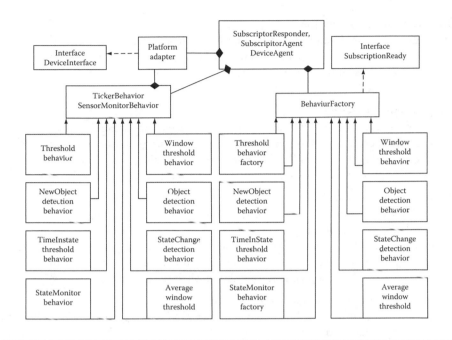

Figure 20.1 Core class diagram for sensor-monitoring module.

and Figure 20.1). There is a one-to-one mapping between the specific abstract events and behaviors.

20.4.3.1.4 Event-Based Publish/Subscribe System

A yellow page interface and a publish/subscribe system is proposed based on the three types of sensors categorized. A quantitative RFID/sensor produces and publishes numerical data measuring some phenomenon, like a temperature sensor reporting degree Celsius. Alternatively, a temperature sensor may provide state data, hot or cold, based on the sensor's interpretation of the condition. Similarly, an identification-based sensor, like an RFID reader, returns the identification of a sensed object.

20.4.3.2 Application Development Process

One of the main goals of the system is to remove programming languages from the high-level development of RSN applications, allowing the domain specialists without sophisticated programming knowledge a direct access to the benefits of RSN applications. The foundation for this goal is the extension of the service discovery system provided by JADE into the RSN domain. On top of this foundation, we propose an event composition interface for describing applications, following the steps below. Note that the process not only eases an application development, it also allows the application to be modified and expanded without much difficulty.

Step 1: Application creation
To begin a new application, the developer (domain specialist) first creates the design space for the application. It will provide an area to store links to selected RFID and sensors, event subscriptions, grouped events, and selected actuators and event–actuator relations, as described in the following.

Step 2: Sensor selection
Application developers first submit a query to the system requesting the type of sensors they are interested in. More complex queries involving restrictions on location, sample rate, etc., are possible. A set of RFID and sensors meeting the constraints specified in the query are provided to the developer, who then selects the set of useful sensors.

Step 3: Event subscription
Recall that each type of RFID/sensors has a set of corresponding abstract events (Table 20.2). The developer may then browse and choose useful events. The process of event selection would be simply marking a check box, entering the required inputs to the event generation function, and clicking the subscribe button. A list of the desired events is then stored for subscription invocation when the application is started.

Step 4: Event grouping by finite-state machine
Once interesting events have been identified, they can be grouped to form a composite event. Using events to trigger transitions, the developer can describe a composite event by a finite-state machine (FSM), which represents detailed event correlation, and where reaching the final state results in the generation of a new event [6,8].

For composite event generation, Li et al. have described the absolute validity interval as a way to meet the need for temporal relation of events [8]. We have adopted this approach. When defining the FSM describing the path to composite event generation, the application developer can place a timeout interval on each state, which also triggers transitions. In this way, time restriction on event relation can be expressed accurately. The FSM may then be represented by a behavior that JADE specifically provided for performing the FSM-like actions.

Step 5: Actuator selection
Similar to RFID and sensor selection, a developer submits a query requesting the set of actuators meeting a set of constraints. Constraints may include actuator type, location, etc. The developer then selects actuators desired for use in his application.

Step 6: Event–actuator relation
To trigger an action, an application developer relates an event to an actuator and sets the actuation parameters to be performed. Actuators act on requests; i.e., when an event linked to an actuator is produced, a request for action is sent to the actuator. Depending on the way the action is specified, the result may be a one-time action or a continuous action that terminates only in response to another event.

20.4.3.3 Energy Management

It is well known that energy management is critical in most sensor network applications. In the proposed platform, wherever possible, energy-consuming tasks are kept to a minimum. In the following we discuss two approaches in this regard.

20.4.3.3.1 Minimizing Communication by Local Computation

It is well known that communication is much costly than computation in terms of energy consumption. Because this is an event-driven system, radio communication is necessary only to subscribe to sensor events and then to notify observers about the event occurrence when conditions are met. By creating one device agent (DeviceAgent) locally on each device (refer to Figure 20.1 and Section 20.4.2), the proposed platform keeps most of the computations local, and minimizes both sensor sampling and radio communication, thus reducing energy consumption.

20.4.3.3.2 Controlling Sensor Sampling and Reporting Rate

In the publish/subscription model, it is vital to maintain a minimum message-passing overhead. One measure is the control of sample rates. Because each sensor has a minimum sample interval that determines the maximum sample rate, it is important to carefully control these minimum sample intervals. Many applications do not require a sensor to sample and report at its maximum rate. Allowing the sample rate to be specified in subscription messages effectively controls sensor data communications, and thus reduces energy consumption.

20.4.4 Prototype Implementation

We have developed a simple, functional prototype as a proof of concept. The prototype is made up of three modules. The first module is an agent-based sensor-event registration and monitoring environment built on top of JADE. The second is a GUI for developing applications on top of the first module. The third is an experiment environment where observable phenomena, devices, and sensors can be arranged to form environments for testing. These three modules are described, followed by a discussion of an application implemented in the prototype.

20.4.4.1 Core Module: Registration and Monitoring

The main functional portion of the platform is an agent-based RFID/sensor-event registration and monitoring environment; a class diagram of core components is shown in Figure 20.1. Everything is centered on the DeviceAgent class, and there is a one-to-one relationship between DeviceAgents and devices. Every DeviceAgent in turn has a link to a PlatformAdapter, which is a generic interface, implemented by each hardware platform to be used. In this prototype, we have simulated underlying hardware; an extension is planned to include actual hardware. By identifying a standard interface, new hardware platforms may be added without any change to the core module.

The abstract PlatformAdapter class provides a number of methods for interaction with devices. To keep the interface simple but flexible, a SensorProfile class is created to encapsulate all the details of any RFID/sensors attached to a device. When created, a device agent will call the getSensorProfiles() method of its PlatformAdapter, and in return will receive an array of SensorProfiles, one for each RFID or sensor. The PlatformAdapter also provides a method to set a sensor profile, allowing properties of each RFID or sensor to change over time. In addition, it provides two methods for RFID or sensor data to be retrieved by the SensorMonitorBehaviors of the DeviceAgent: either to read only the most recent piece of data taken or to retrieve all the data taken since some point in time, returned as an iterator pointing to a list of SensorData. Note that SensorData is another abstract class whose subclasses are broken into QuantityData, StateData, and IdentityData.

Once received by the DeviceAgent, the properties of the SensorProfile are registered with the DF (the yellow pages service) as a service provided by the agent representing the associated device. The DF provides efficient methods for searching and selecting services of interest to an application developer. The developer can query the DF for devices with a particular type of sensor, and then register with the DeviceAgent to monitor the data that is produced by that sensor. Each DeviceAgent implements the SubscriptionManager interface defined by JADE, and in turn has many SubscriptionResponder behaviors running within its thread.

Referring to Figure 20.1, there is a one-to-one relationship between Subscription-Responder behaviors running on the agent and BehaviorFactories known by the agent. Behavior factories are used to create the behaviors that actually monitor the sensor data and return sensor events of interest to the subscribers. To initiate a behavior, the application developer sends a subscription message to the DeviceAgent of interest. Included in the subscription message is the sensorID of the target sensor, the eventID that links the subscription message to a SubscriptionResponderBehavior (and subsequently to the correct behavior factory), and several other data items particular to the event generation behavior of interest, such as sample rate, threshold, window size, state name, and object ID. The application developer would be able to determine what type of information is needed for each event generator by examining the information provided in the DF.

20.4.4.2 Application Development GUI

Applications can be easily developed by utilizing the FSM behaviors provided by the JADE platform. An application is simply a set of states with various actions associated with the entrance to a new state. (Actions could be tied to the state exit in future extensions.) Recall that JADE has a subclass of Behavior that mimics FSM behavior. Each state of a JADE FSM contains another functional JADE behavior (possibly even another FSM behavior allowing for fairly complex applications). State transitions are triggered by the returned values that a behavior generates at the completion of its life cycle. Each state of applications is a StateMonitorBehavior or some combination of many behaviors. The returned values that trigger transitions are the event IDs returned by these behaviors when either event conditions are met, timeouts occur, or failures are detected.

We have created a simple application-specific GUI for the example application (to be described in the next section). The GUI is for selecting RFID and sensors and for altering properties related to the predefined application-specific FSM. Note that a generic application-development GUI is also possible and is left for future extension.

20.4.4.3 An Experimental Environment

An experimental environment has been built, which includes observable phenomena, RFID, sensors, and other wireless devices. As RFID and sensor services are grouped into three categories, three types of phenomena are also created: quantity-based, state-based, and identity-based.

The environment is managed through a simple GUI, which manages the environment, including observable simulated devices, runnable RFID and sensors, and runnable and observable phenomena. Each device then interacts with the DeviceAdapter, which interacts with the DeviceAgent through the PlatformAdapter, as described in Section 20.4.4.1.

20.4.4.4 An Example Application

To further illustrate the proposed development platform, in this section we describe an example of applying the platform to a health-care service. It is based on a previously implemented hardware prototype, a medication-assistant system for elderly patients, as reported in Section 20.3.3 [5]. The proposed framework may be shown to be useful in this illustration.

Consider an elderly patient's home, outfitted with sensors that monitor the patient's health and behavior. RFID readers are placed in several doorways throughout the patient's home to monitor movement. A special scale with a weight sensor replaces the patient's current scale. Sound sensors throughout the home are used to detect a patient's movement, activity, or cries for help. A light sensor in the refrigerator is used to infer how often a patient is eating.

Using our system a health-care provider would develop an application to monitor a patient's routine, and define actions based on detection of significant events. An identification-based event, signaled by the RFID reader every time the patient goes to or gets out of bed, is used to track sleep patterns and to signal potential illness when the interval between expected detections goes too long. The scale is used to detect any drastic changes in weight that can also signal a potential illness. An overly long interval between refrigerator light events triggers an event suggesting the patient is not eating. A sudden spike in sound triggers an event signaling a possible falling or cry for help. Depending on the event generated, the application then takes actions, including paging a health-care provider of unhealthy change, calling or emailing a neighbor or relative for a welfare checkup, and under extreme conditions requesting immediate medical assistance.

20.4.5 Summary

In this section, we have demonstrated an RFID and sensor network application development platform for health care, building on top of the JADE middleware. It provides a simple, generic, application-independent set of RFID/sensors and events, from which health-care providers may easily construct and augment simple, efficient applications. Future work may include an extension of the platform to other non-Java platforms using an interface like Java native interface (JNI), and development of a generic, top-level GUI for application development and maintenance. For actuation implementation, an interface similar to the PlatformAdapter may be developed to add hardware and software actuation capabilities. Finally, a facility for data-driven actuation may be added to complement the event-driven system.

20.5 Conclusion

RFID and sensor networks are both exciting areas; integrating the two would provide much more potentials in the near future. This chapter has described several health-care applications using the combined technologies. As their research and development efforts are still growing strong, and the worldwide aging population is steadily increasing, we expect to see more thrilling proposals applying these two and other emerging wireless technologies onto health care and other areas of medical science.

References

1. Bellifemine, F., Caire, G., Poggi, A., Rimassa, G., JADE A While Paper, http://exp.telecomitalialab.com, September 2003.
2. Bellifemine, F., Caire, G., Trucco, T., Rimassa, G., JADE Programmer's Guide version 3.3, http://jade.tilab.com/doc/programmersguide.pdf, March 2005.
3. Bonnet, P., Gehrke, J., Seshadri, P., Querying the physical world, *IEEE Personal Communications*, 7(5), 10–15, October 2000.
4. Heinzelman, W.B., Murphy, A.L., Carvalho, H.S., Perillo, M.A., Middleware to support sensor network applications, *IEEE Networks Magazine*, 18(1), 6–14, January/February 2004.
5. Ho, L., Moh, M., Walker, Z., Hamada, T., Su, C.F., A prototype on RFID and sensor networks for elder healthcare: Progress report, *Proceeding of the 2005 ACM SIGCOMM Workshop on Experimental Approaches to Wireless Network Design and Analysis*, Philadelphia, PA, 2005.
6. Kaston, O., Romer, K., Beyond event handlers: Programming wireless sensors with attributed state machines, *Fourth International Symposium on Information Processing in Sensor Networks* (*IPSN 2005*), Los Angeles, CA, pp. 45–52, April 2005.
7. Kim, D.-S., Yoo, S.K., Kim, H.O., Chang, B.C., Bae, H.S., Kim, S.J., Location based blood bag management using active RFID and ubiquitous sensor network, *6th International Special Topic Conference on ITAB*, Tokyo, pp. 320–322, 2007.
8. Li, S., Lin, Y., Son, S., Stankovic, J., Wei, Y., Event detection services using data service middleware in distributed sensor networks, *Information Processing in Sensor Networks Workshop*, Pao Alto, CA, 2003.
9. Madden, S., Franklin, M.J., Hellerstein, J.M, Hong, W., The design of an acquisitional query processor for sensor networks, *Proceedings of the 2003 ACM SIGMOD International Conference on Management of Data*, San Diego, CA, pp. 491–502, 2003.

10. Mainland, G., Kang, L., Lahaie, S., Parkes, D.C., Welsh, M., Using virtual markets to program global behavior in sensor networks, *Proceedings of the 11th ACM SIGOPS European Workshop*, Leuven, Belgium, 2004.
11. Moh, M., Ho, L., Walker, Z., Moh, T.-S., A prototype on RFID and sensor networks for elder health care, in *RFID Handbook: Applications, Technology, Security and Privacy*, M. Ilyas and S. Ahson, Eds., CRC Press, Boca Raton, FL, pp. 311–328, 2008.
12. Romer, K., Mattern, F., The design space of wireless sensor networks, *IEEE Wireless Communications*, 11(6), 54–61, December 2004.
13. Souto, E., Guimaraes, G., Vasconcelos, G., Vieira, M., Rosa, N., Ferraz, C., A message-orienter middleware for sensor networks, *Proceedings of the 2nd Workshop on Middleware for Pervasive and Ad-hoc Computing*, Toronto, Canada, pp. 127–134, 2004.
14. Steffan, J., Fiege, L., Cilila, M., Buchman, A., Scoping in wireless sensor networks, *Proceedings of the 2nd Workshop on Middleware for Pervasive and Ad-hoc Computing*, Toronto, Canada, pp. 167–171, 2004.
15. Sun SPOT System: Turing Vision into Reality, http://research.sun.com/spotlight/SunSPOTSJune30.pdf, June 2005.
16. Teaw, E., Hou, G., Gouzman, M., Tang, K.W., Kesluk, A., Kane, M., Farrell, J., A wireless health monitoring system, *Proceedings of the 2005 IEEE International Conference on Information Acquisition*, Hong Kong and Macau, pp. 247–252, 2005.
17. Walker, Z., Moh, M., Moh, T.-S., A development platform for wireless sensor networks with biomedical applications, *Proceedings of 5th IEEE Consumers Communication Networks (CCNC)*, Las Vegas, NV, January 2007.
18. Wu, B., Liu, Z., George, R., Shujaee, K.A., eWellness: Building a smart hospital by leveraging RFID networks, *2005. 27th Annual IEEE International Conference of the Engineering in Medicine and Biology Society 2005*, Shanghai, China, pp. 3826–3829.
19. Wu, B., George, R., Shujaee K., Architecting an event-based pervasive sensing environment in the hospital, *3rd International IEEE Conference on Intelligent Systems*, September 2006, London, U.K., pp. 273–277.
20. Xiao, Y., Shen, X., Sun, B., Cai, L., Security and privacy in RFID and applications in telemedicine, *IEEE Communication Magazine, Special issue on Quality Assurance and Devices in Telemedicine*, April 2006, 64–72.
21. Yoneki, E., Bacon, J., A survey of wireless sensor network technologies: Research trends and middleware's role, Technical Report UCAM-CL-TR-646, Cambridge University, September 2005.

Chapter 21

Integrated RFID and Sensor Networks for Structure Monitoring

Jiming Chen, Santiago Pujol, David K. Y. Yau, and Di Miao

CONTENTS

Automated structural "health" monitoring (SHM) seeks to assess continuously the integrity of building structures using a set of embedded sensors. In this chapter, we discuss the use of resistance-based strain gages in SHM applications, where the gages are used to measure changes in one or more dimensions of a critical structural element. We also present a binary gage designed to detect unequivocally cracks of widths exceeding a predetermined threshold. The binary gages can be used to map damage directly without complex data analysis and correlations between the data reports, structural analysis results, or information on structural drawings. In terms of network requirements, binary gage data reports have a low data rate and are naturally resilient to data loss. We discuss how commercial off-the-shelf (COTS) data acquisition and communication devices can be used to build a multi-hop wireless SHM network. The SHM network is low cost, configurable, and amenable to flexible and clutter-free deployments.

21.1   Introduction

Had the World Trade Center Towers been instrumented with temperature sensors, part of the tragedy of September 11, 2001 could have been avoided. The engineering community seems to have reached the consensus that the collapse of the towers was triggered by reductions in load-carrying capacity caused by high temperature. Had the temperature inside the building been known to engineers trying to coordinate rescue operations, the decision to send personnel in could have been reversed. With advances in sensing and communication technologies, arrays of sensors to monitor building structures continuously and proactively become feasible. During extreme events such as earthquakes, storms, or explosions, damage information reported by sensor arrays will inform rescue/response operations and stakeholders quickly. Without these arrays, damage detection has to be attained by visual inspection. The process is very time consuming and expensive because it requires experienced personnel.

The goal of automated structural "health" monitoring (SHM) is to assess continuously the integrity of a structure using a set of embedded sensors, and to report the results in real time to a remote control center. The control center analyzes the data reports to detect changes in the signals from sensors that indicate damage or other safety issues. Physical properties that are relevant to structural health include strain, stress, accelerations (related to vibrations of the structures), and crack widths. SHM networks based on vibration measurements by accelerometers mounted on the structures of interest include [KPC07,XRC04]. These networks infer damage by detecting changes in the stiffness of a structure. Relating stiffness to damage compromising the safety of the structure is not trivial because structures are designed to yield when subjected to large loads. A structure undergoing a change in stiffness may simply be responding according to the design. On the other hand, local failures, failures of individual elements forming part of a large structure, may not produce large changes in the stiffness of the full structure, and may therefore escape detection. In terms of the network requirements, SHM systems based on vibration measurements require tight time synchronization between data reported by the different sensors. High-resolution accelerometer data needs to be sampled at 100 Hz or faster. The required network should therefore have a high data rate.

In this chapter, we discuss various uses of resistance-based strain gages in SHM. Strain gages are used to measure changes in one or more dimensions of a critical structural element. We also present a binary gage designed to detect unequivocally cracks of widths exceeding a predetermined threshold. These gages can be used to map damage directly without complex data analysis and correlations between the data reports, structural analysis results, or information on structural drawings. In terms of the network requirements, binary gage data reports have a low data rate and are inherently resilient to data loss. We discuss how COTS data acquisition and communication devices can be used to build a multi-hop wireless SHM network. The SHM network is low cost, configurable, and amenable to flexible and clutter-free deployments.

21.2 Background on Electrical Resistance-Based Sensors

Electrical resistance-based sensors work by relating changes in dimensions and changes in electrical resistance. The resistance of a conductor with cross-sectional area A, length L, and specific resistance ρ is

$$R = \frac{\rho \cdot L}{A}. \tag{21.1}$$

If the conductor is stressed in its axial direction, its length L changes in direct proportion to the applied strain. The cross-sectional area A also changes with axial strain. The area decreases if the applied strain is tensile and it increases if the applied strain is compressive. For small strains, the relative change in area is directly proportional to the Poison ratio of the material. Both the relative change in length and the relative change in area result in a change in resistance. The strain also affects the specific resistance of the material, ρ. The specific resistance increases with tensile strain and decreases with compressive strain.

For small strains (smaller than 8 percent [DAL93]), the total change in resistance, ΔR, associated with a strain can be expressed as

$$\Delta R = S \cdot \varepsilon \cdot R, \tag{21.2}$$

where S is a proportionality constant. For certain nickel-based alloys (Constantan, Karma, Nickel–Chromium, and Isoelastic) and other alloys (Armour D and Platinum–Tungsten), S varies from 2 to 4 [DAL93]. These alloys have high specific resistance, and wide ranges in which the relationship between the change in resistance and the strain remains linear.

21.3 Electrical-Resistance Strain Gages

Equation 21.2 implies that the change in resistance associated with a given strain is larger for longer conductors with a large resistance R. To increase the length and ease handling and use, conductors are photo-etched into a grid that is mounted onto a thin backing material, as shown in Figure 21.1.

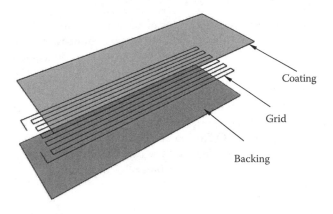

Coating

Grid

Backing

Figure 21.1   Components of a strain gage.

The device shown in Figure 21.1 is called an electrical-resistance strain gage, or simply a strain gage. Strain gages are produced in a number of configurations each having different geometries, alloys, and backing materials. Strain gages are carefully adhered to the media where strains are to be measured. The adhesive, the backing material, the alloy, the configuration, and the geometry of the gage are all factors affecting the sensitivity of the gage, i.e., the ratio of relative change in resistance to strain. The proportionality constant between the relative change in the resistance of the gage and strain is therefore different from S:

$$\Delta R = S_G \cdot \varepsilon \cdot R. \qquad (21.3)$$

Commercially available strain gages have a sensitivity S_G of approximately 2.0 [MG07a].

Because of the way the conductor is configured in a strain gage, a finite length is oriented in the direction perpendicular to the direction of the strains of interest. This fact makes strain gages sensitive to transverse strains [MG07b]. Commercially available strain gages are calibrated in strain fields in which the transverse strain is equal to approximately -0.3 times the axial strain. Strain measurements made under different conditions require correction of the gage factor, S_G. For common gages, the error incurred by neglecting transverse sensitivity is not likely to exceed 5 percent for ratios of transverse to axial strain between -1 and 0.5. Measurements made with strain gages are also sensitive to temperature changes, changes in the resistance of the wires used to connect the gage to the signal-conditioning system, noise, and moisture (which may affect the stability of the adhesive).

Strain gages are designed to measure local strains, usually not exceeding 10 percent, caused by mechanical actions (forces and moments). If used to measure strain directly, the gages are attached to the structure being monitored. Strain gages are also used in a variety of applications in which they are part of transducers that convert the mechanical actions applied to the structure into electrical signals. In these applications, the gages are attached to an element that deforms under the mechanical actions to be monitored. Transducers that use strain gages include load cells and pressure transducers. Strain gages are also used in extensometers to measure deformations over finite gage lengths.

21.4 Signal Conditioning for Electrical-Resistance Strain Gages

In common applications using strain gages it is usually preferred to measure changes in voltage rather than changes in resistance. A circuit called Wheatstone Bridge is used to convert changes in resistance to changes in voltage. The circuit is shown in Figure 21.2. It consists of four resistors (R_1, R_2, R_3, R_4) connected in series to form a loop. The connection between two resistors is called a node. A stable power supply is used to create a potential differential V_s between two opposite nodes of the bridge. The potential differential across the remaining pair of opposite nodes of the bridge is V_o. If the product $R_1 \times R_3$ is equal to $R_2 \times R_4$, V_o is zero.

As the resistors in a Wheatstone Bridge change, because of applied strains or because of other factors, the output V_o deviates from zero. The change in voltage caused by changes in the resistance of each resistor in the bridge is

$$\Delta V_o = \frac{r}{(1 + r)^2} \cdot \left(\frac{\Delta R_1}{R_1} - \frac{\Delta R_2}{R_2} + \frac{\Delta R_3}{R_3} - \frac{\Delta R_4}{R_4} \right) \cdot (1 - \eta) \cdot V_s, \qquad (21.4)$$

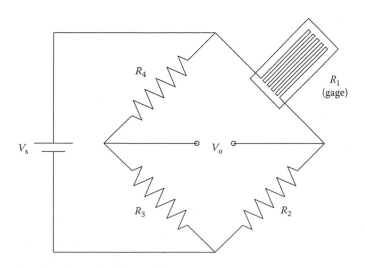

Figure 21.2 Wheatstone Bridge.

where

r is the ratio R_1/R_2, R_i is the change in resistance associated with resistor i ($i = 1, 2,$ 3, and 4)

V_s is the excitation voltage

η is a nonlinearity factor

For most applications, η does not exceed 1.25 times the applied strain. For bridges with a single resistor subjected to strain, η is approximately equal to the applied strain.

Note that if all resistors experience the same change in resistance, the output is zero because the signs associated with changes in resistance in adjacent arms of the bridge are opposite. This fact can be used to filter noise and changes in resistance not associated with the mechanical action being monitored. But before we discuss this in detail, let us consider the order of magnitude of the signals the Wheatstone Bridge produces. In civil engineering, strains of interest are seldom smaller than 0.0001. For $r = 1$, the voltage output at a strain of 0.0001 for a single active gage (the "quarter bridge" configuration) is of the order of

$$\Delta V_o = \frac{1}{4} \cdot (2 \cdot 0.0001) \cdot V_s = 0.00005 \cdot V_s. \tag{21.5}$$

For excitation voltages as high as 10 V, we would need to measure accurately voltages as low as 0.0005 V. Ideally, the resolution of the equipment required to measure voltage should be one hundredth of the desired measurement: 5×10^{-6} V. We conclude that we need to amplify the signals produced by the bridge, and we need to use stable and accurate equipment and resistors. The need for accurate and stable monitoring equipment and circuitry constitutes a challenge in the implementation of mobile, compact, and easily deployable strain-gage networks. Other challenges involved in the use of traditional strain-gage technology are

■ Sensitivity to temperature
■ Sensitivity to moisture
■ Noise
■ Wiring
■ Cracking

Sensitivity of the system to temperature is related to (1) changes in the sensitivity of the gage itself, (2) changes in the resistance of the circuitry, and (3) differences in the volumetric changes induced by temperature in the structure and the gage. These differences take place if the coefficient of thermal expansion of the structure is different from that of the gage. Gages can be selected so that the difference in thermal expansion coefficients is small [MG07c]. The sensitivity to temperature can also be reduced by using the configuration of the Wheatstone Bridge in Figure 21.3.

In the configuration shown, two resistors are subjected to strains and two are not. The two resistors that are not subjected to strains form part of the signal-conditioning equipment. The other two resistors are strain gages. One is attached to the structure and responds to stresses in the structure. The other gage (usually called "dummy gage") is attached to an element made with the same material as the structure but not subjected

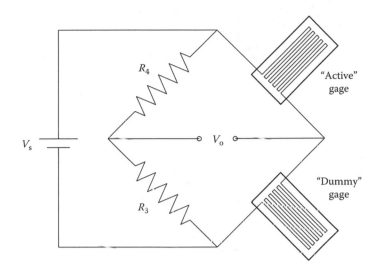

Figure 21.3   Wheatstone Bridge with "temperature-compensating" or "dummy" gage.

to stress. Because the signals in adjacent resistors in a Wheatstone Bridge cancel one another, changes in temperature, if they do not induce stress in the structure, would produce no output. If the strain in the gage attached to the structure comprises both strains that are due to changes in temperature and strains associated with stresses in the structure, the component of the signal related to temperature would be cancelled by the signal in the adjacent gage attached to an unstressed element.

In many applications, the sensing gage is located away from the signal-conditioning system. The gage and the signal-conditioning system are connected by lead wires. Changes in temperature can cause changes in the resistance of these wires. These changes in resistance can cause a spurious output. But the same idea behind the use of a "temperature-compensating" or "dummy" gage can be used to reduce the effects of changes in resistance of lead wires. To do so, and if a single gage is available, the configuration shown in Figure 21.4 is used to connect the gage to the signal-conditioning circuit.

In the configuration shown, equal relative changes in the resistance of wires 1 and 2 cancel one another. Changes in the resistance of wire 3 do not affect the output if the instrument used to measure the output has high impedance.

Lead wires create another challenge in that they affect the sensitivity of the system. They do not respond to stresses in the structure. But they do affect the denominator in the ratio $\Delta R/R$ (Equation 21.4). In essence, instead of R being equal to the resistance of the strain gage, R is equal to the resistance of the gage plus the resistance of the wire, R_{wire1}:

$$\frac{\Delta R}{R} = \frac{\Delta R}{R_{gage} + R_{wire1}}. \tag{21.6}$$

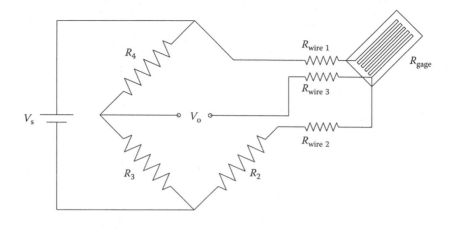

Figure 21.4 Three lead-wire connection.

The change in resistance is still

$$\Delta R = S_g \cdot \varepsilon \cdot R_{gage}. \tag{21.7}$$

From Equations 21.6 and 21.7, and reorganizing terms,

$$\frac{\Delta R}{R} = S_g \cdot \varepsilon \cdot (1 - L), \tag{21.8}$$

where

$$L = \frac{R_{wire1}}{R_{gage} + R_{wire1}}. \tag{21.9}$$

L represents the "loss" in the sensitivity. If the resistance of the lead wires is known, the output of the system can be corrected by using an adjusted gage factor equal to $S_g \cdot (1 - L)$. If the resistance of the lead wires is not known, the system can be calibrated by connecting a resistor in parallel with R_{gage} or R_2 (Figure 21.5). Connection of this resistor causes a change in resistance Delta$R$ similar to the change in resistance caused by strain. If the resistor is so selected that it produces a known change in resistance, then the sensitivity of the system can be so adjusted that the output strain is equal to the strain associated with the induced change in resistance: $\Delta R/(R_{gage} \times S_g)$. This procedure is called "shunt" calibration.

21.5 Large-Strain Binary-Output Resistance-Based Sensors

The Wheatstone Bridge is very versatile and can be used in a number of configurations depending on the sensing application. As discussed, it has limitations but many of them

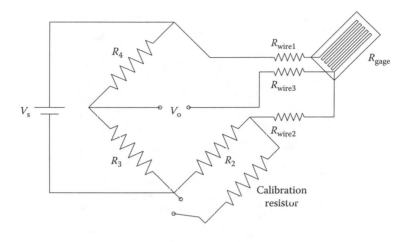

Figure 21.5   Shunt calibration.

can be dealt with by careful configuration or calibration. But cracking in the structure does represent a critical challenge. A crack traversing a strain gage would render the gage, and the Wheatstone Bridge it is part of, useless. Disruption of the circuitry of the Wheatstone Bridge would cause a large change in its output. This is a disadvantage in that the magnitude of the deformations causing the crack cannot be estimated. But it is also an advantage if all that is of concern is simply the detection of cracks because the change in output associated with formation of a crack is easily identifiable. This idea prompted the development of a different type of gage. These gages are intended solely to detect cracks, and not to estimate strain. If the only intent is the detection of cracks, the requirements for system accuracy can be relaxed and the process of data acquisition can be simplified.

Wood and Neikirk [WN01] used EAS stickers and Morita and Noguchi [MN06] used RFID tags for crack detection. Both systems are based on the simple idea that conductors adhered to the surface of a structural element can fracture if a crack forms in the element. These conductors are used as switches in the inductance circuits that power either EAS stickers or RFID tags. Toggling the switch can render the RFID tags unable to communicate with an external reader or, in the case of the EAS stickers designed by Wood, it can change the resonance frequency of the circuit. The information produced by each sensor is binary and indicates whether the conductors have fractured. These techniques have been proven effective for detection of relatively narrow cracks (of thickness not exceeding 2 mm) in structural elements. There are applications in which detection of cracks of small thickness is not necessary or even desired. In reinforced concrete in particular, narrow cracks are inevitable and do not indicate that the integrity of the structure has been compromised. What is of relevance in reinforced concrete is the detection of wide cracks and large discontinuities (Figure 21.6).

Chin et al. [CRM08] have developed a sensor that is triggered by a large change in one of the dimensions of a structural element. The sensor is shown schematically in

Figure 21.6   Damage to RC column (Pisco, Peru, 2007).

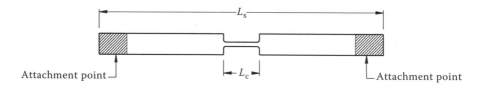

Figure 21.7   The CRM gage.

Figure 21.7 and is referred to as "CRM gage." The CRM gage can be used to detect cracks with the thickness exceeding a predetermined threshold. This threshold can be selected depending on the level of the damage that can be tolerated. With lengths exceeding 12 in., the gage can cover an area large enough to allow reliable assessment of the state of the structural element.

The CRM gage consists of a conductive laminate with a narrow and thinner "neck." This neck can be manufactured from a different material or the same material as the rest of the laminate. The total length of the gage, L_s, determines how large an area can be monitored by a single gage. The length of the neck is referred to as the gage length, L_c. The gage is attached to the structural element at each of its two ends using standard techniques to mount electrical-resistance strain gages. Deformations taking place between the two attachment points concentrate in the neck. These deformations are related to strains and cracks in the structural element. The gage length, L_c, is so chosen that the gage fractures if the total deformation taking place between attachment points exceeds a predetermined value. The deformation that causes fracture of the neck is correlated to L_c experimentally. The gage can be used as a switch enabling or disabling a passive RFID tag (Figure 21.9) or a switch in a sensing circuit (Figure 21.8).

If embedded into a passive RFID tag (Figure 21.9), the status of a CRM gage can be queried by a nearby RFID reader. The RFID tag consists of an antenna loop and a small microchip containing a small amount of information (an ID number associated, for instance, with the location of the gage in the structure), which is transmitted from the tag to the reader when the tag is excited by the reader. Notice that, in this approach,

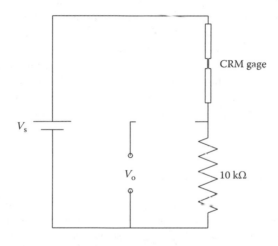

Figure 21.8   Circuit for use with CRM gage.

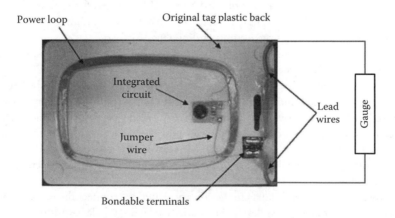

Figure 21.9   Embedding the gage in a passive RFID tag.

the tag obtains energy from the reader, and itself requires no power to operate. The gage is used as a switch between the antenna loop and the microchip. If the gage fractures, the reader cannot communicate with the microchip, indicating the presence of cracks and deformations resulting in a total change in dimension exceeding the selected threshold.

The CRM gage has also been used with the sensing circuit shown in Figure 21.8. In this configuration, the output is equal to V_s before the gage fractures. After fracture, the output is equal to zero indicating the presence of excessive cracking or deformation.

21.6   Data Acquisition and Communication

Embedded gages in a building can give an accurate view of the health of the building, and of any structural damage that has occurred. The status of the gages must be continually monitored and converted into a digital format suitable for computer processing. This requires the use of data acquisition devices that can easily integrate with the gages. Further, the acquired data must be communicated to a remote control center for timely analysis. For the communication, we seek a flexible, clutter-free wireless solution, because buildings that are monitored, or their located areas, may not have existing wireline network connections, and installing an extensive wireline network can be expensive and disruptive.

21.6.1   Passive RFID Design

The CRM gage designed to report cracking (Section 21.5) can be supported by a passive RFID. In this case, the status of a gage is a binary value: broken or not broken. We can embed the gage in a passive RFID tag, as shown in Figure 21.9. The embedding integrates the gage into the tag's antenna loop and includes it as a segment necessary to complete the data report circuit. When the gage is broken, the circuit is open and the tag will not respond when it is queried by an RFID reader. Otherwise, the tag will respond with its tag ID when queried. Cracks in the monitored building can be surveyed by mapping the locations of the nonresponding tags.

Passive RFID has the advantage that the tags have an extremely low cost (well below $1 per piece), especially when ordered in large volumes. Additionally, they do not require power to run, because they operate by reflecting the energy from the RFID reader in their response. A limitation is that a passive RFID has a relatively short reading range compared with, say, an active RFID. To amortize the reader cost ($50 and more) by using one reader to control a set of tags, all the tags must be close to each other. Further, the reader/tag implementation must support a MAC protocol, e.g., the slotted termination adaptive collection protocol [MIT03], to prevent signal interference between the readers and tags.

To disseminate the RFID data to the control center, we connect the reader's communication interface to a wireless device. For initially relaying the data within the building, i.e., an indoor environment, technologies such as 802.11 Wi-Fi or Bluetooth can be used. Both communication methods have decent ranges of tens to hundreds of meters such that a fully connected network is expected to form wirelessly without too many communication nodes. Moreover, their signals can go through obstacles such as walls and furniture, which further obviates concerns for the loss of indoor connectivity. In the unusual case that the readers by themselves will form network partitions, wireless relay nodes can be used to connect the partitions, although this would require advance planning or a site survey. Notice that certain readers may not be equipped with a wireless interface. For example, the Phidgets reader used in [CRM08] has a USB port for communication. In that deployment, up to 4 Phidgets readers are connected to a Keyspan US-4A USB server with an Ethernet port, which in turn interfaces with a Linksys Wi-Fi router for wireless connectivity. Hence, the implementation details become more involved but the conceptual design remains unchanged.

Data forwarded inside the building will eventually need to go out of the building for transport to the control center. In the outdoor environment, a Wi-Fi mesh network, whose nodes are typically fitted with directional or omni-directional antennas for an extended range, is a ready and relatively well-tested technology to provide wireless transport over the wide area. Such a network runs 802.11x in an ad hoc mode. Without a fixed supporting infrastructure, it employs an ad hoc routing protocol, e.g., AODV, with wireless specific routing metrics (e.g., ETX) to discover high-quality data paths between the peering access points.

21.6.2 Mote-Based Design

MICA2 motes can also be used to query the status of gages. The motes can be purchased with a variety of commodity sensors, such as accelerometers, thermometers, magnetometers, and inclinometers. In our case, however, the sensor is the CRM gage. A data acquisition board, e.g., the MDA320CA or the ADAM-6017, is needed in this case to connect to the processor and radio module of the MICA2. In the case of the MDA320CA, up to 8 gages can be connected to its ADC and digital I/O, respectively. This way, the gages are powered by the 2.5 V sensor excitation output of the MDA320CA and the mote can read the status of the gages through the acquisition board. Note that because the gages are powered in this case, both CRM gages and general electrical-resistance strain gages can be supported.

Figure 21.10a shows the wiring of four CRM gages to the single-ended voltage ADC inputs of the board. The E2.5 terminal denotes the 2.5 V excitation channel on the board, and this terminal is connected to one end of each gage. The other end of the gage is connected to the input of an ADC channel and to an array of 10 kΩ pull-down resistors. The pull-down resistors are used to prevent the ADC inputs from staying in a floating state in which the voltage is unknown and subjected to environmental interference. This wiring gives a 2.5 V reading when the gage is not broken and 0 V when the gage is broken. Figure 21.10b shows the wiring of the gages to the digital channels. The wiring connects one end of the gage to the digital input and the other end to electrical ground. This wiring gives a low reading (GND) when the gage is not broken (because the gage closes the circuit to GND), and a high reading (VCC) when the gage is broken, due to internal pull-up resistors in the data acquisition board.

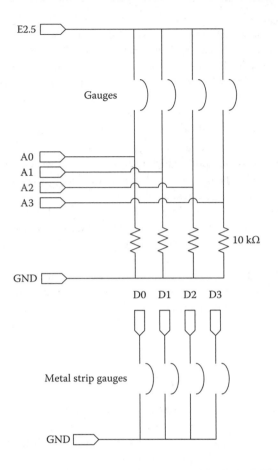

Figure 21.10   Wiring the gages to the data acquisition board.

A two-tiered wireless network can be used for data communication to the control center [CRM08]. The lower tier is used for multi-hop relaying of the gage data reports indoor to a mote network gateway. The indoor communication occurs over the 900 MHz radio interface of the MICA2 in which the motes form a sink tree terminated at the mote network gateway. The routing protocol gathers data from all the data sources (i.e., the motes with gages attached) and forwards the data to the sink. In general, the sink tree may have internal nodes that are MICA2 motes used purely for data forwarding (i.e., they do not have attached gages).

Data collected from the sink tree is in turn forwarded by the mote network gateway to the upper tier 802.11x wireless mesh network for outdoor, wide-area transport to the control center. To bridge between the RF sink tree network and the 802.11x mesh network, the mote network gateway is implemented as a computing node with both a MICA2 mote and an 802.11x router attached. The 802.11x mesh network has an identical operation as the mesh network described for the passive RFID-based design in Section 21.6.1.

21.7 Control Software

We have discussed the deployment of sensors with supporting data acquisition and communication devices in the building being monitored. The local devices can then be configured and controlled by a remote SHM software system to form an integrated deployment with given monitoring parameters. The component-based SHM software in [CRM08] is designed to be portable and configurable, and can accommodate heterogeneous hardware and ensure their interoperability. In particular, it supports both the passive RFID-based and mote-based deployments in Section 21.6.

21.7.1 *Installation and Sensor Configuration*

The SHM software is distributed as a single package, and can be installed on common Windows, OS/X, Linux, and UNIX platforms running the .NET 2.0 framework. Once installed, the system can be configured for specific applications. This is done by modifying an XML sensor configuration file, SensorConfig.xml, located in the installation directory. Figure 21.11 shows an example configuration file. The configuration is structured as a tree of XML elements. The root element is sensorconfig, whose children are the sensor elements for the specific sensor types. Each sensor

```
<?xml version="1.0" encoding="utf-8" ?>
<sensorconfig>
  <sensor id="200" type="phidgetRFID">
    <description>Phidgets RFID Reader (17959)</description>
    <hardwareid>4627</hardwareid>
  </sensor>
  <sensor id="100" type="MDA320CA">
    <description> Crossbow MDA320 Data Acquisition Board
    </description>
    <port>TCP:192.168.1.6:10002</port>
    <channel id="0">
      <description> MDA320 Digital Channel #1</description>
    </channel>
    <channel id="1">
      <description>MDA320 Digital Channel #2</description>
    </channel>
  </sensor>
  <sensor id="300" type="ADAM6017">
    <description>ADAM6017 Data Acquisition Module
    </description>
    <port>TCP:192.168.1.160:502</port>
    <pollinterval>1000</pollinterval>
    <channel id="0">
      <description>ADAM6017 Channel #0</description>
      <range min="-5" max="5"/>
    </channel>
    <channel id="1">
      <description>ADAM6017 Channel #1</description>
      <range min="-5" max="5"/>
    </channel>
  </sensor>
</sensorconfig>
```

Figure 21.11 An example sensor configuration file.

element specifies the sensor type, and multiple instances of sensors of the same type can be distinguished by the unique `id` attribute.

The detailed operations of a sensor are configured by the `sensor` element, for example, in the case of a gage connected through a data acquisition board, the protocol and port for network communication; the board's data channel to which the gage is connected; the polling intervals for data reports; and the interpretation of readings returned in the reports. In general, XML provides a semantic-based description of each sensor, where the semantics can be defined in a sensor-specific manner.

21.7.2 Configuration of Experiment

The core of the SHM software is a set of software components for device/communication control and subscription for sensor events, and data collection, logging, selection, and presentation. An SHM deployment, called an experiment, is then configured as a graph of the software components used. The components interact through data/control flows controlled by well-defined data/control interfaces.

An example SHM experiment is shown in Figure 21.12. It has three layers: the application layer, the generic sensor layer, and the instantiated sensor layer. The application layer interacts with the end user by accepting user commands to control the experiment, and displaying updated views of the building status in a visual interface. The generic sensor layer provides an abstract view of a sensor in terms of the sensor itself and the sensing channel of the sensor (e.g., an MDA320CA data channel, as described in Section 21.6.2). Example methods in the abstract interfaces include opening and closing a sensor, obtaining an identification of the sensor, querying the data/error status of the sensor, reading data from the sensor, setting the polling schedule of the sensor, etc. The instantiated sensor layer provides the actual implementation of the generic sensor and sensing-channel classes. It realizes the low level communication protocols needed for the sensor hardware to perform the high-level tasks defined by the generic

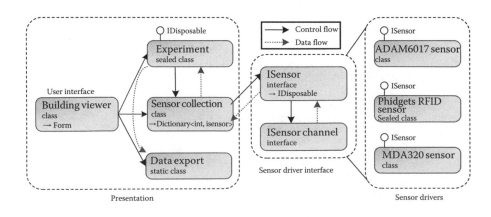

Figure 21.12 An example software configuration.

interfaces. For example, the MODBUS protocol is often used as a de facto standard of communication with a wide range of industrial electronic devices, including a MICA2 data acquisition board. The SHM software implements the control-center side of MOD-BUS query/response messages that communicate with MODBUS sensing devices over TCP/IP. The instantiated sensor layer cleanly encapsulates the driver details of the sensor hardware, and hides these details from the user.

21.7.3 Data Logging and Presentation

In an experiment, all the data and events collected from the sensors are logged in a `.edf` log file. The log file contains detailed information about the experiment, including the sensor configurations, the date/time when the experiment started/ended, all the sensor events/data reports received including their source and time, and any error that occurred during the experiment.

Not all the information in the detailed log may be of immediate interest to the user. Hence, the `DataExport` component in Figure 21.12 provides functions to filter the logged data according to user requests, and to format the data in a user-readable format. Because the log file is written in XML, `DataExport` can implement either a customized program or an XSD transformation of an XQuery. This enables end users to develop their own data export functions without programming knowledge, and DataExport is fully compatible with popular software such as Microsoft Excel.

21.8 CRM Gage Functionality Test

The feasibility of a monitoring system based on CRM gages was evaluated for a particular configuration of the CRM gage. In this configuration, the gage consisted of two thin steel strips of length $L_s/2$ each and a copper tape spanning from one strip to the other (Figure 21.13). L_s was selected to be 16 in. The steel strips measured 3/8 in. in width and 0.005 in. in thickness. The copper tape used was GC Tool Pure Copper Circuit Tape and had a width of 1/8 in. and a thickness of 0.002 in.

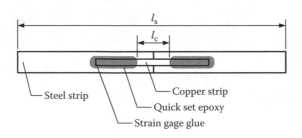

Figure 21.13   CRM gage used in the functional test.

The gage shown in Figure 21.13 was used to detect cracks on a column subjected to constant axial load and cyclic lateral loads. The column was tested at the Nagoya Institute of Technology (NIT) in Nagoya, Japan (Figure 21.14).

Two of the four vertical faces of the column specimen were instrumented with four gages each: two with a gage length (L_c) of 1/8 in. and two with a gage length of 3/8 in. The locations of the two types of gages used were alternated along each face. The gages were oriented as shown in Figure 21.14 (perpendicular to the longitudinal axis of the

Figure 21.14 Experiment setup in the functional test: The silver strips are the CRM gages being deployed.

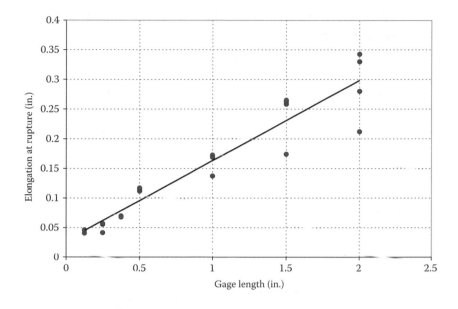

Figure 21.15 A plot of elongation at rupture against gage length, L_c.

column). This orientation enables the gages to detect cracks forming at an angle with respect to the longitudinal axis of the column. This type of crack has been observed repeatedly to lead to structural failures during past earthquakes.

Prototype gages were calibrated before the test at NIT to estimate the relationship between deformation at rupture and gage length (Figure 21.15). Calibrations were made by attaching gages with different lengths to two 2 in. diameter concrete cylinders, with each end of the gage mounted on a different cylinder. One reinforcing steel bar was cast in each cylinder. The bars were used to mount the cylinders on a universal testing machine. With this machine, the cylinders were displaced with respect to one another at a constant speed. The total separation between the cylinders at the time of rupture was recorded. Rupture was detected with the help of a voltage signal sent through the sensor. The signal was monitored using a data acquisition system. Rupture was sensed by an abrupt variation in the voltage between the ends of the sensor.

21.8.1 Test Results

The widths of cracks crossing the gagues tested were measured with a crack width comparator. During the test, six of the eight gages detected deformations large enough to cause rupture of the gage before the loading was completed. The gages with a gage length of 1/8 in. were expected to rupture, in average, at a total deformation of 0.05 in. The gages with a gage length of 3/8 in. were expected to rupture, in average, at a total deformation of 0.07 in. Table 21.1 lists the observed total width of cracks that crossed each gage and caused rupture. The measured total widths of cracks match (on average) the expected deformations at rupture.

Table 21.1   Stretch at Breakage or at the End of the Functional Test

| Gage ID | Gage Length (in.) | Stretch at End/Breakage | | Broke? |
		Design (in.)	Experimental (in.)	
1	1/8	0.050	0.050	Y
2	1/8	0.050	0.050	Y
3	3/8	0.070	0.075	Y
4	3/8	0.070	0.065	Y
5	1/8	0.050	0.050	Y
6	1/8	0.050	0.060	Y
7	3/8	0.070	0.070	N
8	3/8	0.070	0.080	N

21.9   Full-Scale Deployment of CRM Gage

An experimental evaluation of the CRM gage on a full-scale three-story reinforced concrete building structure has been reported in [CRM08]. The structure was built by researchers at Purdue University [Fick2008] as part of an investigation on the response of flat-plate structures to earthquakes.

The total height of the structure was 30 ft. Six reinforced concrete columns, 18 × 18 in. in cross section, supported a total of three 7 in. thick flat slabs. Each slab was 50 ft. by 30 ft. in plan. The experimental structure is shown in Figure 21.16. The structure was subjected to 12 psf of superimposed dead load applied with water-filled barrels. Alternating cyclic lateral loads were applied at each slab with hydraulic actuators. These loads caused cracks in the slabs (Figure 21.17). The purpose of our experiment was to evaluate the potential of the CRM gage to detect these cracks. A set of 4 CRM gages were deployed around 3 of the columns on each floor, for a total of 12 gages. Each gage was configured to detect cumulative crack widths exceeding 0.03 in.

At different stages during the test, the status of each gage as reported by the SHM network, and the locations and widths of cracks were recorded. The records show that the monitoring network achieved a 100 percent detection rate in the sense that it reported reliably the presence of cracks crossing a gage with total widths exceeding the preselected threshold. There was no false positive in the sense that no gage spanning cracks, with a cumulative crack width smaller than the threshold crack width, ruptured.

The network performance varied depending on whether passive RFID or mote-based data acquisition was used. In the case of passive RFID, a reader reported only when it lost contact with an RFID tag because the tag's data circuit was broken by cracks. Reliable reporting of the tag lost event was assured by the use of TCP/IP. Hence, the network data

Figure 21.16   Three-story reinforced concrete building structure in the full-scale experiment.

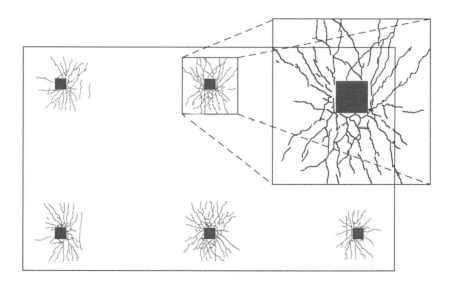

Figure 21.17   Plan views showing cracks in the slab.

rate was extremely low. In the case of mote-based data acquisition by the ADAM-6017, the status of a gage was reported every 2 seconds, resulting in a data rate of 1.208 kb/s per acquisition module. In this case, the SHM network data reports were naturally resilient to loss, because a broken gage is a stable event: once the gage is broken, it remains broken. Hence, a later sensor data report subsumed an earlier data report in the case that the earlier report was lost.

21.10 Conclusion

We have discussed various uses of the Wheatstone Bridge and resistance-based strain gages for SHM. Strain gages exhibit changes in electrical resistance as a response to the applied strain. The Wheatstone Bridge and electrical-resistance strain gages can be used to obtain accurate measurements of strains in continuous media. This use requires availability of a stable and accurate data acquisition system. It is limited to cases where cracks do not form across the strain gages. We then presented the design of a large-strain binary-output gage, the CRM gage, for the specific challenge of detecting large defomations and cracks in critical structural elements. The CRM gage is based on the ideas of Wood and Neikirk [WN01] and Morita and Noguchi [MN06], and has been designed to create a disruption of the circuitry of the sensing hardware to produce a large change in output that can be associated with damage unequivocally and without the need for analysis.

To support the function of remote structural monitoring, we have presented the design and implementation of a multi-hop wireless SHM network using low-cost COTS data acquisition/communication devices. We have also presented a component-based SHM control software for configuring and controlling heterogeneous sensing/communication hardware remotely, and supervising the data reporting by the deployed sensors. The control software components can themselves be composed in a plug-and-play manner to realize different sensor applications. They allow the user to select interesting data reports and format the data for effective presentation.

We have reported a functionality test for the CRM sensor to show that it can be manufactured to detect reliably cracks exceeding a configurable threshold crack width. In addition, we have reported results from a deployment of 12 CRM sensors on a full-scale three-story reinforced concrete building structure. These results show again that the sensors can reliably detect cracks of widths exceeding a preselected threshold in scenarios resembling actual field conditions. The simplicity of the CRM sensor reports allows effective real-time structural monitoring at a low network bandwidth and with a high resilience against packet loss.

Acknowledgments

This work was supported in part by the U.S. National Science Foundation under grant numbers CNS-0305496 and CMS-0443148. The authors are indebted to Dr. S. C. Liu, CMS program director, for his patience and help. This work was also supported in part by the Joint Funds of NSFC-Guangdong under grant number U0735003, the Natural Science Foundation of China under grant numbers 60604029 and 60702081, the Natural Science Foundation of Zhejiang Province under grant number Y106384, the Science and Technology Project of Zhejiang Province under grant number 2007C31038, and the Scientific Research Fund of Zhejiang Provincial Education Department under grant number 20061345.

References

[CRM08] Chin J. C., Rautenberg J. M., Ma C. Y. T., Pujol S., and Yau D. K. Y., A low-cost, low-data-rate rapid structural assessment network: Design, implementation, and experimentation. In *Proceedings IEEE International Conference on Mobile Ad-hoc and Sensor Systems*, Atlanta, GA, September 2008.

[DAL93] Dally J. W., Riley W. F., and McConnell K. G., *Instrumentation for Engineering Measurements*, 2nd edn., Wiley, New York, 1993.

[Fick08] Fick D., Testing and structural evaluation of a large-scale three-story flat plate, Doctoral dissertation, Purdue University, West Lafayette, IN, April 2008.

[KPC07] Kim S., Pakzad S., Culler D. E., Demmel J., Fenves G., Glaser S., and Turon M., Health monitoring of civil infrastructures using wireless sensor networks. In *Proceedings ACM/IEEE Information Processing in Sensor Networks*, Cambridge, MA, 2007.

[MG07a] Measurements Group Inc., Strain gage selection: Criteria, procedures, recommendations, Tech. Note TN-505-4, 16p, 2007.

[MG07b] Measurements Group Inc., Errors due to transverse sensitivity in strain gages, Tech. Note TN-509-4, 9p, 2007.

[MG07c] Measurements Group Inc., Strain gage thermal output and gage factor variation with temperature, Tech. Note TN-504-1, 13p, 2007.

[MG07d] Measurements Group Inc., Errors due to Wheatstone Bridge nonlinearity, Tech. Note TN-507-1, 5p.

[MG07e] Measurements Group Inc., Shunt calibration of strain gage instrumentation, Tech. Note TN-514, 19p, 2007.

[MIT03] MIT Auto-ID Center, 3.56 MHz ISM band class 1 radio frequency identification tag interference specification: Candidate recommendation, verson 1.0.0. Technical report MIT-AUTOID-WH-002, 2003.

[MN06] Morita K. and Noguchi K., Crack detection sensor using RFID-tag and electrically conductive paint. Building Research Institute, Tsukuba, Japan, 2006.

[WN01] Wood S. L. and Neikirk D. P., Development of a passive sensor to detect cracks in welded steel construction. U.S.–Japan Joint Workshop and Third Grantees Meeting, Seattle, WA, 2001.

[XRC04] Xu N., Rangwala S., Chintalapudi K. K., Ganesan D., Broad A., Govindan R., and Estrin D., A wireless sensor network for structural monitoring. In *Proceedings ACM Sensys*, New York, 2004.

Index

A